Infrastructure in the Anthropocene

SECOND EDITION

Infrastructure in the Anthropocene

SECOND EDITION

Mikhail Chester
Braden Allenby

Metis Center for Infrastructure and Sustainable Engineering
Arizona State University
Tempe, Arizona, USA

Infrastructure in the Anthropocene: Second Edition

June 2025

To my father, Dr. Mitchell Chester, who imparted upon me the value of perspective change that comes from curiosity, and a commitment to future generations. My dad used to ask me after school whether I had any homework and if I answered no he'd suggest I assign myself some. This book is one of those assignments.

To my family Becca, Ethan, Abbie, and Norah, who help me to see the world in all its colors and potential.

Mikhail Chester

To my mother and father, who made me the adult I try to be, and to my wife and children, who help me be the imaginative child I need to be.

Braden Allenby

Contents

Forward

At the dawn of the Anthropocene the world appears to be much more complex than in the past, and our infrastructure systems appear unable to respond at pace and scale. This is the motivation for this book. This book is designed to help bridge the gap between what our infrastructure systems can do and what we need them to do. It consists of concepts, framings, and new mental models that challenge normative assumptions of infrastructure, that these systems will continue to be reliable and meet our needs, despite growing evidence that the environments that they must function in as well as the systems themselves are rapidly changing and becoming far more complex.

The focus of this book is on a particularly important set of engineered infrastructure systems that are designed, built, and managed to deliver basic and critical services, such as energy, water, mobility, and information. We acknowledge that the term "infrastructure" has many meanings and can refer to physical or non-physical (e.g., governance) systems. We view engineered infrastructure as sociotechnical systems that are composed of physical assets and the institutions that manage, govern, finance, and regulate them, as well as the massive educational enterprise that creates, stores, and disseminates knowledge to infrastructure engineers and managers. Infrastructure are not purely physical systems and of the ownership of engineers. Our perspective is that these systems are a tripartite (physical assets, governance, and education) and these domains independently and together include many non-engineering actors including planning, financial, cybertechnology, computer science, government professionals, and perhaps most importantly a broad community of consumers of infrastructure services that rely on the systems for their well-being.

Our perspective is often on U.S. or modern infrastructure, but we acknowledge that regions of the world are in various stages of infrastructure development. Regardless of where a region is in deploying infrastructure, we believe the principles presented in this book are valuable.

While much of the book focuses on infrastructure as engineered systems that often frames from a perspective of legacy infrastructure (Sections 1-4), we also introduce challenges and opportunities for infrastructure as they hybridize with cybertechnologies and larger digital and information ecosystems (Section 5). To ignore the rapid integration of cybertechnologies into legacy infrastructure and the increasing access to information about and created by infrastructure and their users is unsagacious.

We use the term *infrastructure* in the plural. In the U.S. the term infrastructure is commonly plural, and we recognize that in other parts of the world the term *infrastructures* is common. Dictionaries list infrastructure as both a countable and uncountable noun so the term can be used as plural. We recognize that this may be unconventional for some readers.

This is the second edition of this book, and the content is considerably expanded and different from the first edition (ISBN: 0999587781; ISBN-13: 978-0999587782). The first edition was published in 2021 through Arizona State University's Consortium for Science, Policy, and Outcomes under their *The Rightful Place of Science* book series. The book was short, and each chapter discussed particular challenges or theories without significant depth. Over the years as we've accumulated insights into these challenges and theories, we felt it prudent to publish a second edition to expand and formalize our perspectives. The second edition is published under Arizona State University's Metis Center for Infrastructure and Sustainable Engineering. As such we have dropped "The Rightful Place of Science" from the title.

RECOGNITION

This book includes new writing by the authors as well as adapted content from our previous works. The previous works were often led by one or both of us, and sometimes led by our colleagues with our involvement and often advising. If the content is adapted then we have secured permissions from the publisher and lead author, and provide a citation to the original work. We are grateful to these lead author colleagues for allowing us to adapt the past work into this book. They are Thomaz Carvalhaes, Alysha Helmrich, Ryan Hoff, Yeowon Kim, Samuel Markolf, and the adapted works are cited as footnotes with the chapter. As such, we view the collective narrative of this book not as ours but of a community of researchers that have pushed conventional thinking of how we should approach infrastructure.

BOOK ORGANIZATION

The book is organized into six sections. Each section is a theme and we intend for the sections to be read consecutively for a coherent perspective.

In Section 1 we argue that the scale and scope of human activities have grown so large that the dichotomy between infrastructure and the environment is shrinking and often nonexistent; reductionist approaches that approach infrastructure as hardware problems are insufficient. We conclude Section 1 with the perspective that infrastructure are becoming less about physical assets and more so a process for navigating wicked and complex challenges.

Section 2 focuses on the capabilities and opportunities for navigating infrastructure through increasingly complex environments. We start by reconceptualizing what infrastructure need to do and then describe tenets of agility and flexibility to match the pace of change surrounding infrastructure.

We turn our attention to the importance of innovative governance and leadership in Section 3. We start by describing how infrastructure governance in its current form came to be and how that model thrived during a long period of stability. However, we then describe how the governance models that dominated infrastructure today are ill-suited for complex and chaotic conditions. We describe governance models that can pivot between efficiency and exploration. We conclude with a discussion of the need to be able to pivot what infrastructure we consider critical during times of crisis.

Section 4 focuses on the challenge of changing climatic conditions and how they undermine normative assumptions about how we plan and manage infrastructure. In this section we introduce novel approaches for infrastructure to confront nonstationary conditions, namely decision-making under deep uncertainty, safe-to-fail, and biomimicry.

The rapid implementation of cybertechnologies, novel data streams, and communications technologies into legacy infrastructure represent not simply new capabilities but more so an irrevocable integration into a broader cognitive ecosystem. This is the focus of Section 5 which challenges how we see the nature of infrastructure, and what this means into the future. We introduce the concept of the cognitive ecosystem –as an ecology of massive data flows, artificial intelligence, and connected technologies – and what this means for infrastructure. We then discuss how these technologies are resulting in infrastructure becoming battlespaces in cyber and asymmetric warfare. We conclude this section with discussions on the challenges of managing infrastructure as cybertechnologies, and the potential for artificial intelligence to disrupt how we understand and control infrastructure.

Section 6 concludes the book with a synthesis of our arguments and perspectives into novel first principles for infrastructure in the Anthropocene. We view these first principles as our charge to the infrastructure community, a roadmap for operationalizing our perspectives.

Section 1

THE ANTHROPOCENE: THE WORLD AS INFRASTRUCTURE

The environments that infrastructure professionals are expected to function in are rapidly changing and markedly more complex than in the past. At the dawn of the Anthropocene humans are increasingly hybridizing infrastructure, controlling planetary systems, and delegating decision-making to software (i.e., artificial intelligence). Whereas in the past the dichotomy between infrastructure systems, or an infrastructure system and the natural environment was somewhat clear, boundaries that we've taken as normal for nearly a century are rapidly dissolving. Are electric vehicles transportation systems or hybrid transportation-building-energy systems increasingly steered by third parties?

Discussion about the Anthropocene has so far largely focused on changes to planetary systems. In Section 1 we contextualize the implications of the Anthropocene on how we *conceive* the environments in which infrastructure must function. We begin to challenge the notion of what infrastructure are, away from legacy mental models that position infrastructure as purely engineered and siloed systems, towards systems that are increasingly hybridized and in control of regional to global systems, including water supplies and the atmosphere. In Section 1 we rethink the relationships between infrastructure and their environments, and in Section 5 we delve further into the implications of these changing relationships including governance, cybertechnology integration, and ultimately the integration of legacy infrastructure into cognitive ecosystems defined by monumental leaps in data collection and storage, sensing, analytics, and artificial intelligence.

Throughout Section 1 we challenge the prevailing mental models of what infrastructure are and do, and we often situate the discourse considering technologies, governance, and education. These three domains together define the capabilities and limitations of our infrastructure systems as physical, social and ultimately knowledge processes. Section 1 should not only establish that infrastructure capabilities are decoupling from the increasingly complex environments that they need to function in, but also that they need to become wicked and complex environments that warrant requisite complexity in the approaches used to navigate them in the Anthropocene.

1.1

WELCOME TO THE ANTHROPOCENE

Infrastructure appears to be caught between the past and the future, and when it comes to these obdurate systems, the present is not a good place to be. Technological evolution is accelerating, politics and society are fragmenting, climate and the natural world are changing, and infrastructure (at least in the developed world) is in dire need of modernization. As countries around the world deploy new infrastructure, the initiatives largely perpetuate the physical forms that we've seen for decades and reflect the same financing structures, management structures, and relationships with the surrounding natural environment as in the past.

Today's design principles seem to reflect those of the last century, when the types of services we demanded of our infrastructure were stable—just like the climate and technological conditions under which infrastructure had to be reliable. But stability is no longer the norm. Furthermore, many forces critical to infrastructure design and performance (such as climatic conditions, advanced technologies such as artificial intelligence, and, less obviously, security protections from today's conflicts) appear to be changing in unpredictable ways. The inflexible and long-lasting nature of infrastructure is directly at odds with these dynamic forces.

1.1.1. INTRODUCTION

Evidence has been accumulating for decades that we are at the dawn of major shifts in the relationships among humans, social systems, the environment, and technology. This has profound implications for how we design and manage infrastructure, which long-time policymakers and the public have been mostly able to ignore. But no more.

Hyperconnectivity and the embedding of new technologies into legacy systems, artificial intelligence managing how we understand and demand services from infrastructure, and destabilizing climate extremes represent just a few of the emerging realities that infrastructure managers will face. The rate at which we are deploying new technologies embedded within infrastructure appears to be outpacing the infrastructure itself. A single roadway intersection may have experienced a progression of control systems that started with a traffic officer, progressed to loop detectors, became traffic cameras with image recognition, and is now driven by cloud-based systems such as Google Maps.

Several trends are emerging that affect how we think about and manage infrastructure into the future. First, while change has occurred across social, environmental, and technological systems over the history of humankind, the acceleration of change appears to be taking off: technological development in the coming century is expected to be greater than the past several thousand years. This has profound implications for the planet and human-managed systems.

Second, "nonstationarity," meaning that past trends are no longer good predictors of future conditions, has emerged as an important concept in climate science. This concept can also be applied across a number of systems that affect infrastructure.

Third, infrastructure and the systems that support them have in many ways become so complex that their emergent characteristics — i.e., what we expect them to do when perturbed — are no longer predictable.

Fourth, infrastructure has become a battleground between adversaries in new forms of war that pose huge challenges but are not yet perceived, much less understood, by either the public or infrastructure managers. Yet the training that educational institutions deliver for future infrastructure managers focuses mainly on elements of the system and their predictability (tools for a complicated, not complex, system).

These forces are combining to create conditions of un-predictability that are inimical to the approaches that we've used to design and manage infrastructure in the past. As these forces emerge concurrently, they represent an environment of destabilization that requires new thinking and competencies for how we approach infra-structure into the future. We summarize these trends in Table 1.1.1.

Table 1.1.1: Trends Affecting Infrastructure Across Domains

INFRASTRUCTURE. <u>Acceleration</u>: Quickening integration of cyber technologies into legacy infrastructure, and introduction of information sense-making organizations (Uber, Lyft, Tesla, Google Nest, etc.) that distribute control and affect service consumption. <u>Nonstationarity Emergence</u>: Design norms focused on stability increasingly insufficient with unpredictable climate-related extreme events. Increasing complexity due to layering of technologies (accretion, interactions, edge cases, and common rarities).

CYBER-TECHNOLOGIES. <u>Acceleration</u>: Acceleration of technologies where cycle times exceed that of infrastructure. Decreasing of technology costs, increasing data processing capabilities, increasing of communication capabilities. <u>Nonstationarity Emergence</u>: Cybersecurity in an age of civilizational conflict, where infrastructure is a primary target. Artificial intelligence; social media tribe formation; China's social credit system.

BIOTECH. <u>Acceleration</u>: Integration of AI/big da-ta/analytics, development of accurate gene editing techniques, and increasing importance of biological data for national security purposes and state AI capabilities significantly accelerate state of art. <u>Nonstationarity Emergence</u>: As humans become a design space, they are both the designers and the designed, creating a rapid reflexivity that affects all hu-man systems dramatically and unpredictably. The human in all domains – physical, psychological, identity, community – becomes a design space.

POLITICS. <u>Acceleration</u>: Integration of information and communication technology and AI/big data/analytics with political activities dramatically increases speed of political change. Increasing ideological polarization stalls infrastructure policy. Increasing contingency of world order and Western Universalism morality significant source of new complexity. <u>Nonstationarity Emergence</u>: Rise of non-traditional political entities leads to growth of neomedieval governance structures. End of traditional Enlightenment politics leading to unpredictable shifts in geopolitical power, possible ascendency of soft authoritarianism over classic Western pluralism.

FINANCING. <u>Acceleration</u>: New players investing in and building their own systems to circumvent the legacy systems; e.g., Amazon and drone delivery. <u>Nonstationarity Emergence</u>: Increasing needs for infrastructure financing and restructuring of financing, yet large uncertainty about consistent infrastructure investment. Increasing tying up of infrastructure financing with other goals.

EARTH SYSTEMS, INCLUDING CLIMATE. <u>Acceleration</u>: Accelerating effects of human activity on water, nutrient, resource, and climate systems, both unintentionally and intentionally. Nonstationarity in environmental conditions (water, temperature, fire, etc.) that threatens the basic design assumptions of infrastructure and their reliability. <u>Nonstationarity Emergence</u>: Feedback loops that future climate change will reduce the efficiency of the Earth system to absorb anthropo-genic carbon. Emergence of integrated systems that are not well managed under current governance regimes.

SOCIAL. <u>Acceleration</u>: Changing demographics. Software increasingly steering decision-making. Rise of China and Confucianism challenge post-WWII Western universalist ethics and institutions. Success of Enlightenment has created complexity that in turn undermines Enlightenment concepts, institutions, and governance mechanisms. <u>Nonstationarity Emergence</u>: Civilizational conflict results in tribal narratives and individual identity becoming both design spaces and battlespaces. Weaponized narrative and manipulation of individuals evolves into new normal for politics.

CULTURAL. <u>Acceleration</u>: Changing values, customs, and beliefs. Changing patterns of war and conflict. Cultural shifts as demographics change (the old get older, and the poor get poorer). <u>Nonstationarity Emergence</u>: Cultural homogeneity at globalist level replaced by regional cultural hegemons as new social structures evolve, but destabilize existing social and cultural practices.

For infrastructure designers, managers, and even users, it's time to re-think the relationship with the core systems that serve as the backbone for virtually every activity and service that society demands. It's time to come to grips with the reality that the complexity of infrastructure is exploding, that emerging and disruptive technologies are accelerating in ways that are antithetical to current infrastructure's core design principles, and that education on these issues is insufficient. New accelerating and interactive forces are defining what infrastructure can and should do, and how it should function on a planet dominated by human systems. To understand these accelerating forces and complexity, it's helpful to start in the past.

1.1.2. A BRIEF HISTORY OF INFRASTRUCTURE

Humans have been changing the Earth for millennia. The development of agriculture some ten thousand years ago reflected drastic technological, cultural, economic, and social change, and over time supported a significant jump in global human population from perhaps a million to hundreds of millions of people. It also affected natural systems, from water and land use patterns to biodiversity, to patterns of nutrient and material flows, including those of many metals. It began a process of urbanization that has continued to characterize the human species—and change the planet in numerous ways.

But it was the Industrial Revolution in Europe in the 1700s, and concomitant changes in integrated and co-evolving human, natural, and built systems, that marked the real emergence of the Anthropogenic Earth. The shift from hunter-gatherer to agricultural lifestyles trans-formed local and regional systems and generated new patterns of resource consumption and energy and waste flows; but the magnitude of those impacts increased exponentially with the Industrial Revolution.

The result over a period of a few short centuries was the terraforming of the planet. Not only did human systems—from economics to technology to culture—change in unpredictable and fundamental ways, but the dynamics of virtually all major natural systems were increasingly affected by human activity. The Industrial Revolution fostered a rapid acceleration in the growth of energy and water use, environmental impacts of all kinds, human population levels and urbanization, economic growth, technological complexity, and the built infrastructure to support it all. These patterns show few signs of slowing.

This acceleration includes not only global population growth, from roughly 450 million people in 1500 to over 7.5 billion today, but also economic growth. Between 1500 and 1800, the global economy grew by a factor of almost three, but between 1500 and today, it has grown by a factor of 12, and much of that growth has been uneven. Per-capita gross domestic product (GDP) grew from 1500 to today by a factor of 10. And with these population and GDP explosions came exponential growth in technologies, resource use, infrastructure, and environmental impacts.

Such growth rates, which are continuing, imply dramatically expanded interactions between human and natural systems, frequently based on fundamentally new technology systems such as railroads, electricity, and the internet. They also explain the shift from a planet where humans are but one species among many to a world increasingly shaped by the activities of, and for the purposes of, a single species. In such a world, natural cycles and systems transmute from exogenous conditions into infrastructure components.

The growth experienced during this Industrial Era can be framed from a complex adaptive systems perspective. Economies grew and reorganized naturally, and infra-structure grew to support those activities. Energy is sometimes used as the core resource to describe the economic growth of complex adaptive systems. Early humans had few technologies to harness the abundance of fossil and renewable energy around them. The development of technologies and specializations that grew rapidly during the agricultural era and exploded during the industrial era to access and better use energy resulted in increasing complexity in human societies. Agricultural societies relied largely on free solar energy; in the Industrial Era society transitioned to fossil fuels.

Ultimately, as the cost of local energy escalates, eco-nomic activity switches to less energy-intensive services and relies on manufacturing and resources from other countries. Throughout these transitions, infrastructure are designed and deployed to provide services that are driven by these activities, many of which rely on the direct consumption of resources (energy, water, materials, etc.). Where abundant resources exist, infrastructure is often "grown" to consume them. Indeed, the Industrial Revolution and subsequent global economic and human population growth reflects the greater availability of energy, which is both an infrastructure and a resource issue. Thus, for example, in the United Kingdom waterpower was the original energy source in the early Industrial Revolution, but it was rapidly augmented, and then replaced, by fossil fuel use—first coal and then petroleum.

Looking at the processes of growth of complex adaptive systems through the lens of resource exploitation and the managing of unintended consequences has led some scholars to hypothesize about the inevitability of collapse. Nonetheless, although the environmental and energy implications of industrialization and global development have encouraged such dystopian scenarios, economic and demographic growth is a mixture

of human institutions and cultural factors; environmental and resource issues and constraints; and interdependent, co-evolving technological and
economic factors. It is not clear what "collapse" means in the context of
such evolving systems.

For example, it is now accepted that the "fall of Rome" was not the
collapse of Western civilization at the hands of barbarian tribes, but a shift
in form that enabled the subsequent rise of modern civilization. Infrastructure systems are complex precisely because single-domain determin-
isms and ideological certainties fail; in the end, it is not the belief system,
but whether the infrastructure works, that marks an engineering and institutional success. And, especially in the future, the ability to integrate
across human, natural, and built systems, with all the concomitant complexity, requires a fundamental shift in how we view the relationship between human and environmental systems.

The rapidly increasing pace of technological change, human culture,
and built environments, coupled with their interactions with natural systems, has produced novel and highly complex emergent behaviors that
require us to think differently about infrastructure. This complexity is the
result of rapidly coevolving human and natural systems. The changes in
technological, human, built environment, and natural systems, and the
complex outcomes they produce are not intentional. Changes in climate,
biodiversity, nutrient cycles, and microbial evolution are unplanned dynamics, the result of technology, policy, and cultural shifts that have been
accumulating for generations. But given the emergence and global scale
of environmental challenges, the notion that infrastructure should be limited to local engineered systems must be challenged.

Reflecting this shift to a planet dominated by human impacts and activities, scientists have proposed the term "Anthropocene," from the
Greek for "human" ("Anthropo-") and "new" ("-cene"), as an appropriate
name for the current geologic period. This is not a new idea: while the
term was popularized in an article in 2000 entitled "The 'Anthropocene,'"
the concept underlies earlier texts such as Man's Role in Changing the
Face of the Earth (1956) and The Earth as Transformed by Human Action
(1993). Indeed, as early as 1873 the Italian priest, geologist, and paleontologist Antonio Stoppani used the term "anthropozoic era."

The challenges and opportunities of the Anthropocene require the development of new institutions and frameworks that are beyond today's
traditional disciplinary structures and reductionist approaches. Our current infrastructure institutions compartmentalize knowledge and emphasize disciplinary expertise with a focus on components. In contrast, the
anthropogenic Earth will be characterized by rapidly evolving technolo-

gies, human and natural systems, and information access. This will require us to develop new capabilities to analyze, design, engineer, construct, maintain, manage, and reconstruct infra-structure. The increasing complexity and rates of change require new approaches and sophistication for infrastructure. Sustainable engineering and associated domains; industrial ecology and associated methodologies such as life-cycle assessment (LCA) and materials-flow analysis; systems engineering; adaptive management; as well as relevant parts of the urban planning and sociology of technology are each positioned to support this complexity.

The institutions and engineering disciplines that today are responsible for infrastructure are still appropriate for many challenges. In this book we use the term "infrastructure" to encompass a plurality of parts (both physical and institutional) that provide services that enable human capabilities. The discussion largely focuses on infrastructure such as physical systems. The competencies for designing and building a bridge, jet turbine, or microchip are successfully taught and practiced across engineering. However, when it comes to complex and integrated systems, our infrastructure institutions and training are ill-prepared.

At the beginning of the Anthropocene our systems are becoming increasingly integrated, complex, and defined by a greater number of stakeholders with competing priorities. We find ourselves with little training or capability to perceive or parse such systems to design, build, operate, refurbish, or retire infrastructure. This inability is tied to complex governance, fiscal, and educational systems (to name a few) with substantial inertia over time. Codes, conventions, processes, and procedures get developed and are embedded in legal processes and instruments that take considerable time and effort to change. Moreover, different groups (professional associations, government bureaucrats, engineering firms, and university courses) are invested in particular ways of doing things. The complexity of the effects of engineered systems is becoming apparent, even if we don't have the training to fully understand it.

Transportation infrastructure development, for instance, must consider shared, electric, and connected vehicles; power infrastructure, new battery storage technology and rapid advancement of renewables; and water infrastructure and climate change. Additionally, considerations for social equity, cultural impacts, aging populations, terrorism, and the increasing connectedness of hardware and embedding of software must be included.

As natural systems increasingly become design spaces for infrastructure, new approaches are needed for planning, constructing, operating, rehabilitating, and decommissioning infrastructure. As the scale, scope, and technologies of human activities have accelerated, the reductionist approach when assessing the relationship between infrastructure and the environment is no longer acceptable.

We posit that fundamental challenges to managing infrastructure exist in both education and design, arguing that in the short-term training should include competencies in complex systems, big data, artificial intelligence, and cybersecurity. In the medium-term new approaches are needed that emphasize agility and flexibility. And in the long term we must embrace the complexity inherent in the infrastructure-environment interface and evolve our infrastructure institutions to embrace this complexity.

It is not our intention to exhaustively explore the ways in which infrastructure design and management should change, but instead to establish the need to rethink the relationship between infrastructure and the natural world, and the fundamental challenges for meeting human needs in the coming century. We do not take the position that understanding leads to control. On the contrary, with complex adaptive systems, it is the inability to control the system that is the problem — one that requires the embrace of uncertain emergent behaviors.

1.1.3. RECOMMENDED READING

Allenby, B. (2007) 'Earth Systems Engineering and Management: A Manifesto', *Environmental Science & Technology*, 41(23), pp. 7960–7965. Available at: https://doi.org/10.1021/es072657r.

Allenby, B. *et al.* (2014) 'Anthropocene', in *Ethics, Science, Technology, and Engineering: A Global Resource*. Farmington Hills, MI: Gale, Cengage Learning, pp. 690–690.

Allenby, B. (2019) '5G, AI, and big data: We're building a new cognitive infrastructure and don't even know it', *Bulletin of the Atomic Scientists*, 19 December. Available at: https://thebulletin.org/2019/12/5g-ai-and-big-data-were-building-a-new-cognitive-infrastructure-and-dont-even-know-it/ (Accessed: 13 October 2023).

Allenby, B.R. (2012) *The Theory and Practice of Sustainable Engineering*. Hoboken, NJ: Upper Saddle River: Pearson Prentice Hall.

Allenby, B.R. (2015) 'The paradox of dominance: The age of civilizational conflict', *Bulletin of the Atomic Scientists*, 71(2), pp. 60–74. Available at: https://doi.org/10.1177/0096340215571911.

Arbesman, S. (2016) *Overcomplicated: Technology at the Limits of Comprehension*. New York, NY: Penguin Publishing Group.

Beinhocker, E.D. (2006) *The Origin of Wealth: Evolution, Complexity, and the Radical Remaking of Economics*. Cambridge, MA: Harvard Business Press.

Bendix, R. (1977) *Max Weber: An Intellectual Portrait*. Berkeley, CA: University of California Press.

Berkes, F., Folke, C. and Colding, J. (2000) *Linking social and ecological systems: management practices and social mechanisms for building resilience*. Cambridge, UK: Cambridge University Press. Available at: http://www.gbv.de/dms/bowker/toc/9780521591409.pdf (Accessed: 24 November 2023).

Bijker, W.E., Hughes, T.P. and Pinch, T.J. (1997) *The Social Construction of Technological Systems: New Directions in the Sociology and History of Technology*. Cambridge, MA: MIT Press.

Birdzell (1986) *How The West Grew Rich*. New York, NY: Basic Books.

Carse, A. (2016) 'Keyword: infrastructure: How a humble French engineering term shaped the modern world', in A. Morita, C.B. Jensen, and P. Harvey (eds) *Infrastructures and Social Complexity*. Oxford, UK: Routledge, pp. 45–57. Available at: https://doi.org/10.4324/9781315622880-11 (Accessed: 15 October 2023).

Carse, A. and Lewis, J.A. (2017) 'Toward a political ecology of infrastructure standards: Or, how to think about ships, waterways, sediment, and communities together', *Environment and Planning A: Economy and Space*, 49(1), pp. 9–28. Available at: https://doi.org/10.1177/0308518X16663015.

Chester, M. and Allenby, B. (2018) 'Reconceptualizing Infrastructure in the Anthropocene', *Issues in Science and Technology*, 34(3), pp. 58–63.

Chester, M.V. and Allenby, B. (2018) 'Toward adaptive infrastructure: flexibility and agility in a non-stationarity age', *Sustainable and Resilient Infrastructure*, 4, pp. 1–19. Available at: https://doi.org/10.1080/23789689.2017.1416846.

Chester, M.V. and Allenby, B. (2019) 'Infrastructure as a wicked process', *Elementa: Science of the Anthropocene*. Edited by A. Iles and M.E. Chang, 7(1), p. 21. Available at: https://doi.org/10.1525/elementa.360.

Clark, S.S., Seager, T.P. and Chester, M.V. (2018) 'A capabilities approach to the prioritization of critical infrastructure', *Environment Systems and Decisions*, 38(3), pp. 339–352. Available at: https://doi.org/10.1007/s10669-018-9691-8.

Crutzen, P. and Stoermer, E. (2000) 'The Anthropocene', *Global Change Newsletter*, 41, pp. 17–18.

11

Friedlingstein, P. *et al.* (2006) 'Climate–Carbon Cycle Feedback Analysis: Results from the C4MIP Model Intercomparison', *Journal of Climate,* 19(14), pp. 3337–3353. Available at: https://doi.org/10.1175/JCLI3800.1.

Geiger, A. (2017) *The Partisan Divide on Political Values Grows Even Wider.* Washington, DC: Pew Research Group. Available at: https://www.pewresearch.org/politics/2017/10/05/the-partisan-divide-on-political-values-grows-even-wider/ (Accessed: 13 October 2023).

Graedel, T.E. and Allenby, B.R. (2010) *Industrial Ecology and Sustainable Engineering.* Upper Saddle River, NJ: Pearson.

Grübler, A. (1998) *Technology and Global Change.* Cambridge, UK: Cambridge University Press.

Gunderson, L.H., Holling, C.S. and Light, S.S. (1995) *Barriers and Bridges to the Renewal of Regional Ecosystems.* New York, NY: Columbia University Press.

Hall, P. (1998) *Cities in Civilization.* New York, NY: Pantheon Books.

Hughes, T.P. (2000) *Rescuing Prometheus: Four Monumental Projects that Changed Our World.* New York, NY: Vintage Books-Knopf Doubleday Publishing Group.

Kissinger, H. (2014) *World Order.* New York, NY: Penguin Publishing Group.

Kossiakoff, A. (2011) *Systems engineering: principles and practice.* 2nd ed. Hoboken, NJ: Wiley (Wiley series in systems engineering and management). Available at: http://uclibs.org/PID/290288 (Accessed: 24 November 2023).

Kurzweil, R. (2005) *The Singularity Is Near: When Humans Transcend Biology.* New York, NY: Viking Books.

Lins, H.F. (2012) 'A Note on Stationarity and Nonstationarity', in *Commission for Hydrology, Advisory Working Group. Fourteenth Session of the Commission for Hydrology,* World Meteorological Organization.

McFate, S. (2019) *The New Rules of War: Victory in the Age of Durable Disorder.* New York, NY: William Morrow-HarperCollins.

Rajkumar, R. (2012) 'A Cyber–Physical Future', *Proceedings of the IEEE,* 100(Special Centennial Issue), pp. 1309–1312. Available at: https://doi.org/10.15779/Z38R11D.

Scott, J.C. (2017) *Against the Grain: A Deep History of the Earliest States.* New Haven, CT: Yale University Press.

Syvitski, J. (2012) 'Anthropocene: An epoch of our making - IGBP', *Global Change Magazine*, (78), p. 12.

Tainter, J. (1990) *The Collapse of Complex Societies*. Cambridge, UK: Cambridge University Press.

Thomas, W.L. (1956) *Man's Role in Changing the Face of the Earth: Int. Symp. Organized by the Wenner - Gren Foundation for Anthropological Research at Princeton, June 1955*. Chicago, IL: U. of Chicago Press, for Wenner-Gren Found. for Anthropological Research & Nat. Science Foundation.

Turner, B.L. *et al.* (1993) *The Earth as Transformed by Human Action: Global and Regional Changes in the Biosphere over the Past 300 Years*. Cambridge, UK: Cambridge University Press.

Tverberg, G. (2015) *A new theory of energy and the economy – Part 1 – Generating economic growth, Our Finite World*. Available at: https://ourfiniteworld.com/2015/01/21/a-new-theory-of-energy-and-the-economy-part-1-generating-economic-growth/ (Accessed: 15 October 2023).

Weck, O.L.D., Roos, D. and Magee, C.L. (2011) *Engineering Systems: Meeting Human Needs in a Complex Technological World*. Cambridge, MA: MIT Press.

1.2

EARTH SYSTEMS ENGINEERING
AND MANAGEMENT

The Industrial Revolution led to changes in human demographics, agricultural and technology systems, cultures, and economic systems. A principal result has been the evolution of an anthropogenic Earth in which the dynamics of major natural systems are increasingly affected by human activity. That does not mean deliberately designed by humans, because many things, from urban systems to the internet, are clearly human in origin yet have not been consciously designed by anyone. But it does mean an Earth where human activity increasingly modulates all Earth systems to the point where those things that are not subject to such impact, such as perhaps volcanoes and earthquakes, are increasingly limited and rare.

It is a world characterized by rapidly increasing integration of human culture, built environments, and natural systems to produce novel and complex emergent behaviors that are beyond traditional disciplinary structures and reductionist approaches.

The boundaries reflected in today's engineering disciplinary structures, and indeed in academic systems as a whole, are still appropriate for many problems. But we fail at the level of the complex, integrated systems and behaviors that characterize the anthropogenic Earth. No disciplinary field in either the physical or social sciences addresses these emergent behaviors, and very few even provide an adequate intellectual basis for parsing such complex adaptive systems. This situation has two important implications for infrastructure professionals.

First, it means that infrastructure managers cannot continue to rest on their traditional strengths, which are increasingly inadequate given today's social, economic, environmental, and technological demands. For

This chapter was adapted from the following journal article with publisher permission: Braden Allenby, 2007, Earth Systems Engineering and Management: A Manifesto, *Environmental Science & Technology*, 41(23), pp. 7960-7965, doi: 10.1021/es072657r.

example, a road built into a rain forest to support mineral exploitation be-comes a corridor of development and environmental degradation. Simi-larly, a new airport in a developing country dramatically increases tour-ism and puts pressure on fragile, previously remote, ecosystems. Alterna-tively, planning for urban transportation infrastructure increasingly re-quires understanding the status of the information and communication technology (ICT) infrastructure, because ICT enables virtual work struc-tures that affect potential traffic loading and peak patterns. In every case, traditional infrastructure approaches, although necessary, do not address the systemic impacts of the project. Infrastructure is critical but not neu-tral.

Second, from a proactive viewpoint, the anthropogenic Earth is a highly complex and tightly integrated system that challenges society to rapidly develop tools, methods, and knowledge that enable reasoned re-sponses. Infrastructure professionals must be a critical part of any such response. As problem solvers who create solutions in the real world, they have to understand and appropriately consider this new and more com-plex environment within which we work and create future options for changing ecosystems, built environments, and human culture. The ra-tional and analytical infrastructure culture, along with the role of infra-structure professionals in creating and maintaining the built environment, makes the infrastructure community a necessary partner — indeed, leader — in Earth systems engineering and management (ESEM).

1.2.1. PRINCIPLES OF EARTH SYSTEMS ENGINEERING AND MANAGEMENT

Continued stability of the information-dense, highly integrated hu-man, natural, and built systems that characterize the anthropogenic Earth requires the ability to rationally design, engineer and construct, maintain and manage, and reconstruct such systems — in short, an ESEM capability. Although this is an unprecedented challenge, ESEM can draw on experi-ence from many existing areas of study and practice. From a technical per-spective, these include industrial ecology methodologies such as life-cycle assessment, design for environment, materials flow analysis, resilience en-gineering, and systems engineering. From a managerial perspective, it draws on the literature about learning organizations and adaptive man-agement. Parts of the urban planning, sociology of technology, and social construction literatures are also relevant.

On the basis of these discourses, a tentative and partial, albeit instruc-tive, set of initial ESEM principles can be developed.

Given our current level of ignorance, only intervene when necessary, and then only to the extent required, in complex systems. This follows from the obvious need to treat complex adaptive systems with respect, because their future paths and reactions to inputs can seldom be predicted. It supersedes formulations such as the precautionary principle, which, in holding that new technologies should not be introduced if the risks cannot be known, demands an unrealistic level of knowledge of the future. Moreover, engineers who solve problems in the real world must accept the world as it is — globalizing, growing rapidly economically, with a population of more than 7 billion people, and heavily reliant on technological systems. Intervention is thus not discretionary, as some would rather fancifully wish, but it nonetheless must be careful and rational.

The capability to model and dialogue with major shifts in technological systems should be developed before, rather than after, policies and initiatives encouraging such shifts. Although projections of technological evolution are seldom accurate, we could do much better in developing frameworks, tracking systems (including metrics, especially ones that signal potential danger), and families of scenarios that would help us perceive problematic trends in real time, and perhaps steer technological evolution to in-crease social and environmental benefits. Such systematic tracking capabilities can help avoid some of the costs of premature adoption of emotionally appealing technologies. Recent examples might include the current infatuation with the hydrogen economy or the massive effort by the United States to create a corn-based ethanol energy economy. The point is not, of course, that technology shifts may not be beneficial; the point is to improve their design and management as they evolve within the real world.

A characteristic of complex systems is that the network that is relevant to a particular analysis is called forth by that analysis. Accordingly, it is critical to be aware of the particular boundaries within which one is working and to be alert to the possibility of logical failure when one's analysis goes beyond the boundaries. For example, to perform a study of New York City's water supply by considering only the five constituent boroughs of New York would result in a flawed assessment, because the system being analyzed (water provision to the city) is not mapped adequately by the political boundaries of the city. Similarly, the application of a life-cycle assessment tool that relied heavily on energy consumption as a proxy for environmental damage to a product where toxicity was a primary issue might well result in dysfunctional conclusions. For example, replacing chlorofluorocarbon-cleaning technologies with aqueous ones in electronics manufacturing makes sense from a systems perspective, even though the latter is more energy intensive.

A point that is critical to an understanding of the anthropogenic world is that the actors and designers are also part of the system they are purporting to design, creating interactive flows of information (reflexivity) that make the system highly unpredictable and perhaps more unstable. As scientists develop data on the effects of global climate change, for example, people's perceptions are changed. This, in turn, changes social practices affecting the climate. Thus, activities at the levels of the emergent behaviors of these complex systems must be understood as processes and dialogues, rather than simply problems to be solved and forgotten. This is an issue that bifurcates engineering: most engineering still involves artifacts, but ESEM requires ongoing and highly sophisticated dialogues with the systems at issue.

Implicit social engineering agendas and reflexivity make ethical and value implications inherent in all ESEM activities. To achieve long-term clarity and stable, effective policies, these normative elements must be explained and accepted, rather than hidden.

Conditions characterizing the anthropogenic Earth require democratic, transparent, and accountable governance and pluralistic decision-making processes. Virtually all ESEM initiatives raise important scientific, technical, economic, political, ethical, theological, and cultural issues in an increasingly complex global polity. Given the need for consensus and long-term commitment, the only workable governance model is one that is democratic, transparent, and accountable.

We must learn to engineer and manage complex systems, not just artifacts. An obvious result of the above analysis is that the anthropogenic world—and ESEM as a response—requires that far more attention be paid to the characteristics and dynamics of the relevant systems, rather than just to constituent artifacts. This does not negate the need to design artifacts; ESEM augments, instead of replaces, more traditional activities.

Given the complexity of the systems involved, our relative ignorance, and the recognition of engineering as a process, it follows that continual learning at the personal and institutional level must be built into project and program management. Some experience with this approach already exists. High-reliability organizations, such as aircraft carrier operations or well-run nuclear power plants, usually have explicit learning structures. Similarly, the adaptive management approach to complex natural resource management challenges, such as in the Baltic Sea, the Everglades, and the North American Great Lakes, is heavily dependent on continual learning.

Unlike simple systems, complex, adaptive systems cannot be centrally or explicitly controlled. Accordingly, it's important to understand not just the substance of the system—the biology of the Everglades or the Baltic, for example, or the physics and chemistry of the troposphere—but also inherent systems dynamics. Where in a system do small shifts propagate across the system as a whole, and where are they dampened out? The famous example of the butterfly that flaps its wings and causes a storm elsewhere in the world may be iconic, but what is perhaps forgotten is that millions of butterflies flap their wings thousands of times each day, without causing an ensuing storm. Perhaps the really interesting question, then, is why one flap has such an impact, when the others don't.

Whenever possible, changes should be incremental and reversible, rather than fundamental and irreversible. Accordingly, premature lock-in of system components should be avoided where possible because it leads to irreversibility. In complex systems, practices and technologies can get locked in quickly—that is, coupled to other systems and components in such a way as to make subsequent changes, including reversion to previous states, difficult or impossible. Thus, tightly coupled networks are more resistant to change than loosely coupled networks, an effect that can be reduced by ensuring that, when couplings to other networks do exist, they are designed to be as loose, and as few, as possible. This supports the more general goal of reversibility: under conditions of high uncertainty and complexity, easy reversibility is a desirable option should the system begin to behave in an un-anticipated and undesired way.

ESEM projects should aim for resiliency, not just redundancy, in design. Redundancy provides backup capability in case a primary system fails, and it is commonly de-signed into high-reliability systems such as jet airplanes. Redundancy assumes, however, that the challenge to the system is of a known variety. Resiliency, to the contrary, is the ability of a system to resist degradation or, when it must degrade, to do so safely even under unanticipated conditions.

1.2.2. DEVELOPING AN ESEM CAPABILITY

One way to begin responding to the challenge of the anthropogenic Earth, as well as continuing the process of clarifying and better understanding ESEM, is to develop a model research agenda. The complexity of the challenges does not allow for more than a partial and exploratory exercise at this point, and the examples given below are also idiosyncratic in that they reflect an engineering perspective on ESEM. In addition, it is a legitimate concern that any discipline, including engineering, that attempts to train professionals to design, engineer, manage, and interact with such complex systems is doomed to overreach and fail. Nonetheless, it is also important to remember that these effects, from climate change to massive urbanization, are already occurring. Our failure to accept responsibility for them does not diminish human impacts but is merely an evasion of our ethical duties. Engineering has an important role here: its projects are frequently the vehicle by which these large and complex systems are affected, and engineering education — rational, quantitative, problem-oriented, systems-based, and pragmatic — is a solid base upon which to build the required expertise and insight.

Accordingly, in addition to its specific research goals, any ESEM research agenda should aim to support the development of highly transdisciplinary research programs capable of looking at Earth systems at emergent levels (including, importantly, the social science dimensions; ideology and politics are often as important as any physical feature of the system). It should also support an overarching program that mines specific research areas for general principles and learning that over time can be leveraged into development of a rational, responsible, and ethical ESEM framework.

1.2.3. INTEGRATED URBAN INFRASTRUCTURE SYSTEMS

Given accelerating urbanization, increasing urban vulnerability to natural disaster or deliberate attack, and the complexity of urban systems, the emergent domain of urban infrastructure systems as comprehensive wholes is underappreciated. Yet, at this point no U.S. government agency, research funding organization, or engineering discipline has the mission or research support for understanding urban systems as integrated systems.

This is a near-term concern because of the increasing demand for new and replacement infrastructure. At the same time, the nature of urban systems is changing profoundly as ICT capability is increasingly integrated into all levels of urban functionality: sensor systems, smart materials, smart buildings, smart infrastructure, and the like. Especially as ICT systems are redesigned to be autonomic — virtualized, self-defining, self-monitoring, self-healing, and learning-capable at all scales from chip to computer to regional and global grids — the implications for urban system design, performance, and behavior accelerate in complexity. Moreover, the increasing role of urban systems as nodes in energy, financial, and virtual information networks adds many layers of information complexity to urban environments. Such research should contribute to a new infrastructure competency in urban-scale systems design and management.

1.2.4. SUSTAINABLE INFRASTRUCTURE

Growing populations, economic development, accelerating technological change, urbanization, and aging and failing infrastructure systems are increasing the need for sustainable infrastructure systems. Although ESEM provides some conceptual basis for developing such systems, clearly the translation of social interest in sustainability to the implementation of sustainable engineering of any type has just begun. It's currently marked by an intellectually confused jumble of superficial, ideological, and heuristic approaches.

Accordingly, a research program to help define sustainable infrastructure and to develop appropriate methodologies, analytical methods, and tools is needed. This is urgent because the time to understand and deploy sustainable infrastructure is now, instead of after newly built environments with decades of active life are constructed.

1.2.5. TECHNOLOGICAL CONVERGENCE

A number of authors, from the dystopian Bill Joy to the techno-optimist Ray Kurzweil, have written about the subject of technological convergence, generally understood as including the accelerating development of the fields of nanotechnology, biotechnology, information and communication technology, applied cognitive science, and robotics as well as their mutually reinforcing integration. These converging technologies constitute major Earth systems in their own right, and their complexity and challenging philosophical, religious, ideological, and economic implications are just beginning to be recognized. However, some of the major arenas where effects of technological convergence can first be seen are in areas familiar to engineering professionals. These include urban and regional integrated infrastructure design, energy technology and infrastructure design, and the like.

What is most challenging, perhaps, about technological convergence is not just its effect of turning natural systems — from the carbon and climate cycles to biology at all scales — into design spaces (and commodities). Rather, as humans gain the tools to design biological and cognitive systems, it also turns the human into a self-reflexive design space. In doing so, the feedback systems, and concomitant increases in system complexity, become truly daunting. Engineering traditionally has been based on the assumption (unspoken because it was so clearly fundamental and valid) that the environment must be designed and built for the human. As both parts of that assumption become design spaces, and thus interact in new and dynamic ways, engineering becomes new, more complex, and ethically challenging in ways that have never before been part of our professional experience. Although research programs designed to respond to this unique challenge do not lie entirely within engineering's ambit, we can bring significant skills to transdisciplinary research efforts.

1.2.6. RESILIENCE OF COMPLEX EARTH SYSTEMS

That complex natural Earth systems are vulnerable to destabilization as a result of human activity is evident from the depletion of stratospheric ozone by chlorofluorocarbons or from the global climate change dialogue. But the vulnerability is also apparent with anthropogenic systems: recent years have seen major challenges to social stability and order, ranging from extreme weather events to terrorist attacks to cyber and asymmetric warfare to substantial cultural conflict. Although each incident is unique and too often tragic, the key to understanding and responding to these constellations of challenges is to recognize that although each is expressed uniquely, they all represent emergent characteristics of the anthropogenic Earth—including, critically, information and cultural networks—at unfamiliar scales and levels of complexity. Thus, although immediate responses have necessarily relied primarily on specific engineering, institutional, and policy responses to particular incidents, the range of challenges, their systemic nature, and the practical impossibility of adequately addressing each one individually argue for adopting a more comprehensive systems perspective. This should be based on the principles of enhancing infrastructure, social, and economic resiliency; meeting security and emergency response needs; and relying to the highest extent possible on dual-use technologies that offer societal benefits, even if anticipated disasters never occur.

Patterns of the built and human environments play an important role in vulnerability. Thus, for example, the damage and disruption from weather events such as hurricanes or from natural disasters such as tsunamis are more disruptive and extensive than in the past because of changing demographic patterns (urbanization, for example) and the relocation of economic activity near riskier areas, such as geologically active Pacific Ocean coastlines. Disease epidemics and their associated economic and social effects are more challenging given the modern transportation infrastructure and globalized patterns of commerce and travel. Terrorism is not new, but terrorist access to weapons of mass destruction is. Cultural conflict is as old as historical records, but the internet and social media create an environment where a few cartoons in a small northern European country can ignite global unrest.

Engineering professionals have important roles in virtually all of these examples, including designing adequate levees; hardening buildings and infrastructure against attack and enabling rapid restoration of services and the built environment; and constructing energy, transportation, and ICT infrastructure that have profound and varied effects across regional and global natural systems. We should thus be leaders in enabling systemic understanding and enhancement of resilience across not just the built environment but also Earth systems as a whole. This is a substantial challenge: how can we, as the engineering community, begin the complex process of building the transdisciplinary capabilities necessary to understand, work, and live rationally, ethically, and responsibly in the world that we have already created?

These research challenges, and many others that undoubtedly come to mind, are "wicked" problems, because they are irreducibly complex and highly transdisciplinary, and require substantial changes in the way we think about infrastructure and engineering in general. Learning to work across the disciplinary divides involved will be exceedingly difficult and personally challenging for many individuals. Many engineers are not accustomed to accepting a leadership role in such a difficult task, but our age has its own imperative.

Activity in each area will be complicated because not enough trained individuals are available to begin many programs in these areas. Peer review will also be a challenge, both because finding appropriate panels will be nontrivial and because that process tends to be highly conservative when faced with profoundly transdisciplinary proposals. In many cases, ideological and even religious feelings run high.

But against all of these barriers lies one fact: we do not have a choice. These emergent behaviors are here, now. We only can decide as engineers and professionals whether to respond to these behaviors rationally and ethically or to ignore them, retreating to wishful fantasy and evading our professional and, indeed, personal responsibility to ourselves and the future.

1.2.7. RECOMMENDED READINGS

Allenby, B. (2000) 'Earth Systems Engineering and Management', *IEEE Technology and Society Magazine*, 19(4), pp. 10–24. Available at: https://doi.org/10.1109/44.890078.

Allenby, B. (2005) *Reconstructing Earth: Technology and Environment in the Age of Humans*. Washington, DC: Island Press.

Allenby, B., Allen, D. and Davidson, C. (2007) 'Sustainable Engineering: From Myth to Mechanism', *Environmental Quality Management*, 17(1), pp. 17–26. Available at: https://doi.org/10.1002/tqem.20148.

Allenby, B. and Fink, J. (2005) 'Toward Inherently Secure and Resilient Societies', *Science*, 309(5737), pp. 1034–1036. Available at: https://doi.org/10.1126/science.1111534.

AT&T (2005) *AT&T Network Continuity Overview*. Dallas, TX: AT&T, p. 19.

Berger, P.L. and Luckmann, T. (1967) *The Social Construction of Reality: A Treatise in the Sociology of Knowledge*. New York, NY: Anchor.

Bijker, W.E., Hughes, T.P. and Pinch, T.J. (1997) *The Social Construction of Technological Systems: New Directions in the Sociology and History of Technology*. Cambridge, MA: MIT Press.

Castells, M. (2000) *The Rise of the Network Society*. 2nd edn. Oxford, UK: Blackwell Publishers.

Graedel, T.E. and Allenby, B.R. (2002a) *Industrial Ecology*. Subsequent edition. Upper Saddle River, N.J: Pearson College Div.

Graedel, T.E. and Allenby, B.R. (2002b) *Industrial Ecology, Second Edition*. 2nd edn. Hoboken, NJ: Pearson Prentice Hall.

Gunderson, L.H., Holling, C.S. and Light, S.S. (1995) *Barriers and Bridges to the Renewal of Regional Ecosystems*. New York, NY: Columbia University Press.

Habermas, J. (1975) *Legitimation Crisis*. Boston, MA: Beacon Press.

Hughes, T.P. (1998) *Rescuing Prometheus*. New York, NY: Pantheon Books.

IBM (2005) *An Architectural Blueprint for Autonomic Computing*. IBM. Available at: https://www.semanticscholar.org/paper/An-Architectural-Blueprint-for-Autonomic-Computing./54bc974d3e9ebc3c0adbce99665cdcf2a94b27f1 (Accessed: 13 October 2023).

Joy, B. (2004) 'Why the Future Doesn't Need Us', *WIRED*, 1 April. Available at: https://www.wired.com/2000/04/joy-2/ (Accessed: 13 October 2023).

Kurzweil, R. (2005) *The Singularity Is Near: When Humans Transcend Biology*. New York, NY: Viking Books.

NRC (2003) *Cities Transformed: Demographic Change and Its Implications in the Developing World*. Washington, DC: National Academies Press, p. 550. Available at: https://doi.org/10.17226/10693 (Accessed: 13 October 2023).

Pool, R. (1997) *Beyond Engineering: How Society Shapes Technology*. Oxford, UK: Oxford University Press.

Rorty, R. (1989) *Contingency, Irony, and Solidarity*. Cambridge, UK: Cambridge University Press.

Senge, P.M. (1990) *The Fifth Discipline: The Art and Practice of the Learning Organization*. 1st edn. New York, NY: Doubleday/Currency.

1.3

INFRASTRUCTURE IN A RAPIDLY CHANGING WORLD

Any investment in infrastructure must be viewed through the lens of a rapidly changing world, where in the future climate, technologies, politics, and basic needs will be very different than today. But to constrain our definition of infrastructure to a narrow historical understanding that emphasized hardware, such as bridges and roads, over the software — including the institutions that govern infrastructure and the people whom infrastructure serves — is to miss a critical opportunity to invest in the services and needs to ensure future generations thrive. Much as telephone wires have been replaced by wireless cell towers, and banking is increasingly shifting from brick-and-mortar buildings to the internet, what constitutes "infrastructure" is evolving rapidly. Any policy position to frame infrastructure must recognize that the systems that deliver basic and critical services, that give us our economic strength, must be able to respond to a rapidly changing world.

To understand why infrastructure is often so poorly positioned to adapt to a rapidly changing world it's helpful to revisit its history. The core technologies and bureaucratic structures that define US infrastructure have persisted for decades. Some systems have been around for a century or more.

Today's infrastructure was built around the technologies and goals of the past. Twentieth century engineers and planners operated in an environment in which the climate was considered relatively stable and information technologies were in their nascent stages. They managed the financial risk of large capital investments by building infrastructure assets with very long lifetimes, based on the assumption that long-term demand and growth would produce sound investments.

This chapter was adapted from the following article with publisher permission: Mikhail Chester, 2021, Can Infrastructure Keep Up With a Rapidly Changing World?, *Issues in Science & Technology*.

Because of this historical set of preferences, most people today think of infrastructure as large, rigid systems, comprising purely physical assets: roads, rail networks, pipes, powerlines, pumps, and bridges. But these components cannot be separated from the "soft" processes that manage our infrastructure. Those bureaucratic structures that define the governance of infrastructure today also reflect those historical assumptions about stable operating conditions. The bureaucracy that emerged to manage that infrastructure was excellent for standardized goals (e.g., meeting performance metrics for roadway maintenance). But today, the governance models that largely steer infrastructure appear increasingly unable to navigate the growing complexity of the world. Preparing for the deep uncertainty associated with climate change, protecting the agency and users from cyberattacks, managing demand amidst growing numbers of services that steer demand (Uber, Lyft, Tesla, Google, Amazon drone delivery), making decisions that require consensus building among increasingly polarized voices requires new expertise, new ways of thinking about what infrastructure are, and new ways of doing infrastructure.

As new volatility confronts our old infrastructure, a gap is emerging between what our systems need to do and what they can do. Many infrastructure and their governance systems simply can't keep up. Transformational changes are coming on fast, and they are combining with other shifts, such as the integration of information technologies and artificial intelligence. Whereas in the past people exclusively mediated their relationships with themselves and the environment, humans are now increasingly ceding control to software. The rigidity inherent in legacy infrastructure is at odds with these changing conditions, and vicious feedbacks are beginning to develop.

Infrastructure's inability to keep pace with these changes can give new actors the opportunity to assert control. Amazon and Google are creating a new infrastructure of drone delivery that allows them to circumvent the risk of aging roadways. Tesla is increasingly managing energy use and storage. Google and Apple direct drivers using a navigational cognition of cities that travelers and transportation agencies are incapable of. Whereas city planners once decided which streets were major thoroughfares and supplied those roadways with resources for control and maintenance, smartphones now route drivers down formerly sleepy side streets in the name of efficiency.

Whereas in the past users decided how they consumed infrastructure services (e.g., transportation or energy), software informed by massive data streams are increasingly in control. Thus, the nature of what infrastructure are and who controls them is changing. As we experience the very nascent stages of this shift, infrastructure managers must modernize how we understand infrastructure, what we think they should do, and how they are governed.

In the future, we need to do a better job of aligning the systems that provide basic and critical services with the complexity of the world we're learning to live in. We must focus on "loose-fit" infrastructure solutions — assets that are modular, multifunctional, and scalable, with the ability to flex and change as society and the environment place new and unanticipated demands on them. In order to build these new infrastructure, we also need new ways of understanding them.

First, we should recognize that the scale and scope of human activities has grown so large that the dichotomy between infrastructure and the environment is shrinking. Human activities enabled by infrastructure have increasingly large and planetary effects, which in turn create increasingly complex conditions for those infrastructure. Rather than imagining that we are building one road or one power line, we need to recognize that we are building complex, integrated systems that require the negotiation of tradeoffs across social, environmental, and infrastructural dimensions. Gone are the days when a transportation agency could view its purview as simply roads and traffic. Now that portfolio contains a complex network of private, public, and shared technologies, operating between negotiated physical and virtual access, increasingly steered by cloud-based software, subject to increasing disturbances, such as extreme weather events and cyberattacks, viewed as a design space for mitigating climate change.

Second, we should transform our notion of infrastructure from one focused exclusively on hardware to one that reckons with the wicked, complex processes that require negotiation and adaptive management. The wicked complexity emerges from the grand problems — such as sustainability, resilience, and equity — that infrastructure must address, coupled with growing technical and social complexity. The technical complexity comes from decades of infrastructure buildouts layered on top of one another and newly integrated with information technologies. The social complexity derives from the decentralizing of control, growing number of players who decide how services are used, and the polarization that increasingly influences infrastructure decision-making. Understood in their full complexity, these challenges show that hardware alone cannot be the sum total of tomorrow's infrastructure.

Finally, we can no longer ignore that fact that infrastructure are increasingly connected and are themselves an information conduit. The rapid integration of cybertechnologies into infrastructure creates new capabilities and efficiencies, but it also produces vulnerabilities. Smart and connected technologies in infrastructure create opportunities for safety and cost improvements. At the same time, they create opportunities for foreign adversaries and criminal hackers to interrupt daily life in surprising ways. War is shifting where the whole of society can be targeted through our increasingly connected systems. Infrastructure that once seemed quite safe has become a battlespace.

Infrastructure must be able to adapt and transform to keep up with society's demands as well as shifts in the environment. Simply making legacy infrastructure do more work or work faster is not the solution. Nor is installing more rigid infrastructure. Instead, we must recognize the shifting nature of what infrastructure is, what society needs it to do, and who controls it. These new concerns necessitate new ways of approaching infrastructure.

Infrastructure investment at any level, whether it's federal spending or municipal budget allocations for local systems, must let go of the notion that the infrastructure of the past is appropriate for the accelerating and increasingly uncertain conditions of the future. Investing in the future means creating the conditions for infrastructure agencies to adapt and transform, integrate cybertechnologies, actively anticipate climate change and other extreme events, and restructure themselves to meet the challenges and needs of the next century. This will require an imagining of what we'll need in the future and how services will differ from today. Fundamentally, we must question whether today's infrastructure as they're structured and governed are appropriate for the future, and if not what new technologies and governance models are needed.

1.4

INFRASTRUCTURE AS A
WICKED COMPLEX PROCESS

Herbert Simon, a pioneer of decision-making theory, posited that given the vast information that is needed to completely understand how to maximize one's benefit from a particular course of action, people instead "satisfice." That is, they will make a decision about "which is good enough, rather than the absolute best (Simon, 1947, 1957; Brown, 2004; The Economist, 2009)." Satisficing, a combination of satisfy and suffice, describes the process by which, in situations of wicked complexity where optimization techniques fail, we often settle on a course of action that is good enough and "deals with a drastically simplified model of the confusion that constitutes the real world (Simon, 1957; Brown, 2004)."

We use the term "optimization" to describe traditional engineering decision making approaches that emphasize quantification of performance with a focus on maximizing performance or efficiency while minimizing costs and meeting some desired level of service. Simon initially developed the concept of satisficing at a time when commercial, industrial, and public sector organizations were restructuring post World War II to peacetime services. It had profound implications on how we view decision making, not as a process that uses complete and consistent information and preferences but instead as a "rational behavior that is compatible with the access to information and the computational capacities that are actually possessed by organisms (Simon, 1957)." With what appears to be growing complexity in the constraints imposed on how our infrastructure systems are designed and operated, it's necessary to examine how the role of engineering is changing from optimizing to satisficing.

This chapter was adapted from the following journal article with publisher permission: Mikhail Chester and Braden Allenby, 2019, Infrastructure as a Wicked Complex Process, *Elementa*, 7(21), pp. 1-11, doi: 10.1525/elementa.360.

Several long-term trends appear to be creating complexity for our infrastructure systems. First, our ability to change infrastructure at a meaningful pace appears to be constrained. The infrastructure that we rely on today shares many of the core design features from when these systems were initially conceived decades ago (Chester and Allenby, 2018). Because demand for the core services that infrastructure deliver hasn't changed for many civil systems, the physical forms that infrastructure takes and the institutions that support those forms have been arranged and connected with each other over such a long history that shifting to new forms of infrastructure appears a monumental challenge—we describe this effect as *lock-in*.

Second, as the scale and scope of human activities have grown, infrastructure systems have incorporated more and more of "nature." The dichotomy between the human-built and natural worlds is shrinking, thereby greatly increasing complexity, both technically and culturally (Chester and Allenby, 2018; Chester, Markolf and Allenby, 2019).

Third, increasingly rapid changes in technology are driving our infrastructure to deliver services that they weren't designed to deliver—a trend that will only increase in the future. This means that we need faster cycle times. Yet as per the first trend, our infrastructure appear to be increasingly locked-in, slowing down our ability to change them quickly. As the need for more rapid and complex infrastructure upgrades, redesigns, and construction accelerates, lock-in becomes a significant constraint and likely long-term trend.

Fourth, because civil infrastructure have persisted for so long there has been significant accretion—i.e., the accumulation, layering, and interconnectedness of technologies, institutions, rules, and policies—so that understanding the emergent behavior of infrastructure (particularly when perturbed) is now challenging if not impossible. Significant change now requires action across many different subsystems and their governing agencies. For example, modernizing the U.S. power grid requires the coordinated deployment, financing, and permitting of hardware and software across hundreds if not thousands of companies and agencies.

The combination of these four factors raises serious questions about whether rapidly changing demands, technologies, and perturbations (such as climate change or terrorist attacks) will affect our infrastructure's capacity to provide services. It is these concurrent trends that also demand a shift from optimization to satisficing as the critical engineering competency.

At the dawn of the Anthropocene—where rapid environmental, technological, social, political, and even cultural changes, increasingly at regional and global scales, are likely to put new demands on our infrastructure—our infrastructure remains mostly rigid, unable to change quickly (Chester and Allenby, 2018). Making infrastructure agile and flexible for the Anthropocene will require us to acknowledge and work with the fact that infrastructure are now wicked complex systems, and that satisficing has become the new normal for designing and managing these systems.

This has profound implications for infrastructure managers who have been trained to assess bounded design problems that can be optimized, rather than wickedly complex problems that must be satisficed. In this chapter we first explore how infrastructure have become wicked complex systems. We then discuss how this wicked complexity has led to a shift in how we design and operate infrastructure and the challenges this approach introduces when large and fast changes are needed. Lastly, we propose that engineering (including education) embraces the wicked complexity of infrastructure and the increasing satisficing that is needed to affect change.

1.4.1. INFRASTRUCTURE DESIGN THROUGH THE AGES

Modern infrastructure systems are several millennia in the making. The Neolithic revolution saw many human cultures shift from hunter gatherer societies to farming and agriculture. During the Neolithic period (approximately 10000 to 2000 BC), deforestation, terraforming, crop domestication, and irrigation resulted in densely populated settlements. The era also witnessed increased specialization in activities, the establishment of social and sexual roles and hierarchies, and new outlooks on ownership.

During this time, in addition to farming and irrigation, dirt road and durable good technology developed (Bogucki, 2011). Town systems by were common by the end of the Neolithic paving the way for the Classical era (800 BCE to 600 CE) in Europe, and significant scientific, technological, and cultural advances in the Chinese and Islamic civilizations. During this time the Greco-Roman world developed paved roadways, aqueducts and water mills, and metal use for building construction, communications, and energy use. Driven by extensive trading networks, transportation technologies advanced in Chinese and Islamic empires. And in Asia there was considerable development of agricultural technologies and practices.

Through the Middle Ages (500 to 1500 CE), infrastructure and associated technology development accelerated. The horse harness, watermills, heavy plow, metallurgy, gunpowder, gothic architecture, and the caravel developed in Europe, much of which was borrowed and modified from elsewhere (Gies and Gies, 1995). By the Industrial Era, infrastructure and technologies were rapidly changing.

The rate of change of infrastructure and technologies over the ages provides an important perspective on how much faster human systems are changing today than in the past. In the past century alone, the level of social, cultural, financial, and political complexity has exploded to the point where the rules by which we deploy infrastructure have drastically changed (Birdzell, 1986; Kurzweil, 2010; Marchant, Allenby and Herkert, 2011; Allenby, 2012).

Today's cities are integrative technologies in that they reflect the engineering, culture, and infrastructure of their various periods. Medieval cities, for example, had narrow, crooked streets crowded with non-standardized buildings that reflected the dominant transportation mode (walking and the occasional horse), and a lack of zoning or building regulations. The streets doubled as sewage infrastructure; the lack of toilets meant that chamber pots were used, and were simply emptied out of windows (Hall, 1998). Slow modernization of urban systems, from sanitation to public transit, marked the eighteenth and nineteenth centuries, culminating in "high modernism."

High modernism as practiced by planners such as Robert Moses in New York, and Le Corbusier in urban design, was marked by a brutal technocratic elitism which knowingly ignored local context and culture in the interest of implementing "universalist" and "scientific" principles in urban design. As Robert Moses declared, "When you operate in an over-built metropolis, you have to hack your way with a meat ax…. I'm just going to keep right on building. You do the best you can to stop it (as quoted by the philosopher Marshall (Berman, 1982a))" (Berman, 1982b). Critics such as Jane Jacobs and Charles Jencks derided high modernism as anti-human and hubristic, and favored "post-modern" city planning and architecture (Jacobs, 1961; Jencks, 1977).

And indeed top down high modernism was replaced by postmodernist bottom-up community activism, seen in such controversies as the defeat of the proposed Greenpoint-Williamsburg waste incinerator in Brooklyn and the London Docklands proposal, both in the 1990s (Hall, 1998; Gandy, 2003). The resulting power of activists and communities, while it accords with modern sentiment in many ways, makes many large projects such as Boston's "Big Dig" major political efforts, and can allow small but vocal minorities to stifle projects and infrastructure that the larger public good or urban community requires. In other words, the postmodern political environment of today makes wicked complexity integral to virtually any significant infrastructure project, a point that today's engineers—and the professors that teach them—cannot ignore.

Modern infrastructure is thus not a technical problem, it's a wickedly complex problem. Going forward, we explore this wicked complexity by first examining what is causing it and then discussing how conceptualizing the process of infrastructure change as satisficing numerous stakeholders and objectives, rather than simply optimizing a technical solution, is an increasingly important engineering framework, and hence professional skill.

1.4.2. WICKED COMPLEXITY

The process by which we design, build, operate, manage, rehabilitate, and decommission infrastructure is a wickedly complex problem. From an engineering perspective, infrastructure is often thought of and taught as a physical end-product, an agglomeration of specialized technological and built artifacts that provide services. However, as technologies and the demand for services change more and more rapidly, thinking of infrastructure as a physical, cultural, and institutional process providing an ever-changing function becomes important.

This means that while an individual piece of physical infrastructure can be optimized for its function within an existing built system, infrastructure design and modernization necessarily becomes wickedly complex. Wicked complexity is the result of three competing forces that are, given the current approaches for designing and managing infrastructure, inimical to rapid and sustained change of infrastructure in a future marked by uncertainty. These forces are 1) wicked problems, 2) technical complexity and lock-in, and 3) social complexity, which work together to fragment our capacity to manage change (see Figure 1.4.1).

There is a dearth of work characterizing engineered infrastructure as wicked complex systems, but it is becoming more widely accepted that this is the case. The complexity of infrastructure has often been defined in terms of physical structure, or technical complexity (Brown, Beyeler and Barton, 2004; Findlay and O'Rourke, 2007). The researchers Edward J. Oughton and Tyler (2013) characterize infrastructure complexity by contrasting the properties of general and complex adaptive systems at the interface of supply and demand (Oughton *et al.*, 2018). They posit that given emergent behavior and self-organization, instability and robustness, dynamics and evolution, adaptiveness, and agent diversity, infrastructure exhibit the key features of complex adaptive systems.

Figure 1.4.1: Wicked Complexity as a Product of Wicked Problems, Technical Complexity, and Social Complexity

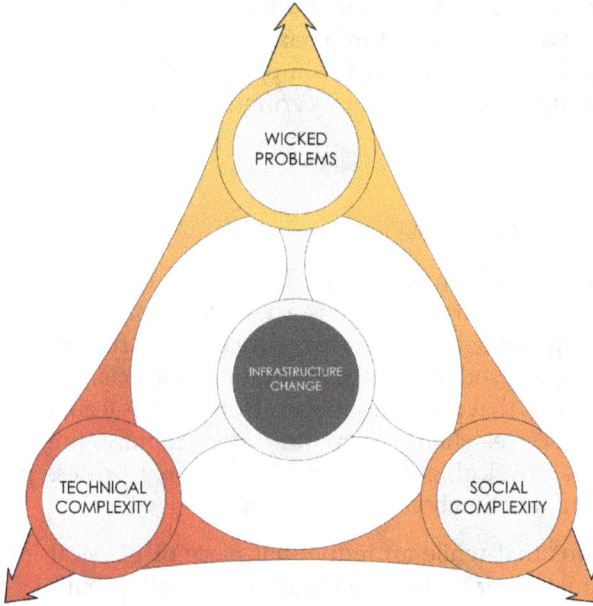

Adapted from J. Conklin, Dialogue Mapping: Building Shared Understanding of Wicked Problems (Hoboken, NJ: Wiley, 2006).

Complexity scientist Samuel (Arbesman, 2016a) describes technical complexity as resulting from several dominant technological forces, including accretion (the accumulation and layering of technologies and assets over long periods of time), interaction (an often-indeterminate number of components interact in ways that we no longer fully comprehend), and edge cases (we sometimes design hardware and processes outside of standard design and operating rules) (Arbesman, 2016b). The combination of these forces, and equally complex social, financial, cultural, technological, and management forces that affect infrastructure at every scale, results in a situation where the emergent behavior, particularly when perturbed, is no longer predictable. Defining infrastructure complexity in terms of physical structure is still appropriate for many purposes, of course, but rapidly becomes inadequate beyond the artifactual and short term, and especially when planning for new technologies.

There are many complexity definitions, often specific to the field and system or organization being analyzed. The Cynefin framework, created by management consultant Dave Snowden in 1999, classifies systems as simple (the domain of best practice), complicated (the domain of experts), complex (the domain of emergence), chaotic (the domain of rapid response), and disorder (Snowden and Boone, 2007). Knowing whether you're working in the complicated or complex domain when it comes to infrastructure is critical because each domain requires fundamentally different approaches.

The complex domain (the realm of unknown unknowns) is one where the inability to predict emergent behaviors means that we can only understand what happened after the fact. Emergence (the ability of individual components of a large system to work together to give rise to dramatic and diverse behavior) is often central to characterizing a system as complex (Cilliers, 2002; Martin and Helmerson, 2014). Although emergence is front and center, complexity is determined from a number of characteristics, including the number of elements, interactions that are dynamic, rich, non-linear and short range, feedback loops, system openness, non-equilibrium behavior, histories, and (related to emergence) elements which are ignorant to the behavior of the whole system (Cilliers, 2002).

The critical boundary for engineers is between complicated and complex. At this transition, optimization tools become obsolete as social, institutional, political, and economic forces come into play and together render insufficient quantitative analyses as the dominant approach for addressing the challenges. Organization scholar Jeffrey (J. Conklin, 2006) describes social complexity as a force of fragmentation that results from the number and diversity of players involved in a project (wickedness is another source of fragmentation in decision making) (E. J. Conklin, 2006). The more parties involved and the more different the stakeholders are, the greater the diversity of perspective they bring, and the more socially complex a decision-making process becomes, thereby making collective intelligence a challenge and consensus virtually impossible to achieve (E. J. Conklin, 2006).

Figure 1.4.2: Cynefin Framework Applied to Infrastructure

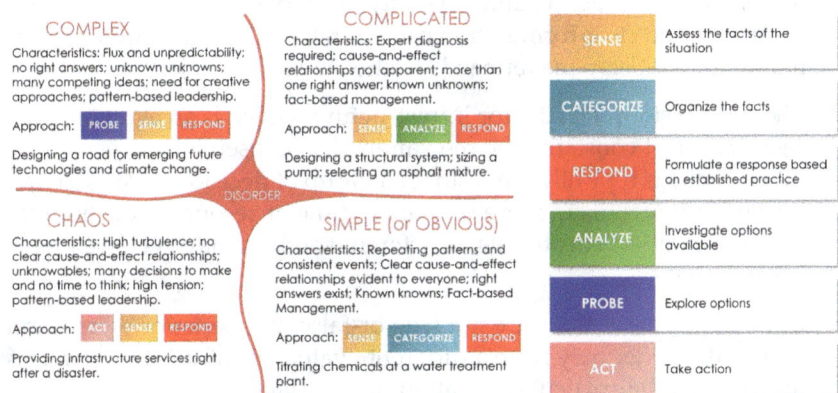

COMPLEX
Characteristics: Flux and unpredictability; no right answers; unknown unknowns; many competing ideas; need for creative approaches; pattern-based leadership.

Approach: PROBE SENSE RESPOND

Designing a road for emerging future technologies and climate change.

COMPLICATED
Characteristics: Expert diagnosis required; cause-and-effect relationships not apparent; more than one right answer; known unknowns; fact-based management.

Approach: SENSE ANALYZE RESPOND

Designing a structural system; sizing a pump; selecting an asphalt mixture.

CHAOS
Characteristics: High turbulence; no clear cause-and-effect relationships; unknowables; many decisions to make and no time to think; high tension; pattern-based leadership.

Approach: ACT SENSE RESPOND

Providing infrastructure services right after a disaster.

DISORDER

SIMPLE (or OBVIOUS)
Characteristics: Repeating patterns and consistent events; Clear cause-and-effect relationships evident to everyone; right answers exist; Known knowns; Fact-based Management.

Approach: SENSE CATEGORIZE RESPOND

Titrating chemicals at a water treatment plant.

SENSE	Assess the facts of the situation
CATEGORIZE	Organize the facts
RESPOND	Formulate a response based on established practice
ANALYZE	Investigate options available
PROBE	Explore options
ACT	Take action

The somewhat ambiguous boundary between highly complicated technological artifacts or subsystems (for which optimization techniques are often appropriate) and wicked complexity (where satisficing techniques are necessary) occurs when operational requirements mean that infrastructure cannot simply be defined in terms of physical assets, but must be viewed as a process. Sometimes, after all, one is simply adding a component where the design is significantly constrained by existing function and assets — a tower to an existing urban mobile device network, or a new water feed to a subdivision.

But especially as human, natural, and built systems integrate in new ways at virtually all scales, the fundamental goal of infrastructure — to provide the resources and services needed to allow people the capabilities that build adaptive capacity and improve human well-being (Clark, Seager and Chester, 2018) — becomes more explicit. At this point, wicked complexity begins to dominate engineering decisions, from design to operation to maintenance to system evolution, and the lens of the engineer must broaden to consider not only the physical structures but also the institutional, cultural, financial, and other forces that affect infrastructure, both in the immediate environment and as the infrastructure process evolves over time.

The pioneering work of Horst W. J. (H. Rittel and Webber, 1973) provides the foundational criteria for wicked problems that for the most part endures today (H. W. J. Rittel and Webber, 1973). Instead of simply reiterating their 10 criteria, we'll instead describe them in the context of infrastructure:

1. Wicked problems have no definitive problem formulation. Providing services at reasonable costs, given financial constraints, competing stakeholder views, environmental and social implications of various options, and with agility and flexibility to adapt to changing conditions means that no clear formulation can be stated containing all the information the decision maker needs to choose a single objective solution.

2. There is no stopping rule for infrastructure, which is an evolving process that must constantly be re-examined. A chosen infrastructure solution today is typically the result of a group of decision makers saying it's good enough. Had there been more time or resources, or different social or political constraints, a different solution may very well have been chosen.

3. Infrastructure solutions are not true-or-false, but better or worse, and each stakeholder may have a different evaluation of what these terms mean given the particular situation. It is usually not the case in these situations that a choice will be regarded as optimal by any party; rather, the best choice may be one that meets the minimal requirements

of most stakeholders (in other words, the one that satisfices, rather than optimizes, from their perspective).

4. There is no immediate and no ultimate test of an infrastructure solution. The implementation of infrastructure will lead to direct and indirect consequences over an extended time period.

5. Given the sheer scale and reach of infrastructure, implementation is often a one-shot operation. There is no opportunity to learn by trial and error, and every attempt counts significantly. This does not mean, however, that engineers and engineering institutions such as universities and professional organizations should not explicitly embrace continual learning processes, because the effectiveness of a particular one-shot solution may well inform better practices elsewhere. For example, the Big Dig in Boston is a one-shot solution in that it is not significantly modifiable at this point, but the learning from what went well and didn't, and what performed socially and environmentally in the ways anticipated, can be applied to similar initiatives in, say, New York City or Toronto or Brasília.

6. There are no criteria by which we can say that all infrastructure solutions to a problem have been identified. As such, judgement dictates whether one should enlarge the set of solutions being considered.

7. Despite having implemented infrastructure solutions in similar forms for centuries, there is almost always a new set of distinguishing features for the next problem that makes it unique. This is especially true because, while purely technological elements of the system may be similar to predecessors, it is seldom the case that the environmental, social, political, and economic dimensions are the same. Rittel and Webber use a subway as an example. Using a subway system design that worked for one city as the template for another is likely to be disastrous, as differing commuter preferences and urban form may outweigh similarities in subway network structure (H. W. J. Rittel and Webber, 1973).

8. Change in infrastructure is often incremental, with the hope that each small step contributes to systematic improvement. However, this can be dysfunctional: tackling the need for change on such a small level can, by avoiding foundational improvement, even create more systemic inertia against fundamental, non-incremental change. Thus, for example, incremental improvement in the U.S. air traffic system has kept an obsolete infrastructure functioning, but the cost is continuing and growing inefficiency, and a more complex upgrade process when it is finally necessary.

9. The modes of reasoning for why we pick infrastructure solutions often far outweigh the choices available through scientific discourse. The problems that these solutions are attempting to address are often based on the most powerful decision makers' world views. When engineers and decision makers are not trained in satisficing techniques, or in recognizing the impact of their own worldviews, decisions are often costly and socially damaging (this was one of the critiques of the high modernists such as Robert Moses).

10. The goal of infrastructure is typically not to find some truth, but instead to improve some characteristic of the world. The problems that infrastructure managers must deal with don't have boundaries and their causes are often obfuscated.

The challenges associated with affecting change in infrastructure that are wicked and complex are exponentially greater and require fundamentally different approaches than what we've historically used. The technical and social forces of complexity combined with wicked infrastructure problems (including access to water, affordable housing, safe, efficient, and affordable mobility, and public health) result in a fragmentation of priorities driven by increasing numbers of stakeholders that render traditional approaches obsolete.

In the Anthropocene it appears that infrastructure are complex adaptive systems that engage wicked complexity on a routine basis. As such, agility and flexibility are needed because of unpredictable emergent properties, with wicked complexity adding the need for satisficing.

1.4.3. INFRASTRUCTURE SATISFICING

To understand how wicked complexity has emerged around infrastructure it's necessary to review the cultural history surrounding development. The modern era emerged at the end of the Renaissance and proceeds through the twentieth century (Hugill, 1995). It is largely associated with the Enlightenment: the development of scientific methods and an emphasis on empirical observation over reason and innate knowledge (Landes, 1998). It is closely associated with the development of individualism, capitalism, expansion of world trade, urban development, and rapid technological progress (Findlay and O'Rourke, 2007; Morris, 2010). So-called high modernism took root after World War II, emphasizing scientific and technological progress delivered by scientists, engineers, and bureaucrats attempting to master nature and largely ignoring the complex social, political, and economic underpinnings in development (Berman, 1982b). Development often focused on spatial ordering and disregarded history, geography, and social context (Jacobs, 1961).

The high modernist period was the domain of the classically trained engineer and planner. Dominance of the political process and the top-down elitist approach characterized by Moses and Le Corbusier meant that integrated design optimized by the engineer was the model for successful infrastructure projects (Jencks, 1977; Berman, 1982b; Le Corbusier, 1987). But high modernism failed, both because it didn't adequately serve an increasingly complex urban culture, and because it was too rigid and inflexible to meet the demands of increasingly vocal community and issue activists, such as environmentalists (Jencks, 1977; Hall, 1998).

In today's postmodernist environment, where many conflicting demands must be integrated in infrastructure projects, satisficing is the only viable methodology. This can be seen not just in infrastructure, but in the "adaptive management" methods that are becoming popular for managing "natural" infrastructure such as the Everglades (Berkes, Folke and Colding, 1998; Allenby, 2012; NASEM, 2018). The role of the infrastructure planner and designer has shifted from prescribing a professional solution to a given situation—a high modernist framing of the professional's role—to being the expert facilitator of the emergence of a design that satisfices enough stakeholders, along with the legacy systems and natural infrastructure components, to be both stable and viable (Allenby, 2012).

What the infrastructure manager needs to do today is very different than in the past. Tribalism is the new norm, where solutions are dictated not by technical performance measures but instead by "acceptable enough" to all parties (Chua, 2018; Hawkins *et al.*, 2018). While this approach has done a better job at bringing more voices to the table, it is inimical to flexibility and agility, the ability to change infrastructure quickly thereby creating adaptive capacity.

This is not to say that having community and multi-stakeholder perspectives involved in infrastructure decision making is necessarily a bad thing (this is crucial to ensuring that no party asserts their will over others). Instead, we must recognize that the fragmentation that comes with this tribalism means that there are competing narratives for what infrastructure should and should not do. A wicked problem is one where you don't understand the problem until you have a solution.

This social complexity creates situations where different stakeholders think that their problem is the only and right problem (e.g., the result of where they're from or the mission of their organization), and as such collective inquiry into the problem is prevented (E. J. Conklin, 2006). The stronger the network of stakeholders, the less agile and flexible the system. The challenge is not to deny the involvement of stakeholders but to recognize that taken as a whole they are a significant design constraint that must be considered, and engineers must try to design for agility and flexibility in spite of it.

1.4.4. MANAGING INFRASTRUCTURE IN THE ANTHROPOCENE

The processes of infrastructure design, delivery, management, and change will need to be as fast as the relevant changes in demand, technologies, demographics, and Earth systems in the Anthropocene. Optimizing, which is the current dominant paradigm, is wholly insufficient for the future. The fragmentation that has resulted from the wicked complexity that has emerged in the contemporary infrastructure process produces disincentives to embrace change and plan for uncertainty. In an increasingly divided and tribal society, even something as fundamental to infrastructure design and management as problem definition becomes ambiguous and contested.

Thus, rather than simply designing a solution to a given problem, the engineer will often find themself needing to satisfice both on problem definition and design solution. Wicked problems can fragment direction and mission since agreeing on the problem to develop a solution is challenging. Furthermore, social complexity fragments stakeholder unity through competing interests and identities. Given these challenges, we must seriously question whether optimizing as the dominant approach is sufficient in terms of significant infrastructure transformation for the Anthropocene. Even if we sought a different approach, the organizational competencies needed to address complexity in infrastructure institutions are not, for the most part, present.

A major challenge that infrastructure systems face in adapting is that their managing institutions are largely configured around bureaucracies designed to address complicated rather than complex systems. What's more, they typically encompass single disciplines (e.g., transportation, water, power, etc.).

Bureaucracies—characterized by hierarchical organizations, jurisdictional responsibilities, intentional and abstract rules, production and administration belonging to the office, appointed officials, and career employment—became commonplace during the industrial era (Constas, 1958; Weber and Parsons, 2008; Diefenbach, 2012). Industrialization (in addition to representative governance and nationalism) added many new tasks to bureaucracies, transforming their role in the modern era. Public services, and their associated infrastructure, became necessary for democratizing governments during the colonial era, and bureaucracies became commonplace in the management of these services, many emerging during industrialization (Riggs, 1997).

The modern infrastructure system is mostly managed by bureaucratic structures that are designed for the industrial era, both in terms of the problems that we were facing and the intellectual needs to address those problems. The engineering education process largely reflects this. With the industrial era came a need to develop competencies around complicated problems.

Production became increasingly concentrated at factories (instead of agriculture) and the rapidly developing technologies and machinery of mass production needed workers who could understand and work with machines (Nason, 2017). Hierarchies—the staple structure of bureaucracies—were used to organize tasks vertically within organizations establishing functional and social relationships (Max Weber, T. Parsons, and R.H. Tawney, 1930). Expertise was compartmentalized, access to information constrained, and resources earmarked for particular classes of workers (Diefenbach, 2012).

These hierarchies work against creativity: they emphasize efficiency of the status quo process. Ideas become stranded between levels of management within the hierarchy—each of which has competing priorities and doesn't necessarily have the authority to implement new ideas. And given the separation of problems, management often doesn't understand the value of the idea.

This organizational structure in many ways persists in current infrastructure management institutions, as well as in the educational structures that feed them. Compounding the traditional challenges is wicked complexity; not only are these bureaucracies expected to efficiently deliver public services but they must do so given the wicked problems, technical complexity, and social complexity that has become endemic.

Adaptive management starts at infrastructure institutions that embrace the competencies needed to deal with complexity, emphasize agility and flexibility of operations and physical hardware, and recognize that infrastructure as an ever-changing process. This kind of management will be necessary for the non-stationarity norms and rapidly changing needs of the Anthropocene.

1.4.5. ADAPTIVE MANAGEMENT: COMPLICATED TO COMPLEX COMPETENCIES

Infrastructure management in the Anthropocene will require new competencies that give systems the capacity to quickly adapt to deliver resources in non-stationarity and ever-more complex environments. The models of adaptive management and other successful strategies that have been developed for managing complexity should serve as a foundation for transforming the processes of infrastructure design, planning, operation, maintenance, and evolution.

Here we don't pretend to have a final vision of how this process is explicitly structured. On the contrary, we view the process of infrastructure as one that will need to evolve, and the forms that it takes at any given time in the future will need to reflect the complex environments and challenges of the time, most of which cannot be forecast, certainly not with any detail. However, the competencies associated with adaptive management strategies to address wicked complexity can serve as a guidebook.

There are lessons that can be learned for hard infrastructure systems from the practices developed by socio-ecological system (SES) practitioners. SES practitioners have been developing best practices for ecosystem resource management to coordinate resources in the face of complexity and uncertainty (Armitage *et al.*, 2009; Chaffin, Gosnell and Cosens, 2014). These adaptive management practices produce a "flexible system of resource management, tailored to specific places and situations, supported by, and working in conjunction with, various organizations at different scales."

The SES literature abounds with successful implementations of adaptive management for complex and rapidly changing social-ecological systems. Successes include the management of the Kristianstads Vattenrike region in Sweden, which balanced development, conservation, and ecological services to address water quality and biodiversity collapse; preservation of the Great Barrier Reef Marine Park in Australia, which was experiencing coral bleaching, overfishing, and eutrophication; and protection of fishing in the Southern Ocean surrounding Antarctica, which was experiencing fish stock collapse. The common thread in these examples was a looming crisis that triggered a few individuals to build trust and knowledge, connect networks, and develop a shared system vision (Schultz *et al.*, 2015).

It is useful, for example, to compare recent adaptive management approaches to the Florida Everglades (challenged by phosphate mining, agricultural interests, and population growth and developers) with the experience of the Aral Sea (the Soviet Union diverted much of its inflow to growing cotton after World War II). Both regions are highly complex economically and culturally, but the use of adaptive management principles in the Everglades has prevented the economic, social, and ecological collapse that the Aral Sea region has suffered. A primary reason is that in the case of the Everglades, the wickedly complex nature of regional management was recognized, while in the case of the Aral Sea, the system was optimized for cotton production for export purposes (Allenby, 2012;

NASEM, 2018). It's critical that engineers and infrastructure managers develop competencies to be able to support and facilitate adaptive management practices.

First and foremost, infrastructure managers will need to be able to distinguish between complicated—even chaotic —but deterministic problems, and those characterized by wicked complexity. Distinctions will have to be made between systems where wicked complexity is relatively low, and systems where it is high. Engineers have in recent history largely operated in and been trained for the complicated domain—that of known unknowns—where management processes focus on unique skills associated with diagnosing and optimizing cause and effect relationships (e.g., what types of bridges could a city build across a canyon given geologic conditions, traffic requirements, and a budget). The complex domain is unique in that it's characterized by unpredictability and flux, which lead to emergence that cannot be predicted (unknown unknowns) (Snowden and Boone, 2007). We won't know what will work and have to accept that the best that we'll be able to do is generate educated guesses and commit to a dynamic process where we constantly reassess what's happened and adjust course.

Examples of complex infrastructure challenges that face engineers are plenty. What is the right solution to a new tourism road through a Peruvian cloud forest for access to Incan ruins? The engineer must consider highly sensitive ecosystem impacts, preserving cultural heritage, design for climate change and possibly more extreme rainfall events, autonomous vehicles, among other factors. The likely answer is we don't know. Sure, we could build a typical asphalt road for modern automotive technology. But this road—given the unpredictability of the factors mentioned—may be insufficient for the conditions, technologies, and needs two decades from now. Instead, infrastructure processes that embrace complexity would focus on experimental management practices and a flexibility in understanding what does and does not work, and commit to trying something different later on once new information emerges.

Additionally, it's reasonable to expect infrastructure managers to have to operate in the domain of chaos more frequently in the future. Catastrophic events like climate change-driven natural hazards and terrorist attacks (particularly cyberattacks, given how interconnected physical infrastructure is becoming with information and communication technologies) are already becoming more common (Singer and Friedman, 2014; USGCRP, 2017; Roser, Nagdy and Ritchie, 2018)—and revealing how little training our infrastructure managers have for working through these situations. In chaotic situations, infrastructure managers will need to be trained to help establish order, sense where stability is and is not present, and then help support transitions back to the complexity domain (Snowden and Boone, 2007). This will require training and practice that is often absent from traditional university curricula.

Managing complex infrastructure systems will require new approaches to training, education, and practice. The conventional infrastructure management organization is structured as a top-down hierarchy where expertise and resources are compartmentalized. Consider a hypothetical state department of transportation that has a leader at the top (often politically appointed), division directors that oversee various domains of the system (e.g., infrastructure delivery, planning, operations), and groups within those divisions that carry out the mission. In this structure, the sharing of knowledge and resources across groups to address interdisciplinary challenges is typically infeasible, and solutions to challenges are often prescribed with little opportunity for deviation. This structure emphasizes compartmentalization of knowledge and efficiency in solutions. More fundamentally, if a problem requires coordination beyond the transportation domain—say, implementing a tax structure that will encourage home office work rather than commuting to an office on bad air quality days—it is essentially impossible with the stove-piped structure of today's engineering management and education institutions.

There is an obvious need to develop organizational structures that 1) emphasize a diversity of ideas and perspectives, and offer frequent opportunities for exchanging them; 2) implement infrastructure systems as an experiment and dynamic reassessment exercise where patterns are allowed to emerge, are then studied, and new infrastructure is then implemented; and 3) emphasize the creation of new ideas over historical models. This organization would, for example, allow emerging technologies to be tested in practice, pit interdisciplinary teams against each other on the same problem, and provide workers discretionary time to pursue outside-the-box ideas as they relate to the agency's mission (Snowden and Boone, 2007).

Combining the emerging SES, management, and infrastructure concepts on wicked complexity, several core competencies emerge.

- *Shared Understanding*: Prior to managing wicked complex problems, there needs to be shared understanding of the problems and a willingness to address the problems collectively. Shared understanding is the process by which stakeholders are made aware of each other's goals and concerns; it is not consensus building. Shared understanding can lead to shared commitments on a project's directions, goals, emerging solutions, group decisions, and actions, and ultimately collective intelligence (E. J. Conklin, 2006). The processes by which infrastructure decisions are made, both internally and externally to the organization, should be radically altered toward collective intelligence. This is of course easier said than done, as a plethora of barriers exist that lock-in current practice and work against radical change. Nonetheless, as the forces increase that dictate how our future infrastructure can and cannot be in the Anthropocene, current practices that attempt to balance different constraints and perspectives without building collective intelligence appear to be increasingly insufficient.

- *Manage, Not Solve*: Infrastructure systems are too often treated as physical assets designed and operated to solve a problem, whether the facilitation of a volume of vehicle traffic or a minimum water pressure and volume. Yet in the face of increasing change, the solution will need to constantly evolve. As such, infrastructure managers will need to adopt a perspective that embraces change — that a system is temporary, the right fit for a short period of time, and likely unacceptable for the not-too-distant future. Management then focuses on the changing conditions that infrastructure needs to adapt to. This requires making decisions under deep uncertainty. In the past, infrastructure could be framed as artifactual; going forward, it must be reconceptualized as a process (Allenby, 2012).

- *Try, Learn, Adapt*: Complex environments are not conducive to command-and-control, top-down strategies; there's just too much distributed information, changing too rapidly, for a centralized approach to work. Instead, an approach that emphasizes experimentation as a process of learning to better enable adaptation capacities is needed. Here infrastructure organizations, the general public, and financiers need to allow for some level of failure. For example, a region could establish that a portion of transportation infrastructure funds be routinely used for testing for new technologies, such as alternative materials or traffic management, with the recognition that many of these technologies will not be implemented at scale but will provide valuable insights into what the agency needs to do to prepare for disruptive conditions. In addition, those responsible for infrastructure must learn to distribute management out to the network; decisions should be made at the most local level possible (a principle that in politics has become familiar as "subsidiarity," particularly in the European Union).

- *Complexity Mindset*: The training that many infrastructure managers receive focuses on managing complicated problems, often through the use of a predefined set of options or processes, and with explicit or implicit reference to optimization techniques. A complexity mindset accepts that complexity exists, that it needs to be managed differently, and that there are limitations on what a manager can control (Nason, 2017). A complexity mindset focuses on what can be over what is, and relies on satisficing, not optimizing, mental models and methods. Planning is different in each case. With complicated systems, planning is done to determine paths which are then followed. With complexity, it is the process of constantly planning that informs action; the plans themselves are seldom, if ever, implemented. As General (later President) Eisenhower observed, "In preparing for battle I have always found that plans are useless, but planning is indispensable (Nixon, 2013)." It isn't the plan that's important; it is the act of planning — thinking about the possibilities, the interconnectedness of problems, and emerging properties of the system — that's not just necessary but a critical part of adaptive management.

Every indication is that the competing forces that now define how infrastructure is implemented, and that are driving increasing complexity today, will continue if not accelerate. Unpredictable and accelerating evolution is occurring across the entire technological frontier. Social and political fragmentation and complexity is growing, and the stability of a world order which has prevailed for over three centuries is failing (Fukuyama, 2014; Kissinger, 2014). Environmental pressures and associated perturbations in human demographics and migration patterns, as well as unpredictable shifts in natural cycles and systems, are set to accelerate. Urbanization is accelerating, especially in developing countries.

There is thus little question that demands on technologists, managers, and engineers working on infrastructure systems are also going to grow. New approaches are needed to create coherence. For example, if water demand is increasing, then a treatment plant will need to be created based on the forecasted demand; funding may need to be secured from a new tax measure; political and social factors will need to managed; and water rights may need to be secured. Acceptance by local communities, activist organizations (which often have different if not mutually exclusive agendas), and regional water users and managers will be required. Designs that are flexible enough to adapt to unpredictable changes in supply and demand because of climate change must be developed, and metrics for tracking performance of the infrastructure — and the systems context of the infrastructure — must be developed and institutionalized. The complicated task of designing the water treatment plant is possibly the simplest exercise in the whole process, and is perhaps the only one for which the infrastructure manager has been trained.

1.4.6. CONCLUSION

In the Anthropocene, infrastructure is an integration of co-evolving human, built, and natural systems, and inevitably is characterized by wicked complexity. As such, engineers and infrastructure managers will need to

Infrastructure in the Anthropocene: Second Edition

shift their mindset away from optimizing and toward satisficing. They will need to shift their thinking about infrastructure from the delivery of physical assets that meet a defined and relatively stable need to a process by which physical systems are designed, built, and operated in response to shifting priorities, new technologies and social practices, and irreducible uncertainty.

1.4.7. REFERENCES

Allenby, B.R. (2012) *The Theory and Practice of Sustainable Engineering.* Hoboken, NJ: Upper Saddle River: Pearson Prentice Hall.

Arbesman, S. (2016a) *Overcomplicated: Technology at the Limits of Comprehension.* Penguin Publishing Group.

Arbesman, S. (2016b) *Overcomplicated: Technology at the Limits of Comprehension.* New York, NY: Penguin Publishing Group.

Armitage, D.R. *et al.* (2009) 'Adaptive co-management for social–ecological complexity', *Frontiers in Ecology and the Environment*, 7(2), pp. 95–102. Available at: https://doi.org/10.1890/070089.

Berkes, F., Folke, C. and Colding, J. (1998) *Linking Social and Ecological Systems: Management Practices and Social Mechanisms for Building Resilience.* Cambridge, UK: Cambridge University Press.

Berman, M. (1982a) *All that is solid melts into air: the experience of modernity.* New York, NY: Simon and Schuster.

Berman, M. (1982b) *All that is solid melts into air: the experience of modernity.* New York, NY: Simon and Schuster.

Birdzell (1986) *How The West Grew Rich.* New York, NY: Basic Books.

Bogucki, P. (2011) 'How Wealth Happened in Neolithic Central Europe', *Journal of World Prehistory*, 24(2), pp. 107–115. Available at: https://doi.org/10.1007/s10963-011-9047-5.

Brown, R. (2004) 'Consideration of the origin of Herbert Simon's theory of "satisficing" (1933-1947)', *Management Decision*, 42(10), pp. 1240–1256. Available at: https://doi.org/10.1108/00251740410568944.

Brown, T., Beyeler, W. and Barton, D. (2004) 'Assessing infrastructure interdependencies: the challenge of risk analysis for complex adaptive systems', *International Journal of Critical Infrastructures*, 1(1), p. 108. Available at: https://doi.org/10.1504/IJCIS.2004.003800.

49

Chaffin, B., Gosnell, H. and Cosens, B. (2014) 'A decade of adaptive governance scholarship: synthesis and future directions', *Ecology and Society*, 19(3). Available at: https://doi.org/10.5751/ES-06824-190356.

Chester, M., Markolf, S. and Allenby, B. (2019) 'Infrastructure and the environment in the Anthropocene', *Journal of Industrial Ecology*, 23(5), pp. 1006–1015. Available at: https://doi.org/10.1111/jiec.12848.

Chester, M.V. and Allenby, B. (2018) 'Toward adaptive infrastructure: flexibility and agility in a non-stationarity age', *Sustainable and Resilient Infrastructure*, 4, pp. 1–19. Available at: https://doi.org/10.1080/23789689.2017.1416846.

Chua, A. (2018) *Political tribes: group instinct and the fate of nations*. New York, NY: Penguin Press.

Cilliers, P. (2002) *Complexity and postmodernism: understanding complex systems*. London, UK: Routledge. Available at: http://site.ebrary.com/id/10070596 (Accessed: 9 November 2023).

Clark, S.S., Seager, T.P. and Chester, M.V. (2018) 'A capabilities approach to the prioritization of critical infrastructure', *Environment Systems and Decisions*, 38(3), pp. 339–352. Available at: https://doi.org/10.1007/s10669-018-9691-8.

Conklin, E.J. (2006) *Dialogue mapping: building shared understanding of wicked problems*. Chichester, UK: J. Wiley. Available at: http://catdir.loc.gov/catdir/toc/ecip0513/2005013962.html (Accessed: 5 November 2023).

Conklin, J. (2006) *Dialogue Mapping: Building Shared Understanding of Wicked Problems*. Wiley.

Constas, H. (1958) 'Max Weber's Two Conceptions of Bureaucracy', *American Journal of Sociology*, 63(4), pp. 400–409.

Diefenbach, T. (2012) 'Bureaucracy and Hierarchy — What Else!?', in R. Todnem (ed.) *Reinventing Hierarchy and Bureaucracy: From the Bureau to Network Organizations*. Bingley: Emerald Publishing Limited, pp. 1–27.

Findlay, R. and O'Rourke, K.H. (2007) *Power and plenty: trade, war, and the world economy in the second millennium*. Princeton, NJ: Princeton University Press (Princeton economic history of the Western world). Available at: http://swbplus.bsz-bw.de/bsz274538768inh.htm (Accessed: 11 November 2023).

Fukuyama, F. (2014) *Political Order and Political Decay: From the Industrial Revolution to the Globalization of Democracy*. New York, NY: Farrar, Straus and Giroux.

Gandy, M. (2003) *Concrete and Clay: Reworking Nature in New York City*. Cambridge, MA: The MIT Press.

Gies, F. and Gies, J. (1995) *Cathedral, forge, and waterwheel: technology and invention in the Middle Ages*. New York, NY: HarperCollins.

Hall, P. (1998) *Cities in Civilization*. New York, NY: Pantheon Books.

Hawkins, S. *et al.* (2018) *Hidden Tribes: A Study of America's Polarized Landscape*. New York, NY: More in Common, p. 159. Available at: https://hiddentribes.us/media/qfpekz4g/hidden_tribes_report.pdf.

Hugill, P.J. (1995) *World Trade Since 1431: Geography, Technology, and Capitalism*. First Edition. Baltimore, MD: Johns Hopkins University Press.

Jacobs, J. (1961) *The Death and Life of Great American Cities*. New York, NY: Vintage Books.

Jencks, C. (1977) *The Language of Post-modern Architecture*. New York, NY: Academy Editions.

Kissinger, H. (2014) *World Order*. New York, NY: Penguin Publishing Group.

Kurzweil, R. (2010) *The Singularity Is Near: When Humans Transcend Biology*. Ebook. London, UK: Duckworth Overlook.

Landes, D.S. (1998) *The wealth and poverty of nations: why some are so rich and some so poor*. 1st ed. New York, NY: W.W. Norton. Available at: http://www.h-net.org/review/hrev-a0a1j7-aa (Accessed: 11 November 2023).

Le Corbusier (1987) *The city of To-morrow and its Planning*. Translated by F. Etchells. New York, NY: Dover. Available at: https://ebookcentral.proquest.com/lib/londonmet/detail.action?docID=1897253 (Accessed: 11 November 2023).

Marchant, G.E., Allenby, B.R. and Herkert, J.R. (2011) *The Growing Gap Between Emerging Technologies and Legal-Ethical Oversight: The Pacing Problem*. New York, NY: Springer Science & Business Media. Available at: https://doi.org/10.1007/978-94-007-1356-7.

Martin, A. and Helmerson, K. (2014) *Emergence: the remarkable simplicity of complexity, The Conversation*. Available at: http://theconversation.com/emergence-the-remarkable-simplicity-of-complexity-30973 (Accessed: 9 November 2023).

Max Weber, T. Parsons, and R.H. Tawney (1930) *The Protestant Ethic and the Spirit of Capitalism*. New South Wales, AU: G. Allen & Unwin.

Morris, I. (2010) *Why the West rules-- for now: the patterns of history, and what they reveal about the future.* 1st ed. New York, NY: Farrar, Straus and Giroux (Business book summary).

NASEM (2018) *Progress Toward Restoring the Everglades: The Seventh Biennial Review - 2018.* Washington, DC: National Academies Press. Available at: https://doi.org/10.17226/25198.

Nason, R. (2017) *It's Not Complicated: The Art and Science of Complexity in Business.* Toronto, OT: University of Toronto Press.

Nixon, R. (2013) *Six Crises.* New York, NY: Simon and Schuster.

Oughton, E. and Tyler, P. (2013) *Infrastructure as a Complex Adaptive System.* Oxford, UK.

Oughton, E.J. *et al.* (2018) 'Infrastructure as a Complex Adaptive System', *Complexity*, 2018, p. 11. Available at: https://doi.org/10.1155/2018/3427826.

Riggs, F.W. (1997) 'Modernity and Bureaucracy', *Public Administration Review*, 57(4), pp. 347–353. Available at: https://doi.org/10.2307/977318.

Rittel, H. and Webber, M. (1973) 'Dilemmas in a general theory of planning', *Policy Sciences*, 4(2), pp. 155–169. Available at: https://doi.org/10.1007/BF01405730.

Rittel, H.W.J. and Webber, M.M. (1973) 'Dilemmas in a General Theory of Planning', *Policy Sciences*, 4(2), pp. 155–169.

Roser, M., Nagdy, M. and Ritchie, H. (2018) 'Terrorism', *Our World in Data* [Preprint]. Available at: https://ourworldindata.org/terrorism.

Schultz, L. *et al.* (2015) 'Adaptive governance, ecosystem management, and natural capital', *Proceedings of the National Academy of Sciences*, 112(24), pp. 7369–7374. Available at: https://doi.org/10.1073/pnas.1406493112.

Simon, H. (1947) *Administrative Behavior: A Study of Decision-making Processes in Administrative Organization.* Basingstoke, UK: Macmillan Company (Original from University of Minnesota).

Simon, H.A. (1957) *Models of Man: Social and Rational; Mathematical Essays on Rational Human Behavior in Society Setting.* Berkeley, CA: Wiley.

Singer, P.W. and Friedman, A. (2014) *Cybersecurity: What Everyone Needs to Know.* New York, NY: Oxford University Press USA.

Snowden, D.J. and Boone, M.E. (2007) 'A Leader's Framework for Decision Making', *Harvard Business Review*, 1 November, pp. 68–76, 149.

The Economist (2009) 'Herbert Simon', *The Economist*, 20 March. Available at: https://www.economist.com/news/2009/03/20/herbert-simon (Accessed: 9 November 2023).

USGCRP (2017) *Climate Science Special Report: Fourth National Climate Assessment*. Washington, DC: U.S. Global Change Research Program, pp. 1–470. Available at: https://science2017.globalchange.gov/ (Accessed: 12 November 2023).

Weber, M. and Parsons, T. (2008) *The theory of social and economic organization*. Reprint. New York, NY: Free Press.

Section 2

NAVIGATING COMPLEXITY

With increasingly complex environments infrastructure will need to develop new capabilities to engage with internal and environment complexity. Section 2 frames *resilience* as effectively navigating increasingly complex conditions. Through Section 2 we present context and capabilities needed to position infrastructure for the Anthropocene. Central to the discussion is the need to open up infrastructure technologies and governance processes to engage at pace and scale with changes in their environments. Flexibility and agility are foundational as they represent the ability to speed up infrastructure sensemaking and ultimately the technologies, governance, and educational processes appropriate for the environment. The capabilities described in Section 2 support infrastructure adaptation as an alignment between environment complexity and infrastructure complexity (i.e., requisite complexity). Currently we argue there is a decoupling between growing environment complexity and the capabilities of infrastructure to respond.

Central to resilience is the critical examination of structure and function of infrastructure systems. We discuss how centralized and decentralized structures of technologies and governance support or limit the ability of the system to engage with complex conditions. Furthermore, we describe how cybertechnologies are increasingly networking assets towards distributed models of control.

Additionally, Section 2 explores the concepts of self-organization of infrastructure in the face of disruption (autopoiesis). For infrastructure to adapt and thrive in the face of disruption they needs to be able to generate a repertoire of responses at least as large as the repertoire of disturbances the environment can unleash (requisite variety or complexity). We argue that novel infrastructure knowledge- and sensemaking techniques are necessary to produce requisite complexity.

2.1

RECONCEPTUALIZING INFRASTRUCTURE IN THE ANTHROPOCENE

A fundamental shift is afoot in the relationship between human and natural systems. It requires a new understanding of what we mean by infrastructure, and thus dramatic changes in the ways we educate the people who will build and manage that infrastructure. Similar shifts have occurred in the past, as when humanity transitioned from building based on empirical methods developed from trial-and-error experience, which was sufficient to construct the pyramids and European cathedrals, to the science-based formal engineering design methods and processes necessary for the electric grid and jet aircraft. But just as the empirical methods were inadequate to meet the challenges of nineteenth- and twentieth-century developments, today's methods are inadequate to meet the needs of the twenty-first century. Now that we have entered the Anthropocene period in which human activities affect natural systems such as climate, engineers face far more complex design challenges.

The durability of the pyramids and cathedrals is evidence that the inherited wisdom of experience can be adequate for many tasks, but we can also see the disadvantages of working without the formal design tools based on advanced mathematics and engineering science. The choir vault and several buttresses of the Beauvais Cathedral in France collapsed in 1284, only 12 years after its partial completion. Retrospective analysis suggests that the problem was resonance under wind load, the type of stress that could be anticipated only with the formal design methods that were yet to be developed. And when the first iron bridge was built at Coalbrookdale, England, in 1781, the engineers used the same design they would have used for a masonry bridge. Within a hundred years, however, structures such as James Eads's 1867 bridge over the Mississippi River at

This chapter was adapted from the following article with publisher permission: Mikhail Chester and Braden Allenby, 2019, Toward Adaptive Infrastructure: Flexibility and Agility in a Non-stationarity Age, *Sustainable and Resilient Infrastructure*, 4(4), pp. 173-191, doi: 10.1080/23789689.2017.1416846.

St. Louis and Gustave Eiffel's 1884 Garabit Viaduct were being formally designed using scientific methods and quantitative design processes that made it possible to predict the performance of the new materials. Personal experience and historical heuristics were replaced by scientific data and quantitative models. Opportunities for infrastructure evolution that would have been impossible without scientific understanding suddenly became available, leading to more efficient, effective, and safer infrastructure, and eventually to entirely new forms of infrastructure such as air travel and electrification.

The engineering tools, methods, and practices necessary for a practicing professional; the engineering education necessary to prepare such a professional; and the way engineering is performed institutionally have all changed fundamentally throughout history as the context within which humans live and prosper has changed. We are now in a period of rapid, unpredictable, and fundamental change that is literally planetary in scope. To date, neither our ideas regarding infrastructure, nor our educational and institutional systems for conducting engineering, have reflected this striking evolution in context. We are well past the point where adaptation is pragmatically, and ethically, required.

It is important to remember that this doesn't mean that all current practice and educational methods are suddenly obsolete. Rather, it means that they are increasingly inadequate, especially when applied to projects that involve complex social, economic, and cultural domains. Designing a water treatment plant will remain well within the capability of today's field of environmental engineering; designing the nitrogen and phosphorus cycles and flows in the Mississippi River basin to reduce the life-depleted "dead zone" in the Gulf of Mexico cannot be done so simply.

2.1.1. THE PLANET AS INFRASTRUCTURE

That we now live on a terraformed planet is not a new idea. The term "Anthropocene" was popularized in an article in 2000 by Paul Crutzen and Eugene Stoermer, but as early as 1873 the Italian priest, geologist, and paleontologist Antonio Stoppani used the term "anthropozoic era," and others have used similar language throughout the twentieth century. The force driving the birth of a new era is obvious. A planet that supported roughly 450 million humans in 1500 now supports over 7.5 billion, and because of industrial progress and economic growth, each of those humans has a vastly more consequential impact.

In a blink of geologic time, then, humans left a planet where they were but one species among many and built a world increasingly shaped by the activities of, and for the purposes of, a single species — themselves. It is not that the planet has been deliberately designed by humans, but many human-built systems from transportation to communications have resulted in global changes that were not consciously designed by anyone. Moreover, that human activity is clearly a significant contributor to current dynamics of such natural phenomenon as climate change, biodiversity, nitrogen and phosphorous cycles, microbial evolution, and regional ecosystem parks such as the Everglades does not imply deliberate design in such cases either. Rather, the world we find ourselves in is one where deliberate human activity, itself a complex network of functions, interacts with a highly complex planetary substrate to create unpredicted, often challenging, emergent behaviors, of which climate change is but one example.

These emergent properties would not exist but for humans; the design process involved, however, is not the explicitly rational and quantitative method that we are used to. Rather, a planet characterized by rapidly increasing integration of human culture, built environments, and natural systems producing novel, highly complex, rapidly changing emergent behaviors challenges all our existing ideas about design, operation, and management of infrastructure. Indeed, it challenges the very idea of infrastructure as limited to local, highly engineered systems. For in such a world, "natural" cycles and systems transmute from exogenous conditions into infrastructure components, a process implicitly recognized by, for example, the substantial literature on ecosystem services and earth systems engineering and management.

Consider urban water supply. Replacing a water pipe in New York City, albeit expensive and complex, is well within current professional and institutional capabilities. On the other hand, to maintain that water supply means designing, through engineering and land use regulation, a continuously monitored network of 19 reservoirs in a roughly 2,000-square-mile watershed. Similarly, most people know that Arizona's population is concentrated in several large urban areas in semiarid desert regions of the state. Contrary to popular belief, however, Arizona's water supply is robust. It is very diverse, with 17% from in-state rivers and associated reservoirs, 40% from groundwater, 3% reclaimed, and 40% imported from the Colorado River, which channels water from seven states. Flexibility and resilience in operations is enhanced because imported water is largely banked in underground reservoirs, while wastewater is highly managed and either returned to the ground where it can be accessed later or sent to power generation facilities. Similarly, water in California is tightly designed. Mountain snow releases water as it melts, and that runoff is heavily managed for multiple uses as it moves toward the ocean. As in many other regions, natural flows from precipitation are now highly managed and there is increasing recognition that historical flow patterns are becoming less and less relevant for predicting future flows.

In short, watersheds are now so highly managed that the environment has in effect become the water infrastructure.

As these examples suggest, the advent of the Anthropocene requires the development of new institutions and frameworks that are beyond today's traditional disciplinary structures and reductionist approaches. Continued stability of the information-dense, highly integrated, rapidly evolving human, natural, and built systems that characterize the anthropogenic Earth requires development of new abilities to rationally analyze, design, engineer and construct, maintain and manage, and reconstruct and evolve such systems at local, regional, and even planetary scales. It is not that current institutions and disciplinary boundaries are completely obsolete; rather, it is that increasing complexity and rates of system evolution require a new and more integrative level of sophistication in infrastructure conceptualization, design, and management. This can draw on a number of relatively new fields of study, including sustainable engineering, industrial ecology and associated methodologies such as life-cycle assessment and materials flow analysis, systems engineering, adaptive management, and parts of the urban planning and sociology of technology literatures.

The boundaries reflected in current engineering institutions, disciplines, and educational practices and structures are still appropriate for many problems. But today, at the dawn of the Anthropocene, they increasingly fail at the level of the complex, integrated systems and behaviors that characterize the anthropogenic Earth. No disciplinary field in either the physical or social sciences addresses these emergent behaviors, and very few even provide an adequate intellectual basis for perceiving, much less parsing, such complex adaptive systems. Engineering cannot continue to rest on traditional curricula and strengths, which are increasingly inadequate given today's social, economic, environmental, and technological demands. Planning for urban transportation infrastructure, for example, is inadequate unless it includes consideration of the implications of such varied developments as hybrid, electric, and autonomous vehicle technologies, potential climate change effects on mobility, aging populations, and terrorism reduction and mitigation. We cannot stop building infrastructure, but especially as many forms of the built environment outlast the context within which they were built—think of coal-fired power plants being built today that will be operating for decades—we can try to design infrastructure to be more resilient to change even when we cannot predict the particulars of such change, to be more adaptable and agile as those changes occur, and to be more aware of obvious effects of our designs that may have heretofore been ignored.

In a world where natural systems are for many purposes becoming simply another element of regional and global infrastructure systems, society in general and engineers in particular are challenged to rapidly develop tools, methods, and intellectual frameworks that enable reasoned responses. The reductionism inherent in pretending the challenge is just climate change or new transportation systems or lower-carbon energy production and distribution systems or water quality and quantity, although appropriate in certain cases, can no longer be defended as an acceptable disciplinary or institutional approach to infrastructure design, operation, and management. Rather, engineers, engineering educators, and institutions that support and manage infrastructure need to create solutions in the real world, and to embrace the implications of the Anthropocene. Bluntly, the world is a design space, and absent catastrophic collapse, there is no going back.

2.1.2. NEW ENGINEERING FOR A NOVEL PLANET

The reality of the terraformed Earth, whatever one wants to call it, requires some significant changes in the way technology and infrastructure are designed, implemented, understood, and managed. Although the obvious target disciplines are in engineering, such a focus would be inadequate. Increasingly, private firms develop and manage emerging technologies that shape the integrated human/built/natural infrastructure that are the defining characteristics of the Anthropocene, so business education writ broadly must be part of the shift. Moreover, as systems that were previously considered external to human design, such as biology and material cycles, increasingly become design products, many aspects of science segue into engineering fields, requiring different mental models and tools. This implies that many fields of science also require some grounding in technology, design, and engineering. Against this background, and recognizing that any suggestions for reform at today's early stage will necessarily be partial and somewhat arbitrary, we can nevertheless make some recommendations for change. For simplicity, we will group these suggestions into three categories: short term, medium term, and long term.

Short term. In all relevant disciplines, including at least all engineering and business programs, at least three areas of learning should be mandatory for all students, undergraduate and graduate, and professional accreditation institutions such as the Accreditation Board for Engineering and Technology (ABET), which approves engineering programs, should set an explicit timetable for achieving this. Whether these areas of expertise are introduced as modules in existing curricula or as independent courses is not as important as ensuring that some level of substantive knowledge is acquired by all students. In order to reinforce this, professional exams, such as the Fundamentals of Engineering, and Principles and Practice of Engineering exams given to professional engineers, should be expanded to include these subject areas. These broad areas of competence are:

1. Technological, social, and sustainable systems theory and principles, including concepts such as wicked complexity and satisficing versus optimizing system performance. This category of knowledge includes understanding the difference between simple and complex system behavior, and designing and managing integrated human/built/natural systems.

2. Big data/analytics/artificial intelligence (AI) functions and systems. Any business or engineering student who graduates without knowing something about these fields is verging on incompetent.

3. Cybersecurity and cyberwar operations from a defensive and offensive perspective. Today, business managers and engineers are trained to rely on, and design and deploy, "smart" systems at all scales; indeed, in many cases this is a performance and business imperative. On the other hand, virtually none of these professionals are taught anything about cybersecurity or cyberwar, a gaping competency chasm when both Russian and Chinese strategic military doctrines have shifted toward deployment of unrestricted warfare / hybrid warfare / cyber warfare initiatives. Inviting an adversary to subvert all your critical infrastructure and systems, which the U.S. educational system now does, is simply stupid, especially since these adversaries are very engaged, and quite competent, in such activities.

These changes in curricula are a one-time improvement and by themselves will quickly become inadequate. At least as regards engineering, two further steps should immediately be taken. It is, for example, sadly the case today that too many engineering students are still being graduated with skill sets that prime them for replacement by expert systems. Thus, the first step should be to review the entire engineering curriculum for each discipline, identifying those elements of existing courses that software, AI systems, and changes in technology have made obsolete. A thorough analysis of the content of existing courses, coupled to modernization of ABET criteria, will help create educational programs that prepare students for the future, not the 1950s.

This should be combined with implementation of a change that engineering schools and those who hire engineers have been talking about for many years but have yet to implement: making engineering a graduate-level professional degree. Uniquely among professions, engineering remains, at least in theory, an undergraduate professional degree: neither law nor medicine nor any other major profession does so. This anachronism today works only by impoverishing the scope and scale of the education provided to undergraduate engineering students, who are then thrust into a world that demands ever more operational sophistication from them—something they cannot provide because there's no room for anything but engineering-related courses in their undergraduate programs.

Medium term. The Anthropocene is characterized by unpredictable change with far shorter time cycles than those common to most infrastructure systems. An important emphasis, therefore, must be on developing frameworks, tools, and methods that enable more agile, adaptable, and resilient infrastructure design and operation even in the face of fundamental uncertainty about virtually every important challenge infrastructure systems will face.

Although specifics are difficult and remain to be worked out, there are a number of ideas and models across existing engineering systems that can be generalized toward the goal of agile, adaptive, and resilient infrastructure. For example, the entire domain of consumer computational technology is characterized by many different technologies that must converge in working systems that are robust and simple enough in operation for average consumers to understand. One of the basic mechanisms for this process is "roadmapping," where modularity of technology—printers are different from data storage devices are different from input devices—is combined with robust module interfaces to enable both module and overall product evolution that is at once unpredictable, functional, and economic. Another example from consumer electronics is the substitution of software for hardware functionality whenever possible. Assuming appropriate hardware design, software can easily be upgraded, whereas hardware is far more difficult to change out (imagine if every improvement in software security or function required a new motherboard!). And programming offers a third model: the core of C++, for example, has remained fairly stable for years, but software evolution based on that core has been explosive, unpredictable, and very rapid. This suggests a "hub-and-spoke" framework: those elements of an infrastructure system that are not likely to change rapidly can be built with longer time frames and be more robust, whereas the spokes and edge elements, connected with stable interfaces, can evolve unpredictably and much more rapidly. Other characteristics might include planned obsolescence, resilience thinking that includes safe-to-fail and extensible design that can be easily updated, and increased compatibility and connectivity of hardware. These and similar models should be applied to infrastructure design generally.

Because of their importance to society, infrastructure systems tend to be designed to be robust and long-lasting. But especially as "natural" systems increasingly become a part of large infrastructure design and planning, it may be necessary to identify infrastructure where context may be changing rapidly, and "down-design" it: that is, design it to be less robust, more temporary, and capable of being replaced inexpensively and much more quickly. This could be combined with a major hubs/minor hubs/links design, where robust major hubs last for long times, minor hubs are down-designed, and structural links are robust in the face of unpredictable challenge.

Long term. The longer-term situation is more inchoate and more complex. Three fundamental shifts, one conceptual, one systemic, and one educational, are required, but none of them can be planned with any specificity. Accordingly, an overriding requirement is to learn how to implement continuous dialog and learning with the many systems within which we are embedded so that we can adjust our mental models, institutions, and educational practices in response to real time changes in system state. Policy and practice become processes of adaptive evolution rather than static responses to a stable world, and technocrats, managers, and engineers must learn to focus on ethical and responsible processes, and less on content, which will be continually and unpredictably changing.

The conceptual shift sounds simple. The fundamental implication of the Anthropocene is that everything from the planetary to the human body itself become design spaces: the world has become infrastructure. Especially given the complexity of the systems involved, this doesn't mean design in the sense of a completely understood controlled system such as a toaster. The world of simple, optimizable design is gone. Rather, it means that in a world of radical human life extension, integrated AI/human cognition, ecosystems as infrastructure, and re-wilding using previously extinct species in accord with wilderness chic design aesthetics, everything is subject to human intervention and intentionality. The Everglades is a design choice and exists today in its current state because we choose to have it exist in that state, even if it obviously has biological and ecological complexity beyond our current understanding. The simple truth of a terraformed planet is difficult for many people to accept, for reasons ranging from fear of the responsibility to ideological commitment to archaic visions of the sanctity of nature, but it is a necessary basis for technocratic and policy professionals if they wish to act ethically and responsibly under today's extremely challenging conditions.

Systemically, institutions need to evolve to mirror the complexity and interdisciplinarity of the networks with which they are interacting. Importantly, this means that institutions must become multi-ideological if they want to manage regional and planetary infrastructure. If they do not, they will be ineffectual. Thus, for example, the United Nations Framework Convention on Climate Change, signed in 1992, prioritized environmental ideology over other values, with the result that implementation has been sporadic, contentious, and ultimately unsuccessful. The shift involved here is also conceptual, and deceptively simple: institutions driven by activist stakeholder and dominant ideological demands must instead begin to frame themselves as problem-solvers, with the desired goal being not the advance of a particular agenda, but the creation of a politically and culturally stable solution to a particular challenge. Moreover, the institutions that typically govern infrastructure are mechanistic, characterized by hierarchical structure, centralized authority, large numbers of difficult to change rules and procedures, precise division of labor, narrow spans on control, and formal means of coordination. In contrast, industrial organizations in sectors exposed to rapid and unpredictable change, such as Silicon Valley technology firms, are far more organic, with the emphasis on moving fast rather than the lumbering pace imposed by formal structure. Whether current infrastructure institutions are capable of transitioning their management structures is an open question, given their legalistic and bureaucratic incentive structures; if they cannot, they must be replaced.

The educational shift is not unlike the institutional shift. Engineering education in particular combines quantitatively complicated but structured models and learning with creativity, design, operation, and problem-solving capability. The former is rapidly becoming mechanized, and the latter is increasingly done in a rapidly changing social, economic, and technological environment. Because of this, it will not be enough to simply reform engineering education in the short term: rather, engineering education must become a constantly evolving process and curriculum that changes at the same pace as its context. Engineers, who today often resemble expert systems—highly competent in one specific area such as chemical or civil engineering, but with knowledge that rapidly scales off at the edge of that domain because they don't have time to gain much education outside their particular discipline—must instead become constantly learning quantitative problem solvers, able to integrate across many different domains, as well as the still yawning gap between the science and technology fields and the social sciences and humanities.

These changes sound challenging, and indeed they are. But the Anthropocene is not a choice that we can reject; it is a reality that we have often unintentionally been creating since the Neolithic. We are now at the point where it must be embraced in all its complexity if we intend to respond rationally, ethically, and responsibly to its challenges.

2.1.3. RECOMMENDED READINGS

Ahern, J. (2011) 'From fail-safe to safe-to-fail: Sustainability and resilience in the new urban world', *Landscape and Urban Planning*, 100(4), pp. 341–343. Available at: https://doi.org/10.1016/j.landurbplan.2011.02.021.

Allenby, B. (2007) 'Earth Systems Engineering and Management: A Manifesto', *Environmental Science & Technology*, 41(23), pp. 7960–7965. Available at: https://doi.org/10.1021/es072657r.

Allenby, B.R. (2012) *The Theory and Practice of Sustainable Engineering*. Hoboken, NJ: Upper Saddle River: Pearson Prentice Hall.

Chester, M.V. and Allenby, B. (2018) 'Toward adaptive infrastructure: flexibility and agility in a non-stationarity age', *Sustainable and Resilient Infrastructure*, 4, pp. 1–19. Available at: https://doi.org/10.1080/23789689.2017.1416846.

Costanza, R. *et al.* (2017) 'Twenty years of ecosystem services: How far have we come and how far do we still need to go?', *Ecosystem Services*, 28, pp. 1–16. Available at: https://doi.org/10.1016/j.ecoser.2017.09.008.

Turner, B.L. *et al.* (1993) *The Earth as Transformed by Human Action: Global and Regional Changes in the Biosphere over the Past 300 Years*. Cambridge, UK: Cambridge University Press.

2.2

FLEXIBILITY AND AGILITY

A 1963 quotation by management professor Leon Megginson—one that's often misattributed to Charles Darwin—states that "It is not the most intellectual of the species that survives; it is not the strongest that survives; but the species that survives is the one that is able best to adapt and adjust to the changing environment in which it finds itself (Megginson, 1963)." The concept of adaptation and the complex principles that support it have been the focus of researchers in the fields of biology, but with significant application in other fields including business (Brennan, Turnbull and Wilson, 2003), management (Chakravarthy, 1982; Hrebiniak and Joyce, 1985), and computer science (Garlan *et al.*, 2004).

Adaptation is perhaps one of the most fundamental and powerful explanatory concepts for the changes in complex systems, in that it provides an explanation for the persistence of successful systems in the face of significant changes in internal and external environments. In biology, adaptation is a trait maintained by natural selection that enhances fitness and survival. More broadly, the concept characterizes the capability of organisms, complex systems, businesses, or institutions to change their organizing principles, structure, and behaviors to succeed in changing environments. How success is measured differs across disciplines, with biology focused on reproduction and business focused on maintaining growth and ultimately profitability (Chakravarthy, 1982; Grisogono, 2006).

Yet when it comes to infrastructure, the systems that we've deployed and continue to maintain—the backbones of our cities, economies, and overall well-being—there appears to be limited capabilities to adapt. This raises serious questions about their ability to provide services in a future with changing demands, population, climate, security challenges, and environmental conditions.

This chapter was adapted from the following article with publisher permission: Mikhail Chester and Braden Allenby, 2019, Toward Adaptive Infrastructure: Flexibility and Agility in a Non-stationarity Age, *Sustainable and Resilient Infrastructure*, 4(4), pp. 173-191, doi: 10.1080/23789689.2017.1416846.

The infrastructure that supports our societies provides untold benefits. Infrastructure are socio-technical systems composed of physical assets and the institutions that manage, govern, finance, and regulate them. The services they provide deliver resources such as energy, water, and information, and move and process waste. These services are not purely physical. While our focus on infrastructure is primarily on hard (or gray) systems—roads, buildings, power, water, etc.—we will also examine the role of soft (i.e., institutions) infrastructure and its relationship with hard systems.

Transportation infrastructure provides mobility and ultimately access to people, goods, and services. Buildings provide shelter for people, businesses, and services. Hard infrastructure can be characterized as services that either produce or deliver resources directly (such as energy, water, waste, or information) or provide mechanisms for resource consumption (such as buildings). In the United States, critical infrastructure is defined as chemical, commercial facilities, communications, critical manufacturing, dams, defense, emergency services, energy, financial, food and agriculture, government facilities, healthcare and public health, information technology, nuclear, transportation, water, and wastewater (CISA, 2017). People's daily needs are typically met by municipal infrastructure and in many cases private energy companies. Supply chains for food, fuel, and materials are generally supported by institutions at larger scales—regional, state, and federal. As such, the funding and planning for adaptive infrastructure must recognize the client. More broadly, infrastructure facilitates derived demands; we don't usually demand the resource or service that infrastructure provides, but instead what that resource or service enables, in other words, the utility that it provides.

While the extent of infrastructure, what it delivers, and how it is used are somewhat quantifiable, the benefits of infrastructure ultimately are in its functioning as an engine of social well-being, which can be characterized through economic growth, health, quality of life, etc. To communicate the value of infrastructure, efforts have been made to monetize this well-being, both for gray infrastructure (ASCE, 2016) and even ecological infrastructure (Costanza *et al.*, 2017).

In the developed world, the core physical structures that define our infrastructure have often not changed in decades, sometimes centuries, from roads to water delivery to power generation and transmission. These infrastructure have certainly seen the implementation of new technologies (e.g., sensors and computing, automation, more efficient components) in support of the services delivered, but the core structures that have been used for decades (if not longer), from roadways to centralized fossil-based electricity generation to water distribution networks, are the cornerstones of the systems that we rely on today. Some are old and in need of rehabilitation or replacement. Some are new and likely to last a long time, making change difficult. And some are yet to be built, with the opportunity to affect their design.

Furthermore, infrastructure has often been built in support of the dominant technologies at the time it was conceived — not just in physical manifestations, but also in the rules, financing, and governing of the institutions that manage the infrastructure. This becomes a problem when the demands that we ask infrastructure to satisfy change and that infrastructure cannot change quickly enough to meet these demands.

In the past century we've seen the design of many technological and institutional forces that lock-in infrastructure systems. Prioritization of funds to roadways after World War II and minimum off-street parking standards are perhaps the most dominant forces for automobile-centric transportation development in the United States (Pollard, 2003; Shoup, 2011). The commitment of manufacturers through the investment of resources, labor, and manufacturing plants supported several waves of technological innovation that used polyphase electric supply (Hughes, 1993). The rapid growth of water utilities and distribution systems in the late 1800s was largely driven by concerns for public health (through the emergence of bacteriology) and safety, to protect growing populations against disease and fires. Centralized water distribution systems grew exponentially between 1880 and 1895 (from roughly 600 to 3,000 in the United States), while at the same time regulatory agencies emerged to ensure provision of services and affordability (NRC, 2002).

As new technologies come online or as demand for services changes, our infrastructure (both hard and soft) may be unable to adapt, raising questions about how quickly they can change given new societal needs or threats. Given that our infrastructure tend to persist for long time periods, are they agile? Can they adapt to changing conditions? Why do our infrastructure need to be adaptable? Why and how should we design our infrastructure to be adaptable?

There are fundamental reasons why these questions arise at this time. There is a critical category distinction between physical infrastructure designed to be part of an overall infrastructure that is intended to last many decades, and the shorter and more abrupt changes in economic, technological, social, and institutional systems that are coupled to infrastructure. If the rate of change of these latter systems is relatively slow, as it has been for most of our history, infrastructure with half-lives of decades is not a problem.

If, however, the rate of change of the system accelerates, we reach a point where the cycle time of infrastructure change decouples from the increasingly rapid social systems which they serve. We have seen this happen in other long-lived institutions such as law (Marchant, Allenby and Herkert, 2011). For example, the shift to autonomous local vehicle service from owned automobiles is happening much faster than in historical periods, yet we have just begun to think about the implications for urban and transport infrastructure design.

Here we attempt to answer the aforementioned questions. We start by identifying several major challenges that have created a crisis for current infrastructure. We then attempt to unpack the design principles for infrastructure in the past century and how these principles constrain our ability to adapt infrastructure to challenges. We then characterize—based on evidence from industries that have successfully deployed adaptable infrastructure—the novel design and operation principles for infrastructure for a future in which demands on our systems are changing rapidly and there is heightened unpredictability across a number of domains.

2.2.1. INFRASTRUCTURE CHALLENGES IN THE NON-STATIONARITY AGE

Infrastructure systems are facing several major challenges that threaten their performance, the services they deliver, and ultimately the well-being of the societies that rely on them. The confluence of these challenges can be described as a *crisis*. This is especially true in the United States, where significant attention is now focused on the state of disrepair of many major infrastructure systems (ASCE, 2017), but is also true in many other developed regions of the world. We posit that these challenges are 1) inflexibility; 2) funding; 3) maturation; 4) utilization; 5) interdependencies; 6) earth systems changes, most immediately climate change; 7) designing for social and environmental well-being; 8) transdisciplinary practices and processes; and 9) geopolitical security.

These challenges are interrelated and several produce so-called non-stationarity effects. We define non-stationarity loosely on statistical definitions as the unpredictability of future conditions based on past trends. There is a rich discourse around how climate change produces non-stationarity. For instance, precipitation and rates of discharges of rivers are becoming increasingly difficult to predict (Milly *et al.*, 2008).

We argue that funding for public infrastructure (namely transportation and water) also now exhibits non-stationarity. This is the result of policies (at federal, state, regional, and municipal levels) and financial planning that now inconsistently allocates funding (partly the result of escalating rehabilitation and maintenance costs) and creates significant uncertainty as to how much funding will be available for upkeep. This non-stationarity combined with the other challenges creates a crisis that must be imminently addressed to ensure that we are able to adapt infrastructure for the future.

1) *Inflexibility*: A unique characteristic of hard and the soft infrastructure that support them is that they provide services for demands that are difficult to change except incrementally, even in the long-term. Some exceptions exist, notably information and communication technology (ICT) services, which have changed radically over short time periods. Inflexibility, which we'll explore in more detail later, emerges partly because physical infrastructure don't need to significantly change form as the services they've delivered have remained fairly consistent for long periods of time. Unlike microchip fabrication, automobile manufacturing, and ICT services, the demands that infrastructure facilitate are relatively consistent on decadal scales. Electricity consumption, mobility (particularly by automobile), water use, and waste management demands are similar to demands 10, 30, 50, even 80 years ago in many mature areas of the U.S.. There have, of course, been efficiencies and technological improvements implemented in these infrastructure systems, and some, such as ICT, may change more rapidly than others, but their core physical structure has not changed dramatically in the long-term.

2) *Funding*: While there has been much attention focused on the state of disrepair of infrastructure in the United States, we argue that a major challenge is the sustainability of funds, particularly for long-term rehabilitation and technological improvement. Funding sustainability challenges result from two major forces: 1) many infrastructure were deployed in the middle of the last century and are now in need of major rehabilitation; and 2) there remains significant uncertainty about the availability of funds for this rehabilitation. The explosive growth of hard infrastructure in the United States with the New Deal, but more substantially after World War II, continued through the latter parts of the twentieth century. The Silent Generation, born between the 1920s and 1940s, experienced heavy capital investment (as a percentage of gross domestic product) in new infrastructure, which continued through the 1970s (Davis, 2017).

As infrastructure built in the middle of the twentieth century began to reach the end of its service life, new pressures emerged to rehabilitate these systems. Many public agencies find themselves with insufficient funds to cover maintenance activities as these rehabilitation demands have grown (ASCE, 2017). Municipalities are forced to triage their limited rehabilitation funds, deciding which components of infrastructure get rehabilitated while delaying others (Menendez *et al.*, 2013).

In the United States a compounding challenge is the uncertainty of federal funds. The Highway Trust Fund, for example, is supported by federal fuel taxes that have not been increased since 1993 and is not indexed to inflation ('The Status of the Highway Trust Fund and Options for Paying for Highway Spending', 2016). Furthermore, many states use their fuel tax for purposes other than transportation (Paletta, 2014) .There have been instances where the fund was projected to become insolvent. Major challenges also exist for water, electricity, aviation, waste management, rail, and other infrastructure (AWWA, 2012; ASCE, 2016).

3) *Maturation*: Some infrastructure in developed regions of the world has grown to a point where substantial expansion no longer takes place. This infrastructure is mature in the sense that cumulative increases in physical infrastructure and its capacity have leveled off over time, and funding priorities are shifting from capital investment in new infrastructure to increased investment in maintenance and rehabilitation of existing infrastructure (Fraser and Chester, 2016).

This maturation can occur for several reasons. First, there are limits to the outward extent that infrastructure can grow, not necessarily physically, but more so practically. Resource, demand, and budget constraints, including travel-time budgets, natural boundaries (such as oceans, bays, mountains, and protected land), and growth boundaries, can limit how far outward people and services will be—and ultimately how infrastructure is deployed (Kornai, 1979; Garreau, 2011). There are certainly many places where infrastructure is still being deployed outward; however, infrastructure grows where people demand services. If demographics, costs, or other barriers reduce or constrain that demand, then growth will slow or stop.

Related to maturation is geographic scale. In many countries, and particularly the United States, infrastructure exists on such large scales that meaningful and timely changes may require herculean efforts. Take for example the U.S. roadway, railway, power, and pipeline networks (Figure 2.2.1), which when visualized show the boundaries and urban centers of the country. Replacing or enhancing infrastructure at large scale means a big change in physical assets and across many geographies. Even if unlimited funds were available, physical resource, time, and manpower constraints likely exist, and non-technical barriers may be so large that the rate of change is severely limited.

4) *Utilization*: Long-term infrastructure capacity planning remains a major challenge for financing and upkeep given the centralized nature of systems, lack of modularity, and resulting inflexibility. Infrastructure capacity is often planned on decadal scales, with forecasting of traffic de-

mands, water consumption, and power consumption, for example, developed with increasingly sophisticated models. Yet accurate forecasting for roughly 30 or more years out remains elusive given the increasing uncertainty associated with the multitude of variables that drive infrastructure use: population, socioeconomics, climate, technologies, economics, and activities.

Figure 2.2.1: Geographic Extent of U.S. Infrastructure

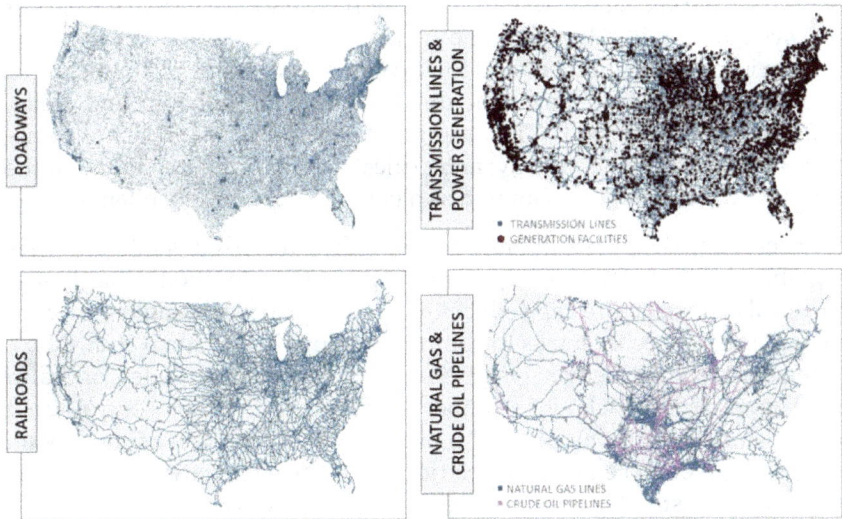

With largely centralized and inflexible infrastructure, managers will ultimately be confronted with the challenge of infrastructure that is either under- or oversized, sometimes grossly. This is evident in the oversizing of infrastructure after the population collapses of Detroit and New Orleans, or undersizing in the cases of cities that have experienced rapid population growth, such as Phoenix and Las Vegas. While oversizing is apparent through cries for more funding to maintain underutilized systems or derelict structures, undersizing is not usually as obvious as short-term policies to meet, say, rapid changes in population growth are quickly established to deploy tried-and-true technologies.

5) *Interdependencies*: Infrastructure are becoming increasingly interdependent with other hard infrastructure, with managing institutions, and with information (more and more delivered digitally). Imagine early instances of shared public hard infrastructure, systems that were deployed on small scales that in no significant way relied on other infrastructure. The earliest roads didn't have electronic traffic control, nor did they have power lines above them or water lines beneath them. By the late 1800s the

Edison Illuminating Company had deployed a number of electricity generating facilities in the Northeast United States and connected them to nearby neighborhoods, each disconnected from the other. Early water conveyance and distribution systems exclusively relied on gravity. Electrical pumps didn't appear until the early 1900s (Walski, 2006).

Today, vast and centralized infrastructure systems are deeply connected with each other. Infrastructure can be interdependent in several ways: geographic (co-location or in close proximity); physical (output of one system is an input into another); cyber (data or information from one system is input for another); and logical (the social, financial, political, etc. relationships between infrastructure) (Rinaldi, Peerenboom and Kelly, 2001). Power, water, and ICT share space with roadways; pipelines sometimes follow rail rights-of-way; and critical systems are often found at the same spatial location, also known as a geographic interdependency.

Because investments in hard infrastructure are fixed, sunk, and irreversible, they are a large risk. Sharing (in terms of co-location or hardware) physical infrastructure can reduce the costs of entry, making it easier for new players or technologies to compete in a market (Chanab *et al.*, 2007). Infrastructure that deliver resources (such as energy or water, and also ICT) serve as the backbone of other infrastructure. Traffic control, train propulsion, water pumping and treatment, and communications rely on electricity. Wet-cooled thermoelectric facilities rely on water systems. Virtually all infrastructure needs transportation services to move people and goods. And the digital age has shifted mechanical controls to digital and introduced remarkable opportunities for generating, transmitting, and processing digital information — processes that are now deeply embedded in many infrastructure processes (cyber interdependencies).

The degree to which this embrace of ICT has introduced new vulnerabilities and unpredictabilities into infrastructure systems is underappreciated and poorly addressed in most infrastructure systems (Amoroso and Vacca, 2013). The interface of hard infrastructure with the institutions that manage them produces logical interdependencies that define the rules, policies, and norms for how they are designed and operated, and how quickly they can change (more on this later).

These tightly coupled interdependencies are a challenge because they introduce complexity at scales and with outcomes that we poorly understand. A perturbation — or even worse, a failure — in one infrastructure can cascade to other infrastructure, leading to service interruptions. The complexity of these interconnected systems, the emergent behaviors of infrastructure when one is shocked, is largely unknown, and represents a critically important area of study when financial, security, or climate change disturbances are introduced.

6) *Earth Systems Changes, Including Climate Change*: There is increasing evidence that critical earth systems are becoming destabilized due to human activity. While climate change is receiving more and more attention, it is likely that we will need to manage other systems, including nitrogen, phosphorous, and water, going forward (Vitousek *et al.*, 1997; Vörösmarty and Sahagian, 2000; Childers *et al.*, 2011). Infrastructure creates a human-made world in place of a naturally evolving one, and at the decadal level the dynamics of changing Earth systems becomes important for engineers. Climate change is likely the most immediate and direct Earth systems hazard that we're confronting. As such, we focus on climate change as a case study that illustrates a fundamental challenge of infrastructure design and management.

Some weather-related extreme events are occurring with greater frequency and intensity (U.S. National Climate Assessment, 2014), and infrastructure, typically designed based on historical conditions, are vulnerable to both extreme and gradual perturbations. Infrastructure are the front line of defense against climate change. The services that they provide are critical during storms, heat, flooding, wildfires, and cold, in terms of the resources they deliver and their direct protection against exposure.

Infrastructure are typically designed against return periods, the frequency that the infrastructure will experience a particular intensity. For example, a bridge over a wash might be designed to maintain structural integrity for a 100-year return period, meaning a flow rate of water through the wash that is experienced on average every 100 years. Two major challenges exist. First, much of our existing infrastructure has been designed for return periods that are likely to change under climate change forecasts. A storm that has historically occurred at a particular intensity once every 100 years may now occur every 20 years (Gilroy and McCuen, 2012; Tramblay *et al.*, 2013).

Second, codes require that designs be based on historical weather conditions that are no longer valid. Those who design infrastructure have not used climate forecasts and even if they were to, they would need different design processes that embrace the uncertainty associated with climate forecasts. We can expect indirect effects on infrastructure from climate change as well, including new conflicts, mass migration, and disease. How these effects will impact infrastructure remain largely unexplored. They nonetheless present serious risk to the reliability of infrastructure services and challenges for delivering services in a future marked by these events.

7) *Social and Environmental Awareness*: Gone are the days when infrastructure could be designed without serious considerations for social well-being and adverse environmental effects. The last half century has produced a mountain of knowledge about how the design, construction, and

use of infrastructure affect people and the environment. Some of this knowledge has affected regulatory processes that require environmental assessments. In addition to the National Environmental Policy Act process requiring more and more disclosures through special-purpose laws, research on social equity and the rapid incorporation of sustainability principles has created novel thinking about how infrastructure should be deployed.

That's not to say that the deployment and use of infrastructure does not produce social and environmental impacts, but that we are more aware of these impacts and some measures have been put in place to reduce them. When deploying new infrastructure, the knowledge that has been generated from this past half-century of study is much more likely to be known by engineers, designers, and managers, as well as the general public, which is able to participate by voting, providing public comments, and protesting. While we are much more aware of social and environmental impacts, infrastructure designers and managers do not necessarily have the flexibility and resources to avoid them, and more broadly balance social, environmental, economic, and technical costs and benefits in a holistic but rigorous manner. Policies, financing, and codes may perpetuate existing practices despite evidence of negative outcomes.

8) *Transdisciplinary Practices and Processes*: Integration of disciplinary and institutional practices and processes is needed to reflect the interdependencies in not just physical infrastructure, but in the institutions and cultures within which they are embedded. As infrastructure have become increasingly interconnected and as our knowledge of the complex systems in which these infrastructure function, provide services, and result in unintended tradeoffs grows, our traditional disciplinary boundaries are no longer sufficient. Infrastructure are typically designed, funded, and managed by a multitude of players, sometimes private and sometime public. They are governed and owned by different asset management systems, standards, businesses, and funding mechanisms. To effectively acknowledge and work within these complex arrangements, transdisciplinarity will be required.

9) *Geopolitical Security*: Several fundamental trends in geopolitical and military doctrine and strategy have come together to make security challenges a critical challenge to infrastructure. The first is a rise in non-state-actor violence, often in the guise of terrorism, against communities and societies. Because infrastructure systems are increasingly reliant on cyber for connectivity and software that can be hacked for operational capacity, deliberate attacks against infrastructure are ever more tempting for those seeking soft targets.

The second is a shift in military strategy by major adversaries of the United States and Europe, especially Russia and China, toward "hybrid warfare" and "unrestricted warfare," which reframe military confrontation as a conflict across all social and cultural systems, including infrastructure (Qiao and Wang, 1999; Galeotti, 2014). Along these lines, it is notable that at a particularly fraught moment in the Ukrainian-Russian conflict, Ukraine's electric system was hacked and taken down, substation by substation, on December 23, 2013 (Zetter, 2016).

Finally, the extent of the Russian attack on American and European social and structural systems is just becoming apparent, and is far more significant than most professionals, embedded in their daily routine, realize. Indeed, a leading NATO analyst has voiced what many have concluded: "Recent Russian activities in the information domain would indicate that Russia already considers itself to be in a state of war (Giles, 2016)." No infrastructure design that isn't hardened against deliberate information attacks can be considered resilient; failure to design security into infrastructure from the beginning is a major source of fragility and vulnerability. And given that cities may have decades- or century-old infrastructure, there may need to be prioritization of assets when hardening (FEMA, 2017).

10) *Wicked Complexity*: Interdependent and even independent infrastructure are dominated by nonlinear interactions, emergent and self-organizing behavior, and distributed control: the key properties of complex systems (Oughton and Tyler, 2013). These properties are defined by physical and non-physical factors and result in limitations on our ability to understand the emergent behavior of infrastructure systems, where the interactions at one level produce unanticipated phenomena at another (Arbesman, 2016). Interdependencies explode this complexity.

Consider the 2003 North America blackout. What started as a single downed power line resulted in a cascading failure throughout the Northeast United States and Canada that left 55 million people without electricity, some for up to two weeks (NERC, 2004). Beyond the power system, outages were experienced in the water, transportation, communications, and industrial systems.

Technical complexity results from several forces: accretion, interaction, and edge cases (Arbesman, 2016). Accretion describes how infrastructure have accumulated and layered technologies over long time periods, to the point where it is no longer apparent how controls work (consider the use of 1980s IBM mainframes by the U.S. Federal Aviation Administration). The ease of interconnections coupled with accretion leads to interactions that over time and scale become so numerous that testing and understanding their behavior becomes challenging. Lastly, edge cases — exceptions to standard design and operating rules—introduce additional layers that obfuscate our ability to understand system behaviors.

Given the obdurate nature of infrastructure, their scale, and ubiquitous use, it can be argued that the systems that we so critically rely on naturally tend toward complexity. Complexity is not strictly the result of technical variables. The increased fragmentation of organizations that have some say in infrastructure, and the processes associated with accommodating different perspectives on how infrastructure is designed or managed has contributed to wicked complexity (Conklin, 2006).

How to build and operate infrastructure is a wicked problem. Wicked problems are characterized as 1) not understanding the problem until a solution is developed (implementation of infrastructure provides new insights into the problem); 2) having no stopping rule (once infrastructure are deployed you often continue to modify them based on changing needs); 3) having solutions that are not right or wrong (there are generally multiple ways to deploy infrastructure, e.g., route alternatives); 4) having novel problems (the multitude of technical and social considerations means that how infrastructure are deployed and operated for a particular circumstance are unique); 5) solutions are a one-shot operation (as per design theorist Horst Rittel: "One cannot build a freeway to see how it works"); and 6) having no alternative solutions (there may be no way to meet the need, or there may be many potential solutions but no single solution) (Rittel and Webber, 1973; Conklin, 2006). The combination of these factors means that infrastructure are wicked complex systems. It has become extremely difficult, if not impossible, both to predict how systems will behave across space and time when perturbations occur and to change systems toward future goals.

As we've transitioned into the twenty-first century, we will likely find our infrastructure increasingly defined by these challenges. Several of these challenges manifested during the latter part of the last century and combined with emerging challenges—specifically the non-stationarity introduced by climate change and financing—meaning that new models of infrastructure design, construction, operation, and use will be needed.

As services and technologies change, the demands that we place on infrastructure will also change. To meet these changing demands, infrastructure will need to be *agile* (in the face of both predictable and unpredictable challenges) and *flexible*, preconditions for *adaptability*. Yet when it comes to hard infrastructure, we have not seen a system that has these characteristics. In the following sections we explore systems that have successfully implemented these characteristics to meet rapidly changing demands and environmental conditions. We identify the design principles and operating conditions that enable these systems to behave with these characteristics and discuss the changes that are needed in hard infrastructure systems so that they can meet rapidly changing demands in the twenty-first century.

2.2.2. DESIGNING AND PLANNING PRINCIPLES

Successful infrastructure in the twenty-first century will need radically different design principles. Engineers will need to be part of a process that reconceptualizes infrastructure from the purely physical to a system that includes institutional components and knowledge as integral parts. Infrastructure will need to change their structure, behavior, or resource use as demands change. In doing so they will need to support the rapid deployment and growth of nascent technologies (such as renewable electricity generation, microgrids, gray water systems, material reuse, and autonomous and electric vehicles) as well as technologies that we haven't yet begun to envision. These technologies are liable to change not just the physical operation, but the mathematics, of the underlying systems in unpredictable ways.

A deeper challenge is that we're not just operating on the level of infrastructure itself, but at the implicit models of operation that we sometimes haven't revisited in decades, requiring us to expand how we think about the institutional and disciplinary ways that we think about infrastructure. They will need to create opportunities for embedding digital sensing, data processing, and data analysis (so-called "smart") technologies that improve our understanding of how interdependent and complex infrastructure behave; provide us with protective measures against vulnerabilities and added security; and enable a new understanding of how our built environment functions to improve well-being.

These infrastructure will need to be able to meet these characteristics with unpredictability around financing and in the face of both extreme events and gradual changes in climate and the challenges that result from climate instability — itself a proxy for the challenge of designing and building infrastructure in a world where human activities increasingly impact all environmental systems. And they will usher in new metrics of infrastructure success that measure the ability to meet rapidly changing needs and respond to perturbations. This will require a fundamentally new paradigm in how we design, build, and operate our infrastructure.

Flexibility and *agility* will need to be at the core of this new paradigm. In the context of hard infrastructure, we distinguish between flexibility and agility based on changing demands and non-stationarity. With rapid changes in technology and ultimately the services that our systems provide, infrastructure will need to be flexible to changing demands. Infrastructure will also need to be agile in that its physical structure, along with the rules, policies, norms, and actors who manage and operate it, will need to be able to maintain function in a non-stationarity future. This includes planning and responding to unpredictable events, such as extreme weather or budget shortfalls, disease events, security challenges such as cyberattacks and physical terrorism, population migrations, and other phenomena.

Infrastructure managers can expect that at some point in the future these events will occur but cannot easily or accurately predict when or to what extent, or for that matter how perturbations in underlying human and natural systems will manifest themselves. The combination of flexible and agile design and operation characteristics are the preconditions for *adaptability*. These challenges are monumental, but they are not without precedent. Success in implementing flexible and adaptable infrastructure for rapidly changing demands in other industries can offer invaluable insight into the processes that need to shift the design paradigm of civil systems.

How do we design our infrastructure to be adaptable? We don't know the ultimate forms of infrastructure that enable flexibility and agility, as it is likely that these forms have not yet been identified, developed, tested, or implemented. Also, we're not training engineers and planners to function in the integrated infrastructure systems of the future. (Allenby, 2012)

However, we can identify the characteristics of flexibility and agility that have been successfully implemented in other infrastructure and their processes and describe how they may translate to civil systems. These characteristics do not strictly belong to the physical world; they must also exist in the organizations and institutions that manage and govern physical systems. We synthesize characteristics of flexibility and agility from several industries that have successfully changed their organizations and physical processes to meet rapidly changing demands and respond to unpredictability. We focus largely on ICT and manufacturing. We also explore the shifting of technological functions within automobiles and automobile travel to characterize the substitution of functionality from physical to digital and efficiency gains within technological and infrastructural constraints. These industries have infrastructure that have evolved in response to economic and competitive pressures that are not usually felt strongly by either the professionals that design infrastructure or the public planners and managers who operate it.

Through this review we identify several characteristics that enable flexibility and agility. Given the rapidly changing demands for services that commercial sectors must often meet and the structure of public institutions that typically manage civil infrastructure, we question where best practices are most likely to arise. Finally, we attempt to organize these characteristics into a structured framework, identifying drivers and characteristics that produce competencies for flexibility and adaptability. This framework is shown in Figure 2.2.2.

An adaptive infrastructure is one that has the capacity to perceive and respond to perturbations in such a way as to maintain fitness over time. Adaptive infrastructure have the capacity to recognize that stimuli or changes in demand are occurring or will occur, including the effects of these stimuli, and have the socio-technical structures in place to change quickly enough to meet future demands.

Figure 2.2.2: Stimuli, Properties, and Competencies for Adaptive Capacity

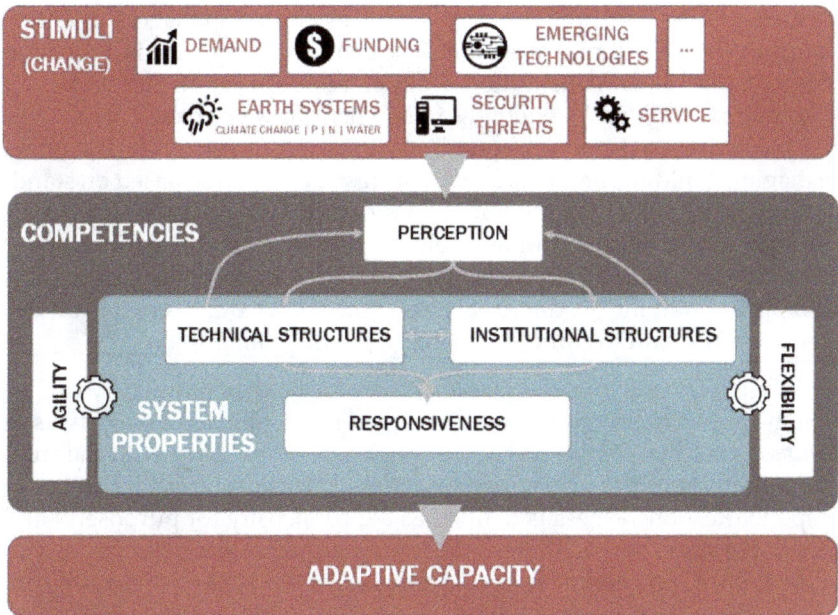

Stimuli can take many forms; we focus on those related to the aforementioned challenges. They describe a direct stress (e.g., climate change, extreme events, or inadequate funding for maintenance), change in demand (e.g., a rapid change in population resulting in more or less need for infrastructure services), or change in service (e.g., a technology or behavioral shift that brings about less-to-no need for an infrastructure). Other stimuli also exist, including emerging technologies and physical and cyber threats.

Competencies for adaptive capacity include agility and flexibility, but are preceded by the ability to perceive stimuli and how they will affect the system and ultimately the infrastructure. There is a rich body of study on how individuals and organizations perceive risk (Mitchell, 1995; Slovic, 2016). Infrastructure managers must be able to recognize that stimuli are or will occur and understand how they will affect the system. Beyond supervisory control and data acquisition systems, they must have physical and informational sensing capabilities that provide insight into the behaviors of complex and interconnected systems (both in terms of the infrastructure and its use). This is related to sensing and anticipating in resilience frameworks (Hollnagel, Pariès and Wreathall, 2011; Park *et al.*, 2013).

Knowledge is a critical aspect of perception, which we argue is currently insufficient to deal with the aforementioned challenges facing infrastructure. As such, the capacity for infrastructure managers to anticipate the effects of stimuli is lacking. The capacity to perceive is a function of the technical and institutional structures supporting infrastructure (Hommels, 2005), and frequently fails when rates of change, or system complexity, exceed normal bounds. Institutions tasked with designing, managing, building, and maintaining infrastructure do so based on standard practices, codes, and methodologies reinforced by disciplinary expertise, training, and organizational culture (Bijker, Hughes and Pinch, 1987; Carse and Lewis, 2017). In periods of rapid, non-incremental change, this disciplinary training produces barriers to the knowledge needed to understand and respond to stimuli, to perceive, to maintain fitness.

While infrastructure can take on many forms defined by network typology, public to private management, and national to local scale, the *system properties* that define physical configuration combined with the rules and objectives of the managing institutions ultimately affect its ability to respond. Responsiveness is defined as the propensity for purposeful and timely behavior change in the presence of stimuli (Santos Bernardes and Hanna, 2009). The definitions of responsiveness, flexibility, and agility are often conflated, and there have been few efforts to differentiate. Following research by Ednilson Santos Bernardes and Mark D. Hanna, in the context of infrastructure adaptation we differentiate "responsiveness" as being the propensity for timely behavior change and "agility and flexibility" as being associated with reconfiguration of the system.

Competencies and appropriate system properties enable *adaptive capacity*, the ability of infrastructure to respond to inevitable and unexpected stimuli. Adaptive capacity has largely been defined by sociological-ecological systems researchers (Meerow, Newell and Stults, 2016). The dominant approach for designing infrastructure systems is that of risk management, that is, sizing infrastructure to be able to withstand an event of a particular magnitude and frequency (the 100-year return period is often chosen). This approach leads to large gray infrastructure that favor designs that keep hazards away (e.g., levees) or can continue operating during the hazard. Yet as infrastructure becomes larger and more permanent, the consequences of failure increase (Park, Seager and Rao, 2011). Furthermore, infrastructure becomes less adaptive when it has more legacy components that impede efficient system evolution in response to unanticipated stresses.

This approach, while robust in protecting against particular shocks, ultimately is highly inflexible for changes in demand beyond what was forecast; it can result in major consequences when failure occurs; and it is generally unable to cope with unforeseen stimuli. Adaptive capacity approaches are inimical to risk-based approaches in that they focus on maintaining capacity in the face of stimuli; minimize the consequences of stimuli instead of minimizing the probability of damage; privilege the use of solutions that maintain and enhance services; design autonomous management schemes instead of hierarchical; and encourage interdisciplinary collaboration and communication (Möller and Hansson, 2008; Ahern, 2011; Park *et al.*, 2013).

We contend that the competencies and system properties needed to achieve adaptive capacity will require transformational shifts in the way that we build, operate, and perceive system purpose and function, and the educational and organizational institutions that we have historically relied on to design and manage infrastructure systems.

2.2.3. ADAPTIVE INFRASTRUCTURE

The competencies and system properties that can help enable adaptive capacities require novel planning techniques, technical and institutional structures, and integration of education and interdisciplinary practices across the lifecycle of infrastructure. Drawing on past successes from other industries, recommendations can be made for civil infrastructure systems. Following from Figure 2.2.2, the competencies and system properties associated with current infrastructure are contrasted with those of successful adaptive systems, based on the following discussion. These are summarized in Table 2.2.1 and discussed in detail as follows.

Table 2.2.1: Competencies for Adaptive Systems

Competency	Current	Adaptive
Perception & responsiveness	Prioritizes perpetuation of existing designs	Roadmapping
Perception, responsiveness, & technical structure	Obdurate design	Design for obsolescence
Technical structure	Hardware-focused	Software-focused
Technical & institutional structures	Risk-based	Resilience-based
Technical structure	Incompatibility	Compatibility
Technical structure	Disconnected	Connectivity
Technical structure	Non-modular design	Modularity
Institutional structure	Mechanistic	Organic
Institutional structure	Culture of status quo	Culture of change
Perception & responsiveness	Discipline-focused education	Transdisciplinary education

For each of the four driving competencies identified in Figure 2.2.2, the current approaches and exploration of adaptive approaches are shown.

ROADMAPPING AND PLANNED OBSOLESCENCE

Industry roadmaps have proven to be valuable for enabling radical innovation and evolution by aligning common goals across a number of different domains. We use the term "roadmap" to describe the development of a model or structure that allows multiple organizations with competing goals operating at many different levels of a technology system to plan together to enable the rapid evolution of systems and manage uncertainty.

In the 1990s during the early stages of development of much of today's ICT backbone and associated computational technologies and tools, roadmapping emerged as a valuable process for prioritizing technologies and identifying infrastructure gaps, suggesting robust interconnections between modules within which powerful innovation was occurring, and creating standards and business practices to meet these challenges across many (often competing) organizations (Pedersen, 1995; Rae, 2017). For example, electronic industry roadmaps enabled constant improvement in computer performance at the user level even as disruptive innovation characterized component subsystems (e.g., portable storage devices evolved from large floppy disks to hard disks to thumb drives).

At the institutional level, roadmapping includes creating industry committees and associations, which subsequently use conferences, workshops, less formal collaborative practices, and other activities to create institutional structures that support constant innovation and communication in often highly competitive environments with significant antitrust and other legal constraints. Given that right now funding is often spent on expensive failures instead of preventative maintenance, roadmapping, when using measures such as return-on-investment, could help identify lower-cost pathways in addition to necessary technological change. Without roadmapping, the combination of unpredictable and disruptive innovation with smooth system-level evolution that characterized the growth of electronics and communications technologies across the entire spectrum of the ICT sector would have been impossible.

These roadmapping techniques suggest several general principles supporting agile and adaptive design. Most importantly, at an institutional level they suggest that a complex combination of competition, innovation, and collaboration can be managed through sophisticated use of modular design. They also suggest that rapid cycles of innovation and obsolescence within modules cannot just be tolerated, but encouraged, even as the overall system remains stable. In ICT, they helped to enable rapid and unpredictable development in ICT subsystems while maintaining a framework that ensured the overall technologies remained operational (Pedersen, 1995). They managed unpredictable and important innovation within the module, yet maintained interconnection at the systems level.

Applying a roadmap model to infrastructure becomes a way to enable radical innovation within systems, while supporting a high level of inter-domain communication and constantly improving product, providing continuous and uninterrupted functionality to users. Consortia including cities and firms can generate such a roadmap that not only crosses engineering domains (e.g., energy, water, ICT, and transport, at a minimum), but policy domains (e.g., tax policy and transportation management), and institutional domains (e.g., the city government with all its silos and the critical private firms in each sector) to facilitate the planning and operation of next-generation infrastructure in non-stationarity conditions, and to encourage continuing innovation and efficiency in provision of services without disruption and at low lifecycle cost.

Roadmapping can be valuable for shifting design considerations from obdurate paradigms to planned obsolescence. For many components of infrastructure systems, managers favor designs and assets that can last a long time. This paradigm is problematic in that it locks users into technologies for the long-term, constraining the ability to modernize systems for changes in demand. *Obduracy* in infrastructure persists because managing institutions are constrained in their ways of thinking and changing one component requires multiple changes; it is therefore easier to maintain old configurations than introduce innovation (Hommels, 2005; Bulkeley, Castán Broto and Edwards, 2014).

Alternatively, *planned obsolescence*, that is, planning for changes regarding function, profitability, and other dimensions of performance, can result in greater capacity to substitute infrastructure components and technologies to more efficiently meet changes in demands. (Lemer, 1996) Access to infrastructure is important when replacing hardware. Much of our infrastructure is buried underground where access is costly and conditions are often unknown. This inaccessibility encourages waiting for failure. Accessibility will be critical for ensuring quick and frequent replacement or upgrades. While infrastructure obsolescence mostly has been studied from a physical asset perspective, the integration of computing and substitution of software for hardware function creates new opportunities for shifting functions.

SOFTWARE-FOR-HARDWARE SUBSTITUTION

With the increasing availability of sensors, processors, and data analytical tools at decreasing costs, there is a growing substitution of software for hardware, transitioning physical to digital processes that increase the flexibility and adaptive capacity of technologies while improving their efficiencies. These technologies are sometimes collectively referred to as "smart" technologies or systems. They are increasingly replacing or being used to augment the capabilities of physical processes. With their use, industries are finding that core business practices can be shifted (Lohr, 2016).

Sensors can predict the structural health of hardware, notifying operators of needed maintenance (Lynch and Loh, 2006), measure fluctuations in manufacturing processes, and adjust inputs to improve production and reduce costs (Frankowiak, Grosvenor and Prickett, 2005). They can also provide real-time information to users or software to adjust operations, likely avoiding the need for manual labor and associated resources. Fuel sensors in automobiles can adjust the air-to-fuel ratio, optimizing combustion and emissions. Digital technologies now allow industrial manufacturers to configure processes virtually before changing or upgrading equipment to proactively identify potential incompatibilities in parts or processes (Resnick, 2016). Variable frequency drives that use electronics to monitor motor performance and load requirements to optimize work by pumps, eliminating the need for smaller pumps and control valves in applications such as water distribution (Roethemeyer and Yankaskas, 1995; Neuberger and Weston, 2012). Traffic camera software is now smart enough to identify cars, pedestrians, and bicycles, reducing the need for in-pavement loop detectors and the associated asphalt impacts (Kenny, 2004). Implementing wireless communications instead of landlines reduces the need for wiring.

In addition to the efficiencies that are gained in substitution, there is likely less waste when you upgrade via software instead of hardware. Additionally, software-driven functions are more agile when faced with unpredictable and rapid change than hardware-driven functions that require physical alterations for upgrades. As software has progressed, the prior practice of upgrading via physical media—sending a disk with the upgrade through snail mail, for example—has become largely obsolete, replaced by the use of online software fixes that are more efficient and produce less waste. This is an important evolution, improving the agility of embedded software systems. Real-time fixes are necessary in an environment where malware and viruses are instantaneous; the next step is to have artificial intelligence responding to challenges and changes as they occur on a network-wide basis.

The implementation of smart technologies within existing infrastructure and technologies can create efficiencies within the constraints of inflexible systems. They can also help improve our understanding of the increasing complexities of infrastructure. Take, for example, the use of GPS, smart phones, and navigation software in personal automobile travel. Initially the automobile consisted of mechanical components with little to no interaction with the environment except through the driver's decisions. With the advent of electronics, automobile systems began working together, communicating information to each subsystem so that subsystem-specific adjustments could be made, thereby increasing the efficiency of the overall vehicle. Currently smart technologies are introducing new efficiencies. Within the confines and rules of the roadway system, sensing technologies and software are now able to communicate to drivers the shortest paths, including routes that avoid traffic, thereby possibly saving fuel and time (Gonder, Earleywine and Sparks, 2012).

These technologies introduce agility within the constraints of the current infrastructure, which is now up to a century old. There are of course limitations to the benefits that these technologies bring. The inefficiencies or poor condition of old infrastructure may prove to be a limiting factor in how much improvement smart technologies can provide. In the future it is conceivable that smart technologies will know the condition of infrastructure and reroute flows or traffic away from vulnerable links, or prevent component failures from cascading through or across infrastructure. They may provide us with insight into the increasing complexities of infrastructure.

Another benefit of hardware-to-software substitution is the integration of modules into a larger system or with the environment, effectively changing the scale and scope of the system that can then be optimized. With a mechanical system, efficiency improvements are largely confined to particular modules. Prior to the integration of software and smart technologies into automobiles, it wasn't possible to optimize the performance of the vehicle in real time, much less reach beyond it. But with sensors and ICT, you can not only optimize the automobile in real time, but make it part of a much larger system that includes other vehicles, infrastructure conditions, and traffic controls. This hardware-to-software substitution shifts not only the underlying technologies, but also the larger governing rules of travel; today navigation software can provide improvements across the entire transportation system and onboard software can optimize how your vehicle is performing in real time.

RISK-TO-RESILIENCE THINKING

Infrastructure will need to operate in natural environments defined by non-stationarity. For example, infrastructure systems are already experiencing more frequent extreme weather events, raising questions of whether traditional risk-based approaches to design are adequate. Many infrastructure (or their components) are designed for specific return periods; for example, a 100-year precipitation event that characterizes, in this case, an event with a particular intensity that has a 1% chance of occurring annually.

However, there is growing evidence that these events will become more frequent and unpredictable, raising questions of which return periods to design to, the affordability of larger designs, and whether people want to live near bigger structures. This risk-based approach, which focuses on the risk triplet of *threats* × *threat probability* × *consequences* (Kaplan and Garrick, 1981), is often based on historical data and results in large gray infrastructure with low probabilities of failure, long lifetimes, and oversizing (e.g., levees or retention basins).

The problem is that the risk management approach does not incorporate an understanding of what may happen when the infrastructure itself fails. Larger and more permanent infrastructure tend to be associated with greater damages when they fail (Park, Seager and Rao, 2011). Climate change necessitates new approaches to infrastructure design — approaches that recognize risks but are adaptable, in that they do not compromise the entire system upon failure. More broadly, infrastructure will need to be designed with new rules that recognize the non-stationarity in Earth systems created by human activities.

COMPATIBILITY, CONNECTIVITY, AND MODULARITY

ICT, particularly those technologies developed in the internet age, are designed to meet rapid changes in demand and types of services offered. Early work in ICT recognized the necessity of flexibility as both a system attribute and core competency (Chung, Rainer and Lewis, 2003). The characteristics of flexibility are defined as compatibility, connectivity, and modularity (Duncan, 1995).

Compatibility is the ability to share information across different technological components, involving integration rules and access standards that affect shareability and reusability. *Connectivity* is the ability of any technology to communicate with components inside and outside of the system. It is a measure of the number of processes that are able to interact. Connectivity enables shareability, which is central to flexibility in that it allows resources to be used for new functions (Duncan, 1995).

Modularization in the manufacturing and computing industries has helped manage complexity, enabled parallel work, and accommodated future uncertainty (Baldwin and Clark, 2006). *Modularity* is the ability to add, modify, or remove components easily, without needing to change other modules and subsystems, achieved through standardization (Duncan, 1995). Integrated systems that lack modularity have fixed processes embedded within the structure that interact, but cannot be easily removed or reconfigured. Likewise, these systems are not capable of easily adding new processes. Systems become more manageable when processes are modularized — that is, processes are designed with standards for information and hardware interactions, and the core routines are compartmentalized. They can be designed independently and used in a variety of situations or systems. Modern computing uses modular design in both software and hardware to enable rapid responses to changing customer demand and manufacturing processes more adaptive (Baldwin and Clark, 2006).

The explosive reach of the internet both in terms of information and hardware embodies these three characteristics. Standards for information transfer, such as web languages (e.g., HTML and FTP), transmission and information control protocols (i.e., TCP/IP), and the heavily modularized use of hardware and software, have enabled the internet to grow at a pace never before seen with other technologies (Freeman and Louçã, 2001).

FLEXIBLE MANAGEMENT

Successful organizations reflect the complexity of the environment in which they operate (Hatch, 1997; Vecchio, 2006). Contingency theories state that organizations must be analyzed as open systems that directly interact with their environment, and as such, in order to be effective, must be able to adapt to changing contingencies (Donaldson, 2001; Sherehiy, Karwowski and Layer, 2007). Organizations that can successfully operate in unstable, changing, and unpredictable environments have organic design characterized by less precise division of labor, wider span of control, more decentralized authority, fewer rules and procedures, and more personal means of coordination (Sherehiy, Karwowski and Layer, 2007).

This is in contrast with the typical mechanistic design of organizations that manage infrastructure. They are characterized by highly hierarchical structures, formal management with a centralized authority, a large number of rules and procedures, precise division of labor, narrow span of control, and formal means of coordination. Table 2.2.2 contrasts these characteristics in the context of infrastructure.

The mechanistic form persists because of historically relatively stable and predictable demands and environments. The ways in which water, electricity, and mobility are demanded have not changed significantly in the past century. The mechanistic approach has been shown to be most effective in environments that require routine operation and little change. In these environments, high-level management possesses the appropriate amount of knowledge to make decisions and organize work.

However, when the environment becomes unstable, management cannot acquire all the knowledge associated with the changing environment, and distributing the knowledge and decision making at the bottom of the hierarchy becomes more effective (Sherehiy, Karwowski and Layer, 2007). This is because in order for one system to be able to understand and manage another, it needs to be of the same or greater complexity (Ashby, 1960). Organic structures allow for more internal specialization to respond to changing environments, thereby increasing responsiveness (Lawrence, Garrison and Lorsch, 1967).

The scholar Bohdana Sherehiy, citing a large body of literature, argues that flexible and adaptable organizations have fewer regulations of job description, work schedules, organization policies, and power differentials (e.g., titles). They have fewer levels of hierarchy, informal and changing lines of authority, open and informal communication, loose boundaries among function and units, distributed decision making, and fluid role definitions. Furthermore, authority is tied to tasks instead of positions, and shifts as task shifts (Weick and Quinn, 1999).

Table 2.2.2: Infrastructure Management Structure Characteristics

	Current (Mechanistic)	Adaptive (Organic)
Authority	• Hierarchy	• Less adherence to authority and control
Communication	• Hierarchical	• Networked
Knowledge	• Centralized	• Decentralized
Loyalty	• Organization	• Project
	• High degree of formality	• High degree of flexibility and discretion
Coordination	• Formal and impersonal	• Informal and personal
Rules and Procedures	• Many	• Few
Tasks	• Specialized	• Contribution to common tasks

Adapted from B. Sherehiy, W. Karwowski, and J. K. Layer, "A review of enterprise agility: Concepts, frameworks, and attributes," International Journal of Industrial Ergonomics 37, no. 5 (2007): 445–460.

Significant questions remain as to whether the public institutions that manage infrastructure have the flexibility to change from mechanistic to organic cultures: do laws and policies constrain their organizational form? Given that public institutions are directly beholden to taxpayers, can they take on new organizational forms or change how infrastructure performance is measured? This should not be taken as an argument that infrastructure should be privatized, but instead as a challenge to conceptualize new management structures that embrace organic characteristics.

Taxonomies of organizational flexibility tend to focus on a few key factors to meet changing demands, which may be helpful when developing new organizational structures for managing infrastructure (Dastmalchian and Blyton, 1993). They alter the number of employees and hours through employing part-time, temporary, or short-term contracts, or by changing working times. They create opportunities for changing workforce skills to accomplish a wider range of tasks. And they recommend financial flexibility through pay-for-performance and profit-sharing plans (Dastmalchian and Blyton, 1993; Kalleberg, 2001; Sherehiy, Karwowski and Layer, 2007). Whether these factors are feasible for public institutions in the context of infrastructure management remains an open question.

CULTURE OF CHANGE

Related to flexibility in management is a culture of change, an organization supportive of experimentation, learning, and innovation, and one that is aware of changes in the environment (Sherehiy, Karwowski and Layer, 2007). Whether infrastructure organizations are capable of achieving these remains an open question as there are a lack of incentives (related to what has historically been consistent market demand), legal and regulatory requirements (such as adherence to codes and regulations), safety requirements, reliability requirements, and constraints (e.g., missions that are tied to public goods, or earmarked funding).

Given that infrastructure is typically associated with the provision of public goods, many constraints exist to ensure that resources are delivered reliably, fairly, and at the lowest cost. These constraints may be inimical to rapid experimentation, innovation, and change. We can even question whether it makes any sense to reimagine public institutions that manage infrastructure as reflections of private enterprises that must persist in very different environments.

However, there remains a pressing need for infrastructure organizations to be able to change to ensure reliability and foster new activities and technologies into the future. Change-oriented cultures are partly the result of education, which can take several forms. Organizations can directly support research activities and structurally integrate research into decision making at all levels. They can budget for experimentation or the testing of new and emerging infrastructure designs and management strategies. They can consider changing organizational and individual responsibilities toward anticipated stressors (and ultimately solutions) and away from structures that are purely disciplinary and focused on process or challenges within those disciplines—toward competencies that span multiple disciplines. Part of this responsibility falls to those educating the next generation.

EDUCATION

Institutional design is partially an artifact of training, and infrastructure education (largely engineering) continues to emphasize knowledge and problem solving within single domains. Relatedly, infrastructure continues to be planned, designed, and operated as rigid silos with little to no understanding of the complexity that emerges from the inherent interdependencies of systems. The result is that each system, to the extent it does try to optimize, does so within subsystems.

Integrated education at the university level, and integrated planning in practice, is almost the opposite of what educators and engineers do today. But it's necessary for understanding the impacts of stressors and opportunities for developing strategies to handle stressors going forward. The true complexity of mitigating the infrastructure crisis includes challenges not only in physical infrastructure and the institutions that manage them but at all levels of knowledge production, starting with education and training. Engineers must be able to think about institutional design in these complex environments, as well as technical design. And engineers don't necessarily need all of the expertise themselves; they could be part of teams that design, deploy, and operate systems.

2.2.4. LOCK-IN AND PATH DEPENDENCY

Transitions toward agile and flexible infrastructure will require the identification of, and strategies to overcome, barriers that perpetuate current infrastructure forms despite a need to change. These barriers will need to be catalogued and associated with the actors and forces (rules, policies, norms, financing, etc.) that support them. There is a long history of describing how major changes are needed for infrastructure as well as the technologies that use these infrastructure, often expressed as scenario analyses showing how things can be given some monumental aligning of forces (McCollum and Yang, 2009; Delucchi and Jacobson, 2011; Jacobson and Delucchi, 2011).

Although some infrastructure have changed quickly (e.g., the shift from landlines to wireless technologies in ICT), when it comes to many core civil systems (i.e., water, electricity, transport) large-scale transitions have not happened. This is because many barriers exist that prevent these transitions, including financial (lack of funding for capital investments or earmarking of funds for particular purposes), political (limited political will and "not in my term of office" mentality), codes (such as minimum parking requirements), social (communities may not see the value of redirecting resources from an established technology to an alternative), cultural (for example, consumers' unwillingness to consume treated wastewater), and technological forces.

We describe *lock-in* as the inability to change infrastructure due to these barriers and their often-synergistic interactions with other infrastructure (i.e., interdependencies where we cannot radically change the structure or function of one infrastructure because another relies on that structure or function). Many infrastructure designs have persisted for so long that other infrastructure and institutions have become interdependent, leading to additional barriers and complications for making transitions.

Furthermore, given that demand for infrastructure services doesn't change quickly, managers and policymakers end up prioritizing low-cost rehabilitation and supporting established technologies. Despite the dire state of infrastructure (ASCE, 2016), as long as people get basic services cheaply, the impetus for major reform will not exist. New funds tend to go to existing infrastructure and technologies because we know how to do them inexpensively and we've codified so much for them. Also, regulated infrastructure is often permitted to expend capital for variable costs, but not for fixed costs or capital improvements; American railroads are a case in point (Wolmar, 2012). While the American Society of Civil Engineers has done a good job of communicating the state of infrastructure to policymakers and engineers, they do not engage directly with the general public — the group that will need to pay to avoid aging and failing infrastructure.

The persistence of forces that maintain lock-in into the long term creates *path dependency*, a characteristic of all complex adaptive systems whereby past conditions significantly impacts possible future trajectories. In the case of long-lived infrastructure in a period of rapid change, path dependency can lead to the perpetuation of infrastructure and the technologies, activities, and behaviors that rely on them, despite alternative futures being preferable. These alternative paths may describe futures with lower user and public agency costs, reduced energy use and greenhouse gas emissions, or more socially equitable outcomes.

As infrastructure and the technologies they support continue into the long term and interdependencies are established with other systems, there exists the possibility that the achievement of benefits becomes limited. Learning how to design infrastructure in such a way as to enable more desirable future states given the reality and complexity of path dependency thus becomes a necessary future competency.

We characterize the confluence of constraints (barriers) that prevent us from achieving a more desirable state as a *limit*. In mathematics, limits are values that functions approach as the inputs approach some value. In the case of infrastructure, limits can be thought of as the practical achievable futures given technical and nontechnical barriers. The concept of a limit is important because it recognizes the reality that the infrastructure that we've integrated into all facets of society may not allow us to reach a desirable future, that there's only so much better we can make it or the activities that use it.

For example, given the preference and incentivizes for automobile-focused roadway infrastructure, there may not be a pathway where greenhouse gas emissions can be reduced fast enough to avoid some significant climate change threshold. Or consider that the design and deployment of housing stock (specifically the materials, building technologies, and form) may limit our ability to reduce building energy use beyond a certain threshold (Nahlik and Chester, 2015). As such, our current infrastructure and the forces that maintain their persistence are limiting our ability to achieve desired goals.

More subtly, so is our educational system. Educating engineers to think of systems only in terms of traditional technological frameworks, such as energy, transportation, or information, makes it difficult to achieve integrated design and management of infrastructure, and thus impedes the ability to reduce the deleterious effects of lock-in. And yet engineering education has its own path dependency, reflecting such constraints as meeting requirements for professional and school accreditation.

Understanding how lock-in with various coupled systems is acting in a particular design space can help identify the barriers that need to be overcome to create opportunities for agile and flexible infrastructure. There is a long history of scenario analysis that shows that it is technologically feasible to reach more preferable future states. Research is needed that identifies the barriers (financial, political, social, technological, institutional, educational, etc.) to achieving these states and competencies needed for overcoming these barriers. These competencies will likely center on changing social and cultural norms — and current educational practices and institutions — that drive the formalization of codes and financial structures, and the institutional practices that drive infrastructure growth and management.

2.2.5. A CENTURY OF CHANGE

A century from now we may look back and ask ourselves what the costs were of not transitioning our antiquated infrastructure fast enough. These costs would have taken many different forms, from missing economic benefits to environmental losses to possibly even loss of life (consider infrastructure not sufficiently protecting us from climate change).

Or maybe we will look back at the monumental success that was the rapid transition of physical infrastructure and their managing institutions, and the accumulating benefits that it afforded. In this future, pipes may still deliver water, lines may still deliver power, and roads may still move people and goods, but infrastructure (as socio-technical systems) will have the competencies and structure to be able to meet changes in service demand in environments of unpredictability.

Today, as we debate what the next infrastructure component should be, we should fundamentally question whether new infrastructure should be more of what we already have or something that doesn't exist yet. Until we get to that point, we will maintain lock-in and perpetuate systems that we know may already be obsolete.

2.2.6. REFERENCES

Ahern, J. (2011) 'From fail-safe to safe-to-fail: Sustainability and resilience in the new urban world', *Landscape and Urban Planning*, 100(4), pp. 341–343. Available at: https://doi.org/10.1016/j.landurbplan.2011.02.021.

Allenby, B.R. (2012) *The Theory and Practice of Sustainable Engineering*. Hoboken, NJ: Upper Saddle River: Pearson Prentice Hall.

Amoroso, E.G. and Vacca, J.R. (2013) *Cyber attacks: protecting national infrastructure*. Waltham, MA: Butterworth-Heinemann.

Arbesman, S. (2016) *Overcomplicated: Technology at the Limits of Comprehension*. New York, NY: Penguin Publishing Group.

ASCE (2016) *Failure to Act: Closing the Infrastructure Investment Gap*. Reston, VA: American Society of Civil Engineers, p. 32. Available at: https://www.infrastructurereportcard.org/wp-content/uploads/2016/10/ASCE-Failure-to-Act-2016-FINAL.pdf.

ASCE (2017) *2017 Infrastructure Report Card: A Comprehensive Assessment of America's Infrastructure*. Reston, VA: American Society of Civil Engineers, p. 112. Available at: https://www.infrastructurereport-card.org/wp-content/uploads/2016/10/2017-Infrastructure-Report-Card.pdf.

Ashby, W.R. (William R. (1960) *Design for a Brain: the origin of adaptive behaviour*. 2nd ed. rev. London, UK: Chapman & Hall : Science Paperbacks.

AWWA (2012) *Buried No Longer: Confronting America's Water Infrastructure Challenge*. Denver, CO: American Water Works Association, p. 37. Available at: https://www.awwa.org/Portals/0/AWWA/Government/BuriedNoLonger.pdf?ver=2013-03-29-125906-653.

Baldwin, C.Y. and Clark, K.B. (2006) 'Modularity in the Design of Complex Engineering Systems', in D. Braha, A.A. Minai, and Y. Bar-Yam (eds) *Complex Engineered Systems: Science Meets Technology*. Berlin, Heidelberg: Springer (Understanding Complex Systems), pp. 175–205. Available at: https://doi.org/10.1007/3-540-32834-3_9.

Bijker, W.E., Hughes, T.P. and Pinch, T. (1987) *The Social construction of technological systems: new directions in the sociology and history of technology*. Cambridge, MA: MIT Press.

Brennan, D.R., Turnbull, P.W. and Wilson, D.T. (2003) 'Dyadic adaptation in business-to-business markets | Emerald Insight', *European Journal of Marketing*, 37(11/12), pp. 1636–1665. Available at: https://doi.org/10.1108/03090560310495393.

Bulkeley, H., Castán Broto, V. and Edwards, G.A.S. (2014) *An urban politics of climate change: experimentation and the governing of social-thenical transitions*. London, UK: Taylor et Francis (Climate change, urban studies, environmental studies). (Accessed: 5 November 2023).

Carse, A. and Lewis, J.A. (2017) 'Toward a political ecology of infrastructure standards: Or, how to think about ships, waterways, sediment, and communities together', *Environment and Planning A: Economy and Space*, 49(1), pp. 9–28. Available at: https://doi.org/10.1177/0308518X16663015.

Chakravarthy, B.S. (1982) 'Adaptation: A promising metaphor for strategic management', *Academy of Management Review*, 7(1), pp. 35–44.

Chanab, A. *et al.* (2007) *Telecom Infrastructure Sharing: Regulatory Enablers and Economic Benefits*. Booz Allen Hamilton.

Childers, D.L. *et al.* (2011) 'Sustainability Challenges of Phosphorus and Food: Solutions from Closing the Human Phosphorus Cycle', *BioScience*, 61(2), pp. 117–124. Available at: https://doi.org/10.1525/bio.2011.61.2.6.

Chung, S., Rainer, R. and Lewis, B. (2003) 'The Impact of Information Technology Infrastructure Flexibility on Strategic Alignment and Application Implementations', *Communications of the Association for Information Systems*, 11(1). Available at: https://doi.org/10.17705/1CAIS.01111.

CISA (2017) *Critical Infrastructure Sectors, Cybersecurity & Infrastructure Security Agency*. Available at: https://www.cisa.gov/topics/critical-infrastructure-security-and-resilience/critical-infrastructure-sectors (Accessed: 4 November 2023).

Conklin, E.J. (2006) *Dialogue mapping: building shared understanding of wicked problems*. Chichester, UK: J. Wiley. Available at: http://catdir.loc.gov/catdir/toc/ecip0513/2005013962.html (Accessed: 5 November 2023).

Costanza, R. *et al.* (2017) 'Twenty years of ecosystem services: How far have we come and how far do we still need to go?', *Ecosystem Services*, 28, pp. 1–16. Available at: https://doi.org/10.1016/j.ecoser.2017.09.008.

Dastmalchian, A. and Blyton, P. (1993) 'The concept of organizational flexibility: exploring new direction', in *Flexibility in organizations: proceedings of a colloquium*. Victoria, BC: School of Business, University of Victoria.

Davis, J. (2017) 'Infrastructure Week Preview – Federal Infrastructure Grants: How to Get Back to "Average?"', *The Eno Center for Transportation*, 11 May. Available at: https://enotrans.org/article/infrastructure-week-preview-federal-infrastructure-grants-get-back-average/ (Accessed: 5 November 2023).

Delucchi, M.A. and Jacobson, M.Z. (2011) 'Providing all global energy with wind, water, and solar power, Part II: Reliability, system and transmission costs, and policies', *Energy Policy*, 39(3), pp. 1170–1190. Available at: https://doi.org/10.1016/j.enpol.2010.11.045.

Donaldson, L. (2001) *The contingency theory of organizations*. Thousand Oaks, CA: Sage Publications (Foundations for organizational science). Available at: http://catdir.loc.gov/catdir/enhancements/fy0656/00010586-t.html (Accessed: 5 November 2023).

Duncan, N.B. (1995) 'Capturing Flexibility of Information Technology Infrastructure: A Study of Resource Characteristics and Their Measure', *Journal of Management Information Systems*, 12(2), pp. 37–57.

FEMA (2017) *Hazard Mitigation Planning, Federal Emergency Management Agency*. Available at: https://www.fema.gov/emergency-managers/risk-management/hazard-mitigation-planning.

Frankowiak, M., Grosvenor, R. and Prickett, P. (2005) 'A review of the evolution of microcontroller-based machine and process monitoring', *International Journal of Machine Tools and Manufacture*, 45(4), pp. 573–582. Available at: https://doi.org/10.1016/j.ijmachtools.2004.08.018.

Fraser, A. and Chester, M.V. (2016) 'Environmental and Economic Consequences of Permanent Roadway Infrastructure Commitment: City Road Network Lifecycle Assessment and Los Angeles County', *Journal of Infrastructure Systems*, 22(1), p. 04015018. Available at: https://doi.org/10.1061/(ASCE)IS.1943-555X.0000271.

Freeman, C. and Louçã, F. (2001) *As time goes by: from the industrial revolutions to the information revolution*. Oxford, UK: Oxford University Press. Available at: http://www.dawsonera.com/depp/reader/protected/external/AbstractView/S9780191529054 (Accessed: 5 November 2023).

Galeotti, M. (2014) 'The "Gerasimov Doctrine" and Russian Non-Linear War', *In Moscow's Shadows*, 6 July. Available at: https://inmoscowsshadows.wordpress.com/2014/07/06/the-gerasimov-doctrine-and-russian-non-linear-war/ (Accessed: 5 November 2023).

Garlan, D. *et al.* (2004) 'Rainbow: architecture-based self-adaptation with reusable infrastructure', *Computer*, 37(10), pp. 46–54. Available at: https://doi.org/10.1109/MC.2004.175.

Garreau, J. (2011) *Edge City: Life on the New Frontier*. New York, NY: Knopf Doubleday Publishing Group.

Giles, K. (2016) *Handbook of Russian Information Warfare*. Rome, IT: NATO Defense College, Research Division. Available at: https://www.ndc.nato.int/news/news.php?icode=995 (Accessed: 5 November 2023).

Gilroy, K.L. and McCuen, R.H. (2012) 'A nonstationary flood frequency analysis method to adjust for future climate change and urbanization', *Journal of Hydrology*, 414–415, pp. 40–48. Available at: https://doi.org/10.1016/j.jhydrol.2011.10.009.

Gonder, J., Earleywine, M. and Sparks, W. (2012) 'Analyzing Vehicle Fuel Saving Opportunities through Intelligent Driver Feedback', *SAE International Journal of Passenger Cars - Electronic and Electrical Systems*, 5(2), pp. 450–461. Available at: https://doi.org/10.4271/2012-01-0494.

Grisogono, A.-M. (2006) 'Success and failure in adaptation', in. *Sixth International Conference on Complex Systems*, Boston, MA, p. 9. Available at: https://www.researchgate.net/profile/Anne_Marie_Grisogono/publication/228676408_Success_and_failure_in_adaptation/links/54ef1c670cf25238f93bb689.pdf.

Hatch, M.J. (1997) *Organization theory: modern, symbolic, and postmodern perspectives*. Oxford, UK: Oxford University Press. Available at: http://catdir.loc.gov/catdir/enhancements/fy0604/97170130-t.html (Accessed: 5 November 2023).

Hollnagel, E., Pariès, J. and Wreathall, D.D.W. and J. (eds) (2011) *Resilience Engineering in Practice: a Guidebook*. Ashgate Gower (Ashgate studies in resilience engineering). Available at: http://site.ebrary.com/id/10431328 (Accessed: 5 November 2023).

Hommels, A. (2005) 'Studying Obduracy in the City: Toward a Productive Fusion between Technology Studies and Urban Studies', *Science, Technology, & Human Values*, 30(3), pp. 323–351. Available at: https://doi.org/10.1177/0162243904271759.

Hrebiniak, L.G. and Joyce, W.F. (1985) 'Organizational Adaptation: Strategic Choice and Environmental Determinism', *Administrative Science Quarterly*, 30(3), pp. 336–349. Available at: https://doi.org/10.2307/2392666.

Hughes, T.P. (1993) *Networks of Power: Electrification in Western Society, 1880-1930*. Reprint edition. Baltimore, MD: Johns Hopkins University Press.

Jacobson, M.Z. and Delucchi, M.A. (2011) 'Providing all global energy with wind, water, and solar power, Part I: Technologies, energy resources, quantities and areas of infrastructure, and materials', *Energy Policy*, 39(3), pp. 1154–1169. Available at: https://doi.org/10.1016/j.enpol.2010.11.040.

Kalleberg, A.L. (2001) 'Organizing Flexibility: The Flexible Firm in a New Century', *British Journal of Industrial Relations*, 39(4), pp. 479–504. Available at: https://doi.org/10.1111/1467-8543.00211.

Kaplan, S. and Garrick, B.J. (1981) 'On The Quantitative Definition of Risk', *Risk Analysis*, 1(1), pp. 11–27. Available at: https://doi.org/10.1111/j.1539-6924.1981.tb01350.x.

Kenny, R. (2004) 'Replacing In - Pavement Loops with Video Detection', *International Municipal Signal Association*, pp. 32–35.

Kornai, J. (1979) 'Resource-Constrained versus Demand-Constrained Systems', *Econometrica*, 47(4), pp. 801–819. Available at: https://doi.org/10.2307/1914132.

Lawrence, P.R., Garrison, J.S. and Lorsch, J.W. (1967) *Organization and environment: managing differentiation and integration.* Cambridge, MA: Harvard Business School Press (Harvard Business School classics).

Lemer, A.C. (1996) 'Infrastructure Obsolescence and Design Service Life', *Journal of Infrastructure Systems*, 2(4), pp. 153–161. Available at: https://doi.org/10.1061/(ASCE)1076-0342(1996)2:4(153).

Lohr, S. (2016) 'G.E., the 124-Year-Old Software Start-Up', *The New York Times*, 27 August. Available at: https://www.ny-times.com/2016/08/28/technology/ge-the-124-year-old-software-start-up.html (Accessed: 5 November 2023).

Lynch, J. and Loh, K. (2006) 'A Summary Review of Wireless Sensors and Sensor Networks for Structural Health Monitoring', *The Shock and Vibration Digest*, 38, pp. 91–128. Available at: https://doi.org/10.1177/0583102406061499.

Marchant, G.E., Allenby, B.R. and Herkert, J.R. (2011) *The Growing Gap Between Emerging Technologies and Legal-Ethical Oversight: The Pacing Problem.* New York, NY: Springer Science & Business Media. Available at: https://doi.org/10.1007/978-94-007-1356-7.

McCollum, D. and Yang, C. (2009) 'Achieving deep reductions in US transport greenhouse gas emissions: Scenario analysis and policy implications', *Energy Policy*, 37(12), pp. 5580–5596. Available at: https://doi.org/10.1016/j.enpol.2009.08.038.

Meerow, S., Newell, J.P. and Stults, M. (2016) 'Defining urban resilience: A review', *Landscape and Urban Planning*, 147, pp. 38–49. Available at: https://doi.org/10.1016/j.landurbplan.2015.11.011.

Megginson, L.C. (1963) 'Lessons from Europe for American Business', *The Southwestern Social Science Quarterly*, 44(1), pp. 3–13.

Menendez, J.R. *et al.* (2013) 'Prioritizing Infrastructure Maintenance and Rehabilitation Activities under various Budgetary Scenarios: Evaluation of Worst-First and Benefit–Cost Analysis Approaches', *Transportation Research Record*, 2361(1), pp. 56–62. Available at: https://doi.org/10.3141/2361-07.

Milly, P.C.D. *et al.* (2008) 'Stationarity Is Dead: Whither Water Management?', *Science*, 319(5863), pp. 573–574. Available at: https://doi.org/10.1126/science.1151915.

Mitchell, V.-W. (1995) 'Organizational Risk Perception and Reduction: A Literature Review', *British Journal of Management*, 6(2), pp. 115–133. Available at: https://doi.org/10.1111/j.1467-8551.1995.tb00089.x.

Möller, N. and Hansson, S.O. (2008) 'Principles of engineering safety: Risk and uncertainty reduction', *Reliability Engineering & System Safety*, 93(6), pp. 798–805. Available at: https://doi.org/10.1016/j.ress.2007.03.031.

Nahlik, M.J. and Chester, M.V. (2015) 'Policy Making Should Consider Time-Dependent Greenhouse Gas Benefits of Transit-Oriented Smart Growth', *Transportation Research Record*, 2502(1), pp. 53–61. Available at: https://doi.org/10.3141/2502-07.

NERC (2004) *Technical Analysis of the August 14, 2003, Blackout*. Princeton, NJ: North American Electric Reliability Council, p. 124. Available at: https://www.nerc.com/pa/rrm/ea/August%2014%202003%20Blackout%20Investigation%20DL/NERC_Final_Blackout_Report_07_13_04.pdf.

Neuberger, T. and Weston, S.B. (2012) *Variable frequency drives: energy savings for pumping applications*. IA04008002E / Z12581. Cleaveland, OH: Eaton Corporation, p. 4. Available at: https://www.eaton.com/content/dam/eaton/markets/mining-metals-minerals/knowledge-center/white-paper/Variable-frequency-drives-energy-savings-for-pumping-applications.pdf.

NRC (2002) *Privatization of Water Services in the United States: An Assessment of Issues and Experience*. Washington, DC: National Academies Press. Available at: https://doi.org/10.17226/10135.

Oughton, E. and Tyler, P. (2013) *Infrastructure as a Complex Adaptive System*. Oxford, UK: Infrastructure Transitions Research Consortium, p. 25. Available at: https://www.itrc.org.uk/wp-content/PDFs/Infrastructure-as-complex-adaptive-system.pdf.

Paletta, D. (2014) 'States Siphon Gas Tax for Other Uses', *Wall Street Journal*, 17 July. Available at: http://online.wsj.com/articles/states-siphon-gas-tax-for-other-uses-1405558382 (Accessed: 5 November 2023).

Park, J. *et al.* (2013) 'Integrating Risk and Resilience Approaches to Catastrophe Management in Engineering Systems', *Risk Analysis*, 33(3), pp. 356–367. Available at: https://doi.org/10.1111/j.1539-6924.2012.01885.x.

Park, J., Seager, T.P. and Rao, P.S.C. (2011) 'Lessons in risk- versus resilience-based design and management', *Integrated Environmental Assessment and Management*, 7(3), pp. 396–399. Available at: https://doi.org/10.1002/ieam.228.

Pedersen, S.W. (1995) 'Electronics industry environmental roadmap', in *Proceedings of the 1995 IEEE International Symposium on Electronics and the Environment ISEE (Cat. No.95CH35718). Proceedings of the 1995 IEEE International Symposium on Electronics and the Environment ISEE (Cat. No.95CH35718)*, pp. 285–289. Available at: https://doi.org/10.1109/ISEE.1995.514991.

Pollard, T. (2003) 'Follow the Money: Transportation Investments for Smarter Growth', *Temple Environmental Law & Technology Journal*, 22, p. 155.

Qiao, L. and Wang, X. (1999) *Unrestricted Warfare*. Beijing, CN: PLA Literature and Arts Publishing House.

Rae, A. (2017) 'iNEMI Roadmap 2017'. Available at: https://thor.inemi.org/webdown-load/2017/2017_iNEMI_RM_ICSR_061717.pdf.

Resnick, J. (2016) *Virtual assembly lines are making the auto industry more flexible, Ars Technica*. Available at: https://arstechnica.com/cars/2016/09/virtual-assembly-lines-are-making-the-auto-industry-more-flexible/ (Accessed: 5 November 2023).

Rinaldi, S.M., Peerenboom, J.P. and Kelly, T.K. (2001) 'Identifying, understanding, and analyzing critical infrastructure interdependencies', *IEEE Control Systems Magazine*, 21(6), pp. 11–25. Available at: https://doi.org/10.1109/37.969131.

Rittel, H.W.J. and Webber, M.M. (1973) 'Dilemmas in a General Theory of Planning', *Policy Sciences*, 4(2), pp. 155–169.

Roethemeyer, D.E. and Yankaskas, D.R. (1995) 'Evolution of Motor and Variable Frequency Drive Technology', *Summer Study on Energy Efficiency in Industry Conference Proceedings*, pp. 541–552.

Santos Bernardes, E. and Hanna, M.D. (2009) 'A theoretical review of flexibility, agility and responsiveness in the operations management literature: Toward a conceptual definition of customer responsiveness', *International Journal of Operations & Production Management*, 29(1), pp. 30–53. Available at: https://doi.org/10.1108/01443570910925352.

Sherehiy, B., Karwowski, W. and Layer, J.K. (2007) 'A review of enterprise agility: Concepts, frameworks, and attributes', *International Journal of Industrial Ergonomics*, 37(5), pp. 445–460. Available at: https://doi.org/10.1016/j.ergon.2007.01.007.

Shoup, D.C. (2011) *The high cost of free parking*. Updated. Chicago, IL: American Planning Association.

Slovic, P. (2016) *The Perception of Risk*. New York, NY: Taylor & Francis.

'The Status of the Highway Trust Fund and Options for Paying for Highway Spending' (2016). Available at: https://www.cbo.gov/publication/50298.

Tramblay, Y. *et al.* (2013) 'Non-stationary frequency analysis of heavy rainfall events in southern France', *Hydrological Sciences Journal*, 58(2), pp. 280–294. Available at: https://doi.org/10.1080/02626667.2012.754988.

U.S. National Climate Assessment (2014) *Climate Change Impacts in the United States*. Washington, DC: U.S. Global Change Research Program. Available at: https://doi.org/10.7930/J0Z31WJ2.

Vecchio, R.P. (2006) *Organizational behavior: core concepts*. 6th ed. Mason, OH: Thomson/South-Western.

Vitousek, P.M. *et al.* (1997) 'Human Alteration of the Global Nitrogen Cycle: Sources and Consequences', *Ecological Applications*, 7(3), pp. 737–750. Available at: https://doi.org/10.1890/1051-0761(1997)007[0737:HAOTGN]2.0.CO;2.

Vörösmarty, C.J. and Sahagian, D. (2000) 'Anthropogenic Disturbance of the Terrestrial Water Cycle', *BioScience*, 50(9), pp. 753–765. Available at: https://doi.org/10.1641/0006-3568(2000)050[0753:ADOTTW]2.0.CO;2.

Walski, T.M. (2006) 'A history of Water distribution', *Journal AWWA*, 98(3), pp. 110–121. Available at: https://doi.org/10.1002/j.1551-8833.2006.tb07611.x.

Weick, K.E. and Quinn, R.E. (1999) 'Organizational change and development', *Annual Review of Psychology*, 50, pp. 361–386. Available at: https://doi.org/10.1146/annurev.psych.50.1.361.

Wolmar, C. (2012) *The great railroad revolution: the history of trains in America*. 1st US ed. New York, NY: PublicAffairs.

Zetter, K. (2016) 'Inside the Cunning, Unprecedented Hack of Ukraine's Power Grid', *Wired*, 3 March. Available at: https://www.wired.com/2016/03/inside-cunning-unprecedented-hack-ukraines-power-grid/ (Accessed: 5 November 2023).

2.3

TOWARD ADAPTIVE INFRASTRUCTURE: THE FIFTH DISCIPLINE

Successful institutions—public or private—adapt to the changing complexity in the environments in which they operate. Adaptability, the capabilities needed to survive and thrive as the environment in which you function changes, has been a topic of study as it relates to human enterprises for decades. It frequently appears as a concept of academic interest in business, management, and computer science (Chakravarthy, 1982; Hrebiniak and Joyce, 1985; Brennan, Turnbull and Wilson, 2003; Garlan *et al.*, 2004). However, the infrastructure institutions that manage the technologies and systems that deliver critical and basic services have received remarkably little attention when it comes to adaptability. They are possibly a victim of their own success, in the developed world delivering services with the highest reliability, and have been doing so for decades. At the dawn of the Anthropocene the world appears to be accelerating in many destabilizing ways. The Great Acceleration graphs show changes in key global indicators since 1750, and that since around 1950 there has been an acceleration of human activity and associated impacts. Yet the institutions that manage our infrastructure, including the technological solutions they deploy, remain rigid and obdurate, reflecting the relatively stable conditions of the past. We cannot underestimate how destabilizing these forces are predicted to become. While change has always happened, we appear to be entering an era marked by rapid acceleration and unpredictability of social, technical, and environmental variables in ways that we as humans have never experienced, and infrastructure are at the center of these trends (Figure 2.3.1). Infrastructure is a three-part system consisting of physical assets, institutions for governance, and education, and each is failing to be agile enough given the rapid pace of change.

This chapter was adapted from the following article with publisher permission: Mikhail Chester and Braden Allenby, 2020, Toward Adaptive Infrastructure: The Fifth Discipline, *Sustainable and Resilient Infrastructure*, 6(5), pp. 334-338, doi: 10.1080/23789689.2020.1762045.

Figure 2.3.1: Accelerating, Non-Stationary, and Increasingly Unpredictable Emerging Forces Affecting Infrastructure

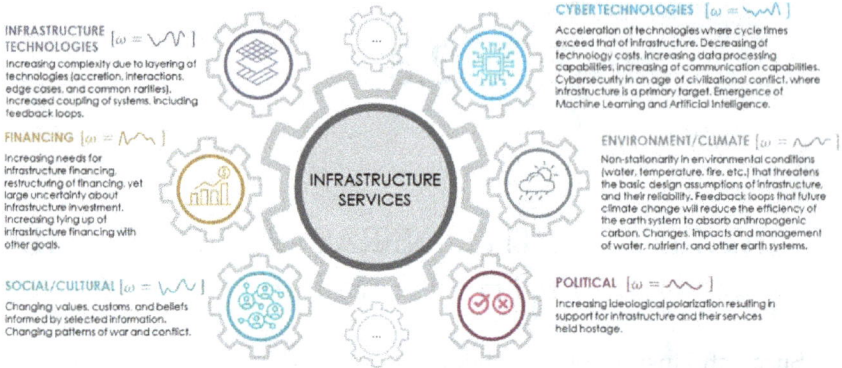

INFRASTRUCTURE TECHNOLOGIES $[\omega = \mathcal{W}]$
Increasing complexity due to layering of technologies (accretion, interactions, edge cases, and common rarities). Increased coupling of systems, including feedback loops.

CYBERTECHNOLOGIES $[\omega = \mathcal{W}\mathcal{M}]$
Acceleration of technologies where cycle times exceed that of infrastructure. Decreasing of technology costs. Increasing data processing capabilities, increasing of communication capabilities. Cybersecurity in an age of civilizational conflict, where infrastructure is a primary target. Emergence of Machine Learning and Artificial Intelligence.

FINANCING $[\omega = \mathcal{M}]$
Increasing needs for infrastructure financing, restructuring of financing, yet large uncertainty about infrastructure investment. Increasing tying up of infrastructure financing with other goals.

ENVIRONMENT/CLIMATE $[\omega = \mathcal{N}\mathcal{N}]$
Non-stationarity in environmental conditions (water, temperature, fire, etc.) that threatens the basic design assumptions of infrastructure, and their reliability. Feedback loops that future climate change will reduce the efficiency of the earth system to absorb anthropogenic carbon. Changes, impacts and management of water, nutrient, and other earth systems.

INFRASTRUCTURE SERVICES

SOCIAL/CULTURAL $[\omega = \mathcal{W}\mathcal{V}]$
Changing values, customs, and beliefs informed by selected information. Changing patterns of war and conflict.

POLITICAL $[\omega = \mathcal{N}]$
Increasing ideological polarization resulting in support for infrastructure and their services held hostage.

Each force is represented as a gear that is moving and affecting the ability to deliver and modernize infrastructure services. These forces are becoming increasingly unpredictable, as represented by different and random rates (ω) by which the gear spins. There are many sources that describe these forces and their trends including: (Friedlingstein *et al.*, 2006; Kurzweil, 2010; Nye, 2011; Kissinger, 2014; Allenby, 2015; Arbesman, 2017; Fukuyama, 2018).

We are already becoming overwhelmed by the growing complexity that is just beginning to emerge at dawn of the Anthropocene (Senge, 1990; Tainter, 1990; Allenby, 2012). It's important to distinguish between complexity and complicatedness prior to discussing rigidity, agility, and flexibility in the context of adaptation. Here we define complexity as it relates to infrastructure based primarily on the inability to predict emergent behaviors. Complexity has emerged from the acceleration and growing uncertainty associated with social, technical, and environmental factors, and the combination thereof, and requires fundamentally new approaches to how we deliver services (Snowden and Boone, 2007; Arbesman, 2017; Helmrich and Chester, 2020). Yet infrastructure management remains largely rooted in principles for complicated systems (Chester and Allenby, 2019). Complexity is about flux and unpredictability, no right answers, and unknown unknowns, thus requiring creative approaches and pattern-based leadership that often test before formalizing solutions. Complicatedness on the other hand is about expert diagnosis to assess multiple right answers, known unknowns, and fact-based management that typically emphasizes data collection and analysis to make decisions (Snowden and Boone, 2007; Chester and Allenby, 2019). Complexity has always existed with human systems in some form, but the rate of change of human activities, technologies, etc. appears to be taking off, resulting in acceleration and scale that is unprecedented (Birdzell, 1986; Kurzweil, 2010; Marchant, 2011; Allenby, 2012; Syvitski, 2012), and producing a new normal where the predictability of what infrastructure will and should do is diminishing. And new dynamics, such as Asymmetric Warfare that leverages the increasingly cyber connected core systems and lax security, make infrastructure a battlefield in civilizational conflict (Allenby, 2015). We argue that infrastructure and the institutions that manage them must adapt by becoming agile and flexible in response to the changing complexity of the world around them. In doing so they must become a Fifth Discipline, organizations that are focused on learning about the rapidly changing environments and demands in which they must deliver services. The Fifth Discipline concept was developed by Senge (1990) to describe the necessity of organizations to learn in complex environments. The disciplines – 1) continual clarity and deepening is needed to see reality objectively, 2) mental models must be challenged; 3) building shared visions is necessary to foster commitment; 4) team learning requires dialogue; and 5) systems thinking is needed to integrate the first four – are competencies to make sense of complex environments, precursors to adaptive and transformative capacity.

The design and management of infrastructure continues to emphasize rigidity through well-established models developed over the past century when conditions were much more predictable. We define rigidity in the context of infrastructure systems as a highly constrained ability to adapt to changing internal and external conditions. In a functioning infrastructure system, it may arise from physical, institutional, political, or economic factors, including lock-in to other systems that prevent responsive change. It may also arise from an inability to perceive or learn rapidly enough to change appropriately. Rigidity occurs for many reasons including governance models that emphasize predictability in environmental conditions and demand, the use of technologies that hedge risk over long periods under stationarity assumptions, and educational norms that emphasize problem solving in the complicated domain. As such, infrastructure management emphasizes risk-based models that assume stationarity, do not consider what may happen when the risk management solutions fail, and large and permanent assets that are prone to greater damages when they fail (Park, Seager and Rao, 2011; Kim *et al.*, 2019). This model is in many ways the result of the prevalent form of government that oversees infrastructure, the divisional bureaucracy. This form of infrastructure bureaucracy emerged in the early 1900s with natural monopolies, first the railroad, and later utilities (Chandler, 1977; Friedlander, 1995a, 1995b, 1996). It silos functional specialization with multiple layers of management that 1) separates strategic visioning from day-to-day operations, and 2) is inimical to interdisciplinary problem solving by creating managerial barriers that work against the exchange of ideas across expertise silos (Chandler, 1977; Edwards, 2003). It was a product of the industrial revolution reflecting the social and technological complexities of the twentieth century. Indeed, the divisional bureaucracy that has managed infrastructure for a century has allowed for some agility and flexibility, but it appears to operate too slow for the accelerating and uncertain change associated with the Anthropocene.

In Chester and Allenby (2018) we identified the growing challenges associated with the rigid institutional forms and associated technologies that drive our infrastructure today. We explored industries that exhibited agile and flexible characteristics, and synthesized these into a set of competencies that we recommend as useful principles for infrastructure going forward. In Gilrein et al., (2019) we used the competencies to identify real-world examples. The competencies and examples covered both centralized to decentralized, green to gray, and even dumb to smart configurations of infrastructure and technologies. Furthermore, we discussed both technological and institutional competencies. We did not advocate for any particular configuration but instead viewed the challenge of implementing agility and flexibility as one that could deploy any number of approaches.

Saxe and MacAskill (2019) provide a thoughtful response to our work, largely arguing that rigidity in some forms may in itself produce agility and flexibility. Their argument appears to be motivated by an urging of caution that we should not aggressively seek to change infrastructure but instead to look carefully at existing rigid systems that have proven successful. In particular, they argue that 1) rigid infrastructure has provided immense value (we agree) that we can learn from and leverage; 2) planned obsolescence — specifically the shortening of lifetimes of many assets and then continuing their use after their design life — has created a paradigm where many assets are in need of major rehabilitation and are designed for past demand assumptions, and longer lifetimes would obviate some of these challenges; and 3) centralized large-scale infrastructure provide stability, a skeleton that can be built upon and utilized for the long-term. We address these counterpoints while expanding our position on the necessity of restructuring infrastructure institutions and the way they deploy technologies at the dawn of the Anthropocene.

We maintain first and foremost that the institutions that manage infrastructure and the technologies they deploy must reflect the rapidly accelerating and uncertain environment in which they operate. As complexity in the world emerges the institutions that operate within it need to change to continue delivering services in those new environments. Infrastructure has for decades, if not centuries, operated in environments that have been relatively stable as compared to today, and the institutions and technologies exemplify this rigidity. Certainly, rigid infrastructure has provided immense value, but this rigidity was able to persist (that is continue delivering services reliably) because the complexity that defined society and technologies was changing at a pace that the infrastructure systems could remain viable. More specifically, changes in demand have for nearly a century been slower than the capacity of the institution to change. As such, the cycle time of infrastructure change and renewal was effectively coupled to the cycle time of change in the external environment. Breakdown in institutional control and effectiveness occurs when the cycle time of the subject matter of the institution exceeds the speed with which the institution can respond (Osinga, 2007). We argue that we are at a point in the acceleration of human systems including technology where infrastructure institutions are going to be unable to keep up to remain viable (both institutionally and in their ability to meet new needs). Similarly, while norms may have shifted to shorter asset lifetimes, and indeed there are benefits to locking in longer lifetime assets, the asset (designed for short or long lifetimes) will only be viable as long as the demands and environment in which it operates remain in a somewhat stable envelope. Evidence mounts that these envelopes are likely to be greatly exceeded, if not become wholly irrelevant into the future (COVID-19 gives us a glimpse into how rigid infrastructure is disrupted when demands change seemingly overnight). Every infrastructure element relies on design constraints and objectives, and many of these are derived from assumptions regarding the state of the external environment within which that infrastructure functions. When those implicit assumptions change, the design becomes obsolete, and sometimes even dysfunctional.

Agility and flexibility differ in their application between physical assets and governance processes. For physical assets they can be met through centralized or decentralized configurations; we imagine future infrastructure having aspects of both. Similarly, we see pathways for increasing agility and flexibility anywhere along the gray to green spectrum, and the physical to cyber spectrum. Indeed, many of the technologies identified by Gilrein et al. (2019) were decentralized, however, centralized configurations—which we would argue today are often configured towards rigidity—can be managed or designed differently to improve their agility and flexibility. The topic of agility and flexibility in infrastructure governance deserves its own deep analysis, however, we think that they are primarily found in the organizational leadership capabilities needed for both stable and unstable environments, and transitioning between. Infrastructure institutions are structured around administrative leadership for stable conditions (the management of bureaucratic function through the structuring of tasks, planning, vision building, resource acquisition, crises management, and organizational strategy). Adaptive leadership describes learning processes (that renegotiates roles, goals, and ideas, sometimes through the clashing of ideas and technologies) while Enabling leadership is the ability to shift between Administrative and Adaptive leadership as environments change, and consists of facilitating the movement of information, creating the pressure to act, and providing resources for creativity (Uhl-Bien, Marion and McKelvey, 2007).

While there are indeed lessons to be learned from cities and infrastructure that have persisted for centuries, we argue that any potential to use these solutions into the future must be rooted in their capacity to address increasing complexity, not simply because evidence suggests they are better than other solutions absent of any consideration of the environment they must function in. Again, we can conceive of pathways where dense urban form that promotes active transport (walking, biking, and transit) and locks in other infrastructure (e.g., power and water) towards agility and flexibility. These pathways must embrace the accelerating climate, technology, and social forces that will drive the viability of infrastructure at the nexus of supply and demand. We urge caution in trying to emulate desirable urban forms that exist elsewhere without consideration of the growing complexity that is the new normal. Simply put, in an era of increasing complexity, we can't only look backwards for answers to what might work for the coming centuries.

Fundamentally, we advocate for changes to infrastructure where the systems (institutions and technologies) respond in pace to the increasing cycle times of technologies, and embrace the wicked complexity that increasingly defines the conditions under which infrastructure must deliver services. We view these institutions primarily as knowledge architects for delivering critical services, testing and employing information capabilities and technologies, i.e., Fifth Disciplines (Senge, 1990). These institutions will likely be structured very differently than those that govern our infrastructure today. They will operate based on new principles that accept uncertainty and rapid change as normal.

REFERENCES

Allenby, B.R. (2012) *The Theory and Practice of Sustainable Engineering.* Hoboken, NJ: Upper Saddle River: Pearson Prentice Hall.

Allenby, B.R. (2015) 'The paradox of dominance: The age of civilizational conflict', *Bulletin of the Atomic Scientists*, 71(2), pp. 60–74. Available at: https://doi.org/10.1177/0096340215571911.

Arbesman, S. (2017) *Overcomplicated: Technology at the Limits of Comprehension.* New York, NY: Penguin Publishing Group.

Birdzell (1986) *How The West Grew Rich.* New York, NY: Basic Books.

Brennan, D.R., Turnbull, P.W. and Wilson, D.T. (2003) 'Dyadic adaptation in business-to-business markets | Emerald Insight', *European Journal of Marketing*, 37(11/12), pp. 1636–1665. Available at: https://doi.org/10.1108/03090560310495393.

Chakravarthy, B.S. (1982) 'Adaptation: A Promising Metaphor for Strategic Management', *The Academy of Management Review*, 7(1), pp. 35–44. Available at: https://doi.org/10.2307/257246.

Chandler, A.D. (1977) *The Visible Hand.* Cambridge, MA: Harvard University Press. Available at: https://books.google.com/books?hl=en&lr=&id=lAI3AwAAQBAJ&oi=fnd&pg=PR9&ots=YWApkToGvQ&sig=fB0xzTg4BxWW6iRUHCzxMJz-H8M#v=onepage&q&f=false.

Chester, M.V. and Allenby, B. (2018) 'Toward adaptive infrastructure: flexibility and agility in a non-stationarity age', *Sustainable and Resilient Infrastructure*, 4(4), pp. 173–191. Available at: https://doi.org/10.1080/23789689.2017.1416846.

Chester, M.V. and Allenby, B. (2019) 'Infrastructure as a wicked complex process', *Elementa: Science of the Anthropocene.* Edited by A. Iles and M.E. Chang, 7(1), p. 21. Available at: https://doi.org/10.1525/elementa.360.

Edwards, P. (2003) 'Infrastructure and Modernity: Scales of Force, Time, and Social Organization in the History of Sociotechnical Systems', in *Modernity and Technology.* Cambridge, MA: MIT Press, pp. 185–225.

Friedlander, A. (1995a) *Emerging Infrastructure: The Growth of Railroads.* Corporation for National Research Initiatives.

Friedlander, A. (1995b) *Natural Monopoly and Universal Service: Telephones and Telegraphs in the U.S. Communications Infrastructure, 1837-1940.* Corporation for National Research Initiatives.

Friedlander, A. (1996) *Power and Light: Electricity in the U.S. Energy Infrastructure, 1870-1940.* Corporation for National Research Initiatives.

Friedlingstein, P. *et al.* (2006) 'Climate–Carbon Cycle Feedback Analysis: Results from the C4MIP Model Intercomparison', *Journal of Climate*, 19(14), pp. 3337–3353. Available at: https://doi.org/10.1175/JCLI3800.1.

Fukuyama, F. (2018) *Identity: The Demand for Dignity and the Politics of Resentment.* New York, NY: Farrar, Straus and Giroux.

Garlan, D. *et al.* (2004) 'Rainbow: architecture-based self-adaptation with reusable infrastructure', *Computer*, 37(10), pp. 46–54. Available at: https://doi.org/10.1109/MC.2004.175.

Gilrein, E.J. *et al.* (2019) 'Concepts and practices for transforming infrastructure from rigid to adaptable', *Sustainable and Resilient Infrastructure*, 6(3–4), pp. 213–234. Available at: https://doi.org/10.1080/23789689.2019.1599608.

Helmrich, A.M. and Chester, M.V. (2020) 'Reconciling complexity and deep uncertainty in infrastructure design for climate adaptation', *Sustainable and Resilient Infrastructure*, 7(2), pp. 83–99. Available at: https://doi.org/10.1080/23789689.2019.1708179.

Hrebiniak, L.G. and Joyce, W.F. (1985) 'Organizational Adaptation: Strategic Choice and Environmental Determinism', *Administrative Science Quarterly*, 30(3), pp. 336–349. Available at: https://doi.org/10.2307/2392666.

Kim, Y. *et al.* (2019) 'The Infrastructure Trolley Problem: Positioning Safe-to-fail Infrastructure for Climate Change Adaptation', *Earth's Future*, 7(7), pp. 704–717. Available at: https://doi.org/10.1029/2019EF001208.

Kissinger, H. (2014) *World Order*. New York, NY: Penguin Publishing Group.

Kurzweil, R. (2010) *The Singularity Is Near: When Humans Transcend Biology*. Ebook. London, UK: Duckworth Overlook.

Marchant, G.E. (2011) 'The Growing Gap Between Emerging Technologies and the Law', in G.E. Marchant, B.R. Allenby, and J.R. Herkert (eds) *The Growing Gap Between Emerging Technologies and Legal-Ethical Oversight: The Pacing Problem*. Dordrecht: Springer Netherlands (The International Library of Ethics, Law and Technology), pp. 19–33. Available at: https://doi.org/10.1007/978-94-007-1356-7_2.

Nye, J.S. (2011) *The Future of Power*. 1st ed. New York, NY: PublicAffairs. Available at: https://library.leeds.ac.uk/site/custom_scripts/authentication/link.cgi?Item=MyiLibrary&targetExtra=309544 (Accessed: 26 November 2023).

Osinga, F.P.B. (2007) *Science, Strategy and War: The Strategic Theory of John Boyd*. London, UK: Routledge.

Park, J., Seager, T.P. and Rao, P.S.C. (2011) 'Lessons in risk- versus resilience-based design and management', *Integrated Environmental Assessment and Management*, 7(3), pp. 396–399. Available at: https://doi.org/10.1002/ieam.228.

Saxe, S. and MacAskill, K. (2019) 'Toward adaptive infrastructure: the role of existing infrastructure systems', *Sustainable and Resilient Infrastructure*, 6(5), pp. 330–333. Available at: https://doi.org/10.1080/23789689.2019.1681822.

Senge, P.M. (1990) *The Fifth Discipline: The Art and Practice of the Learning Organization*. 1st edn. New York, NY: Doubleday/Currency.

Snowden, D.J. and Boone, M.E. (2007) 'A Leader's Framework for Decision Making', *Harvard Business Review*, 1 November, pp. 68–76, 149.

Syvitski, J. (2012) 'Anthropocene: An epoch of our making - IGBP', *Global Change Magazine*, (78), p. 12.

Tainter, J. (1990) *The Collapse of Complex Societies*. Cambridge, UK: Cambridge University Press.

Uhl-Bien, M., Marion, R. and McKelvey, B. (2007) 'Complexity Leadership Theory: Shifting leadership from the industrial age to the knowledge era', *The Leadership Quarterly*, 18(4), pp. 298–318. Available at: https://doi.org/10.1016/j.leaqua.2007.04.002.

2.4

CENTRALIZATION AND DECENTRALIZATION

2.4.1. INTRODUCTION

The environments in which infrastructure need to function, adapt, and thrive appear to be accelerating, becoming increasingly volatile and uncertain, resulting in increasing complexity (Desha, Hargroves and Smith, 2009; Steffen *et al.*, 2015; Chester and Allenby, 2018; Sharma, 2019). Cybertechnologies are being integrated into legacy infrastructure at rapid rates, creating remarkable new opportunities but also vulnerabilities (Rinaldi, Peerenboom and Kelly, 2001; Chester and Allenby, 2020). Climate change is creating deep uncertainty for weather extremes and is threatening to exceed design envelopes of critical systems (Burillo *et al.*, 2017; Underwood *et al.*, 2017; Ayyub, 2018; Bondank, Chester and Ruddell, 2018; Nasr *et al.*, 2019; Chester, Underwood and Samaras, 2020; Helmrich and Chester, 2020). In the United States, aging infrastructure are increasingly confronted with these new challenges and must adapt with scant resources to meet demand and deliver services. Infrastructure is caught between goals, administrative structures, and technologies rooted in the past, and a future defined by complexity, of the infrastructure themselves and the environments they must function in.

The domains of infrastructure—physical networks and governing institutions—are of major importance when examining the capacities of our critical systems to adapt and transform. Pervasive across academic literature and discourse are the concepts of centralized, decentralized, and distributed systems. At first glance, these concepts appear to describe important dimensions of infrastructure. Centralization is often associated with networks distinguished by a small number of providers to a large number of consumers (Figure 2.4.1-A) and a top-down governance model

This chapter was adapted from the following article with publisher and lead author permissions: Alysha Helmrich, Samuel Markolf, Rui Li, Thomaz Carvalhaes, Yeowon Kim, Emily Bondank, Mukunth Natarajan, Nasir Ahmad, and Mikhail Chester, 2021, Centralization and Decentralization for Resilient Infrastructure and Complexity, *Environmental Research: Infrastructure and Sustainability*, 1(2), 021001, doi: 10.1088/2634-4505/ac0a4f..

(Chandler, 1977; Bardhan, 2002; Alanne and Saari, 2006; Wilder and Romero Lankao, 2006; Makropoulos and Butler, 2010; Pagani and Aiello, 2011; Albalate, Bel and Fageda, 2012; Tomlinson *et al.*, 2015; Quezada, Walton and Sharma, 2016; Derrible, 2017; Chester, Underwood and Samaras, 2020; Rodrigue, 2020). Generally, centrality measures the importance of a singular node in a larger network, where increased emphasis on a node produces vulnerability (by, e.g., consumers relying on the operation of a singular producer for a service). A decentralized system (Figure 2.4.1-B), where the ratio of producers to consumers increases, places less reliance on singular nodes, thereby, decreasing vulnerability (Baran, 1964; Freeman, 1978). There does not appear to be a clear definition of distributed in the infrastructure literature, but at times the concept is used synonymously with decentralization (Ackermann, Andersson and Söder, 2001; Alanne and Saari, 2006; Makropoulos and Butler, 2010). Following framings in computer science, a distributed system is one where nodes in a network work toward a common goal (Srinivasa and Muppalla, 2015; Ge, Yang and Han, 2017). A distributed network is a configuration where consumers are provided with a larger variety of service options (producers) and are connected with other consumers to achieve an objective (Figure 2.4.1-C). The configuration of infrastructure networks and governance has taken on a new importance under the emerging field of resilience, a field which studies a system's ability to persist, adapt, and transform to disturbances within and beyond the design envelope. Yet, there does not appear to be a concerted effort to align how centralization, decentralization, and distributed framings are used, and what different configurations mean, when adapting and transforming systems to be able to respond to increasing complexity.

Figure 2.4.1: De/centralization configurations from a network perspective

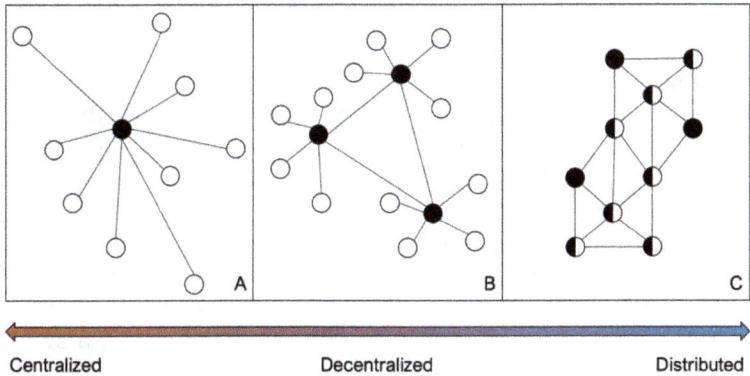

Centralized   Decentralized   Distributed

A) Centralized system: Single producer (black circle) linked to many consumers (white circles). B) Decentralized system: Multiple producers connected together that service specific consumers. C) Distributed system: Producers and consumers coupled together to work toward a common objective (adapted from (Baran, 1964)).

Given a growing interest in configuring infrastructure to better respond to increasingly complex environments (Rinaldi, Peerenboom and Kelly, 2001; Baldwin and Clark, 2006; Tomlinson *et al.*, 2015; Chester and Allenby, 2018; Gilrein *et al.*, 2019), a coherent framing of centralization, decentralization, and distributed (hereby referred to as de/centralization) is needed. A polysemic framing (i.e., carrying multiple meanings) is particularly problematic given the importance of de/centralization within network and governance domains, how systems are physically configured and how authority is structured towards adaptive capacities (Mintzberg, 1979; Hines, Blumsack and Schläpfer, 2015). In this work, infrastructure is defined as human design spaces that produce built environment systems including their technologies and institutions of governance. Water, power, and transportation systems are the primary infrastructure sectors of focus. The upcoming section starts by reviewing key de/centralization framings across infrastructure domains. Following, de/centralization is contextualized within emerging resilient infrastructure theory. Last, a multi-dimensional framing of de/centralization through a coupled network-governance perspective with information as a mediator is proposed, emphasizing the capabilities that infrastructure will need in the Anthropocene.

2.4.2. BACKGROUND

2.4.2.1 PHYSICAL INFRASTRUCTURE AS NETWORKS

Framings of built infrastructure de/centralization often focus on network configuration characteristics that primarily assess whether production of a good or service to a user occurs on a large and isolated (i.e., centralized) or small and integrated (i.e., decentralized) scale (Alanne and Saari, 2006; Makropoulos and Butler, 2010; Pagani and Aiello, 2011; Tomlinson *et al.*, 2015; Derrible, 2017; Rodrigue, 2020). Drawing from a select body of literature that covers a diversity of framings, how de/centralization is specifically used across infrastructure sectors is explored, and whether these framings are consistent. In the power sector, a system is characterized as decentralized when it uses distributed generators of a capacity considerably smaller than those used in larger networks, or centralized when it uses generators located far from the loads (Ackermann, Andersson and Söder, 2001; El-Khattam and Salama, 2004; Luo, 2005). As such, de/centralization in power systems is determined by capacity of distribution and proximity between generation and consumption. Furthermore, the power sector is discussed as becoming increasingly connected (e.g., the integration of renewable technologies at residential properties allows consumers to become small-scale providers) (Pagani and Aiello, 2011). This produces a distributed system, where greater connectivity allows more actors to work toward a goal. Examining the water sector (water supply, wastewater, and stormwater), the presence of site-level technologies that manage lower volumes of water (e.g., point-of-use treatment, septic tanks) are characterized as decentralized while municipal-scale technologies managing large volumes of water (e.g., reservoirs, combined sewer systems) are considered centralized (Makropoulos and Butler, 2010; Larsen, Udert and Lienert, 2013; Quezada, Walton and Sharma, 2016; Gilrein *et al.*, 2019). De/centralization in the water sector appears to be defined by the volume of water managed and the proximity between treatment and use. Finally, in transportation, centralized networks are described as having a singular node with high accessibility where heavy travel navigates through this central node to reach another destination (i.e., hub-and-spoke). Decentralized networks still have a central node, but also have sub-center nodes, providing increased connections, or routing options, between origins and destinations. Lastly, a distributed system is when every node connects with each surrounding node, meaning there is no node that is more accessible than another (Economic Research Centre, 2001; Rodrigue, 2020). De/centralization in the transportation sector can be assessed by the accessibility and, therefore, connectivity, of nodes. Maybe unsurprisingly given the differing nature of infrastructure services, framings differ with some focused on supply and demand capacities, others separated between production and consumption, and others based on pure network characteristics. Yet, there is another critical framing that must be unpacked, that of infrastructure governance configurations.

2.4.2.2. GOVERNANCE AS NETWORKS OF POWER

The governance of infrastructure and its associated organizational and bureaucratic structures has significant impact on system ability to manage changing conditions (Mintzberg, 1979; Uhl-Bien and Arena, 2018; Chester, Miller and Muñoz-Erickson, 2020; Chester *et al.*, 2021). Decentralization of governance can be classified as either vertical through the dispersion of power through a formal chain of command (i.e., strategic leadership to middle managers to front line workers), or horizontal where decision-making power is dispersed across many employees, including those outside of the chain of command, such as analysts and operators (Mintzberg, 1979). Centralization exists when the decision-making power rests with a select few, for example, strategic leadership (Mintzberg, 1979). Governance — a process involving collective action for resource allocation and use across multiple civic and private actors (Kooiman, 1993) — is the balance between the authority, responsibility, and power of management and individuals within an organization through rules, values, and norms (Mintzberg, 1979; Dubois and Fattore, 2009; Faguet, 2014; Chester, Miller and Muñoz-Erickson, 2020). While this definition emphasizes decentralization as a process, the term is also used to reference organizational structure (Siggelkow and Levinthal, 2003; Dubois and Fattore, 2009; Abimbola, Baatiema and Bigdeli, 2019). Infrastructure governance is critical for resilience because human and organizational power structures, if structured appropriately, create knowledge, allocate resources, and form bureaucracies capable of handling both stability and instability.

In water, transportation, and power sectors, the characterization of governance de/centralization depends on the jurisdiction and scale of the organizations which manage services between the producer and consumer. Water (e.g., distribution pipelines, wastewater treatment), power networks (e.g., the grid), and transportation (e.g., roadways and railways) system governance, are often characterized as centralized governance models (Chandler, 1977; Albalate, Bel and Fageda, 2012; Quezada, Walton and Sharma, 2016; Rodrigue, 2020) where divisional bureaucracies emphasize a concentration of power at strategic leadership (i.e., vertical centralization) (Mintzberg, 1980; Chester, Miller and Muñoz-Erickson, 2020). In these centralized structures, the leadership team can assess where to allocate and reallocate resources across their jurisdiction to meet demand through authoritative power. Conversely, water privatization, power generation, and transportation services are characterized by numerous, independent decision makers, and therefore decentralized (i.e., multiple actors (e.g., airlines) set their own policy) (Bardhan, 2002; Wilder and Romero Lankao, 2006; Larsen, Udert and Lienert, 2013; Quezada, Walton and Sharma, 2016; Rodrigue, 2020). In decentralized structures, the decision-makers will be more directly connected with their producer and consumer needs. However, access to resources may vary, which could elevate inequality of supply across marginalized jurisdictions with less power. Each governance structure allows infrastructure managers to address supply and demand within their jurisdiction.

The use of de/centralization across domains and sectors varies widely, leading to a polysemy and inconsistencies. De/centralized networks are characterized by proximity to resources, proximity to treatment, capacity of distribution, volume of product, and number of connections. De/centralization of governance within infrastructure sectors is characterized by the number of actors who hold decision-making power. Centralization is often framed as dominant in infrastructure systems, a remnant of system design emphasizing efficiency for times of stability, but infrastructure systems may take on various network-governance configurations—a system may tend towards centralized-centralized network and governance, de-centralized-decentralized, centralized-decentralized, or decentralized/centralized. Each configuration can be found across infrastructure sectors, and select examples are presented in Figure 2.4.2. There does not appear to be clear boundaries between centralized and decentralized configurations, but instead that a gradient is applicable where the degree of de/centralization is inherently relative to the geographic and operational scale of the infrastructure systems. Reframing de/centralization across networks and governance to support resilience is a necessary and timely endeavor.

Figure 2.4.2: Examples of de/centralization classifications across network and governance domains

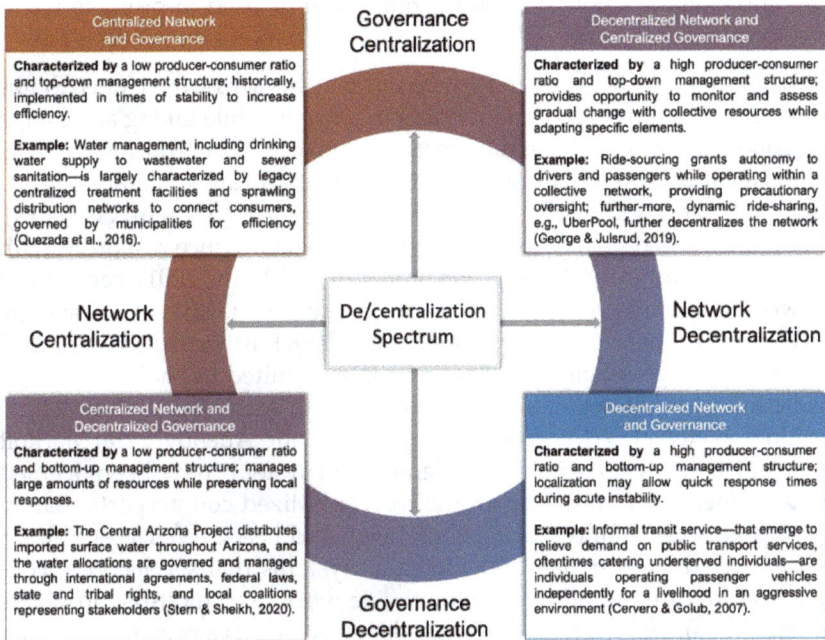

2.4.3. DE/CENTRALIZATION FOR RESILIENCE

A rigorous investigation into the conditions by which de/centralization lead to improved capacities to respond to disturbances remains underexplored and is critical for understanding how to adapt and transform infrastructure for future complexity. To respond to increasing complexity, infrastructure systems will need to be resilient while maintaining services (Park *et al.*, 2013; Woods, 2015). Infrastructure managers can prepare for complexity by 1) recognizing the unpredictability of a future that cannot be based on past events (i.e., non-stationarity) across social, ecological, and technological contexts and 2) understanding that traditional infrastructure design approaches, and thereby much of centralized infrastructure, relied upon assumptions of stationarity (Milly *et al.*, 2008; Tank, Zwiers and Zhang, 2009; Chester and Allenby, 2018; Helmrich and Chester, 2020). By relying on stationarity, it is more likely an infrastructure design envelope will be exceeded, leading to either temporary or catastrophic failure. Given the long design lives of built infrastructure, it is important to highlight governance, which may be more responsive, in periods of instability. Within infrastructure resilience literature, decentralization has been promoted as a tool to increase system resilience (Cascio,

2009; Tomlinson *et al.*, 2015; Meerow and Stults, 2016; Kwasinski *et al.*, 2019). This section reviews uses of de/centralization in resilience literature and examines how de/centralization can be used to promote resilience principles.

Large-scale, centralized infrastructure systems across the United States were created to provide basic services to citizens while taking advantage of economies of scale to reduce costs, contrary to decentralized systems which leverage economies of scope to enable multifunctionality and flexibility (O'Flaherty, 2005; Faguet, 2014; Goldthau, 2014; Gilrein *et al.*, 2019). While centralized governance can coordinate disturbance responses with widespread authority (Rinaldi, Peerenboom and Kelly, 2001), centralized networks, which have a low probability of failure, may have increasingly severe failure consequences (Gilrein *et al.*, 2019). Furthermore, centralized networks and governance are less adaptable, limited by their scale or jurisdiction (e.g., the water sector is divided into wastewater, stormwater, drinking water, etc.) (Mintzberg, 1979; Chester and Allenby, 2018; Markolf *et al.*, 2018; Leigh and Lee, 2019). Centralized systems have been deemed more vulnerable to failure than their decentralized counterparts (Baran, 1964; Ahern, 2011; Gilrein *et al.*, 2019; Leigh and Lee, 2019). Decentralized infrastructure are frequently depicted as systems that service local consumers through redundancy within the network (e.g., multiple providers or pathways), and the network redundancy is seen as a principle of resilience (Ahern, 2011; Zodrow *et al.*, 2017). Decentralization also provides an opportunity for modularity, or the capability to readily adapt or scale a system by reorganizing individual technical or institutional components without significantly disrupting the overall system (e.g., smart grids increase information flows to identify and respond to network component failures (Li *et al.*, 2010)). Decentralized infrastructure systems may limit cascading failures amongst infrastructure sectors by quickly recognizing and isolating the failure (Gleick, 2003; Goldthau, 2014; Zodrow *et al.*, 2017). However, across these studies of de/centralization and resilient infrastructure, there is recognition that, while decentralized networks are frequently promoted for increased resilience, there are situations in which centralization can also be beneficial (e.g., balancing budget, reliability, and network size) (Hines, Blumsack and Schläpfer, 2015; Zodrow *et al.*, 2017; Gilrein *et al.*, 2019; Leigh and Lee, 2019). There is less exploration of de/centralization in governance; however, complexity leadership and organizational science literature assert that the ability to transition between de/centralization, vertically and horizontally, is crucial for long-term viability because this flexibility allows organizations to traverse between modes of exploitation, i.e., business as usual, and exploration, i.e., innovation (Mintzberg, 1979; Siggelkow and Levinthal, 2003; Uhl-Bien and Arena, 2018).

2.4.3.1. INFRASTRUCTURE RESILIENCE AS A DE/CENTRALIZATION SPECTRUM

Leveraging de/centralization as a spectrum across domains increases pathways for infrastructure managers to address growing complexity and develop resilient infrastructure. Biggs et al. (2012) describes seven key principles of resilience: maintain diversity and redundancy, manage connectivity, manage slow variables and feedbacks, foster complex adaptive systems thinking, encourage learning, broaden participation, and promote polycentric governance. Using these principles, and the specific characteristics attributable to each, conceptual linkages between the de/centralization spectrum and resilience are proposed (Figure 2.4.3). These conceptual linkages highlight the importance of recognizing both the network and governance domains of infrastructure systems, as well as the need to consider these domains in the de/centralization spectrum. As illustrated in Figure 2.4.3, both centralization and decentralization can contribute to resilience efforts. Although decentralization appears to be more closely aligned with the resilience principles, there are instances where increased centralization may be warranted. Particularly, increased centralization appears beneficial when circumstance requires greater levels of coordination and/or an understanding of the entire system (rather than sub-systems or individual components). Implementing and achieving all seven of the resilience principles, or even simply a subset, requires focus on both the network and governance domains of infrastructure systems. Specifically, it is important to recognize that there may be instances where resilience efforts necessitate more centralized governance in conjunction with a more decentralized network infrastructure. This dichotomy is likely to have limits, i.e., at some point, a network can become decentralized to the point of being unmanageable by a more centralized governance structure. Therefore, infrastructure managers must reimagine how de/centralization is used and promoted in network and governance domains.

Figure 2.4.3: Examples of how increasing centralization (orange), increasing decentralization (blue), or dynamic combinations of both de/centralization in infrastructure networks and governance can influence resilience principles

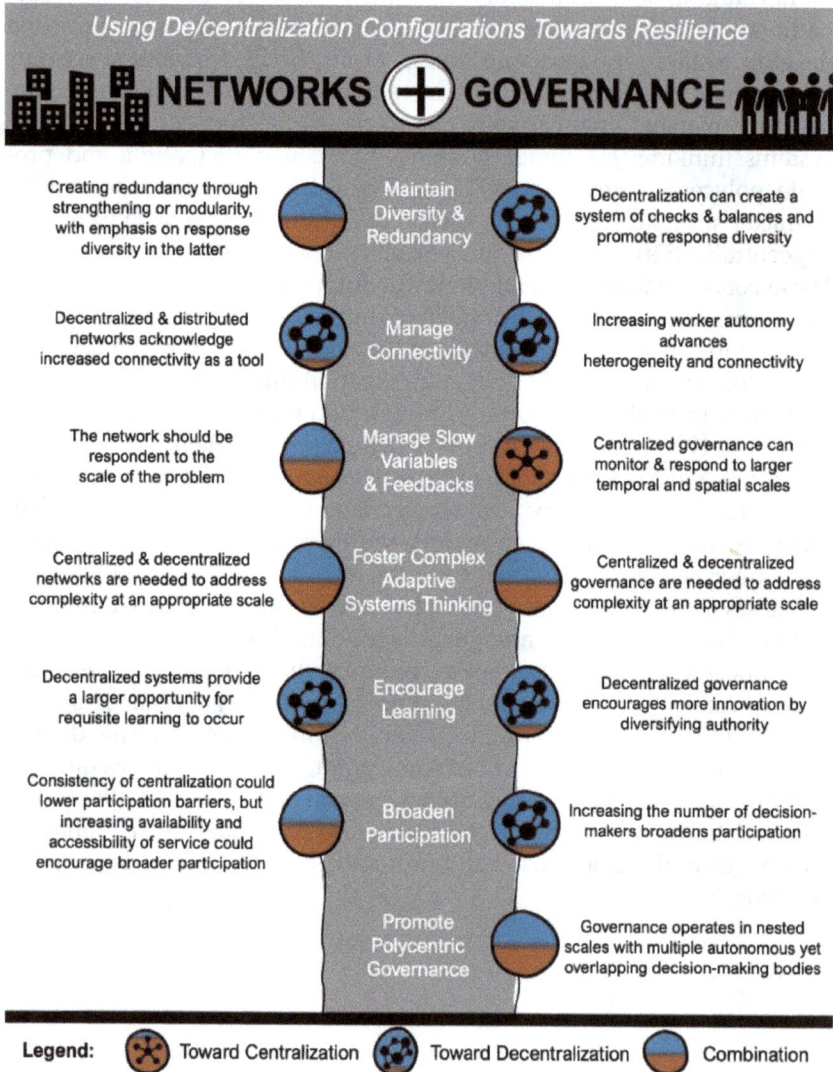

Using De/centralization Configurations Towards Resilience

NETWORKS ⊕ GOVERNANCE

Creating redundancy through strengthening or modularity, with emphasis on response diversity in the latter	Maintain Diversity & Redundancy	Decentralization can create a system of checks & balances and promote response diversity
Decentralized & distributed networks acknowledge increased connectivity as a tool	Manage Connectivity	Increasing worker autonomy advances heterogeneity and connectivity
The network should be respondent to the scale of the problem	Manage Slow Variables & Feedbacks	Centralized governance can monitor & respond to larger temporal and spatial scales
Centralized & decentralized networks are needed to address complexity at an appropriate scale	Foster Complex Adaptive Systems Thinking	Centralized & decentralized governance are needed to address complexity at an appropriate scale
Decentralized systems provide a larger opportunity for requisite learning to occur	Encourage Learning	Decentralized governance encourages more innovation by diversifying authority
Consistency of centralization could lower participation barriers, but increasing availability and accessibility of service could encourage broader participation	Broaden Participation	Increasing the number of decision-makers broadens participation
	Promote Polycentric Governance	Governance operates in nested scales with multiple autonomous yet overlapping decision-making bodies

Legend: ⊛ Toward Centralization   ⊛ Toward Decentralization   ◉ Combination

2.4.4. REFRAMING CENTRALIZATION AND DECENTRALIZATION

A dynamic, multi-dimensional framing of de/centralization through a coupled network-governance perspective with information as a mediator (Figure 2.4.4) may help infrastructure managers achieve resilience across infrastructure sectors. First, it is important to note that the classification of centralized, decentralized, and distributed networks and governance relies heavily on context, specifically scale. Second, this context is further complicated by information flow—coupling legacy infrastructure (water, power, transportation) with emerging communication technologies accelerates the capacity for distributed networks by connecting more nodes and increasing information flow. These information flows must be recognized as increasingly powerful forces that are able to forge new connections between consumers and producers and mediate existing relationships between governing bodies, infrastructure networks, and consumers. By empowering more producers and/or consumers with information (and, therefore, sense-making), as well as decision-making power to act, a system can more readily respond to instances of failure because of increased information sharing.

Figure 2.4.4: Multidimensional framing of de/centralization through a coupled network-governance perspective with information (e.g., SCADA) as a mediator.

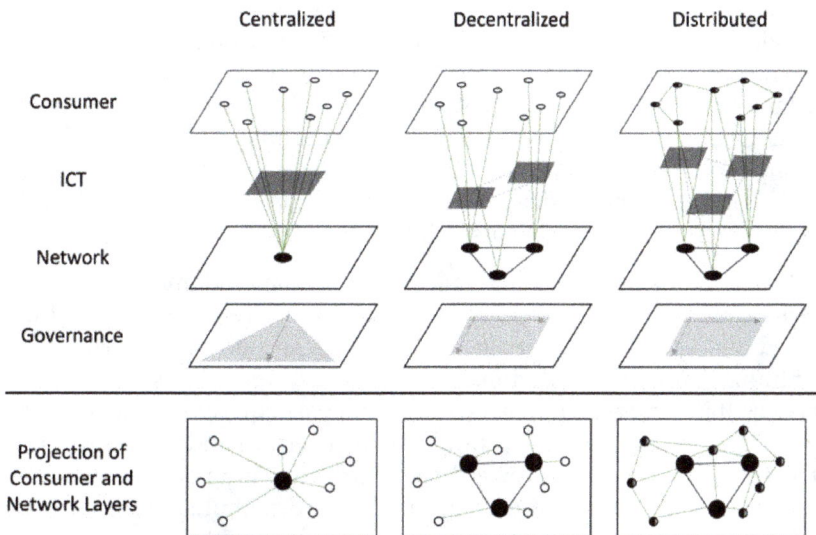

Consumer-producer dynamics are depicted by black (consumer) and white (producer) filled circles.

By recognizing the contributions of both networks, governance, and information toward de/centralization, infrastructure managers increase their opportunities to address instability. While de/centralization configurations have often been applied to infrastructure networks, exploration of de/centralization within the governance domain of infrastructure has largely been ignored. Decentralized governance has generally been shown to increase organizational performance in periods of instability since it brings the decision-making power closer to those experiencing instability or those with expertise on the topic of conflict, while centralized governance typically performs well in stability (Mintzberg, 1979; Andersen, 2004; Uhl-Bien and Arena, 2018). As identified in Complexity Leadership Theory by (Uhl-Bien and Arena, 2018), instability creates a need for organizations to: a) transition between operational (centralized) and entrepreneurial (decentralized) leadership, a skill labelled as enabling leadership and b) recognize a regime shift to initiate this leadership shift. First, enabling leadership must take advantage of vertical and horizontal de/centralization. While vertical decentralization brings the decision-making power closer to the situation, horizontal decentralization diversifies who has the ability to make decisions. Divisional bureaucracies seen in infrastructure systems offer limited horizontal and vertical decentralization by concentrating decision-making and sense-making towards the top of divisional hierarchies (Mintzberg, 1979; Chester, Miller and Muñoz-Erickson, 2020). Second, leadership must identify and have the capacity to respond—individually and as an institution—to shifts in the environment. If an organization continues to operate during instability with governance processes designed for efficiency under stability, they will experience failure (Snowden and Boone, 2007). By recognizing the multidimensional network-governance spectrum of de/centralization, infrastructure managers are provided a more nuanced view of de/centralization that creates more solution pathways to respond to instability.

Increased flexibility of network and governance domains would increase resilience, compared to the systems that are currently rigid and locked into place—whether centralized, decentralized, or distributed. Centralized systems are dominant in infrastructure networks and governances as this configuration has served well, particularly in times of stability. Acknowledging the de/centralization spectrum and the contribution of all three configurations toward resilience, there is an opportunity to integrate more attributes of decentralized and distributed systems. Infrastructure managers must design, create, and maintain infrastructure systems that augment adaptation and respond to disturbances outside their design envelope to continue operating effectively in a world characterized by instability. In terms of infrastructure networks, an immediate action would be to consider 'maintaining diversity and redundancy.' For instance, during COVID-19, passenger airlines began transporting medical supplies as well as people, showing response diversity. This change of function, additionally, required swift institutional action, allowing passenger airlines—not only to transport cargo—but to operate with larger quantities of dry ice than previously allowed in order to maintain extremely low temperatures for vaccine shipments (Kulisch, 2020). This example additionally highlights how the network and governance domains do not operate in isolation—despite being analyzed separately in much literature on de/centralization.

A spectrum of de/centralization across network and governance domains, and with consideration of information flows, provides an abundance of pathways for infrastructure managers to address complexity. Scenarios describing how change might occur can illustrate these pathways. First, there is a scenario of slow, chronic change—much like that seen in climatic conditions across the world. Here, infrastructure systems need to 'manage slow variables and feedbacks' while 'encouraging learning'. This likely requires a decentralized network to allow for experimentation but, at least some degree of, centralized governance to collect and communicate information, such as successful network adaptations or emerging climate science. However, in a second scenario, where there is a sudden, destabilizing event such as an extreme weather event or a cyberattack, the physical network cannot likely be changed in a timely manner. In this situation, decision-making power should have 'broadened participation' to allow local infrastructure managers to respond immediately while 'managing connectivity' to maximize support and provisions to the impacted region. This type of response is likely best achieved through a horizontally decentralized governance structure—a feature found in 'polycentric governance.'

Furthermore, looking across infrastructure sectors, there are likely to be different priorities and definitions of operations and failures. In the power and water sectors, there is a focus on the physical network systems and their ability to maintain constant and consistent services, including within times of failure; while in the transportation sector service is maintained to meet demand and, therefore, may change over time. (Relatedly, information flow also depends on demand—e.g., demand for information increases during times of instability when decision-makers need to respond quicky.) Viewing de/centralization as a spectrum allows infrastructure managers to consider the strengths and weaknesses of each for their particular sector and create a symbiotic relationship between configurations to respond to instability rather than to place one configuration in opposition to the other, encouraging infrastructure managers to 'foster complex adaptive systems thinking.'

2.4.5. CONCLUSION

As infrastructure managers pursue adaptation strategies to respond to increasing complexity in infrastructure systems and environments, they will need to confront existing configurations of networks and governance and critically examine the benefits and trade-offs of de/centralization. While existing framings of de/centralization may be polysemic, reconciling these framings expands infrastructure managers' design space to include both network and governance domains, increasing the opportunities for adaptation and transformation for infrastructure systems. It is critical that governance is observed as a domain of de/centralization. Emerging technologies that allow for rapid information sharing provide an opportunity to empower producers, as well as consumers, with decision-making capabilities to respond with speed and flexibility in instances of failure. This integration of information changes infrastructure systems from centralized or decentralized, to distributed. This increases the number of decision-makers in a system, which adds complexity along with flexibility, and confronts fundamental assumptions of who holds power. Centralization is dominant in today's infrastructure systems, but all three configurations — centralized, decentralized, and distributed — are at times appropriate for supporting resilience capacities because 1) infrastructure sectors have different priorities and, therefore, different solutions will be needed to meet each context, and 2) each configuration has varying abilities to respond to stability and instability. This reframing of de/centralization challenges infrastructure managers to develop new mental models — an evolving understanding of a system that is operational but not necessarily technically correct as it is influenced, oftentimes unconsciously, by an individual's beliefs — and, likewise, institutions to create new collective cognition (Gentner, Stevens and Gentner, 1983; Jones *et al.*, 2011). Reimagining de/centralization as a spectrum may help infrastructure managers navigate infrastructure systems through periods of instability by assessing, and reassessing, de/centralization configurations to increase actionable pathways forward for infrastructure resilience.

2.4.6. REFERENCES

Abimbola, S., Baatiema, L. and Bigdeli, M. (2019) 'The impacts of decentralization on health system equity, efficiency and resilience: a realist synthesis of the evidence', *Health Policy and Planning*, 34(8), pp. 605–617. Available at: https://doi.org/10.1093/heapol/czz055.

Ackermann, T., Andersson, G. and Söder, L. (2001) 'Distributed generation: a definition', *Electric Power Systems Research*, 57(3), pp. 195–204. Available at: https://doi.org/10.1016/S0378-7796(01)00101-8.

Ahern, J. (2011) 'From fail-safe to safe-to-fail: Sustainability and resilience in the new urban world', *Landscape and Urban Planning*, 100(4), pp. 341–343. Available at: https://doi.org/10.1016/j.landurbplan.2011.02.021.

Alanne, K. and Saari, A. (2006) 'Distributed energy generation and sustainable development', *Renewable and Sustainable Energy Reviews*, 10(6), pp. 539–558. Available at: https://doi.org/10.1016/j.rser.2004.11.004.

Albalate, D., Bel, G. and Fageda, X. (2012) 'Beyond the efficiency-equity dilemma: Centralization as a determinant of government investment in infrastructure*', *Papers in Regional Science*, 91(3), pp. 599–615. Available at: https://doi.org/10.1111/j.1435-5957.2011.00414.x.

Andersen, T.J. (2004) 'Integrating Decentralized Strategy Making and Strategic Planning Processes in Dynamic Environments', *Journal of Management Studies*, 41(8), pp. 1271–1299. Available at: https://doi.org/10.1111/j.1467-6486.2004.00475.x.

Ayyub, B.M. (2018) *Climate-Resilient Infrastructure*. Edited by Committee on Adaptation to a Changing Climate. Reston, VA: ASCE Press. Available at: https://doi.org/10.1061/9780784415191.

Baldwin, C.Y. and Clark, K.B. (2006) 'Modularity in the Design of Complex Engineering Systems', in D. Braha, A.A. Minai, and Y. Bar-Yam (eds) *Complex Engineered Systems: Science Meets Technology*. Berlin, Heidelberg: Springer (Understanding Complex Systems), pp. 175–205. Available at: https://doi.org/10.1007/3-540-32834-3_9.

Baran, P. (1964) *On Distributed Communications: I. Introduction to Distributed Communications Networks*. RAND Corporation. Available at: https://www.rand.org/pubs/research_memoranda/RM3420.html (Accessed: 19 October 2023).

Bardhan, P. (2002) 'Decentralization of Governance and Development', *Journal of Economic Perspectives*, 16, pp. 185–205. Available at: https://doi.org/10.1257/089533002320951037.

Biggs, R. *et al.* (2012) 'Toward Principles for Enhancing the Resilience of Ecosystem Services', *Annual Review of Environment and Resources*, 37(1), pp. 421–448. Available at: https://doi.org/10.1146/annurev-environ-051211-123836.

Bondank, E.N., Chester, M.V. and Ruddell, B.L. (2018) 'Water Distribution System Failure Risks with Increasing Temperatures', *Environmental Science & Technology*, 52(17), pp. 9605–9614. Available at: https://doi.org/10.1021/acs.est.7b01591.

Burillo, D. *et al.* (2017) 'Electricity demand planning forecasts should consider climate non-stationarity to maintain reserve margins during heat waves', *Applied Energy*, 206(June), pp. 267–277. Available at: https://doi.org/10.1016/j.apenergy.2017.08.141.

Cascio, J. (2009) 'Resilience', *Foreign Policy*, June, p. 92.

Chandler, A.D. (1977) *The Visible Hand*. Cambridge, MA: Harvard University Press. Available at: https://books.google.com/books?hl=en&lr=&id=lAI3AwAAQBAJ&oi=fnd&pg=PR9&ots=YWApkToGvQ&sig=fB0xzTg4BxWW6iRUHCzxMJz-H8M#v=onepage&q&f=false.

Chester, M. *et al.* (2021) 'Infrastructure resilience to navigate increasingly uncertain and complex conditions in the Anthropocene', *npj Urban Sustainability*, 1(1), pp. 1–6. Available at: https://doi.org/10.1038/s42949-021-00016-y.

Chester, M.V. and Allenby, B. (2018) 'Toward adaptive infrastructure: flexibility and agility in a non-stationarity age', *Sustainable and Resilient Infrastructure*, 4, pp. 1–19. Available at: https://doi.org/10.1080/23789689.2017.1416846.

Chester, M.V. and Allenby, B.R. (2020) 'Perspective: The Cyber Frontier and Infrastructure', *IEEE Access*, 8, pp. 28301–28310. Available at: https://doi.org/10.1109/ACCESS.2020.2971960.

Chester, M.V., Miller, T. and Muñoz-Erickson, T.A. (2020) 'Infrastructure governance for the Anthropocene', *Elementa: Science of the Anthropocene*, 8(1), p. 78. Available at: https://doi.org/10.1525/elementa.2020.078.

Chester, M.V., Underwood, B.S. and Samaras, C. (2020) 'Keeping infrastructure reliable under climate uncertainty', *Nature Climate Change*, 10(6), pp. 488–490. Available at: https://doi.org/10.1038/s41558-020-0741-0.

Derrible, S. (2017) 'Urban infrastructure is not a tree: Integrating and decentralizing urban infrastructure systems', *Environment and Planning B: Urban Analytics and City Science*, 44(3), pp. 553–569. Available at: https://doi.org/10.1177/0265813516647063.

Desha, C.J., Hargroves, K. and Smith, M.H. (2009) 'Addressing the time lag dilemma in curriculum renewal towards engineering education for sustainable development', *International Journal of Sustainability in Higher Education*, 10(2), pp. 184–199. Available at: https://doi.org/10.1108/14676370910949356.

Dubois, H.F.W. and Fattore, G. (2009) 'Definitions and Typologies in Public Administration Research: The Case of Decentralization', *International Journal of Public Administration*, 32(8), pp. 704–727. Available at: https://doi.org/10.1080/01900690902908760.

Economic Research Centre (2001) 'Transport and Economic Development Round Table 119', in. *EUROPEAN CONFERENCE OF MINISTERS OF TRANSPORT*, Paris, FR: OECD Publications Service. Available at: https://www.itf-oecd.org/sites/default/files/docs/02rt119_0.pdf.

El-Khattam, W. and Salama, M.M.A. (2004) 'Distributed generation technologies, definitions and benefits', *Electric Power Systems Research*, 71(2), pp. 119–128. Available at: https://doi.org/10.1016/j.epsr.2004.01.006.

Faguet, J.-P. (2014) 'Decentralization and Governance', *World Development*, 53, pp. 2–13. Available at: https://doi.org/10.1016/j.worlddev.2013.01.002.

Freeman, L.C. (1978) 'Centrality in social networks conceptual clarification', *Social Networks*, 1(3), pp. 215–239. Available at: https://doi.org/10.1016/0378-8733(78)90021-7.

Ge, X., Yang, F. and Han, Q.-L. (2017) 'Distributed networked control systems: A brief overview', *Information Sciences*, 380, pp. 117–131. Available at: https://doi.org/10.1016/j.ins.2015.07.047.

Gentner, Dedre, Stevens, A.L. and Gentner, Debre (1983) *Mental Models*. Hillsdale, N.J: Taylor & Francis Inc.

Gilrein, E.J. *et al.* (2019) 'Concepts and practices for transforming infrastructure from rigid to adaptable', *Sustainable and Resilient Infrastructure*, 6(3–4), pp. 213–234. Available at: https://doi.org/10.1080/23789689.2019.1599608.

Gleick, P.H. (2003) 'Global Freshwater Resources: Soft-Path Solutions for the 21st Century', *Science*, 302(5650), pp. 1524–1528. Available at: https://doi.org/10.1126/science.1089967.

Goldthau, A. (2014) 'Rethinking the governance of energy infrastructure: Scale, decentralization and polycentrism', *Energy Research & Social Science*, 1, pp. 134–140. Available at: https://doi.org/10.1016/j.erss.2014.02.009.

Helmrich, A.M. and Chester, M.V. (2020) 'Reconciling complexity and deep uncertainty in infrastructure design for climate adaptation', *Sustainable and Resilient Infrastructure*, 7(2), pp. 83–99. Available at: https://doi.org/10.1080/23789689.2019.1708179.

Hines, P., Blumsack, S. and Schläpfer, M. (2015) 'Centralizedversus DecentralizedInfrastructure Networks', *ArXiv* [Preprint]. Available at: https://doi.org/10.48550/arXiv.1510.08792.

Jones, N. *et al.* (2011) 'Mental Models: An Interdisciplinary Synthesis of Theory and Methods', *Ecology and Society*, 16(1). Available at: https://doi.org/10.5751/ES-03802-160146.

Kooiman, J. (1993) *Modern Governance: New Government-Society Interactions.* Thousand Oaks, CA: SAGE.

Kulisch, E. (2020) *FAA issues dry ice alert to airlines carrying vaccines, FreightWaves.* Available at: https://www.freightwaves.com/news/faa-issues-dry-ice-alert-to-airlines-carrying-vaccine (Accessed: 19 October 2023).

Kwasinski, A. *et al.* (2019) 'Hurricane Maria Effects on Puerto Rico Electric Power Infrastructure', *IEEE Power and Energy Technology Systems Journal*, 6(1), pp. 85–94. Available at: https://doi.org/10.1109/JPETS.2019.2900293.

Larsen, T.A., Udert, K.M. and Lienert, J. (2013) *Source Separation and Decentralization for Wastewater Management.* London, UK; New York, NY: IWA Publishing. Available at: https://doi.org/10.2166/9781780401072.

Leigh, N.G. and Lee, H. (2019) 'Sustainable and Resilient Urban Water Systems: The Role of Decentralization and Planning', *Sustainability*, 11(3), p. 918. Available at: https://doi.org/10.3390/su11030918.

Li, F. *et al.* (2010) 'Smart Transmission Grid: Vision and Framework', *IEEE Transactions on Smart Grid*, 1(2), pp. 168–177. Available at: https://doi.org/10.1109/TSG.2010.2053726.

Luo, S. (2005) 'A review of distributed power systems part I: DC distributed power system', *IEEE Aerospace and Electronic Systems Magazine*, 20(8), pp. 5–16. Available at: https://doi.org/10.1109/MAES.2005.1499272.

Makropoulos, C.K. and Butler, D. (2010) 'Distributed Water Infrastructure for Sustainable Communities', *Water Resources Management*, 24(11), pp. 2795–2816. Available at: https://doi.org/10.1007/s11269-010-9580-5.

Markolf, S.A. *et al.* (2018) 'Interdependent Infrastructure as Linked Social, Ecological, and Technological Systems (SETSs) to Address Lock-in and Enhance Resilience', *Earth's Future*, 6(12), pp. 1638–1659. Available at: https://doi.org/10.1029/2018EF000926.

Meerow, S. and Stults, M. (2016) 'Comparing Conceptualizations of Urban Climate Resilience in Theory and Practice', *Sustainability*, 8(7), p. 701. Available at: https://doi.org/10.3390/su8070701.

Milly, P.C.D. *et al.* (2008) 'Stationarity Is Dead: Whither Water Management?', *Science*, 319(5863), pp. 573–574. Available at: https://doi.org/10.1126/science.1151915.

Mintzberg, H. (1979) *The Structuring of Organizations: A Synthesis of the Research*. Hoboken, NJ: Prentice-Hall.

Mintzberg, H. (1980) 'Structure in 5's: A Synthesis of the Research on Organization Design', *Management Science*, 26(3), pp. 322–341. Available at: https://doi.org/10.1287/mnsc.26.3.322.

Nasr, A. *et al.* (2019) 'A review of the potential impacts of climate change on the safety and performance of bridges', *Sustainable and Resilient Infrastructure*, 6(3–4), pp. 192–212. Available at: https://doi.org/10.1080/23789689.2019.1593003.

O'Flaherty, B. (2005) *City Economics*. Cambridge, MA: Harvard University Press. Available at: https://www.hup.harvard.edu/catalog.php?isbn=9780674019188.

Pagani, G.A. and Aiello, M. (2011) 'Towards Decentralization: A Topological Investigation of the Medium and Low Voltage Grids', *IEEE Transactions on Smart Grid*, 2(3), pp. 538–547. Available at: https://doi.org/10.1109/TSG.2011.2147810.

Park, J. *et al.* (2013) 'Integrating Risk and Resilience Approaches to Catastrophe Management in Engineering Systems', *Risk Analysis*, 33(3), pp. 356–367. Available at: https://doi.org/10.1111/j.1539-6924.2012.01885.x.

Quezada, G., Walton, A. and Sharma, A. (2016) 'Risks and tensions in water industry innovation: understanding adoption of decentralised water systems from a socio-technical transitions perspective', *Journal of Cleaner Production*, 113, pp. 263–273. Available at: https://doi.org/10.1016/j.jclepro.2015.11.018.

Rinaldi, S.M., Peerenboom, J.P. and Kelly, T.K. (2001) 'Identifying, understanding, and analyzing critical infrastructure interdependencies', *IEEE Control Systems Magazine*, 21(6), pp. 11–25. Available at: https://doi.org/10.1109/37.969131.

Rodrigue, J.-P. (2020) *The Geography of Transport Systems*. 5th edn. London, UK: Routledge. Available at: https://doi.org/10.4324/9780429346323.

Sharma, R. (2019) 'The Straits of Success In a Vuca World', *ABS International Journal of Management*, 7(1), pp. 16–22.

Siggelkow, N. and Levinthal, D.A. (2003) 'Temporarily Divide to Conquer: Centralized, Decentralized, and Reintegrated Organizational Approaches to Exploration and Adaptation', *Organization Science*, 14(6), pp. 650–669. Available at: https://doi.org/10.1287/orsc.14.6.650.24840.

Snowden, D.J. and Boone, M.E. (2007) 'A Leader's Framework for Decision Making', *Harvard Business Review*, 1 November, pp. 68–76, 149.

Srinivasa, K.G. and Muppalla, A.K. (2015) *Guide to High Performance Distributed Computing: Case Studies with Hadoop, Scalding and Spark*. Heidelberg, DE; New York, NY; Dordrecht, UK; London, UK: Springer (Computer Communications and Networks). Available at: https://doi.org/10.1007/978-3-319-13497-0.

Steffen, W. *et al.* (2015) 'The trajectory of the Anthropocene: The Great Acceleration', *The Anthropocene Review*, 2(1), pp. 81–98. Available at: https://doi.org/10.1177/2053019614564785.

Tank, A., Zwiers, F. and Zhang, X. (2009) *Guidelines on Analysis of Extremes in a Changing Climate in Support of Informed Decisions for Adaptation*. WMO-TD No. 1500. Geneva, Switzerland: World Meteorological Organization, p. 55. Available at: http://www.clivar.org/organization/etccdi/etccdi.php.

Tomlinson, B. *et al.* (2015) 'Toward Alternative Decentralized Infrastructures', in *Proceedings of the 2015 Annual Symposium on Computing for Development*. New York, NY: Association for Computing Machinery (DEV '15), pp. 33–40. Available at: https://doi.org/10.1145/2830629.2830648.

Uhl-Bien, M. and Arena, M. (2018) 'Leadership for organizational adaptability: A theoretical synthesis and integrative framework', *The Leadership Quarterly*, 29(1), pp. 89–104. Available at: https://doi.org/10.1016/j.leaqua.2017.12.009.

Underwood, B.S. *et al.* (2017) 'Increased costs to US pavement infrastructure from future temperature rise', *Nature Climate Change*, 7(10), pp. 704–707. Available at: https://doi.org/10.1038/nclimate3390.

Wilder, M. and Romero Lankao, P. (2006) 'Paradoxes of Decentralization: Water Reform and Social Implications in Mexico', *World Development*, 34(11), pp. 1977–1995. Available at: https://doi.org/10.1016/j.worlddev.2005.11.026.

Woods, D.D. (2015) 'Four concepts for resilience and the implications for the future of resilience engineering', *Reliability Engineering & System*

Safety, 141, pp. 5–9. Available at: https://doi.org/10.1016/j.ress.2015.03.018.

Zodrow, K.R. *et al.* (2017) 'Advanced Materials, Technologies, and Complex Systems Analyses: Emerging Opportunities to Enhance Urban Water Security', *Environmental Science & Technology*, 51(18), pp. 10274–10281. Available at: https://doi.org/10.1021/acs.est.7b01679.

2.5

REQUISITE VARIETY TO ENGAGE COMPLEXITY

2.5.1. INTRODUCTION

The external conditions in which our infrastructure needs to function are changing rapidly. Acceleration of climate change (IPCC, 2014), acceleration of the integration of cybertechnologies (and with that security concerns) (Chester and Allenby, 2020), hyperpolarization that routinely stifles innovative reform (Pildes, 2010), distributed control (with third parties increasingly driving service consumption, e.g., phone navigation software and smart home energy technologies) (Poska, Kaplan and Alford, 2021), shifting priorities (e.g., pushes to increase renewables) (The Economist, 2021), and new competition (Amazon drone delivery as infrastructure) (Lohn, 2017), to name a few, represent dramatic shifts in the environments in which legacy infrastructure need to remain viable. We use the term environment in the broadest of senses including technological, political, social, cultural in addition to environmental change. Modernizing infrastructure for any one of these challenges is difficult enough but tackling all at once is necessary and represents accelerating complexity for infrastructure managers and a new frontier of tools. Infrastructure have been designed for variability across some of these external conditions, but 1) are these conditions now or soon going to exceed what the systems are able to respond to, and 2) are infrastructure being adapted too slowly relative to changes in external conditions? If the answer to either question is yes, then we must recognize that there is an infrastructure viability problem.

As the scale and scope of human activities have grown, the boundaries between human, built, and natural systems are becoming porous (Allenby, 2007, 2012; Zalasiewicz *et al.*, 2011; Steffen, Crutzen and McNeill, 2016; Waters *et al.*, 2016). Whereas in the past human systems on smaller

---

This chapter was adapted from the following article with publisher permission: Mikhail Chester and Braden Allenby, 2022, Infrastructure Autopoiesis: Requisite Variety to Engage Complexity, *Environmental Research: Infrastructure and Sustainability*, 2(012001), doi: 10.1088/2634-4505/ac4b48.

scales had clearer bounds, the challenges that have emerged with the Anthropocene appear to be unbounded (Allenby, 2012). Climate change, for example, is an integration of human, technological, social, cultural, economic, and natural system dynamics, and any mitigation or adaptation strategy must embrace this complexity (Hetherington, 2018; Miller, Chester and Munoz-Erickson, 2018). Infrastructure -- design spaces where humans provide basic, lifeline, and critical services through governance structures, physical assets, and mental models informed by educational practices -- appear to be at a critical juncture. Serious questions are emerging around what infrastructure are and should do in the future (Allenby, 2007; Edwards, 2017; Hetherington, 2018; Braden Allenby and Chester, 2021). This is driven in part by the accelerating capabilities and increasingly global scope of human activities (Steffen *et al.*, 2015; Steffen, Crutzen and McNeill, 2016). Infrastructure is generally framed from a local context – city or regional engineered systems that deliver and support services such as water, power, and mobility, in tightly bounded (geography, financing, goals, interdependencies) spaces. But these design spaces (infrastructure) appear increasingly unbounded. For example, carbon capture and storage technologies treat the atmosphere as a managed space, public service and private business operations increasingly rely on data streams hosted and managed by technology firms and warehouses, water provision in the Western U.S. includes the management of continental hydrology, and space (as used by GPS systems) are critical infrastructure for all transportation and routing activities. This deepening integration represents new challenges when contextualized in rapidly changing environments.

Infrastructure adaptation has emerged as a set of responses to rapidly changing environments. Adaptation appears to be heavily rooted in climate preparedness efforts, and for infrastructure conventional risk-based strategies including rebound and robustness are still emphasized (Bassett and Fogelman, 2013; Woods and Hollnagel, 2017) and strategies such as safe-to-fail, design-making under deep uncertainty, and social-ecological-technological capacity planning are just emerging (Kim *et al.*, 2019; Helmrich and Chester, 2020; McPhearson *et al.*, 2021). There appears to be a disconnect in infrastructure planning between adaptation strategies and the fundamental capabilities of infrastructure organizers (Chester, Miller and Muñoz-Erickson, 2020). Adaptation is a set of actions that are predicated on the organization being able to make sense of changing environments (Miller and Muñoz-Erickson, 2018). Making sense of the changing environments requires that the organization be willing to engage with complexity and produce a repertoire of responses commensurate to the complexity it faces (Ashby, 1956; Boisot and Mckelvey, 2011; Naughton, 2017).

Here, we frame the challenges of modernizing infrastructure in the face of growing complexity as a set of processes that can make sense of and appropriately react to the increasing variety produced in the environment. We contend that infrastructure agility and flexibility fundamentally stem from how the organization and its technologies respond to increasing variety. In establishing this context, we draw from the cybernetics, biology, and ecology fields where considerable theory has been developed to describe how systems react to change. We start by establishing the concept of internal-external variety as it relates to infrastructure, as well as how organizations structure themselves to respond to external variety, both sufficiently and insufficiently. Leveraging this theory, we then describe how infrastructure systems should change to be able to respond to increasing complexity in the Anthropocene. Our goal is to help reframe how infrastructure organizations plan for adaptation in the face of unfamiliar Anthropocene conditions. In doing so we position key tenets (sustained adaptation, horizon scanning, horizontal governance, and loose fit design) that infrastructure should advance to engage with complexity.

2.5.2. ONLY VARIETY CAN ABSORB VARIETY

To engage with complexity, infrastructure systems will need to understand the processes by which their organizations make sense of changing environments, and how the organization supports or hinders novel ways of sensemaking (i.e., knowledge generation about the system itself and the environment) and service delivery. "If a system is to be able to deal successfully with the diversity of challenges that its environment produces, then it needs to have a repertoire of responses which is (at least) as nuanced as the problems thrown up by the environment"(Naughton, 2017). Variety is not simply change, but the number of states that a system or its environment can achieve (Ashby, 1956). This concept is referred to as the Law of Requisite Variety (illustrated in Figure 2.5.1) and is a valuable starting point when considering whether systems are able to respond to external pressures. How engineered infrastructure responds to external pressures deserves critical examination.

Figure 2.5.1: Insufficient and Sufficient (Requisite) Variety

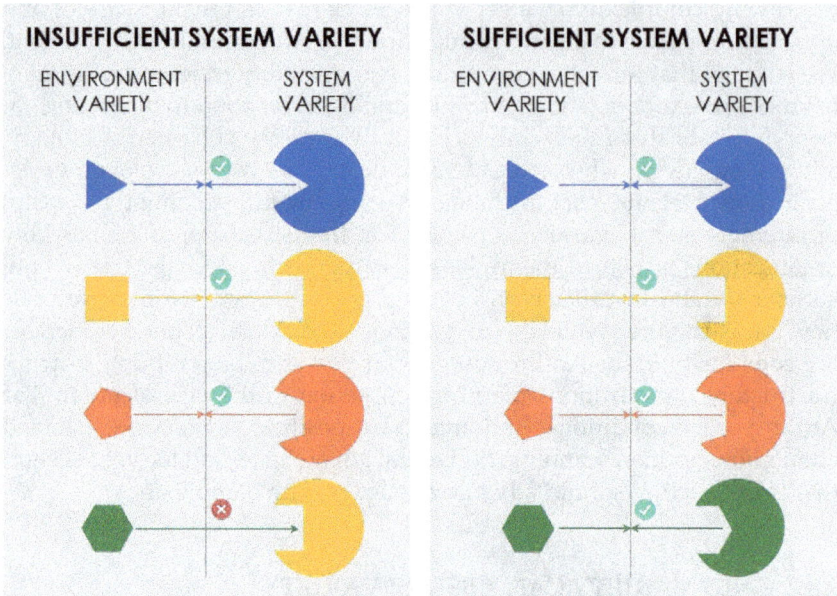

Adapted and reprinted with permission from Norman and Bar-Yam (2018). The left box shows system variety that is unable to respond to environment variety. The right box shows requisite variety, system variety matches that of environment variety.

Infrastructure has been designed with governing principles and bureaucratic processes that emphasize particular bounds around sensemaking. Infrastructure is a product of the historical context inside which they were structured, and the operating goals rooted in that context. This includes the historical normalized goals, cultures, and preferences (Joerges, 1989; Sovacool, Lovell and Ting, 2018), as well as perception of the environment (Chester, Miller and Muñoz-Erickson, 2020; Chester, Underwood and Samaras, 2020). When an infrastructure system was created (such as a transportation or water agency) these goals and perceptions became institutionalized and the technologies to support the service became the backbones of system functionality. We are today reconciling how to modernize infrastructure, at the nexus of past norms and future uncertainty. When environment variety outpaces that of the infrastructure system itself, it's not simply that internal variety must be increased, but more so that the appropriate variety must be generated at least as fast as the environment is changing. It's not about explicitly designing internal system variety, as that implies a state of certainty that we don't have with increasing complexity. Instead, we must position infrastructure systems with the agility and flexibility to organically respond at pace. We refer to this as Infrastructure Autopoiesis which we will elaborate on.

2.5.2.1. INFRASTRUCTURE SENSEMAKING PROCESSES

The organizations that manage infrastructure are designed to amplify and attenuate particular types of information and receive conditioned information. This process drives how infrastructure systems make sense of complexity and their ability to respond, and in many situations remains affected by historical goals. An organization engages with environment complexity by managing how and what types of information flow between decisionmakers (Beer, 1979, 1981, 1985). Consider that infrastructure management (as management and operations) must interact not only with the physical infrastructure but also governance systems and environments consisting of natural systems and non-governmental human systems (Figure 2.5.2). Management is not designed to make sense of Environment and Infrastructure complexity (Chester, Miller and Muñoz-Erickson, 2020); if it did it would be overwhelmed (Beer, 1985). As such, it relies on its operational capabilities, which have been explicitly designed to interact with these systems. Operations make sense of demand, changing conditions in weather and climate, resource availability, and other conditions. Decisions must be made in a timely manner and with limited resources, so Operations are designed to attenuate Environment and Infrastructure information to only what is perceived to be needed (reducing away the complexity). Operational processes will also amplify the effects of management, for example, channelizing rivers to control variability, encouraging particular behaviors, or creating increasingly sophisticated systems in response to perceived needs. A similar set of interactions occur between management and operational processes. Management requires attenuated information from operations to make decisions. If operators updated management with every subtask, activity, and day-to-day goings-on management would be overwhelmed. Note that by the time information about the environment reaches management it is attenuated twice. Management sends effector signals to operations amplifying how it thinks the organization should respond to environmental conditions. In doing so management amplifies what it perceives operations must know and eliminates what it perceives to be irrelevant. Human governance systems are also relevant as they add complexity creating goals, systems, and rule sets, but will receive attenuated information in terms of what the infrastructure system is doing. The relationship between Governance and the Environment is co-evolutionary (red dotted line), i.e., they are interdependent and change in response to each other.

141

Figure 2.5.2: Infrastructure Management Requisite Variety to Engage with Complexity.

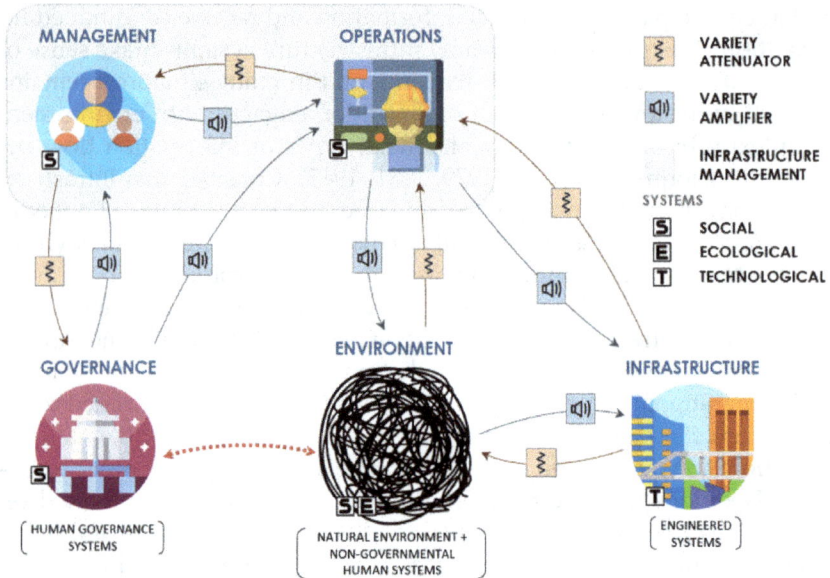

Variety attenuation is when high variety is cut down to the number of states the receiving entity can handle. Variety amplification is when low variety is enhanced to the number of states that the receiving entity needs. The S (social), E (ecological), and T (technological/infrastructure) labels describe the major systems represented in the SETS framework. The figure is detailed from the perspective of infrastructure management (Markolf *et al.*, 2018; McPhearson *et al.*, 2021). Adapted from Beer (1985).

The relationships are critical to understanding how infrastructure organizations understand and respond to complexity. Environment and Infrastructure complexity are intentionally reduced by Operations, and in the long run may be especially problematic, as changing conditions may result in obsolete practices by the organization (Beer, 1985; Hayward, 2004). However, as Management tries to reintroduce complexity, first to operations and then in how Operations respond to the Environment, it may find that its low complexity response is insufficient for the high complexity of the environment (Beer, 1985; Hayward, 2004). This process is by design and includes rules and norms that were intentionally created based on goals that may reflect legacy priorities. They reflect an organization's mental model of how the environment, and the organization's interactions with it, work.

2.5.2.2. REQUISITE COMPLEXITY

Recognizing that organizations can try to manage or reduce external complexity with commensurate internal complexity, Boisot and McKelvey (2011) advance Ashby's work as the Law of Requisite Complexity. They argue that organizations can invest in adaptation in one of two ways: 1) simplify the complexity of incoming stimuli to economize on the resources needed to respond; or, 2) invest extra resources beyond what is deemed necessary to ensure some degree of adaptation. The two options are effectively efficiency versus resilience prioritization, exploitation, and exploration (Papachroni, Heracleous and Paroutis, 2016; Brad Allenby and Chester, 2021). In the first, organizations run the risk of oversimplifying, reducing external complexity by mis-categorizing unfamiliar signals as fitting known patterns. With the second, two risks emerge. The first is associated with expending unnecessarily on complex responses before adaptation occurs (Boisot and Mckelvey, 2011). There is also, however, the risk of applying outdated assumptions to a changing environment. It's important to recognize that the operational space was engineered from how we perceived complexity in the external system relative to its actual complexity (i.e., organizational mental model). If we perceive the complexity of the external system to change then the organization must change how it invests in adaptation. The requisite perception is in fact a creative interaction between institutional cognition and the external, unknowable (because of complexity) environment. Organizations construct a mental model of their reality (i.e., how the environment works and they interact with it) through a dialog between what they desire (as possibly mandated by their mission), and how they perceive their environment. As such adaptation necessitates mental model shifts, an internal renegotiation by the organization of how it perceives the environment.

Exploring complexity within the context of environment variety relative to internal variety is helpful for understanding response strategies. Figure 2.5.3 is the Ashby Space (Boisot and Mckelvey, 2011), where the x-axis represents the variety of system responses and the y-axis variety of environmental stimuli. The identity (1:1) diagonal delineates between two critical regimes: above being perish where the system is producing more variety than the organization can respond to, and below being adaptive where the system is being overwhelmed by environment variety. The chart is separated into three regimes: ordered, complex, and chaotic (Gell-Mann, 2002). In the ordered regime stimuli are relatively unproblematic. In the Cynefin complexity framework, the ordered regime would map to simple and complicated systems, where emergent behaviors are predictable (Snowden and Boone, 2007; Chester and Allenby, 2019; Helmrich and Chester, 2020). In the complex, there is a mix of predictable and unpredictable stimuli, where initiating events, accidents, and non-linear phenomena are obscured by noise. In the chaotic regime extracting useful information appears intractable both because the information and situation are chaotic, making anticipation and prediction impossible. Here operators must either wait for the situation to fully unfold or proceed by trial-and-error. In Figure 2.5.3 the red dotted line indicates the adaptive frontier (or budget) – outside of this curve, the organization does not have the resources or capacity to process external environment input and to make appropriate responses. The ontology is helpful when considering how organizations respond to environmental stimuli, the act of trying to manage the organization towards a response capability (i.e., the requisite variety diagonal). If the organization interprets the stimuli as ordered (low variety stimuli) it will follow a path that emphasizes routine responses associated with stable conditions. The organization's adaptive success will depend on whether the routine tools and processes used during stable times are sufficient for the phenomena. If the organization interprets the stimuli as chaotic (high variety stimuli) and tries to match the response with capabilities that don't exist, the organization will exceed its adaptation budget and collapse. The third approach (strategist) becomes critical as it recognizes that trial-and-error is needed in adaptation. Here, the organization interprets the stimuli and attempts to find patterns (moving down the y-axis), reducing both the range and variety of stimuli (Boisot and Mckelvey, 2011). Where no patterns are found the organization develops new patterns (moving to the right along the x-axis), all while staying within the adaptation budget. Some stimuli will appear ordered and others chaotic, and the better the organization is at discerning between the two the greater chance it has at staying within its adaptation budget. This is because the organization can economize on tackling the ordered challenges and saving limited resources for trial-and-erroring their understanding and response to the chaotic stimuli.

Figure 2.5.3: The Ashby Space.

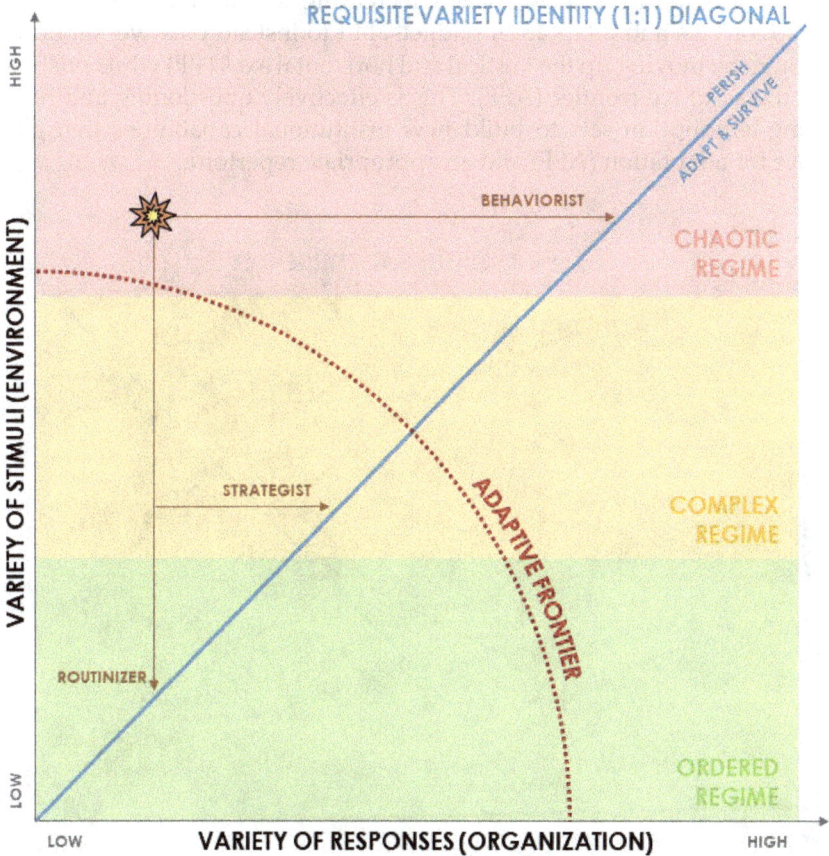

Adapted from Boisot and McKelvey (2011).

Codes, regulations, financing, organizational culture, and priorities define the rules and norms that underpin the mental models that infrastructure organizations follow, effectively informal governance. When should organizations challenge these assumption sets to shift their mental models of how the environment works and how their systems need to change to remain viable? The Ashby Space and Law of Requisite Complexity provides helpful framing. At their inception infrastructure systems may have had a variety of responses greater than that of their environment, as represented in Figure 2.5.4 as Vi. Today, environmental stimuli have increased significantly, and while our response repertoire has not increased commensurate with changing external conditions (VT). We find ourselves in the perish space where there is a decoupling between what infrastructure can do and what we need them to do. Infrastructure systems are now outside of their initial adaptive frontier (AFi) and in a complex regime.

Adaptation to greater environment complexity should not necessarily be met by attempting to make sense of increasing stimuli within existing assumption sets and adaptation budgets, but to instead creatively muddle through by moving up the vertical and horizontal axes (VF) while pushing out the adaptive frontier (AFF). This is effectively questioning and redefining assumption sets to build new institutional capabilities that give space for adaptation (AFF) and an appropriate repertoire.

Figure 2.5.4: Adaptive Frontier

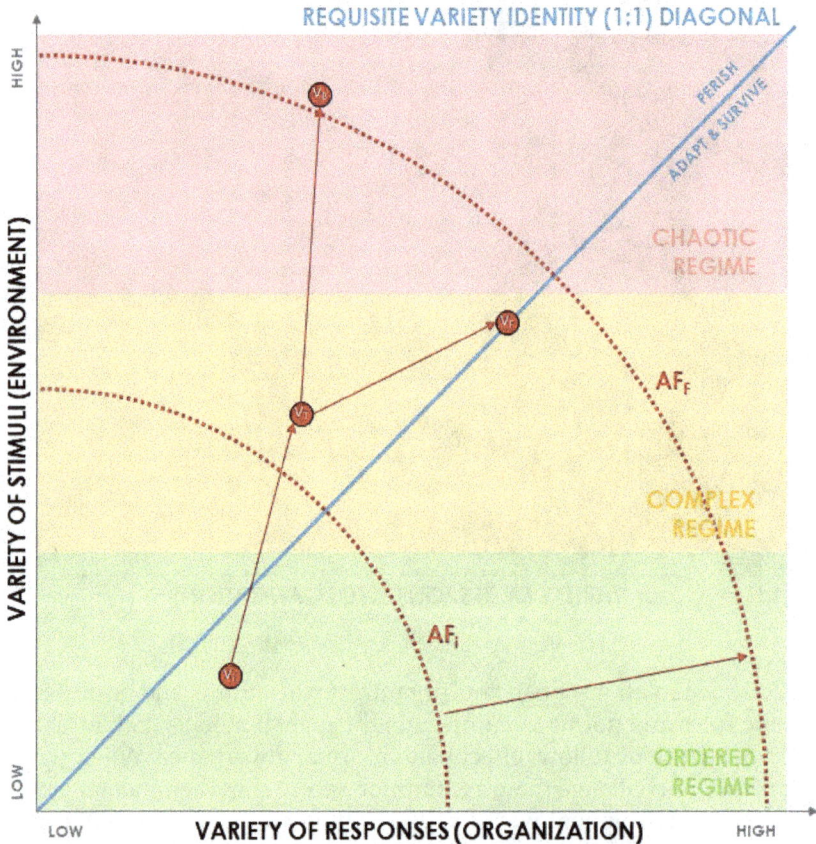

In the Anthropocene, the dichotomy between infrastructure and the environment appears to be shrinking (Chester, Markolf and Allenby, 2019), and as such changes to infrastructure to increase adaptive capacity should be expected to trigger changes in the environment. This in turn demands new capabilities from the infrastructure. As such variety is subject to both infrastructure and environmental co-adaptation over time, a feedback loop that necessitates a process of sustained adaptation.

Infrastructure systems must effectively engage complexity and this will require a restructuring of how they make sense of changing conditions, provide space for the systems to adapt, and restructure their mental models. We call this process Infrastructure Autopoiesis: the ability of infrastructure systems to maintain themselves in the face of growing complexity by creating the knowledge, processes, and technologies necessary to engage environment complexity.

2.5.3. ENGAGING COMPLEXITY

We contend that to meaningfully engage with the growing complexity of the Anthropocene infrastructure systems need to 1) have a sufficient repertoire to respond to growing environment variety, and that 2) necessitates restructuring how organizations generate knowledge and make sense of change. In doing so infrastructure systems will need to adopt autopoietic capabilities, i.e., the ability to self-maintain through restructuring in changing environments. These capabilities will need to be primarily rooted in how the organization generates and acts on knowledge.

We must rethink infrastructure systems as knowledge enterprises with the capabilities necessary to constantly assess change, test paths forward, and restructure leadership, education, and technologies in response to change. We must first recognize that non-competitive environments, while necessary for public service, create disincentives for engaging with complexity.

Infrastructure systems, particularly those aligned with public services, are predicated on conditions associated with non-competitive environments. A city's water utility does not need to concern itself with the prices of competitors or being pushed out of the market by alternatives. The non-competitive environment emerges from regulation and legislation where providers of lifeline infrastructure services are established to provide services that deliver positive externalities, and where determining a fair price is infeasible as no market mechanism exists to determine an individual's willingness to pay (UK Institute for Government, 2015). We argue that this non-competitive environment in which infrastructure as natural monopolies are designed, operated, and governed drives much of our system's rigidity. Without concern that the infrastructure as an organization will fail (e.g., the public cannot do without water), public agencies have fewer incentives to assess their organization's viability into the future. This perspective appears increasingly problematic in a rapidly changing environment, where the conditions of survivability are rapidly changing and the rapid integration of cyber-technologies into legacy infrastructure is creating opportunities for new players to control some aspects of how public services are produced, managed, and consumed.

2.5.3.1. INFRASTRUCTURE AUTOPOIESIS

In the Anthropocene, infrastructure systems will need to be able to make sense of change and restructure to respond. Autopoiesis is generally discussed in the context of natural or biological systems where competition and survivability are contingent on adaptive traits in response to changing conditions (Maturana and Varela, 1980). Infrastructure as a process of transforming, e.g., water supply or energy sources into potable water and electricity, is allopoietic in that the system produces something other than itself. But when governance as knowledge processes is introduced into the infrastructure as a system then we contend that autopoiesis becomes possible. In the Anthropocene, infrastructure must be primarily thought of as knowledge enterprises, and secondarily amalgamations of hardware (Chester, Miller and Muñoz-Erickson, 2020). Sensemaking is an organizational process that underlies infrastructure cognition, the ability to acquire knowledge about changing conditions, what your systems are and can do, and how to pivot in response to change (Miller and Muñoz-Erickson, 2018).

Infrastructure Autopoietic capabilities should be in support of closing the gap between the complexity of the environment and the complexity of the organization's repertoire (adaptive tension) (Maguire and Mckelvey, 1999). At the edge of chaos -- the boundary between order and disorder in a system -- organizations tend to emphasize creativity, agility, flexibility, and innovation (Packard, 1988; Langton, 1997; Levy, 2000; Porter, 2006; Lambert, 2020). This is because to avoid chaos organizational sub-systems must accelerate interactions and restructuring (akin to near-decomposability, discussed later). Infrastructure systems appear to exhibit structures and interactions that assume long-term stability and are poorly suited for increasing complexity (Chester, Miller and Muñoz-Erickson, 2020). The threat of and flirtation with chaos produces self-organizing criticality (Bak, 1996), organizations that are able to reflect on how change is occurring and what they need to do to remain viable.

2.5.3.2. AGILITY AND FLEXIBILITY AS ENGAGING COMPLEXITY

We contend that agility and flexibility are the capabilities of infrastructure systems to adapt and transform by operating at the edge of chaos where they negotiate between exploitative (efficiency) and explorative (resilience) activities to generate an internal repertoire that can respond to changing external complexity. In the past, we've framed agile and flexible infrastructure based on characteristics seen in other industries that have shown a propensity to adapt. These characteristics included technical (compatibility, connectivity, modularity, software-for-hardware substitution), governance (road mapping, design for obsolescence, organic cultures that emphasize change), and education (transdisciplinary) dimensions (Chester and Allenby, 2018). We've described how these characteristics support sensemaking, the ability of infrastructure systems to recognize and keep pace with change (Chester and Allenby, 2021). We argue here that these characteristics fundamentally describe capabilities associated with sensemaking and generating an internal repertoire to adapt; they are an output of an infrastructure system generating complexity. Going forward we discuss what it means for infrastructure to operate at the edge of chaos, in self-organized criticality where the tension between exploitation and exploration of resources is navigated. To do so we frame four tenets that appear frequently in the Complexity literature (Lawrence and Lorsch, 1967; Thompson, 1967; Henderson and Clark, 1990; March, 1991; Carroll and Burton, 2000; Lichtenstein *et al.*, 2007; Sutherland and Woodroof, 2009; Garud, Gehman and Kumaraswamy, 2011; Woods and Hollnagel, 2017). Each of these tenets is described in detail including examples from infrastructure, and Figure 2.5.5 provides an overview.

Figure 2.5.5: Tenets for Infrastructure to Engage Complexity.

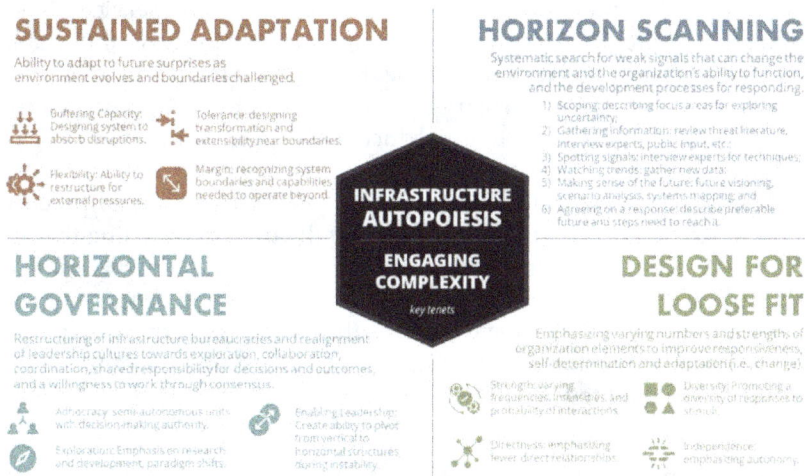

Tenets adapted from (Lawrence and Lorsch, 1967; Carroll and Burton, 2000; Beekun and Glick, 2001; Sutherland and Woodroof, 2009; Woods and Hollnagel, 2017).

2.5.3.3. TENET 1: SUSTAINED ADAPTATION

Sustained Adaptation describes an organization's commitment to change in the face of destabilizing conditions often marked by uncertainty (Woods, 2015), and should become a guiding principle for engaging complexity. Infrastructure as governance and technologies are obdurate, with momentum that carries them along established lines of development (Hughes, 1983; Sovacool, Lovell and Ting, 2018). This path-dependency results from organizational structures that emphasize efficiency over innovation, commitments to technologies with long lifetimes, assumptions of relatively stable operating conditions, and financial and regulatory processes that were structured for past goals (Arthur, 1989; Payo *et al.*, 2016). Massive contingencies are required to disrupt momentum (Hughes, 1983; Sovacool, Lovell and Ting, 2018). On the contrary, sustained adaptation recognizes that over the life cycle boundary conditions will be challenged, conditions and contexts of use will change, adaptation efforts will at times fall short requiring innovation, and systems will need to stretch when their boundary conditions are exceeded (extensibility) (Woods, 2015). The notion that today's infrastructure (a product of legacy technologies and goals) will evolve when needed and will continue to remain viable, without transformative leadership, is foolhardy at best. Instead, we should create the conditions for sustained adaptation to establish as a unifying perspective.

Sustained adaptation will require making space for innovation, and allowing infrastructure governance to renegotiate roles in what the systems do and how they interact with nonincumbent (disruptive) services. It will require a commitment to scanning, asking what if, but most importantly creating the conditions for the organization to shift form and goals. This will require pivoting infrastructure from processes that emphasize reduction of complexity to those that produce a repertoire (variety) to engage with it. Organizations struggle to engage with complexity for several reasons (Garud, Gehman and Kumaraswamy, 2011). Institutional practices can lock people into "thought worlds" by reducing or governing their interactions and dampening creative dialogues (Dougherty, 1992; Kanigel, 2005), emphasize short-term performance metrics at the expense of nurturing long-term ideas, or have cultures that do not benefit from innovative experiences (Tushman and O'Reilly, 1996). Sustained adaptation will require negotiating tensions between innovation and process that attenuate complexity.

Infrastructure leadership for engaging complexity should take the perspective of investing in processes that generate novelty. Early organizational complexity leadership focused on generating innovation during periods of stability (Cyert and March, 1992) or developing innovation in separate organizational units (Tushman and Nadler, 1978; Tushman and O'Reilly, 1996; Benner and Tushman, 2003). These approaches have since been deemed insufficient for environments of constant change where innovation needs to happen continually for organizational survival (Garud, Gehman and Kumaraswamy, 2011). They require organizations to reorient frequently at heavy cost, strain management attention, and as new stakeholders and needs are formed produce a divergence between the change the organization perceives and what is actually happening in the environment (Henderson and Clark, 1990). Instead, organizations should seek to invest in reaching critical thresholds where change must occur, by investing in dynamic activities and teams as well and their connections (Lichtenstein *et al.*, 2007). In doing so the "scaffolding" for innovation is created, and when the organization confronts change its requisite variety is more equipped to match that of the environment. Investments in interactions across the workforce and possibly broader stakeholders are critical, and the rules that govern those interactions should be flexible (Drazin and Sandelands, 1992; Axelrod and Cohen, 2008). The rules that govern interactions should provide space for improvisation and spontaneity (Stacey, 2003). Fundamentally, infrastructure organizations should engage complexity not as something to be analyzed but instead as an unfolding practice that warrants continual investment (Garud, Gehman and Kumaraswamy, 2011). This investment will require a leadership change.

2.5.3.4. TENET 2: HORIZONTAL GOVERNANCE

Sustained adaptation will require the restructuring of infrastructure bureaucracies and the realignment of leadership cultures. Research and development should become central to what infrastructure agencies do, and it should focus not just on normal science (research within current thought paradigms), but on paradigm shifts. Infrastructure agencies should recognize that their legacy roles as natural monopolies are being disrupted and that control of how services are delivered and consumed is rapidly becoming distributed (e.g., solar panels and home batteries, sharing economy services, smart thermostats, big-tech cloud-based navigation), resulting in mismatches between what the organization does and the routines it uses (Henderson and Clark, 1990; Helmrich *et al.*, 2021). Instead of full control, infrastructure agencies should begin restructuring themselves as coordinators of multiple players affecting services and driving appropriate change.

Infrastructure is typically structured as divisional bureaucracies with an emphasis on vertical management (up and down layers of the hierarchy) and control, where expertise is siloed and the organization emphasizes efficiency (exploitation) as assessed by measurable performance goals (Chester, Miller and Muñoz-Erickson, 2020). This rigid structure makes shifting what the organization does challenging (Mintzberg, 1979) and adaptation to disruptive change difficult. To manage in the face of complexity near-decomposability of organizational units becomes critical. The theory of near-decomposability describes how when systems are perturbed, the smallest sub-units in a hierarchy, if given independence, tend to evolve faster and find stable solutions (Simon, 2002). We argue that in the pervasive models that dominate infrastructure bureaucracies today, subunits (e.g., sub-divisions) are not decomposable as they're heavily controlled by management (Chester, Miller and Muñoz-Erickson, 2020). This top-down control limits the ability of the organization to restructure as chaos ensues by retarding the speed at which the organization needs to respond to a fast-changing environment.

Infrastructure organizations should focus on providing flexibility for self-organization. They should restructure bureaucracies and leadership towards a horizontal governance model where teams coalesce around tasks instead of divisional focus, and autonomy is given to the teams with appropriate resources to scan, test, and make decisions. This requires a pivot from the conventional divisional bureaucracy to an adhocracy where differentiation is encouraged (Mintzberg, 1979; Carroll and Burton, 2000; Chester, Miller and Muñoz-Erickson, 2020). Instead of leadership focused on trying to make sense of changing conditions, they should instead focus on providing the integrative pieces for teams (Thompson, 1967; Galbraith, 1977; Gresov, 1989). The separating out of activities means that the organization as a whole is better capable of responding to specific challenges in the environment (Lawrence and Lorsch, 1967). In an adhocracy, teams "comprised of modular units that are partially connected to each other avoid the trap of too much standardization (which creates obstacles to change) and too little coordination (which results in chaos)" (Perrow, 1985; Brown and Eisenhardt, 1998; Carroll and Burton, 2000). The self-organization is loose-coupling (semi-autonomy) that prevents the organization from overreacting to environmental change while increasing the capability of the organization to appropriately and timely react to change (Weick, 1976).

Comprehensive reorganization is likely a longer-term proposition and in the short term, infrastructure organizations can consider structural ambidexterity, a separate structure within an organization committed to exploration (Raisch and Birkinshaw, 2008). The "skunkworks" satellite structure involves an autonomous sub-division tasked with innovating beyond what the mainstream organization is capable of (Donada, Mothe and Vidal, 2021).

U.S. agencies have operated with horizontal governance models, a result of a mandated need for innovation. Examples of public infrastructure agencies operating with horizontal governance models were not identified, likely due to political accountability best served in vertical governance models, and that the divisional bureaucratic form that dominates infrastructure agencies was purposefully adopted from the railroads (Chester, Miller and Muñoz-Erickson, 2020). Created in 1958, the Defense Advanced Research Projects Agency's (DARPA) goal is to conduct revolutionary and high-payoff research with no specific operational mission. To privilege action and exploration over positional authority, DARPA is structured in a horizontal governance model (specifically, an adhocracy), where transient teams and mobile individuals are brought together to assess U.S. security holes, later to be disbanded so that new teams can form around different challenges (Nieto-Gómez, 2011). DARPA accomplishments include major contributions to stealth technologies, the internet, GPS, and unmanned aerial vehicles.

2.5.3.5. TENET 3: HORIZON SCANNING

Infrastructure organizations must restructure how they make sense of information about the environment. As previously discussed, infrastructure organizations take in information about the environment attenuating it based on processes and filters designed for normative and generally legacy goals. The attenuated environmental information is analyzed by operations (people and technology) who pass on a further attenuated information stream to managers to make strategic decisions. If the information processing systems used by the organization are not structured for new and changing conditions then they can be expected to miss or misinterpret critical information about the environment that may affect the viability of the organization. Horizon scanning is the process of systematically searching the environment for opportunities and threats (Sutherland and Woodroof, 2009).

Horizon scanning is not simply future thinking; it is the systematic search for weak signals that can change the environment and the organization's ability to function, and the development processes for responding. Infrastructure organizations today are scrambling to respond to climate change and cyberattacks, two destabilizing challenges that have for decades been slowly boiling. Horizon scanning involves: (i) scoping (describing focus areas for exploring uncertainty); (ii) gathering information (review threat literature, interview experts, public input); (iii) spotting signals (interview experts for techniques); (iv) watching trends (gather new data); (v) making sense of the future (future visioning, scenario analysis, systems mapping); and (vi) agreeing on a response (describe preferable future and steps need to reach it) (Sutherland and Woodroof, 2009). The goal of horizon scanning is to create plans and strategies that are agile and flexible to a variety of future conditions to aid decision making towards organizational change. It is not simply to pontificate about or to try to predict the future.

A horizon scan developed with the stormwater and wastewater management communities revealed poor preparedness for increasing integration of cybertechnologies into assets and how third parties may increasingly use data from public services (Blumensaat *et al.*, 2019). Related, a horizon scanning exercise developed for water resources in Russia that identified trends, weak signals, wild cards, and uncertainties identified potential cross-sector demand conflicts and the need for circular economy planning to conserve resources (Saritas and Proskuryakova, 2017).

2.5.3.6. TENET 4: DESIGN FOR LOOSE FIT

Infrastructure change should embrace loose fit design where assets, processes, and solutions are given the flexibility to work independently or as part of a larger group. Typically, tight fit solutions are deployed for infrastructure where sub-systems have rigidly predetermined functions and interactions with other parts of the system. Put simply, loose fit design enables change, so when the environment or demand changes the asset can pivot to meet the new need. However, underlying loose fit design is the notion of negotiating between exploitation (efficiency) and exploration (innovation) (March, 1991).

Creativity flourishes at the edge of chaos because loose fit provides opportunities for exploration, while not being so unstructured as to leak vital information. Infrastructure systems appear to be organized around principles of tightness; they take comfort in the traditional, there is minimal innovation, and they are slow to adjust to changes in their environments. On the contrary, there are industries that constantly innovate as their markets demand so, e.g., Silicon Valley (Pancs, 2017). Loose and tight coupling is not simply about innovation, but more fundamentally how connections between sub-systems are structured and function (Weick, 1976; Orton and Weick, 1990). Infrastructure are coupled systems with distinct elements (people, hardware, and software) governed by formal and informal rules that govern the responsiveness of elements to each other. Traditionally, infrastructure limits the distinctiveness of elements to pursue standardization; they emphasize tight couplings. Errors of tightness can emerge when organizations constrain their decision-making capacity (and as such choice), which can stem from the specialization that comes with being too large. This can produce organizations that are too tight for their technical and strategic contexts (Tashkun *et al.*, 1998). Looseness emphasizes both responsiveness and distinctiveness, where elements are given room for self-determination and localized adaptation (Orton and Weick, 1990). Looseness is not always preferable. When an organization institutes structures that are too fuzzy then decision making and efficiency can be lost due to power that is too diffuse and nobody with authority to make decisions (Tashkun *et al.*, 1998).

Loosely coupled systems are desirable in certain high-risk infrastructure settings. Perrow (1985) describes how loosely coupled organizational decision-making processes are more likely to contain sensitive areas should an accident take place. In contrast, tight coupling may make things worse as rigid decisions are more likely to be made despite not having a comprehensive view of the nature of the problem.

Loose fit solutions are needed to respond to growing environment complexity. With increased autonomy, loose fit solutions provide greater sensitivity of elements to sense environment changes. With tight fit solutions elements are more constrained by their organizations or by other elements (interdependencies) and their ability to sense is limited (Weick, 1976). The balance between looseness and tightness is driven by Ashby's Law of Requisite Variety; the organization needs sufficient looseness of elements to produce a repertoire that is as large as or exceeds that of the environment.

2.5.4. CONCLUSIONS

The inability to engage with complexity can be expected to result in a decoupling between what our infrastructure systems can do and what we need them to do, and autopoietic capabilities close this gap by creating the conditions where a sufficient repertoire can emerge. This decoupling does not necessarily mean a rapid progression to irrelevance. On the contrary, the obdurate and ossified nature of infrastructure technologies and our normative expectations of the institutions that govern them will support their persistence for some time. However, the decoupling will likely create space for innovation by third parties (often driven by advances in connected technologies), and these parties can be expected to exert more and more control over aspects of services, adding complexity to an increasingly complex environment. Instead of trying to hold back these parties (imagine a transportation agency telling Google or Apple that they can no longer navigate traffic through a road network), infrastructure agencies should embrace the increasingly diverse stakeholder environments, and seek to establish their organizations as consensus builders. And infrastructure agencies should restructure towards a model sustained adaptation, with horizontal governance increasingly used as a bureaucratic structure, with horizon scanning that shifts how they make sense of their environment, and around loose fit design. Fundamentally, infrastructure leadership should recognize that changing conditions into the future are likely to represent a new paradigm for the basic and critical services that future generations will rely upon, and rapid innovation is needed to reposition their organizations to thrive.

2.5.5. REFERENCES

Allenby, B. (2007) 'Earth Systems Engineering and Management: A Manifesto', *Environmental Science & Technology*, 41(23), pp. 7960–7965. Available at: https://doi.org/10.1021/es072657r.

Allenby, Brad and Chester, M. (2021) 'Learning From Engineers', *Issues in Science and Technology* [Preprint]. Available at: https://issues.org/learning-from-engineers/ (Accessed: 12 October 2023).

Allenby, Braden and Chester, M. (2021) *The Rightful Place of Science: Infrastructure in the Anthropocene*. Tempe, AZ: Consortium for Science, Policy & Outcomes.

Allenby, B.R. (2012) *The Theory and Practice of Sustainable Engineering*. Hoboken, NJ: Upper Saddle River: Pearson Prentice Hall.

Arthur, W.B. (1989) 'Competing Technologies, Increasing Returns, and Lock-In by Historical Events', *The Economic Journal*, 99(394), pp. 116–131. Available at: https://doi.org/10.2307/2234208.

Ashby, W.R. (1956) *An Introduction to Cybernetics*. Hoboken, NJ: J. Wiley. Available at: https://doi.org/10.5962/bhl.title.5851.

Axelrod, R. and Cohen, M.D. (2008) *Harnessing Complexity*. New York, NY: Basic Books.

Bak, P. (1996) *How Nature Works: The Science of Self-organized Criticality*. Heidelberg, DE; New York, NY; Dordrecht, UK; London, UK: Springer.

Bassett, T.J. and Fogelman, C. (2013) 'Déjà vu or something new? The adaptation concept in the climate change literature', *Geoforum*, 48, pp. 42–53. Available at: https://doi.org/10.1016/j.geoforum.2013.04.010.

Beekun, R. and Glick, W. (2001) 'Organization Structure from a Loose Coupling Perspective: A Multidimensional Approach*', *Decision Sciences*, 32, pp. 227–250. Available at: https://doi.org/10.1111/j.1540-5915.2001.tb00959.x.

Beer, S. (1979) *The Heart of Enterprise*. Hoboken, NJ: Wiley.

Beer, S. (1981) *Brain of the Firm*. Hoboken, NJ: Wiley.

Beer, S. (1985) *Diagnosing the system for organizations*. Chichester, UK: Wiley (Managerial cybernetics of organization). Available at: https://bac-lac.on.worldcat.org/oclc/300591269 (Accessed: 5 November 2023).

Benner, M. and Tushman, M. (2003) 'Exploitation, Exploration, and Process Management: The Productivity Dilemma Revisited', *The Academy of Management Review*, 28, pp. 238–256. Available at: https://doi.org/10.5465/amr.2003.9416096.

Blumensaat, F. *et al.* (2019) 'How Urban Storm- and Wastewater Management Prepares for Emerging Opportunities and Threats: Digital Transformation, Ubiquitous Sensing, New Data Sources, and Beyond - A

Horizon Scan', *Environmental Science & Technology*, 53(15), pp. 8488–8498. Available at: https://doi.org/10.1021/acs.est.8b06481.

Boisot, M. and Mckelvey, B. (2011) 'Complexity and organization–environment relations: Revisiting Ashby's law of requisite variety', in, pp. 279–298. Available at: https://doi.org/10.4135/9781446201084.n16.

Brown, S.L. and Eisenhardt, K.M. (1998) *Competing on the Edge: Strategy as Structured Chaos*. Harvard Business Press.

Carroll, T. and Burton, R.M. (2000) 'Organizations and Complexity: Searching for the Edge of Chaos', *Computational & Mathematical Organization Theory*, 6(4), pp. 319–337. Available at: https://doi.org/10.1023/A:1009633728444.

Chester, M., Markolf, S. and Allenby, B. (2019) 'Infrastructure and the environment in the Anthropocene', *Journal of Industrial Ecology*, 23(5), pp. 1006–1015. Available at: https://doi.org/10.1111/jiec.12848.

Chester, M.V. and Allenby, B. (2018) 'Toward adaptive infrastructure: flexibility and agility in a non-stationarity age', *Sustainable and Resilient Infrastructure*, 4, pp. 1–19. Available at: https://doi.org/10.1080/23789689.2017.1416846.

Chester, M.V. and Allenby, B. (2019) 'Infrastructure as a wicked complex process', *Elementa: Science of the Anthropocene*. Edited by A. Iles and M.E. Chang, 7(1), p. 21. Available at: https://doi.org/10.1525/elementa.360.

Chester, M.V. and Allenby, B. (2021) 'Toward adaptive infrastructure: the Fifth Discipline', *Sustainable and Resilient Infrastructure*, 6(5), pp. 334–338. Available at: https://doi.org/10.1080/23789689.2020.1762045.

Chester, M.V. and Allenby, B.R. (2020) 'Perspective: The Cyber Frontier and Infrastructure', *IEEE Access*, 8, pp. 28301–28310. Available at: https://doi.org/10.1109/ACCESS.2020.2971960.

Chester, M.V., Miller, T. and Muñoz-Erickson, T.A. (2020) 'Infrastructure governance for the Anthropocene', *Elementa: Science of the Anthropocene*, 8(1), p. 78. Available at: https://doi.org/10.1525/elementa.2020.078.

Chester, M.V., Underwood, B.S. and Samaras, C. (2020) 'Keeping infrastructure reliable under climate uncertainty', *Nature Climate Change*, 10(6), pp. 488–490. Available at: https://doi.org/10.1038/s41558-020-0741-0.

Cyert, R.M. and March, J.G. (1992) *Behavioral Theory of the Firm*. Hoboken, NJ: Wiley.

Donada, C., Mothe, C. and Vidal, J. (2021) 'Managing skunkworks to achieve ambidexterity: The Robinson Crusoe effect', *European Management Journal*, 39. Available at: https://doi.org/10.1016/j.emj.2020.07.008.

Dougherty, D. (1992) 'Interpretive Barriers to Successful Product Innovation in Large Firms', *Organization Science*, 3(2), pp. 179–202. Available at: https://doi.org/10.1287/orsc.3.2.179.

Drazin, R. and Sandelands, L. (1992) 'Autogenesis: A Perspective on the Process of Organizing', *Organization Science*, 3(2), pp. 230–249. Available at: https://doi.org/doi:10.1287/orsc.3.2.230.

Edwards, P.N. (2017) 'Knowledge infrastructures for the Anthropocene', *The Anthropocene Review*, 4(1), pp. 34–43. Available at: https://doi.org/10.1177/2053019616679854.

Galbraith, J.R. (1977) *Organization Design*. Boston, MA: Addison-Wesley Publishing Company.

Garud, R., Gehman, J. and Kumaraswamy, A. (2011) 'Complexity Arrangements for Sustained Innovation: Lessons From 3M Corporation', *Organization Studies*, 32(6), pp. 737–767. Available at: https://doi.org/10.1177/0170840611410810.

Gell-Mann, M. (2002) 'What Is Complexity?', in A.Q. Curzio and M. Fortis (eds). Heidelberg, DE: Physica-Verlag HD, pp. 13–24. Available at: https://doi.org/10.1007/978-3-642-50007-7_2.

Gresov, C. (1989) 'Exploring Fit and Misfit with Multiple Contingencies', *Administrative Science Quarterly*, 34(3), pp. 431–453. Available at: https://doi.org/10.2307/2393152.

Hayward, P. (2004) 'Facilitating foresight: where the foresight function is placed in organisations', *Foresight*, 6(1), pp. 19–30. Available at: https://doi.org/10.1108/14636680410699115.

Helmrich, A. *et al.* (2021) 'Centralization and decentralization for resilient infrastructure and complexity', *Environmental Research: Infrastructure and Sustainability*, 1(2), p. 021001. Available at: https://doi.org/10.1088/2634-4505/ac0a4f.

Helmrich, A.M. and Chester, M.V. (2020) 'Reconciling complexity and deep uncertainty in infrastructure design for climate adaptation', *Sustainable and Resilient Infrastructure*, 7(2), pp. 83–99. Available at: https://doi.org/10.1080/23789689.2019.1708179.

Henderson, R. and Clark, K. (1990) 'Architectural Innovation: The Reconfiguration of Existing Product Technologies and the Failure of Established Firms', *Administrative science quarterly*, 35(1), pp. 9–30. Available at: https://doi.org/10.2307/2393549.

Hetherington, K. (2018) *Infrastructure, Environment, and Life in the Anthropocene.* Durham, NC: Duke University Press.

Hughes, T.P. (1983) *Networks of Power: Electrification in Western Society, 1880-1930.* Baltimore, MD: Johns Hopkins University Press. Available at: https://www-fulcrum-org.ezproxy1.lib.asu.edu/epubs/3j333441h?locale=en#page=23.

IPCC (2014) *Climate Change 2014: Synthesis Report. Contribution of Working Groups I, II and III to the Fifth Assessment Report of the Intergovernmental Panel on Climate Change.* 9789291691432. Geneva, Switzerland: UN International Panel on Climate Change, p. 151.

Joerges, B. (1989) 'Large Technical Systems: Concepts and Issues', in *The Development Of Large Technical Systems.* Oxfordshire, UK: Routledge, pp. 9–36.

Kanigel, R. (2005) *The One Best Way: Frederick Winslow Taylor and the Enigma of Efficiency.* Cambridge, MA: MIT Press.

Kim, Y. *et al.* (2019) 'The Infrastructure Trolley Problem: Positioning Safe-to-fail Infrastructure for Climate Change Adaptation', *Earth's Future*, 7(7), pp. 704–717. Available at: https://doi.org/10.1029/2019EF001208.

Lambert, P. (2020) 'The Order-Chaos Dynamic of Creativity', *Creativity Research Journal*, 32(4), pp. 431–446. Available at: https://doi.org/10.1080/10400419.2020.1821562.

Langton, C.G. (1997) *Artificial Life.* Cambridge, MA: MIT Press. Available at: https://mitpress.mit.edu/9780262621120/artificial-life/ (Accessed: 18 October 2023).

Lawrence, P. and Lorsch, J. (1967) 'Differentiation and Integration in Complex Organizations', *Administrative Science Quarterly*, 12(1), pp. 1–47. Available at: https://doi.org/10.2307/2391211.

Levy, D. (2000) 'Applications and Limitations of Complexity Theory in Organization Theory and Strategy', *Public Adm. Public Policy*, 79, pp. 67–87. Available at: https://doi.org/10.4324/9781482270259-3.

Lichtenstein, B.B. *et al.* (2007) 'Complexity Dynamics of Nascent Entrepreneurship', *Journal of Business Venturing*, 22(2), pp. 236–261. Available at: https://doi.org/10.1016/j.jbusvent.2006.06.001.

Lohn, A.J. (2017) *What's the Buzz?: The City-Scale Impacts of Drone Delivery.* RAND Corporation. Available at: https://www.rand.org/pubs/research_reports/RR1718.html (Accessed: 21 December 2021).

Maguire, S. and Mckelvey, B. (1999) 'Complexity and Management: Moving From Fad To Firm Foundations', *Emergence*, 1(2), pp. 19–61. Available at: https://doi.org/10.1207/s15327000em0102_3.

March, J.G. (1991) 'Exploration and Exploitation in Organizational Learning', *Organization Science*, 2(1), pp. 71–87. Available at: https://doi.org/10.1287/orsc.2.1.71.

Markolf, S.A. *et al.* (2018) 'Interdependent Infrastructure as Linked Social, Ecological, and Technological Systems (SETSs) to Address Lock-in and Enhance Resilience', *Earth's Future*, 6(12), pp. 1638–1659. Available at: https://doi.org/10.1029/2018EF000926.

Maturana, H.R. and Varela, F.J. (1980) *Autopoiesis and Cognition: The Realization of the Living*. Dordrecht, NL: Springer Netherlands.

McPhearson, T. *et al.* (2021) 'Radical changes are needed for transformations to a good Anthropocene', *npj Urban Sustainability*, 1(1), pp. 1–13. Available at: https://doi.org/10.1038/s42949-021-00017-x.

Miller, C.A. and Muñoz-Erickson, T.A. (2018) *The Rightful Place of Science: Designing Knowledge*. Tempe, AZ: Consortium for Science, Policy & Outcomes.

Miller, T., Chester, M. and Munoz-Erickson, T. (2018) 'Rethinking Infrastructure in an Era of Unprecedented Weather Events', *Issues in Science and Technology*, 34(2), pp. 46–58.

Mintzberg, H. (1979) *The Structuring of Organizations: A Synthesis of the Research*. Hoboken, NJ: Prentice-Hall.

Naughton, J. (2017) *Ashby's Law of Requisite Variety, Edge.org*. Available at: https://www.edge.org/response-detail/27150 (Accessed: 8 July 2021).

Nieto-Gómez, R. (2011) 'The Power of "the Few": A Key Strategic Challenge for the Permanently Disrupted High-Tech Homeland Security Environment', *Homeland Security Affairs*, 5 December. Available at: https://www.hsaj.org/articles/50 (Accessed: 9 December 2021).

Norman, J. and Bar-Yam, Y. (2018) 'Special Operations Forces: A Global Immune System?' Edited by A.J. Morales et al., pp. 486–498. Available at: https://doi.org/10.1007/978-3-319-96661-8_50.

Orton, J.D. and Weick, K.E. (1990) 'Loosely Coupled Systems: A Reconceptualization', *Academy of Management Review*, 15(2), pp. 203–223. Available at: https://doi.org/10.5465/amr.1990.4308154.

Packard, N.H. (1988) *Adaptation Toward the Edge of Chaos*. Champaign, IL: University of Illinois at Urbana-Champaign, Center for Complex Systems Research.

Pancs, R. (2017) 'Tight and Loose Coupling in Organizations', *The B.E. Journal of Theoretical Economics*, 17(1). Available at: https://doi.org/10.1515/bejte-2015-0081.

Papachroni, A., Heracleous, L. and Paroutis, S. (2016) 'In pursuit of ambidexterity: Managerial reactions to innovation–efficiency tensions', *Human Relations*, 69(9), pp. 1791–1822. Available at: https://doi.org/10.1177/0018726715625343.

Payo, A. *et al.* (2016) 'Experiential Lock-In: Characterizing Avoidable Maladaptation in Infrastructure Systems', *Journal of Infrastructure Systems*, 22(1), p. 02515001. Available at: https://doi.org/10.1061/(ASCE)IS.1943-555X.0000268.

Perrow, C. (1985) *Normal Accidents: Living With High-risk Technologies*. New York, NY: Basic Books.

Pildes, R.H. (2010) 'Why the Center Does Not Hold: The Causes of Hyperpolarized Democracy in America', *California Law Review*, 99(2), p. 273. Available at: https://doi.org/10.15779/Z38R11D.

Porter, T.B. (2006) 'Coevolution as a Research Framework for Organizations and the Natural Environment', *Organization & Environment*, 19(4), pp. 479–504. Available at: https://doi.org/10.1177/1086026606294958.

Poska, S., Kaplan, A. and Alford, J. (2021) 'Impact of Navigation Applications on Traffic Operations', *Institute of Transportation Engineers (ITE) Journal*, 91(2), pp. 37–42.

Raisch, S. and Birkinshaw, J. (2008) 'Organizational Ambidexterity: Antecedents, Outcomes, and Moderators', *Journal of Management*, 34(3), pp. 375–409. Available at: https://doi.org/10.1177/0149206308316058.

Saritas, O. and Proskuryakova, L.N. (2017) 'Water resources – an analysis of trends, weak signals and wild cards with implications for Russia', *foresight*, 19(2), pp. 152–173. Available at: https://doi.org/10.1108/FS-07-2016-0033.

Simon, H.A. (2002) 'Near decomposability and the speed of evolution', *Industrial and Corporate Change*, 11(3), pp. 587–599. Available at: https://doi.org/10.1093/icc/11.3.587.

Snowden, D.J. and Boone, M.E. (2007) 'A Leader's Framework for Decision Making', *Harvard Business Review*, 1 November, pp. 68–76, 149.

Sovacool, B.K., Lovell, K. and Ting, M.B. (2018) 'Reconfiguration, Contestation, and Decline: Conceptualizing Mature Large Technical Systems'. Available at: https://doi.org/10.1177/0162243918768074.

Stacey, R. (2003) *Complex Responsive Processes in Organizations: Learning and Knowledge Creation*. London, UK: Routledge.

Steffen, W. *et al.* (2015) 'The trajectory of the Anthropocene: The Great Acceleration', *The Anthropocene Review*, 2(1), pp. 81–98. Available at: https://doi.org/10.1177/2053019614564785.

Steffen, W., Crutzen, P.J. and McNeill, J.R. (2016) 'The Anthropocene: Are Humans Now Overwhelming the Great Forces of Nature?', in *THE ANTHROPOCENE: ARE HUMANS NOW OVERWHELMING THE GREAT FORCES OF NATURE?* Berkeley, CA: University of California Press, pp. 440–459. Available at: https://www.degruyter.com/document/doi/10.1525/9780520964297-051/html?lang=en (Accessed: 18 October 2023).

Sutherland, W. and Woodroof, H. (2009) 'The need for environmental horizon scanning', *Trends in ecology & evolution*, 24(10), pp. 523–527. Available at: https://doi.org/10.1016/j.tree.2009.04.008.

Tashkun, S. *et al.* (1998) 'Organizing for Innovation: Loose or Tight Control?', *Long Range Planning*, 31. Available at: https://doi.org/10.1016/S0024-6301(98)00082-X.

The Economist (2021) 'The use of renewable energy is accelerating', 11 May. Available at: https://www.economist.com/graphic-detail/2021/05/11/the-use-of-renewable-energy-is-accelerating (Accessed: 10 December 2021).

Thompson, J.D. (1967) *Organization in Action*. New York, NY: McGraw-Hill.

Tushman, M.L. and Nadler, D.A. (1978) 'Information Processing as an Integrating Concept in Organizational Design.', *Academy of Management Review*, 3(3), pp. 613–624. Available at: https://doi.org/10.5465/amr.1978.4305791.

Tushman, M.L. and O'Reilly, C.A. (1996) 'Ambidextrous Organizations: Managing Evolutionary and Revolutionary Change', *California Management Review*, 38(4), pp. 8–29. Available at: https://doi.org/10.2307/41165852.

UK Institute for Government (2015) *Private vs public markets*. Available at: https://web.archive.org/web/20220121033621/https://www.instituteforgovernment.org.uk/publication/private-vs-public-markets (Accessed: 29 September 2021).

Waters, C. *et al.* (2016) 'The Anthropocene Is Functionally and Strati-graphically Distinct from the Holocene', *Science*, 351(6269), p. 137. Available at: https://doi.org/10.1126/science.aad2622.

Weick, K.E. (1976) 'Educational Organizations as Loosely Coupled Systems', *Administrative Science Quarterly*, 21(1), pp. 1–19. Available at: https://doi.org/10.2307/2391875.

Woods, D.D. (2015) 'Four concepts for resilience and the implications for the future of resilience engineering', *Reliability Engineering & System Safety*, 141, pp. 5–9. Available at: https://doi.org/10.1016/j.ress.2015.03.018.

Woods, D.D. and Hollnagel, E. (2017) *Resilience Engineering: Concepts and Precepts*. Boca Raton, FL: CRC Press.

Zalasiewicz, J. *et al.* (2011) 'The Anthropocene: a new epoch of geological time?', *Philosophical Transactions of the Royal Society A: Mathematical, Physical and Engineering Sciences*, 369(1938), pp. 835–841. Available at: https://doi.org/10.1098/rsta.2010.0339.

2.6

COVID AS A HARBINGER OF
TRANSFORMATION

2.6.1. INTRODUCTION

As COVID-19 propagated rapidly through cities across the world and presented a multitude of public health, social, and economic challenges, a new landscape of issues emerged for infrastructure. Challenges for healthcare systems were more directly evident, but the extended nature of the pandemic changed how cities operated and re-shaped approaches to critical infrastructure (CI) systems (Bliss, 2020b; Florida, 2020). Amid the outbreak, opportunities for reflection, research, and action abounded.

Infrastructure systems have an important role in shaping human activity and supporting public needs during pandemics. Given CI's role in absorbing impacts, maintaining essential services, and facilitating societal adaptation in the face of unforeseen events like COVID-19, how we frame infrastructure resilience is essential. A clear framing of major infrastructure challenges during COVID-19 can help illuminate the research and capacities for resilience in a dynamic and changing world.

Unlike other hazards for which infrastructure managers are more accustomed to planning, pandemics differ significantly mainly in terms of spatio-temporal scale and interdependencies between infrastructure sectors. As opposed to more regionally isolated hazards, modern pandemics can become global-scale phenomena that occur as successive waves unique to different viruses, making them difficult to predict where, how, and to what scale impacts will propagate (Cohen, 2009; Woods, Seager and Alderson, 2020). While there are often no direct physical threats to

This chapter was adapted from the following article with publisher and lead author permissions: Thomaz Carvalhaes, Samuel Markolf, Alysha Helmich, Yeowon Kim, Rui Li, Mukunth Natarajan, Emily Bondank, Nasir Ahmad, and Mikhail Chester, 2020, COVID-19 as a Harbinger of Transforming Infrastructure Resilience, *Frontiers in Built Environment*, 6(148), pp. 1-8, doi: 10.3389/fbuil.2020.00148.

infrastructure, pandemics induce devastating impacts to their sustainment, previously framed as impacts to critical infrastructure workforce (Ryan, 2008; Hessel and Group, 2009; Dietz and Black, 2012). Infrastructure planning must consider sustaining monitoring and response mechanisms for months to years amid economic disruption and workforce challenges (i.e., infected laborers and social distancing), conditions not adequately addressed in national response plans (Dietz and Black, 2012). Measures to "flatten the curve" for the number of cases can increase the capacity for systems to absorb impacts in terms of systems functionality curves (i.e., resilience curves), which differ between sectors (e.g., communications, healthcare, power and water) (Jovanović *et al.*, 2020). While studies have long highlighted alarming gaps in preparedness (Osterholm, 2005; Adalja *et al.*, 2012), hospitals also depend on CI, as for example, all modern medicine depends on electrical systems in some way (Osterholm and Kelley, 2009). Yet, planning often doesn't consider such complexities, interdependencies, and second and third order effects (e.g., supply chains for PPE) (Itzwerth *et al.*, 2006; Huff *et al.*, 2015). CI is not only vulnerable, but responsible for on-going support and adaptation (Hendrickson and Rilett, 2020).

While devastating pandemics have happened in the past, infrastructure impacts during pandemics are not well understood (Williams, 2007; Ryan, 2008; Cohen, 2009). COVID-19 represented an emerging risk (i.e., previously not widely considered), which posed a significant challenge to resilience because knowledge and information were vague or missing, the maturity of risk management was low, and regulatory frameworks were missing or inconsistent (Jovanović *et al.*, 2020). Moreover, these challenges were exacerbated and accelerated by the changing relationship between people and their environments in increasingly complex social (e.g., norms, urbanization, international travel, and trade), ecological (e.g., climate change), political (e.g., public health breakdowns, land-use policy), and built environment systems (Bogich *et al.*, 2012; Bedford *et al.*, 2019). The coupled evolution of human, built, and ecological systems presents new levels of complexity, rapidity, and scales for hazards such as pandemics and their impacts (Chester and Allenby, 2020).

COVID-19 revealed four major themes that warrant examination: (1) Planning for concurrent hazards, (2) Flexibility in how we assess the criticality of infrastructure, (3) Managing trade-offs between efficiency and resilience, and (4) Expanding institutional resilience to include leadership for both stable and unstable conditions. These competencies are in line with broader challenges for infrastructure in the Anthropocene (Chester and Allenby, 2019). We discuss these four themes with the goal of identifying future research pathways and areas for more comprehensive treatment toward guiding resilient infrastructure design and policies for a future characterized by accelerating, increasingly uncertain, and increasingly complex conditions.

2.6.2. THEME 1: PLANNING FOR CONCURRENT HAZARDS

We are entering an era of concurrent crises where global connectivity enables the propagation of shocks through interdependent critical infrastructure systems (Biggs *et al.*, 2011). In Puerto Rico, the pandemic coincided with long-term recovery efforts related to Hurricane Maria, frequent and intense earthquakes due to a newly discovered fault line and an ongoing drought (Rosa and Robles, 2020; USGS, 2020). Unlike past disasters, widespread unemployment and limitations for social aggregation caused by the pandemic undercut capacities for disaster resilience (e.g., income, health insurance, shelters) (Rosa and Robles, 2020). Amid Cyclone Amphan in India and Bangladesh, floods impacted health centers and other CI, while compliance to evacuation protocols was challenged by public fear of COVID-19 (Ellis-Petersen and Ratcliffe, 2020; Gettleman *et al.*, 2020; Okura *et al.*, 2020).

As COVID-19 extended over years, the pandemic increasingly overlapped with other hazards. In summer of 2020 many cities had to manage infrastructure under COVID-19 along with other events that threatened public safety including extreme heat (Anderson *et al.*, 2018; Calma, 2020), wildfires (DOI, 2020), floods (Einhorn, 2020; Zhong, 2020), hail (Cappucci, 2020), and hurricanes (Vann, 2020). Infrastructure in disrepair will fail, and cyberattacks have increased (Chester and Allenby, 2020). Traditional infrastructure responses and operations to extreme events were complicated by the scale and scope of a global pandemic, and recovery efforts were challenged by the shortage of resources and difficulties in safely operating rescue protocols.

Unlike most hazards which are local in nature, COVID-19 was global and presented an opportunity for developing knowledge systems to prepare our critical systems under shared goals (Sarewitz, 2020). Emergency response often assumes that non-affected regions are capable of supporting recovery efforts through the supply chain of goods, backup labor, and mobilization of infrastructure services. However, cooperative recovery efforts were challenged due to the global scale, urgency, and uncertainty of the pandemic (Ryan, 2008; Mogul and Hurt, 2020; Villarreal, 2020). For example, the U.S. wildfire season is usually combated by local, regional, and international firefighting crews. Due to travel restrictions, local crews had to reassess how they prioritized responses including controlled burns and virtual fire risk assessments (Gibbens, 2020; McDowell, 2020).

Cities had to develop creative ways to provide public infrastructure and community services while tackling hazards. Innovative responses highlighted the importance of infrastructure resilience. Cooling centers and congregational spaces for vulnerable heat populations were initially closed to enforce social distancing. As extreme heat events unfolded, cities began leveraging multifunctionality by placing vulnerable populations in hotel rooms, re-opening conventional cooling centers (e.g. libraries, parks, splash pads), and adapting venues (e.g., sporting facilities, stationary buses) (Chilukuri, 2020; City of Chicago, 2020; Flavelle, 2020). Functionally redundant systems were also being implemented such as utility financial assistance, utility shut-down restrictions, and provision of air conditioning units (CDC, 2020a; NYC, 2020). Creatively leveraging capacities that enable infrastructure flexibility can aid in shifting infrastructure functions and extending operability in the face of unprecedented hazards (Gilrein *et al.*, 2019).

Among these capacities is a culture of learning from past failure and success. After Hurricane Maria -- when clinicians in Puerto Rico dealt with treatment interruptions, transportation limitations, and scarce equipment and medicine -- practical emergency measures were developed that paid off in maintaining functions while containing the spread of COVID-19 among vulnerable cancer patients (Gay *et al.*, 2019; Rivera *et al.*, 2020). In facing competing resource-scarcity and disasters, infrastructure agencies pursuing resilience may benefit from adopting multi-hazards approaches (Ryan, 2008), where investments for agility to unforeseen scales, types, and combinations of disasters are emphasized over hazard-specific robustness.

2.6.3. THEME 2: CHANGING NATURE OF CRITICALITY

COVID-19 challenged our industry and defense-based framing around criticality of engineered systems (e.g., energy, healthcare, ICT), in favor of one that considers human capabilities. The Department of Homeland Security (DHS), for example, defines CI as, "systems and assets that are so vital to the United States that their incapacity or destruction would have a debilitating impact on our physical or economic security or public health or safety." Defining which systems are CI results in a prioritization of resources during extreme events (Theoharidou, Kotzanikolaou and Gritzalis, 2009). However, COVID-19 illustrated potential problems with industry-based framings that do not consider differences between hazards, and the interdependencies inherent in solutions (e.g., ventilators were a healthcare, manufacturing, supply chain, and fiscal challenge). Parks, for example, are typically considered a non-essential service. However, during COVID-19, parks proved their value by serving as field hospitals (Fink, 2020), providing alternative shelters for socially vulnerable groups (CDC, 2020a; Welsh, 2020), and promoting physical, emotional, and mental well-being (Friedman, Allen and Lipsitch, 2020; Olin, 2020; Surico, 2020). Criticality varies between events, and the ability to adjust short term resources accordingly as well as long term resource planning is important.

CI definitions should account for the changing services and functions of industries during hazards (see Chapter 3.4 – Dynamic Criticality). The DHS lists commercial facilities as CI, yet many commercial facilities were shut down. Some industries were able to shift production from non-essential to essential products, while others were not. While this strategy provides some revenue, these industries recognized the social value of essential products and adapted accordingly. Perfume and beer bottlers began packaging hand sanitizer, hockey equipment companies began making medical face shields, and vacuum companies began making ventilators (Domonoske, 2020). Flexibility and agility as environments change (Chester and Allenby, 2018) appears to be critical, but is not captured by static CI definitions. In order to address a dynamic definition of CI and better embrace environments of change, organizations will need to implement Enabling (Explorative) Leadership rather than Administrative (Exploitative) Leadership.

Framing infrastructure criticality in terms of human capabilities can enable a more dynamic and effective approach to directing resources. "Capability refers to the set of valuable functionings that a person has effective access to," where functionings are realized uses of resources that infrastructure systems provide (Clark, Seager and Chester, 2018). Infrastructure becomes critical as it enables human capabilities, for example, following Maslow's hierarchy of needs. In the example of industries producing essential products, production chains became critical in order to enhance the capabilities of people to access sanitation products. How we meet sheltering and nutritional needs during a pandemic may be different from other crises like heat waves, hurricanes, or even terrorist attacks. Therefore, treating criticality as dynamic (and maintaining flexibility to define and plan for it accordingly) appears crucial to identifying how to meet basic needs through infrastructure changes as hazards vary (or arise concurrently).

2.6.4. THEME 3: MANAGING THE TRADE-OFF BETWEEN EFFICIENCY AND RESILIENCE

COVID-19 put a spotlight on the challenges that emerge when too much emphasis is placed on efficiency at the expense of resilience (e.g., not having enough ICU beds, testing supplies, staff, etc., in the right places at the right times) (Tenner, 2020; Allenby and Chester, 2021). Efficiency relates to the optimal response to an existing environment (i.e., prioritizing the reduction of waste in terms of time, effort, and resources), whereas resilience relates to the capacity to adapt to disruptive changes in the environment (i.e., increased slack, redundancy, and diversity—features efficiency might consider waste) (Martin, 2019). Thus, there is an unavoidable tension between efficiency and resilience that has in infrastructure historically leaned toward efficiency (Tenner, 2020). While there are certainly limits (i.e., resource constraints) to the amount of 'slack'/redundancies that can be implemented in a system, COVID-19 showed the importance of having both efficiency and resilience within our systems and ensuring that a proper balance between them is maintained. This perspective was summed up nicely by Ridley (2019), "Efficiency is not fragility, nor is resilience wasteful. Rather, these are design choices that we need to be aware of when designing processes and businesses." The need to reconcile the disconnect between efficiency and resilience was further illustrated by COVID-19 guidelines from the United States Centers for Disease Control (CDC) that emphasized "optimizing" the deployment of stockpiled ventilators and supply of personal protective equipment (CDC, 2020a, 2020b).

Whereas a focus on efficiency and optimization is most applicable for operating in stable conditions, it becomes difficult to maintain and effectively apply in conditions of rapid change and widespread uncertainty such as a global pandemic. How organizations responded to the sudden and large-scale impacts of COVID-19 provided valuable insights into what it means to transition between stable (i.e., efficiency) and unstable (i.e., resilience) environments. Natural disasters often play out over the course of days or weeks, and are usually isolated to specific regions. If a crisis affects a particular area for a relatively short period of time, then certain concepts of efficiency may be applicable (e.g., transfer of resources like generators). This approach breaks down when the geographic scale of the disruption develops globally, and the temporal scale extends over years. COVID-19 highlighted the need to allow for some level of 'inefficiency' in the form of redundancy and diversity of services and assets, along with enhancing institutional, knowledge, and leadership capabilities to manage, mobilize, and implement such resilience capacities. However, there are limits to additional capacity; a system needs to have multifunctionality and be able to alter its functionality when needed.

One possibility for implementing these characteristics into systems, communities, and institutions, is to give stronger credence to the idea that resilience is a public good (Galston, 2020). Considering resilience as a common pool resource like air or water, its absence becomes a negative externality (i.e., The Tragedy of the Commons). Externalities are often addressed by government intervention, and in the case of externalities associated with the lack of resilience, entities such as the Federal Emergency Management Agency (FEMA) and the National Flood Insurance Program were created. Although these programs help reduce negative externalities, they exhibit limitations due to their primary focus on post-disaster rebound and recovery. COVID-19 showed the limitations of these reactionary approaches, and illustrated the need for more proactive, large-scale (spatially and temporally), and dynamic efforts to address the fragility-related externalities that permeate our systems. For example, externalities associated with air and water pollution occur across extensive geographic and temporal scales. In response, the Clean Air Act and Clean Water Act were passed by Congress and administered by the Environmental Protection Agency and the States to address related environmental externalities. Perhaps it is time to consider whether a National Resilience Act, Agency, or Department are needed, and what they may entail. Given the stark reality that society will increasingly face multiple risks, considering resilience as a public good can significantly aid toward striking the right balance between efficiency and resilience, and in turn, enhance our ability to navigate stable, chaotic, and complex conditions.

2.6.5. THEME 4: IMPROVING INSTITUTIONAL RESILIENCE THROUGH LEADERSHIP

The rapidity and scale of COVID-19 emphasized the importance of institutional flexibility for infrastructure resilience (see Chapters 3.2 and 3.3 for an in-depth discussion of governance and leadership for stability and instability). Different types of leadership, including some currently lacking in infrastructure management, enable such flexibility. In stable times, Administrative Leadership emphasizes efficiency-focused efforts (formalized bureaucracies, structures, organizations, and roles well-suited for operating in stable conditions), whereas during periods of instability Adaptive Leadership (an emphasis on learning, adaptability, and creativity that enable operation in uncertain and complex conditions) facilitates rapid and appropriate efforts (Uhl-Bien, Marion and McKelvey, 2007). However, Enabling Leadership is also necessary to create flexible knowledge, and the financial and structural conditions needed to alternate between Administrative and Adaptive Leadership modes as conditions shift between stable and chaotic (Uhl-Bien, Marion and McKelvey, 2007). All three leadership models are necessary, but infrastructure agencies have predominantly been modeled around Administrative Leadership.

As COVID-19 shocked infrastructure demands, deficiencies of Administrative Leadership, where management has been designed for stable conditions, emerged. For example, transit financing and operations appear to have been structured around assumptions of a fairly stable demand envelope, and as systems around the world experienced rapid reductions in demand, the viability of public transit models was tested. While Administrative Leadership can typically execute orders quickly due to a clear power structure, it is unable to navigate complex environments effectively (Uhl-Bien, Marion and McKelvey, 2007). Transit agencies – like other infrastructure – scrambled to restructure operations given the rapid changes in demand, a vulnerable workforce, and social distancing guidelines, but there are many indications that lock-in driven by policies, finance, and technology for stable conditions limited their ability to adapt (Bliss, 2020a; Guse, 2020). Adaptive leadership for uncertainty has become critical, as has the ability to shift from leadership in stable to unstable conditions.

Through continuous navigation between Administrative and Adaptive Leadership, an organization's leadership becomes flexible – or Enabling. For example, telecommunications operate in a rapidly changing environment defined by emerging technologies and novel demands (Bourgeois and Eisenhardt, 1988; Vecchiato, 2015). Enabling Leadership allowed video communication technology providers, such as Zoom Video Communications, to change their management approaches and assets to meet exponential demand increases (Gilbert, 2020). There were hurdles along the way, such as "Zoom-bombing" (Bond, 2020), but the ability to satisfice (Chester and Allenby, 2018) between formal top-down decisions (e.g., increasing bandwidth) and creative solutions (e.g., expanding accessibility) to emerging demands allowed Zoom to adapt services. From managing daily operations to redefining CI to enduring crises, Enabling Leadership proactively catalyzes a flexible and effective response by facilitating interaction between Administrative and Adaptive Leaderships (Uhl-Bien, Marion and McKelvey, 2007). COVID-19 has emphasized the importance of flexible leadership in infrastructure resilience to remain effective in a complex and uncertain world.

2.6.6. DISCUSSION

In a future defined by acceleration, increasing uncertainty, and increasing complexity, COVID-19 provided a glimpse of how best practices that were developed under past conditions that focus on efficiency and stability are becoming increasingly insufficient. Consideration of concurrent hazards, the reframing of infrastructure criticality, understanding the balance between resilience and efficiency, and enabling flexible leadership must be addressed together to revise how we govern and build systems that provide critical services. Overarching these themes is the ability to creatively shift between modes, functions, and leadership capabilities. While the right kind of assets are certainly important and necessary to adapt amid pandemics, it is equally (and often more) important to maintain the human capacity and institutional ability to be dynamic and flexible.

Ultimately, infrastructure managers and researchers must recognize the flexibility of physical and institutional aspects of infrastructure required to keep CI operational in times of disturbance. Criticality is dynamic in rapidly developing situations like COVID-19, which showed the importance of Enabling Leadership as it can allow an organization to adapt quickly in volatile contexts. Physical infrastructure is often not as quickly adaptable, underscoring the importance of proactive competencies (e.g., multi-functionality, redundancy, planned obsolescence) that may not always align with traditional emphases on efficiency and Administrative Leadership. Infrastructure design is rooted in stationarity, which decreases the capacity for flexibility. COVID-19 demonstrated how infrastructure demand can violate design envelopes, and how our rigid systems are unable to adapt. Social distancing concurrent with other disturbances emerged and infrastructure institutions found themselves operating with limited crews and remote management that needed to exhibit Enabling and Adaptive Leadership to remain functional. Therefore, adaptive capacities such as culture of change, connectivity, compatibility, modularity, redundancy, multi-functionality, and planned obsolescence are essential for flexibility in the face of disturbance (Chester and Allenby, 2019). While COVID-19 posed many difficulties to our infrastructure, it also created an opportunity to rethink how we approach our basic and critical systems so that they are better equipped to face future challenges.

Another dynamic that deserves attention is the changing relationships between cyber and physical systems (see Section 5 for an in-depth discussion). Many of the themes discussed focus on traditional and largely physical systems, including their governance. But COVID-19 revealed new capabilities at the interface of legacy physical systems and cybertechnologies. Contact tracing through smart networks and phones -- both voluntary and involuntary -- created the possibility of identifying and limiting the spread of the virus, while simultaneously exposing profound privacy challenges (Smith, 2019; Zakrzewski, 2020). Communication systems driven by smart phones rapidly deployed software to help diagnose COVID-19, detect hand washing, and develop image recognition responsive to mask wearing (Perez, 2020). In conjunction with these changes, applications of artificial intelligence directed individuals towards protective behaviors, helped with diagnosis, and even aided the development of a vaccine (Etzioni and Decario, 2020; Fast and Chen, 2020; Kurzweil, 2020; Peckham, 2020). Although these new applications and technologies played an increasingly important role in addressing the pandemic, they are not without drawbacks. For instance, the rapid development and integration of new cyber-physical systems also introduced profound challenges related to privacy (Smith, 2019; Zakrzewski, 2020), equity and fairness (i.e., the growing digital divide within and across nations) (Holpuch, 2020; Milanesi, 2020; Ramsetty and Adams, 2020), cyber-security (Aladenusi, 2020; CISA, 2020), and the preservation of individual freedoms (Funk and Linzer, 2020; Gilmore III, 2020; Nguyen, 2020; UN, 2020). Thus, as novel cyber-physical systems continue to evolve and emerge, it is important to recognize that the capabilities and challenges introduced by these systems will likely have profound impacts that extend far beyond mitigating and managing COVID-19. Although combating the pandemic necessitated a certain degree of urgency and expediency, deliberation and examination of the potential long-term implications of novel cyber-physical systems also was warranted.

COVID-19 was a window of opportunity for laying new foundations for how we design, operate, and manage infrastructure in the Anthropocene. In its early stages, the pandemic shocked infrastructure demand and created tremendous uncertainty about the future. Going forward, infrastructure can be expected to be shocked in new ways that we probably have not yet experienced. At a time when infrastructure agencies are struggling to cope with disrepair, emerging technologies, and climate change, COVID-19 laid bare how difficult some of the challenges will be, and the need for creative new approaches. Agencies that catalyzed around this moment were able to examine how assumptions about stable demand manifest as rigidity in assets and management. They reviewed what competencies are needed for times of stability versus instability, and the governance, management, and financial principles needed to shift between them. They showed and advanced resilience capabilities. During times of crisis, agencies should be supported for not simply restoring and carrying over their pre-crisis mission, but restructuring their organizations (including assets, institutions, and education) towards the future.

2.6.7. REFERENCES

Adalja, A.A. *et al.* (2012) 'The Globalization of US Medical Countermeasure Production and Its Implications for National Security', *Biosecurity and Bioterrorism: Biodefense Strategy, Practice, and Science*, 10(3), pp. 255–257. Available at: https://doi.org/10.1089/bsp.2012.0622.

Aladenusi, T. (2020) *Impact of COVID-19 on Cybersecurity, Deloitte Switzerland*. Available at: https://www.deloitte.com/content/dam/Deloitte/ng/Documents/risk/ng-COVID-19-Impact-on-Cybersecurity-24032020.pdf (Accessed: 16 July 2020).

Allenby, B. and Chester, M. (2021) 'Learning From Engineers', *Issues in Science and Technology* [Preprint]. Available at: https://issues.org/learning-from-engineers/ (Accessed: 12 October 2023).

Anderson, G.B. *et al.* (2018) 'Projected trends in high-mortality heatwaves under different scenarios of climate, population, and adaptation in 82 US communities', *Climatic Change*, 146(3), pp. 455–470. Available at: https://doi.org/10.1007/s10584-016-1779-x.

Bedford, J. *et al.* (2019) 'A new twenty-first century science for effective epidemic response', *Nature*, 575(7781), pp. 130–136. Available at: https://doi.org/10.1038/s41586-019-1717-y.

Biggs, D. *et al.* (2011) 'Are We Entering an Era of Concatenated Global Crises?', *Ecology and Society*, 16(2). Available at: https://doi.org/10.5751/ES-04079-160227.

Bliss, L. (2020a) 'Hit Hard by Covid-19, Transit Workers Call for Shutdowns', *CityLab*, 13 April. Available at: https://www.bloomberg.com/news/articles/2020-04-13/as-transit-workers-get-sick-unions-mull-shutdowns (Accessed: 19 October 2023).

Bliss, L. (2020b) 'Mapping How Cities Are Reclaiming Street Space', *Bloomberg-CityLab*, 3 April. Available at: https://www.bloomberg.com/news/articles/2020-04-03/how-coronavirus-is-reshaping-city-streets (Accessed: 19 October 2023).

Bogich, T.L. *et al.* (2012) 'Preventing Pandemics Via International Development: A Systems Approach', *PLOS Medicine*, 9(12), p. e1001354. Available at: https://doi.org/10.1371/journal.pmed.1001354.

Bond, S. (2020) 'A Must For Millions, Zoom Has A Dark Side — And An FBI Warning', *National Public Radio (NPR)*, 3 April. Available at: https://www.npr.org/2020/04/03/826129520/a-must-for-millions-zoom-has-a-dark-side-and-an-fbi-warning (Accessed: 19 October 2023).

Bourgeois, L.J. and Eisenhardt, K.M. (1988) 'Strategic Decision Processes in High Velocity Environments: Four Cases in the Microcomputer Industry', *Management Science*, 34(7), pp. 816–835.

Calma, J. (2020) *What happens when extreme heat collides with a pandemic?, The Verge.* Available at: https://www.theverge.com/2020/3/27/21197467/extreme-heat-waves-covid-19-pandemic-coronavirus (Accessed: 19 October 2023).

Cappucci, M. (2020) 'A violent hailstorm and flooding struck Calgary, Canada, on Saturday', *Washington Post*, 15 June. Available at: https://www.washingtonpost.com/weather/2020/06/15/calgary-hailstorm/ (Accessed: 16 July 2020).

CDC (2020a) *COVID-19 and Cooling Centers, Centers for Disease Control and Prevention*. Available at: https://www.cdc.gov/coronavirus/2019-ncov/php/cooling-center.html (Accessed: 16 July 2020).

CDC (2020b) *COVID-19: Strategies for Optimizing the Supply of PPE, Centers for Disease Control and Prevention*. Available at: https://www.cdc.gov/coronavirus/2019-ncov/hcp/ppe-strategy/index.html.

Chester, M.V. and Allenby, B. (2018) 'Toward adaptive infrastructure: flexibility and agility in a non-stationarity age', *Sustainable and Resilient Infrastructure*, 4(4), pp. 173–191. Available at: https://doi.org/10.1080/23789689.2017.1416846.

Chester, M.V. and Allenby, B. (2019) 'Infrastructure as a wicked complex process', *Elementa: Science of the Anthropocene*. Edited by A. Iles and M.E. Chang, 7(1), p. 21. Available at: https://doi.org/10.1525/elementa.360.

Chester, M.V. and Allenby, B.R. (2020) 'Perspective: The Cyber Frontier and Infrastructure', *IEEE Access*, 8, pp. 28301–28310. Available at: https://doi.org/10.1109/ACCESS.2020.2971960.

Chilukuri, S. (2020) *City Opens Park Splash Pads, Expands Cooling Centers As Heat Wave Continues, Block Club Chicago*. Available at: http://blockclubchicago.org/2020/07/07/city-offering-socially-distanced-cooling-centers-to-fight-extreme-heat-during-covid-19/ (Accessed: 16 July 2020).

CISA (2020) *UK and US Security Agencies Issue COVID-19 Cyber Threat Update*. Available at: https://www.cisa.gov/news-events/news/uk-and-us-security-agencies-issue-covid-19-cyber-threat-update (Accessed: 16 July 2020).

City of Chicago (2020) *Cooling Centers - Map, Chicago Data Portal*. Available at: https://data.cityofchicago.org/Health-Human-Services/Cooling-Centers-Map/cj7n-sh49 (Accessed: 16 July 2020).

Clark, S.S., Seager, T.P. and Chester, M.V. (2018) 'A capabilities approach to the prioritization of critical infrastructure', *Environment Systems and Decisions*, 38(3), pp. 339–352. Available at: https://doi.org/10.1007/s10669-018-9691-8.

Cohen, J. (2009) 'Past Pandemics Provide Mixed Clues to H1N1's Next Moves', *Science*, 324(5930), pp. 996–997. Available at: https://doi.org/10.1126/science.324_996.

Dietz, J.E. and Black, D.R. (2012) *Pandemic Planning*. Hoboken, NJ: Taylor and Francis Group.

DOI (2020) *Wildfires & COVID-19, US Department of the Interior*. Available at: https://web.archive.org/web/20200609152446/https://www.doi.gov/wildland-fire/wildfires-covid-19.

Domonoske, C. (2020) 'ExxonMobil Starts Making Hand Sanitizer, Following Liquor Companies', *NPR*, 24 April. Available at: https://www.npr.org/sections/coronavirus-live-updates/2020/04/24/844363276/exxonmobil-starts-making-hand-sanitizer-weeks-after-many-liquor-companies (Accessed: 19 October 2023).

Einhorn, E. (2020) *Thousands fled for their lives when two Michigan dams collapsed. Experts warn it could happen again.*, *NBC News*. Available at: https://www.nbcnews.com/news/us-news/thousands-fled-their-lives-when-two-michigan-dams-collapsed-more-n1230841 (Accessed: 20 July 2020).

Ellis-Petersen, H. and Ratcliffe, R. (2020) 'Super-cyclone Amphan hits coast of India and Bangladesh', *The Guardian*, 20 May. Available at: https://www.theguardian.com/world/2020/may/20/super-cyclone-amphan-evacuations-in-india-and-bangladesh-slowed-by-virus (Accessed: 10 July 2020).

Etzioni, O. and Decario, N. (2020) 'AI Can Help Scientists Find a Covid-19 Vaccine', *Wired*, 28 March. Available at: https://www.wired.com/story/opinion-ai-can-help-find-scientists-find-a-covid-19-vaccine/ (Accessed: 16 July 2020).

Fast, E. and Chen, B. (2020) *Can artificial intelligence help us design vaccines?*, *Brookings*. Available at: https://www.brookings.edu/articles/can-artificial-intelligence-help-us-design-vaccines/ (Accessed: 16 July 2020).

Fink, S. (2020) 'Treating Coronavirus in a Central Park "Hot Zone"', *The New York Times*, 15 April. Available at: https://www.nytimes.com/2020/04/15/nyregion/coronavirus-central-park-hospital-tent.html (Accessed: 19 October 2023).

Flavelle, C. (2020) 'Coronavirus Makes Cooling Centers Risky, Just as Scorching Weather Hits', *The New York Times*, 6 May. Available at: https://www.nytimes.com/2020/05/06/climate/coronavirus-climate-change-heat-waves.html (Accessed: 16 July 2020).

Florida, R. (2020) 'We'll Need To Reopen Our Cities. But Not Without Making Changes First.', *Bloomberg-CityLab*, 27 March. Available at: https://www.bloomberg.com/news/articles/2020-03-27/how-to-adapt-cities-to-reopen-amid-coronavirus (Accessed: 19 October 2023).

Friedman, W. 'Ned', Allen, J.G. and Lipsitch, M. (2020) 'Keep parks open. The benefits of fresh air outweigh the risks of infection.', *Washington Post*, 14 April. Available at: https://www.washingtonpost.com/outlook/2020/04/13/keep-parks-open-benefits-fresh-air-outweigh-risks-infection/ (Accessed: 19 October 2023).

Funk, A. and Linzer, I. (2020) 'How the coronavirus could trigger a backslide on freedom around the world', *Washington Post*, 16 March. Available at: https://www.washingtonpost.com/opinions/2020/03/16/how-coronavirus-could-trigger-backslide-freedom-around-world/ (Accessed: 16 July 2020).

Galston, W.A. (2020) 'Efficiency Isn't the Only Economic Virtue', *Wall Street Journal*, 10 March. Available at: https://www.wsj.com/articles/efficiency-isnt-the-only-economic-virtue-11583873155 (Accessed: 19 October 2023).

Gay, H.A. *et al.* (2019) 'Lessons Learned From Hurricane Maria in Puerto Rico: Practical Measures to Mitigate the Impact of a Catastrophic Natural Disaster on Radiation Oncology Patients', *Practical Radiation Oncology*, 9(5), pp. 305–321. Available at: https://doi.org/10.1016/j.prro.2019.03.007.

Gettleman, J. *et al.* (2020) 'Cyclone Amphan Slams India and Bangladesh', *The New York Times*, 20 May. Available at: https://www.nytimes.com/2020/05/20/world/asia/cyclone-amphan-india-bangladesh.html (Accessed: 10 July 2020).

Gibbens, S. (2020) *COVID-19 complicates an already dire wildfire season*, *Science*. Available at: https://www.nationalgeographic.com/science/article/covid-19-complicates-already-dire-wildfire-season (Accessed: 16 July 2020).

Gilbert, B. (2020) *All your friends are using Zoom, the video-chat app that is suddenly dominating competition from Google and Microsoft*, *Business Insider*. Available at: https://www.businessinsider.com/zoom-video-everywhere-google-hangouts-skype-2020-3 (Accessed: 19 October 2023).

Gilmore III, J.S. (2020) *Protecting Human Rights During the COVID-19 Pandemic*, *U.S. Mission to the OSCE*. Available at: https://osce.usmission.gov/human-rights-protections-during-covid-19-pandemic/ (Accessed: 16 July 2020).

Gilrein, E.J. *et al.* (2019) 'Concepts and practices for transforming infrastructure from rigid to adaptable', *Sustainable and Resilient Infrastructure*, 6(3–4), pp. 213–234. Available at: https://doi.org/10.1080/23789689.2019.1599608.

Guse, C. (2020) 'MTA workers dying from coronavirus at triple the rate of agencies that employ NYC first responders', *New York Daily News*, 8 April. Available at: https://www.nydailynews.com/2020/04/08/mta-workers-dying-from-coronavirus-at-triple-the-rate-of-agencies-that-employ-nyc-first-responders/ (Accessed: 19 October 2023).

Hendrickson, C. and Rilett, L.R. (2020) 'The COVID-19 Pandemic and Transportation Engineering', *Journal of Transportation Engineering, Part A: Systems*, 146(7), p. 01820001. Available at: https://doi.org/10.1061/JTEPBS.0000418.

Hessel, L. and Group, T.E.V.M. (EVM) I.W. (2009) 'Pandemic influenza vaccines: meeting the supply, distribution and deployment challenges', *Influenza and Other Respiratory Viruses*, 3(4), pp. 165–170. Available at: https://doi.org/10.1111/j.1750-2659.2009.00085.x.

Holpuch, A. (2020) 'US's digital divide "is going to kill people" as Covid-19 exposes inequalities', *The Guardian*, 13 April. Available at: https://www.theguardian.com/world/2020/apr/13/coronavirus-covid-19-exposes-cracks-us-digital-divide (Accessed: 16 July 2020).

Huff, A.G. *et al.* (2015) 'How resilient is the United States' food system to pandemics?', *Journal of Environmental Studies and Sciences*, 5(3), pp. 337–347. Available at: https://doi.org/10.1007/s13412-015-0275-3.

Itzwerth, R.L. *et al.* (2006) 'Pandemic influenza and critical infrastructure dependencies: possible impact on hospitals', *The Medical Journal of Australia*, 185(S10), pp. S70-72. Available at: https://doi.org/10.5694/j.1326-5377.2006.tb00712.x.

Jovanović, A. *et al.* (2020) 'Assessing resilience of healthcare infrastructure exposed to COVID-19: emerging risks, resilience indicators, interdependencies and international standards', *Environment Systems and Decisions*, 40(2), pp. 252–286. Available at: https://doi.org/10.1007/s10669-020-09779-8.

Kurzweil, R. (2020) 'AI-Powered Biotech Can Help Deploy a Vaccine In Record Time', *Wired*, 19 May. Available at: https://www.wired.com/story/opinion-ai-powered-biotech-can-help-deploy-a-vaccine-in-record-time/ (Accessed: 16 July 2020).

Martin, R.L. (2019) 'The High Price of Efficiency', *Harvard Business Review*, 1 January. Available at: https://hbr.org/2019/01/the-high-price-of-efficiency (Accessed: 19 October 2023).

McDowell, J.D. (2020) 'How COVID-19 Will Change the Way We Fight Wildfires', *Smithsonian Magazine*, 7 July. Available at: https://www.smithsonianmag.com/science-nature/wildfire-season-covid-19-180975250/ (Accessed: 16 July 2020).

Milanesi, C. (2020) 'Digital Transformation And Digital Divide Post COVID-19', *Forbes*, 11 May. Available at: https://www.forbes.com/sites/carolinamilanesi/2020/05/11/digital-transformation-and-digital-divide-post-covid-19/ (Accessed: 16 July 2020).

Mogul, F. and Hurt, E. (2020) 'As Nurses Aid New York, Other States Worry They'll Be Short-Staffed Too', *NPR*, 24 April. Available at: https://www.npr.org/2020/04/24/843529594/as-nurses-aid-new-york-other-states-worry-theyll-be-short-staffed-too (Accessed: 19 October 2023).

Nguyen, A. (2020) 'Vietnam's Government Is Using COVID-19 to Crack Down on Freedom of Expression', *Slate*, 8 May. Available at: https://slate.com/technology/2020/05/vietnam-coronavirus-fake-news-law-social-media.html (Accessed: 16 July 2020).

NYC (2020) *Mayor de Blasio Announces COVID-19 Heat Wave Plan to Protect Vulnerable New Yorkers*, *The official website of the City of New York*. Available at: http://www.nyc.gov/office-of-the-mayor/news/350-20/mayor-de-blasio-covid-19-heat-wave-plan-protect-vulnerable-new-yorkers (Accessed: 16 July 2020).

Okura, Y. *et al.* (2020) *Monsoon, floods and COVID-19: building community resilience in Bangladesh*. Zurich, CH: ZFRA. Available at: https://floodresilience.net/resources/item/monsoon-floods-and-covid-19-building-community-resilience-in-bangladesh/ (Accessed: 10 July 2020).

Olin, A. (2020) 'In the COVID-19 era, a renewed appreciation of our parks and open spaces', *Kinder Institute for Urban Research | Rice University. Urban Edge Blog.*, 10 April. Available at: https://kinder.rice.edu/urbanedge/covid-19-era-renewed-appreciation-our-parks-and-open-spaces (Accessed: 19 October 2023).

Osterholm, M.T. (2005) 'Preparing for the Next Pandemic', *New England Journal of Medicine*, 352(18), pp. 1839–1842. Available at: https://doi.org/10.1056/NEJMp058068.

Osterholm, M.T. and Kelley, N.S. (2009) 'Energy and the Public's Health: Making the Connection', *Public Health Reports*, 124(1), pp. 20–21.

Peckham, O. (2020) *COVID-19 Update: Apple Frees Mobility Data, MIT Predictive ML, IBM Data Challenge, Rolls-Royce & More, EnterpriseAI*. Available at: https://www.enterpriseai.news/2020/04/22/covid-19-update-apple-palantir-rolls-royce-more/ (Accessed: 16 July 2020).

Perez, S. (2020) 'Apple's software updates give a glimpse of software in a COVID-19 era', *TechCrunch*, 23 June. Available at:

https://techcrunch.com/2020/06/23/apples-software-updates-give-a-glimpse-of-software-in-a-covid-19-era/ (Accessed: 16 July 2020).

Ramsetty, A. and Adams, C. (2020) 'Impact of the digital divide in the age of COVID-19', *Journal of the American Medical Informatics Association*, 27(7), pp. 1147–1148. Available at: https://doi.org/10.1093/jamia/ocaa078.

Ridley, M. (2019) 'Blending efficiency and resilience', *Medium*, 3 March. Available at: https://mark-ridley.medium.com/blending-efficiency-and-resilience-1ff876e7f0c9 (Accessed: 19 October 2023).

Rivera, A. *et al.* (2020) 'The Impact of COVID-19 on Radiation Oncology Clinics and Patients With Cancer in the United States', *Advances in Radiation Oncology*, 5(4), pp. 538–543. Available at: https://doi.org/10.1016/j.adro.2020.03.006.

Rosa, A. and Robles, F. (2020) 'Pandemic Plunges Puerto Rico Into Yet Another Dire Emergency', *The New York Times*, 8 July. Available at: https://www.nytimes.com/2020/07/08/us/coronavirus-puerto-rico-economy-unemployment.html (Accessed: 10 July 2020).

Ryan, J.R. (2008) *Pandemic Influenza: Emergency Planning and Community Preparedness*. Boca Raton, FL: CRC Press. Available at: https://www.google.com/books/edition/Pandemic_Influenza/t13C_eWhOX4C?hl=en&gbpv=0.

Sarewitz, D. (2020) 'Pandemic Science and Politics', *Issues in Science and Technology*, 25 March. Available at: https://issues.org/pandemic-science-politics-values/ (Accessed: 19 October 2023).

Smith, T. (2019) *In Hong Kong, protesters fight to stay anonymous - The Verge*. Available at: https://www.theverge.com/2019/10/22/20926585/hong-kong-china-protest-mask-umbrella-anonymous-surveillance (Accessed: 16 July 2020).

Surico, J. (2020) 'The Power of Parks in a Pandemic', *Bloomberg-CityLab*, 9 April. Available at: https://www.bloomberg.com/news/articles/2020-04-09/in-a-pandemic-the-parks-are-keeping-us-alive (Accessed: 19 October 2023).

Tenner, E. (2020) 'Efficiency Is Biting Back', *The Atlantic*, 29 April. Available at: https://www.theatlantic.com/ideas/archive/2020/04/too-much-efficiency-hazardous-society/610843/ (Accessed: 19 October 2023).

Theoharidou, M., Kotzanikolaou, P. and Gritzalis, D. (2009) 'Risk-Based Criticality Analysis', in C. Palmer and S. Shenoi (eds) *Critical Infra-*

structure Protection III. Berlin, Heidelberg: Springer (IFIP Advances in Information and Communication Technology), pp. 35–49. Available at: https://doi.org/10.1007/978-3-642-04798-5_3.

Uhl-Bien, M., Marion, R. and McKelvey, B. (2007) 'Complexity Leadership Theory: Shifting leadership from the industrial age to the knowledge era', *The Leadership Quarterly*, 18(4), pp. 298–318. Available at: https://doi.org/10.1016/j.leaqua.2007.04.002.

UN (2020) *COVID-19 and Human Rights, United Nations*. Available at: https://www.un.org/victimsofterrorism/sites/www.un.org.victimsofterrorism/files/un_-_human_rights_and_covid_april_2020.pdf (Accessed: 16 July 2020).

USGS (2020) *USGS Scientists Find Seafloor Faults Near Puerto Rico Quakes' Epicenters*. Available at: https://www.usgs.gov/news/featured-story/usgs-scientists-find-seafloor-faults-near-puerto-rico-quakes-epicenters (Accessed: 19 July 2020).

Vann, M. (2020) 'FEMA faces multi-front battle on COVID-19 as hurricane season nears', *ABC News*, 16 April. Available at: https://abcnews.go.com/Politics/fema-faces-multi-front-battle-covid-19-hurricane/story?id=70052631 (Accessed: 19 October 2023).

Vecchiato, R. (2015) 'Strategic planning and organizational flexibility in turbulent environments', *Foresight*, 17(3), pp. 257–273. Available at: https://doi.org/10.1108/FS-05-2014-0032.

Villarreal, A. (2020) 'Covid-19: De Blasio urges US enlistment program for doctors and nurses', *The Guardian*, 3 April. Available at: https://www.theguardian.com/us-news/2020/apr/03/de-blasio-new-york-coronavirus (Accessed: 19 October 2023).

Welsh, N. (2020) 'Good News, Bad News for Homeless COVID-19 Response', *The Santa Barbara Independent*, 2 April. Available at: https://www.independent.com/2020/04/02/good-news-bad-news-for-homeless-covid-19-response/ (Accessed: 19 October 2023).

Williams, V.J. (2007) 'Fluconomics: preserving our hospital infrastructure during and after a pandemic', *Yale Journal of Health Policy, Law, and Ethics*, 7(1), pp. 99–152.

Woods, D.D., Seager, T.P. and Alderson, D.L. (2020) 'When Can We Move Forward From COVID-19? When Four Capabilities Are In Action.' Available at: https://doi.org/10.5281/zenodo.3748052.

Zakrzewski, C. (2020) 'The Technology 202: Tech to contain coronavirus on college campuses sparks fresh privacy concerns', *Washington*

Post, 17 July. Available at: https://www.washing-tonpost.com/news/powerpost/paloma/the-technology-202/2020/07/10/the-technology-202-tech-to-contain-coronavirus-on-col-lege-campuses-spark-fresh-privacy-con-cerns/5f077b4a88e0fa7b44f716e8/ (Accessed: 16 July 2020).

Zhong, R. (2020) 'Severe Floods in China Leave Over 106 Dead or Missing', *The New York Times,* 3 July. Available at: https://www.ny-times.com/2020/07/03/world/asia/china-floods-rain.html (Accessed: 19 October 2023).

Section 3

GOVERNANCE AND LEADERSHIP

Positioning infrastructure to engage with Anthropocene complexity necessitates modernization of governance processes including organizational structures and bureaucracies, goals, and leadership. This is the focus of Section 3. We start by describing how and why modern infrastructure are governed the way they are, their efficiency-focused goals and supporting bureaucratic structures that emerged from the railroads over a century ago. We contract these governance structures against alternative forms, namely those that are more adept at creating innovation in the face of chaos. We then provide alternative governance models for infrastructure.

Central to these models is the capability to pivot between efficiency-focused goals and innovation-focused goals. Efficiency (i.e., exploitation) works well during periods of stability, what infrastructure have so far largely been designed around. Innovation (i.e., exploration) becomes necessary before and during surprise. Infrastructure will need to pivot between efficiency and innovation in the Anthropocene and several models are described in Section 3 for alternative infrastructure governance structures.

3.1

INFRASTRUCTURE GOVERNANCE
FOR THE ANTHROPOCENE

3.1.1. INTRODUCTION

Physical infrastructure systems today and the institutions that manage them are facing growing challenges that raise serious questions about their viability in the Anthropocene. Approaches to designing, maintaining, and managing infrastructure systems have remained stubbornly stable for over a century. Both the physical systems and the organizational systems that build and manage them are obdurate, resistant to change (Hommels, 2005). They have been remarkably successful at delivering reliable and affordable critical services—such as power, water, and transportation—and in doing so driving growth and economic stability, and improving well-being. These systems may be victims of their own success. They have become so mundane that they appear taken for granted in the developed world and often viewed as the engineer's domain (La Porte, 1996; Coutard, 2002): we might expect that the delivery of reliable and affordable critical services will continue without question, despite operating environments that are becoming significantly more complex. Infrastructure designed for the past may be problematic for an accelerating, increasingly uncertain, and increasingly complex future. Yet there remains limited insight into how infrastructure is governed, whether these governance models are appropriate for the future, and how governance processes emphasize technologies that may or may not be appropriate for the future.

Changes in the environments in which infrastructure operates, and in infrastructure themselves, will be troublesome for those who manage the systems and need to ensure their ability to meet public needs into the future. In Herbert Simon's *Sciences of the Artificial* a distinction is made between internal and external environments in his design of an intellectual

This chapter was adapted from the following article with publisher permission: Mikhail Chester, Thaddeus Miller, and Tischa Muñoz-Erickson, 2020, Infrastructure Governance for the Anthropocene, *Elementa*, 8(1), 078, pp. 1-14, doi: 10.1525/elementa.2020.078.

structure to characterize natural and artificial phenomena (Simon, 1996). The internal system is an organization capable of attaining goals within some range of environments, while the external system determines the conditions for goal attainment. Infrastructure is designed to deliver services within a somewhat narrow range of environmental conditions, but at the same time contribute to environmental change thereby generating vulnerability (e.g., automobility and air quality; energy and carbon; flood management and unending complexity that can't be controlled). They mediate human-environment interactions in ways and scales that are increasingly difficult to make sense of (Chester, Markolf and Allenby, 2019). This context appears remarkably different than decades ago when our infrastructure systems were designed. A rapidly changing world appears at odds with infrastructure design principles that emphasize consistency, and systems that are instantiated for decades with limited flexibility and agility to transition (Hommels, 2005; Sovacool, Lovell and Ting, 2018).

While there is a growing body of work that examines the challenges associated with agile and flexible infrastructure from a physical design perspective, little work has been done to understand how the structure, functioning, values, rules, norms, and processes of the institutions that manage infrastructure keep infrastructure services obdurate, or create the conditions for transformative change. Indeed a large literature describes the forces that created modern infrastructure systems, and how these socio-technical regimes have changed over time (van der Brugge, Rotmans and Loorbach, 2005; Geels *et al.*, 2016; Desai and Armanios, 2018). This work has shown that a complex governance system, of multiple state and private actors, scales, arrangements, and modes of governing is emerging to address the variety of ownership arrangements, financial constraints, and socio-political pressures exerted on modern infrastructure development (Leach, Scoones and Stirling, 2010; Goldthau, 2014; O'Brien and Pike, 2015). But a systemic review of the bureaucratic structures that persist and the rules, values, norms, and practices that define their operation at the intra-organizational level is not apparent, yet is critical. In this chapter, we endeavor to describe the emergence of the divisional bureaucratic organizational structures of many contemporary U.S. infrastructure, how this organizational form emerged, and what it means to change infrastructure governance for the future. When considering change, we evaluate both organizational leadership and identity to understand across hierarchies the conditions of transformation for both stable and unstable environments. We conclude by focusing on processes of transition towards improving sensemaking of the environment, models of governance that may be more appropriate for an increasingly complex environment, and key factors that support public service organization mission change. We view the whole of the discussion as a treatise on describing how infrastructure governance should transition for the future.

Going forward, we use a lexicon to describe governance concepts that is rooted in governance and sustainability transitions theory (Muñoz-Erickson *et al.*, 2016). Governance is a process involving collective action for resource allocation and use across multiple civic and private actors and not just the state (Kooiman, 1993). Government is to governance as structure is to function, and transitions are the processes that lead to fundamental changes in structure, culture and practices as they relate to a particular goal (Jordan, 2008; Loorbach, 2010; Muñoz-Erickson *et al.*, 2016). Following Ostrom (2008), we discuss institutions as the rules and norms that humans use when interacting within repetitive and structured situations. Governance actions are shaped by institutions. When it comes to organizations and their structures, the term bureaucracy describes the formalization of behavior to achieve coordination, including division of labor, specialization, formalization of behavior, hierarchy of authority, chain of command, regulated communication, and standardization of work processes and of skills (Mintzberg, 1979; Serpa and Ferreira, 2019). Whether public or privately-owned, an organization is bureaucratic if its behavior is standardized (predictable) and specialized (Mintzberg, 1981).

3.1.2. EMERGENCE OF THE INFRASTRUCTURE DIVISIONAL BUREAUCRACY

Infrastructure and the institutions that design, manage and maintain them (as socio-technical systems) emerge due to a diversity of pressures, and bureaucratic structures are a response to the goals, technologies, and cultural preferences of past conditions. Just as physical infrastructure is resistant to change, so are their social and organizational components. Before analyzing the bureaucratic structures that guide infrastructure today, it's important to recognize these pressures. They include the allocation of financial risk (Habib, Brealey and Cooper, 2000), technological emphasis (Kaminsky, 2018), the ability to scale (Edwards *et al.*, 2007), certainty of service delivery (La Porte, 1996), the need to consolidate (Edwards *et al.*, 2007), and the formal and informal rules for operating systems (Ausubel and Herman, 1988). Graham and Marvin, (2001) frame infrastructure as networking activities – water, power, transport, communications, etc. – and their governance as the management of flows across scales. There are myriad pressures that have created the manifestations of infrastructure as socio-technical systems, and recognizing these conditions is critical for shifting the goals and purposes of large engineered systems (Osborne and Brown, 2005; Sovacool, Lovell and Ting, 2018). However, understanding the bureaucratic structures and the embedded knowledge and assumptions that drive (and possibly constrain) infrastructure today is a critical step towards ensuring that they transition to meet our future needs.

The scant evidence suggests that at the end of the twentieth century, the organizations that manage U.S. infrastructure were often structured as a divisional bureaucracy, where divisions reflect departments with focused expertise (Friedlander, 1995a, 1995b, 1996). Transportation agencies may have divisions specialized on pavement construction and rehabilitation, traffic operations, inspections, and environmental planning (USDOT, 2017; ADOT, 2019; Caltrans, 2020; GADOT, 2020; Phoenix, 2020). A water distribution agency may have divisions focused on system design and construction, production, and distribution (CAP-AZ, 2010; EPA, 2020; Long Beach, 2020; USGS, 2020). Power providers often structure themselves according to assets or functions, for example, asset management, field services, and regulatory management (Horan, McGrath and Peterson, 2018). Across governmental and jurisdictional scale, the divisional bureaucracy is present (and worthy of particular focus), and in addition to construction, maintenance, and operational functions, also often includes administrative functions such as communications and outreach. This organizational structure appears to have persisted for the entirety of modern infrastructure systems (Chandler, 1977; Friedlander, 1995b, 1996). As a staple organizational structure, the divisional bureaucracy has undoubtedly delivered tremendous value, but as the challenges around infrastructure in a rapidly changing environment grow, serious questions remain as to whether this form of management is able to handle substantive change and deliver the public service values needed into the future. How did the divisional bureaucracy emerge as a dominant management structure of infrastructure organizations?

3.1.2.1. HISTORY OF THE DIVISIONAL BUREAUCRACY IN INFRA-STRUCTURE

The Industrial Revolution was a key turning point in the scale, scope and rate of human activity, and was associated with immense increases in power and speed enabled by a transition to largely coal-based energy sources (Beniger, 1989). Through the early 1800s, energy use was largely associated with human, animal, and wind power. But the use of coal in the early 1800s, first in ships, and later in manufacturing, contributed to an explosion in the material economy that far outpaced the capacity of the supply and demand landscape (at that time largely small firms) to manage (Beniger, 1989, p. 262). This rapid growth created for the first time in human history a sustained global demand for distribution and control systems including information processing, programming, and telecommunications (Beniger, 1989, p. 185). It led to a "crisis of control", the need for new technologies, processes, and organizational structures that could manage the growing complexity and speed of change of commerce (Edwards, 2003).

Control mechanisms and information flows advanced rapidly to manage the increase in economic production. The rise of organizational hierarchies and associated bureaucracies in the nineteenth century was a direct response to information-handling demands. The American economy until then had been defined by small business with limited to no hierarchy, that relied on communication through market mechanisms (Beniger, 1989, p. 262). Communication through the market for the first time became too slow, and as wholesalers increased in size, they adopted organizational structures (hierarchies and bureaucracies) that could match the market's increasing speed. Bureaucracies flourished because they could yield lower costs, increase productivity, and increase profits relative to other market structures (Beniger, 1989).

America's railroads were at the center of these technology and changing market trends, deploying technological and organizational innovations, that ultimately become a template for large organizations and other infrastructure. Railroads were the first major global system to experience dramatic control problems as the US manufacturing base exploded and population grew during the mid-1800s. The US railroad system had in the early 1800s consisted of smaller carriers, and by the middle of the century was experiencing rapid consolidation. At the end of the nineteenth century, US railroad systems were the largest infrastructure (and business) organizations in the world, in terms of number of people employed, transactions handled, and capital used. In 1891 the Pennsylvania railroad employed 110,000 people, far more than the US armed forces (39,492), and US post office (95,440) which at the time was the largest government agency in terms of personnel (Chandler, 1977, pp. 204–205). The expanding size of the railroads necessitated pioneering in business administration to handle the complexity of their operations.

The consolidation of the railroads brought with it a need for organizational management across geographic and managerial scales (including financing), and two competing models emerged. The decentralized model saw geographic regions of a railway company self-managing, where top managers of the regions worked together to evaluate, coordinate, and allocate resources across the entire network (Chandler, 1977, pp. 185–186). The centralized model used departments focused on functional areas such as traffic, transportation, and finance, making decisions across the entire organization, and remaining independent. Financiers preferred the centralized model as it created fewer managers and thus administrative costs, had all managers in the same location easing communication, and allowed departments to operate autonomously (Chandler, 1977, p. 185). By the end of the 19th century, virtually all railroad systems were using a centralized system. As the railroads consolidated, their size increased, resulting in the need for additional layers of management (hence the middle manager).

The president, vice presidents, and board of directors in these centralized organizations were thus positioned to steer strategic goals. The divisional bureaucracy was born where middle managers controlled operations and top managers allocated resources.

By the early twentieth century the railroads had achieved control over competition. They had consolidated to control large geographic regions and had begun sharing rates with the Interstate Commerce Commission which then handled negotiations between the railroads and shippers. Little competitive pressure meant that there was less need for long-term planning, and coordination of existing activities (Chandler, 1977). Railroads became the administrative model that other natural monopolies adopted: they were highly visible and even low-level managers carried significant status in their communities (Chandler, 1977). Not surprisingly then the railroad was the largest infrastructure in the 19th century and deployed innovations across technologies and organization structure. Organizational innovations included complex administrative structures with multi-layered hierarchies and a large degree of functional specialization (Edwards, 2003).

The innovations necessary for the railroad triggered the development of other critical services including steamships, urban transit, and communications (namely the postal service, telegraph, and telephone). Like the railroad many of these industries operated without competition, and the beginning of the twentieth century saw an explosion of public enterprises that were not regulated by market mechanisms, including lighting, power and heat in cities (Chandler, 1977). These utilities were carried out by a single privately owned enterprise that had no competition and worked with localities to provide services. By the late 1800s railroad managers had become more professional, and were systematically disseminating information about their innovative processes and procedures, including how their use resulted in efficiency gains (Chandler, 1977). Nascent public enterprises adopted the dominant divisional bureaucracy model, with its standardized processes, of the much larger railroads, creating the institutional foundations and specialized structures that are still in use today.

The hierarchy associated with the divisional bureaucratic form of management itself became a source of permanence, power, and continued growth (Chandler, 1977). The enterprises that had existed prior to managerial hierarchies were short-lived. They were based on partnerships between individuals that were easily dissolved, e.g., in the event of retirement or death, or if one businessperson simply decided they'd rather work with someone else. The hierarchies that defined divisional bureaucracies were intrinsically persistent. When a manager left, they were easily replaced by someone with appropriate expertise. The organization's mission and goals persisted despite turnover of managers (Chandler, 1977).

3.1.2.2. DIVISIONAL BUREAUCRACIES FOR STANDARDIZED PRODUCTS

It is helpful to understand the benefits and tradeoffs of divisional bureaucracies based on how they are structured. Organizational design can be classified into several general schema to structure the basic components of organizations: strategic apex (top management), operating core (persons responsible for basic work), middle line (intermediate managers between chief executive and workers), technostructure (personnel who design internal systems for planning and controls), and support staff (personnel who provide indirect services) (Figure 3.1.1) (Mintzberg, 1981). The divisional organization (one such schema) is less an integrated organization and more independent entities (departments) under a loose administration. Each division is treated as an independent entity with its own goals, that get translated down the line into subgoals and standardization of work (bureaucratization of structure). While the divisional organization emerges to improve adaptability -- adding or subtracting divisions in response to new conditions -- evidence suggests that the organizational structure discourages risk taking and innovation (standardized and measurable performance goals work against innovation) (Mintzberg, 1981). It creates hierarchical barriers that make it difficult for innovative ideas at the bottom to reach higher levels of strategic management (Wilson, 1989). Next, the division-specific performance goals can work against cross-division problem solving. The recruiting of employees into a division and building their specialization within the division creates cultural fortresses (Wilson, 1989). And lastly, divisional management structures rely on performance-based goals that are measurable, often economic in nature, and is not conducive to social goals which may be relevant to the broader public that relies on the institution's services (Mintzberg, 1981).

Figure 3.1.1: Organizational Components, Pulls and Forms

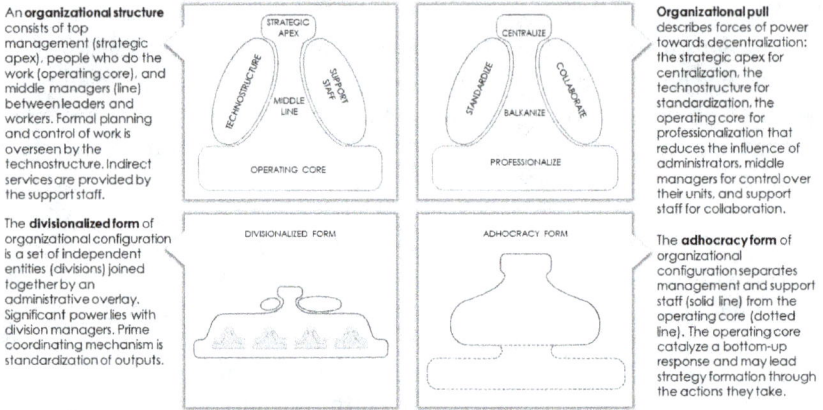

An **organizational structure** consists of top management (strategic apex), people who do the work (operating core), and middle managers (line) between leaders and workers. Formal planning and control of work is overseen by the technostructure. Indirect services are provided by the support staff.

Organizational pull describes forces of power towards decentralization: the strategic apex for centralization, the technostructure for standardization, the operating core for professionalization that reduces the influence of administrators, middle managers for control over their units, and support staff for collaboration.

The **divisionalized form** of organizational configuration is a set of independent entities (divisions) joined together by an administrative overlay. Significant power lies with division managers. Prime coordinating mechanism is standardization of outputs.

The **adhocracy form** of organizational configuration separates management and support staff (solid line) from the operating core (dotted line). The operating core catalyze a bottom-up response and may lead strategy formation through the actions they take.

Four organizational figures are shown: the top row describing structure and tensions, and the bottom row showing the Divisionalized and Adhocracy structural forms. Vertical (delegating decsionmaking down the chain of authority) and horizontal (shifting of power to nonmanagers) dimensions represent control schema. Adapted from (Mintzberg, 1979) and reprinted with permission of Pearson Education, Inc., New York, New York.

Framing infrastructure centralization in terms of how the organization delegates power, instead of the configuration of assets and how they interact, becomes a valuable frame for understanding the capability of different infrastructure governance forms. Centralization and decentralization are often discussed in terms of network typology when it comes to infrastructure (Hines, Blumsack and Schläpfer, 2015). Yet network typology does not sufficiently address power relationships, who controls what, and why that matters as it relates to organization structure. Centralization and decentralization are rooted in power dynamics and the delegation of authority (Mintzberg, 1979). Centralization is the aggregating of power at a single point in the organization, while decentralization is the disbursement of power down the chain of authority (vertical) or out from the chain of authority to non-managers (horizontal). Centralizing power may be necessary for coordination, but for many large organizations simply bringing all relevant information to a central authority is not feasible. Decentralization allows the organization to respond quickly (by avoiding the transmission of information to the center) and is a stimulus for motivation (providing creative space). Vertical decentralization disperses formal power down a chain, from a strategic apex to the middle managers. Horizontal decentralizations disperses decisional power so that nonmanagers are granted control over decision-making (Mintzberg, 1979). We discuss these relationships as we describe the alternatives to the divisional management structure.

The characteristics that define the divisional management structure that is the core arrangement of infrastructure organizations appears inimical to the emerging challenges of the Anthropocene, i.e., growing complexity and uncertainty, and cross-disciplinary efforts to address these challenges through agile and flexible approaches. Serious questions remain as to whether our infrastructure institutions are prepared to change to meet these emerging challenges. We do not believe that an entire restructuring of infrastructure institutions is necessarily needed. Instead, an opening up of how we manage infrastructure is needed that considers both stable and unstable conditions, and an examination of why we allow organizations to focus on performance goals that largely reflect those of the last century, that is, the continued and uninterrupted delivery of services using largely centralized and rigid systems, with what appears to be limited capability of adapting to the accelerating challenges and complexity of the Anthropocene. We now turn to examine infrastructure management within this new landscape, and the paths necessary to ensure that public and critical services are met into the future.

3.1.3. RE-THINKING INFRASTRUCTURE AS KNOWLEDGE OR-GANIZATIONS

If the bureaucratic forms that define infrastructure organizations are insufficient for the future, then what should we do? Approaching increasingly complex systems and environments necessitates new processes that are different from those that dominate infrastructure today. At the macro it requires embracing acceleration and uncertainty where assumptions change based on developments in society, there are multiple accountabilities and the need to balance this power, value is created not as profit but as social welfare, and there is richness of diversity (in people, structure, activities, culture, and processes) where any particular person cannot fully understand systemic interactions and emergent characteristics (Karp and Helgø, 2008). So, what does this mean for institutions and how they're organized? To answer this question, we can look at how successful institutions respond to increasing complexity. Successful institutions create the capacity to generate knowledge about how conditions are changing by shifting power structures to enable those in the organization that are in the best position to sense change, and creating capacities to experiment since certainty about the future is diminished and therefore deterministic recommendations become more problematic.

Prior to assessing institutional change it's important to ask the basic question of what would motivate infrastructure institutions to change. Infrastructure institutions have largely since their inception (generally only 50 to 100 years ago in the U.S.) operated in environments of stability where environmental variables (social, political, financial, climate, technologies) have been relatively stable, and as such, demand has been stable or at least reasonably predictable (Chester and Allenby, 2018). Both internal and external forces are likely to upend this paradigm (Simon, 1996). Infrastructure institutions will most likely have to change out of necessity as they find themselves increasingly unable to match the accelerating change happening in the environment around them, and in the demands they provide.

The infrastructure organization in the Anthropocene will need to reorganize into a knowledge enterprise whose complexity matches that of the environment and knowledge accumulates, is shared, and is used at low cost. Such an organization will need to be capable of anticipating wicked problems (problems that are resistant to a clear definition and agreed solution) (Head and Alford, 2013), and in doing so will shift from knowledge creation and management for control, to navigating the service through the complex environment. Infrastructure organizations are already knowledge organizations, with systems that structure the generation, validation, circulation, and use of knowledge (Miller and Muñoz-Erickson, 2018). However, the knowledge systems of current infrastructure organizations are narrow and specialized to match the expectations for technical standards and service delivery. Whereas these knowledge systems focus on managing for the efficient production and delivery of physical assets, in the Anthropocene infrastructure organizations will need to adapt to produce and share knowledge differently (Boisot, 1998; Schneider, 2002; Uhl-Bien, Marion and McKelvey, 2007). The Law of Requisite Complexity says that to remain effective a system must possess equal complexity to the environment in which it must operate (McKelvey and Boisot, 2003; Uhl-Bien, Marion and McKelvey, 2007). Therefore, to remain effective and relevant, infrastructure organizations need to change their knowledge systems and enable intellectual assets through distributed intelligence rather than relying on upper management, focus on speed, and lead for adaptability, knowledge, and learning, instead of efficiency and control (Uhl-Bien, Marion and McKelvey, 2007; Uhl-Bien and Arena, 2018). In introducing change towards complexity, organizations will need to distribute information, interests, and power (Head and Alford, 2013). They will need to optimize the organization's capacity for learning, creativity, and adaptability (Uhl-Bien, Marion and McKelvey, 2007). We now describe what change means and how it can be incentivized.

3.1.4. TRANSFORMING INFRASTRUCTURE MANAGEMENT

The insights from organizations that have embraced change in the context of rapidly changing environments or in the name of Anthropocene challenges do not often address the foundational questions of what change means to an organization. They instead tend to focus on business acumen and steering employees towards new goals. For example, Lockwood and Papke (2017) describe how Design Thinking as a set of organizational attributes, creates opportunities for innovation. When it comes to infrastructure there remains a dearth of knowledge around what change means to an organization, and why some organizations are able to embrace challenges like sustainability while others are not. Here we explore what change means to public service organizations focusing on a literature base that describes communication and power structure effects in institutions.

3.1.4.1. ORGANIZATIONAL CHANGE AS IDENTITY AND POWER

Leading change is about recognizing the nature of human beings and their reactions to changing structures, processes, routines, and outcomes. It is less about the structures and strategies themselves (Diefenbach, 2007). Organization change has been described as the renegotiation of shared meaning about what is valued, believed, and aimed for (Spencer-Matthews, 2001). Change starts with leaders and managers, who create the conditions for communication and renegotiation of roles. Whereas leaders view change as opportunities to renew the organization and sometimes advance their careers, middle managers and the operating core (those who do the basic work, as per Mintzberg 1981) tend not to seek or welcome change (Karp and Helgø, 2008). Organizational change can happen only when a critical mass of people's own agenda overlap with that of leadership (Karp and Helgø, 2008). As such, organizational change is often described as a socially constructed reality where power relationships are negotiated (Grant *et al.*, 2005). The primary challenge is facilitating a transition where people can associate their new identities and interests in the organization with the goals of leadership, and they do this through communication where they relate to one another and find meaning that is associated with the new goals (Stacey, 2007).

Change is wanted only when people see no other solution to the change (Diefenbach, 2007). Most people don't openly resist change but instead learn to cope with it on a tactical and operational level, often bypassing it in their daily routines (Diefenbach, 2007; Karp and Helgø, 2008).

From a leadership perspective, change involves negotiating between being and not being in control, as the realignment of identities occurs (Streatfield, 2001). The ways of the organization in the status quo allow for plans, processes and tools that allow for predictability in governing. But as change happens, interactions between people and the old structures occur in unpredictable ways as new identities are formed, and leaders have limited control over this process. Furthermore, public agencies are generally formed around models of predictability to hedge risk and thereby resist chaos, complexity and uncertainty (Karp and Helgø, 2008). The primary challenge for the future is restructuring organizations to be welcoming of continued change. Instead of leadership focused on control that seeks to perpetuate the status quo, in complex environments leadership must instead loosen control and focus resources on forming identities to ensure that change is successful (Streatfield, 2001; Griffin, 2003; Stacey, 2007; Karp and Helgø, 2008). The loosening of control allows chaos to occur and removes barriers to reorganization, allowing for self-organization, self-governing, uncertainty, new ideas, sensemaking, and diversity to flourish, necessary catalysts of change (Karp and Helgø, 2008).

3.1.4.2. LEADERSHIP FOR COMPLEX ADAPTIVE SYSTEMS

In the pursuit of organizational change in the name of adaptation, traditional approaches have tended to emphasize fixed boundaries, compartmentalized organizational responses, and simplified coordination and communication that have hampered efforts (Simon, 1962; Cilliers, 2001; Uhl-Bien, Marion and McKelvey, 2007). Infrastructure organizations must emphasize collective thinking and flexibility in viewpoints, towards adaptability that emphasizes the creation and capture of knowledge (Uhl-Bien, Marion and McKelvey, 2007). In managing knowledge production, managers should focus on enabling emergence and coordinating its context. They do this with three leadership functions: Administrative (formal managerial roles to plan and coordinate activities), Adaptive (adaptive, creative, and learning actions that emerge from interactions that drive complexity), and Enabling (creates conditions for Adaptive leadership and manages entanglement -- dynamic relationships between formal top-down hierarchy and informal complex adaptive emergent forces -- between Administrative and Adaptive functions). The formal bureaucracy cannot be disentangled from the complexity. At times the bureaucracy (rationalized structure and coordination) needs to be emphasized (when the environment is stable) and at other times the complexity (when the environment is volatile). Enabling leadership is the management of Adaptive and Administrative emphasis.

In complex environments Administrative leadership is necessary for periods of stability but must exercise authority with consideration of the organization's need for creativity, learning, and adaptation (Uhl-Bien, Marion and McKelvey, 2007). Administrative leadership is the management of bureaucratic functions. It is necessary to ensure the structuring of tasks, planning, vision building, resource acquisition, crises management, and organizational strategy (Yukl, 1981; Mumford, Bedell-Avers and Hunter, 2008). As organizations seek to introduce agility in the face of growing complexity, Administrative leadership will drive the allocation of resources within the organization, and support planning and coordination. As such, how this aspect of the organization supports knowledge building is critical. Institutions that do not relax administrative control during times when strategic realignment is occurring, or the environmental conditions are volatile are going to be less capable of adapting to change.

Adaptive leadership emerges in the movement to change priorities and goals, and is a dynamic rather than a person (Bradbury and Lichtenstein, 2000; Drath, 2001; Lichtenstein *et al.*, 2006). It emerges from the renegotiation of roles, goals, and ideas, and results from asymmetry in authority or preferences. Authority-based asymmetry involves top-down hierarchies that seek change. Preference-based asymmetry describes how the clashing of incompatible ideas, knowledge, or technologies produces new knowledge, creative ideas, learning, or adaptation (Uhl-Bien, Marion and McKelvey, 2007). This innovation occurs when groups debating a particular approach give up the untenable aspects of their respective positions and focus on tenable aspects, or the creation of new ideas, to identify new approaches or thinking. As such, Adaptive leadership is not the result of people but of a process of learning towards the creation of new insights or knowledge (i.e., emergence).

Enabling leadership encompasses the maneuvering of conditions to catalyze change, and middle managers are often in a prime position given their access to resources and their ability to steer production (Jaques, 1989; Osborn and Hunt, 2007; Uhl-Bien, Marion and McKelvey, 2007). Enabling leadership has two primary functions, to: i) manage the conditions in which adaptive leadership occurs, and ii) move innovative knowledge and products of adaptive leadership through the organization (Uhl-Bien, Marion and McKelvey, 2007). In managing the conditions for Adaptive leadership, (1) fostering interactions (facilitating the movement and interplay of information by creating the conditions for self-organizing to occur), (2) fostering interdependency (creating the pressure to act on information, particularly in situations where agents are receiving incompatible information and must adjust to elaborate their information), and (3) creating tension (an imperative to act and elaborate strategy and information, through stimuli such as pressures or challenges that are supported by resources for creativity) are fundamental.

3.1.4.3. ORGANIZATIONAL STRATEGIES

In addition to leadership change several strategies can be employed to increase the capacity of organizations to work within complex environments. A body of research characterizes how public service organizations should approach wicked problems (Head and Alford, 2013). We draw from this work to describe the competencies needed by public service organizations in navigating wicked problems, and increasingly complex environments.

Organizations need thinking that creates greater room for discovering alternative ways of solving problems to elucidate the variables, options, and linkages at the heart of complexity and to advance solutions (Head and Alford, 2013). First, wicked problems are not generally amenable to solutions driven by regulation or appeals to scientific knowledge (Schon, 1995). Instead, they are often grounded in different value perspectives that frame how different parties understand the issue and its solutions. By identifying how these different framings differ, organizations can start to facilitate a dialogue that recognizes the different value perspectives, describe how the implications of these different value perspectives, and address conflicting views (Head and Alford, 2013). Second, by deploying systems thinking, organizations will begin to recognize the inputs, processes, and outputs that are defining the complexity, and that solutions will be context dependent, overcoming some of the command-and-control bureaucratic structures that limit these explorative opportunities (Senge, 1990; Chapman, 2004; Seddon, 2008). Third, analysis can learn from complexity theory which recognizes the interdependencies, feedback loops, emergent features, and surprises that produce chaos and contradictory results (Berkes, Colding and Folke, 2008). Complexity theory can help organizations identify trends in what appears to be noise, and move away from conventional approaches that focus on identifying the more effective solution (Head and Alford, 2013).

Whereas divisional bureaucratic structures limit an organization's ability to sense and respond to complex environments, collaboration is needed to develop a shared understanding, agree on purpose, and establish mutual trust (Huxham and Vangen, 2005). Where there are multiple parties with different knowledge, interests, and values, collaborative relationships can show blind spots in an organization's knowledge system and enhance understanding. It can increase the likelihood that the nature of the problem and its causes are identified, and solutions are found and agreed to by helping diverse parties reach an understanding. Collaborative approaches also help implement solutions as parties are more likely to agree on next steps, and they enable share contributions, coordinated actions, and mutual adjustments as solutions are implemented (Head and Alford, 2013).

The leadership styles previously described tend to focus on organizational agility for power sharing, and should be supported by collaborative leadership for shared power (Crosby and Bryson, 2005). Collaborative leadership is about eliciting cooperation from partners through invitation to the dialogue, framing of the issues, orchestrating of the agenda, recognizing expertise, facilitating win-win negotiation processes, identifying entrepreneurial opportunities, and engaging in diplomacy (Head and Alford, 2013).

Leadership and the capacities to work within and develop solutions for complexity will require organizational structures and processes to change. First, as discussed the traditional divisional bureaucracy tends to silo expertise constraining an organization's ability to collaborate, share ideas, and sense changes in the environment. To alleviate these divisional constraints, assigning staff to a home division but creating opportunities for them to be re-posted to strategic projects improves organizational agility to respond to wicked and complex problems (Head and Alford, 2013). Cross-cutting committees can also be valuable if imbued with decision-making authority. Second, the financial structures and incentives associated with divisional bureaucracies emphasize products and services, but not usually how well the organization is responding to complex challenges. Budgets and financials should consider how to support outcomes, outputs, and processes associated with an organization's knowledge system and its ability to sense and respond to changes in its environment. There should also be opportunities for the pooling of budgets, where incentives are provided for multiple divisions within an organization, or multiple organizations to work together to address multi-faceted problems (Head and Alford, 2013). Similar to financial structuring, performance assessment or knowledge system monitoring should be updated to reflect how well the organization is tackling complexity (Dooren, Bouckaert and Halligan, 2015). Lastly, organizations should shift how they hire, retain, and promote their staff with consideration given to the knowledge, experience, and skills necessary to work in a collaborative environment when addressing complexity (Alford and O'Flynn, 2012).

These leadership and process changes have a commonality of supporting the capacity of the organization to sense and respond to changing conditions at faster rates. As infrastructure organizations are increasingly required to navigate the growing complexity, they will need to transition to accommodate these structural and process differences. How effectively they transition will likely determine their capacity to deliver services in accelerating and uncertain environments, and ultimately their relevance. Next, we reimagine an infrastructure governance model that is capable of transitioning to thrive in the Anthropocene.

3.1.5. TRANSITIONING INFRASTRUCTURE GOVERNANCE

Transitioning infrastructure governance is an effort that will require sustained commitment to taking down forces that lock-in the systems including management, financing, education, and technologies. It will require a recognition that how we currently manage infrastructure and the services it produces is under threat, and a willingness to test what could work without having clear evidence of precisely how to structure governance. Each of these is inimical to the current ways of governing infrastructure, and as such any transition represents a wicked complex process that will require new approaches that most infrastructure managers are not trained for. Following, we first describe what management form infrastructure organizations should consider transitioning to. Next, drawing from literature on how to change public service organizations, we summarize key factors necessary to support the successful transition of infrastructure governance. We then describe the steps necessary to improve the sensemaking capabilities of infrastructure systems, for increasingly complex environments.

Where the divisional bureaucratic form is pervasive, infrastructure organizations should create the flexibility where adhocracy structures can dominate during times of instability. Adhocracy is a structural configuration of management that emphasizes the fusing of experts from different disciplines as ad hoc teams to tackle new and complex problems (Mintzberg, 1979). Whether the divisional bureaucratic structure should remain at all is an open question. Uhl-Bien et al. (2007) contend that the reality is that complete transitions away from the current structure are unlikely and that they may still be needed during periods of stability. However, adaptive leadership for unstable conditions and enabling leadership to shift between stability and instability are critical competencies for public institutions increasingly confronting new and more complex demands and challenges (Uhl-Bien, Marion and McKelvey, 2007). What is clear is that the divisional bureaucratic form is ill-suited for conditions of high complexity and instability. An adhocracy emphasizes a highly organic structure with little rationalization and formalization of behavior (Figure 3.1.1). It involves a grouping of specialists in functional units but with the capacity to deploy them to market-based teams as needs arise, a reliance on a liaison to communicate and coordinate across ad-hoc teams, and selective decentralization of teams and their strategic locating across the organization (Mintzberg, 1979). The adhocracy structure works against the organization relying on any form of standardization for coordination. It not only imbues experts with the authority to study external conditions and drive decisions, but also emphasizes the building of new knowledge and skills. Mutual adjustment becomes the prime coordination mechanism, unlike a divisional bureaucracy where coordination is left to a few managers. Managers do exist in the adhocracy, but instead of focusing on standardized procedures and goals, they will lead small groups and help coordinate across groups by liaising and negotiating (Mintzberg, 1979). Mintzberg's adhocracy model may have been describing Simon (2002)'s near-decomposability feature of systems, i.e., subunits that have autonomy but maintain connections and coordinate activities (Teece, 2007). Near-decomposability describes how subunits at the lowest levels in a system find equilibrium solutions faster than higher levels, due to rates of interaction and therefore greater ability to change (Simon, 2002; Teece, 2007). To shift from a divisionalized form to an adhocracy, the power structures and identities within organizations must be shifted.

Motivating change is less about convincing the organization of a better future, but more about a bad or dangerous present (Diefenbach, 2007). Nevertheless, the organizational and leadership strategies we have described enable the sensemaking necessary to recognize that a new vision is needed prior, or at least desired, prior to a crisis and avoiding disaster conditions if possible. Changing the identities of individuals in the organization through a shared vision requires sustained and sufficient efforts. In reviewing the factors relevant to public service organization change, several commonalities exist (Armenakis and Bedeian, 1999; Fernandez and Rainey, 2006). Summarizing the work by Fernandez and Rainey (2006), we describe key aspects of these factors in the context of infrastructure organizations.

• Leaders must ensure that change is needed by convincing individuals of the need for change. This involves crafting a compelling vision that is easy to communicate, that members of the organization find appealing, and that provides a direction for change. Windows of opportunity such as natural disasters can be excellent times to introduce new visions (Judson, 1991; Kotter, 1995; Nadler, 1997; Abramson and Lawrence, 2001; Carnall, 2007).

• The vision must be transformed into a course of action with goals, plans, and new expectations and rules for achieving it. Plans serve as road maps that provide direction to a preferred end state, identify barriers, and recommend options for overcoming the barriers.

• Internal and widespread support for change must be garnered to overcome resistance. A crisis or shock often results in reduced resistance as individuals view change as inevitable. The shock can be real or manufactured (Thompson and Fulla, 2001; Laurent, 2003). To overcome change resistance, several strategies exist including threats, criticism, persuasion, rewards, compromises, guarantees against personal loss, psychological support, building loyalty, and recognition of the value of past practices (Judson, 1991).

- Upper management support for change is necessary. Coalitions can be important for steering resources and support to organizational members.

- External support from political overseers and external stakeholders is needed to ensure that change resources are appropriated to the organization and reduce barriers.

- Change is expensive, and underfunding can lead to weak implementation including stress on individuals and neglect of the core organizational functions. Appropriate and sustained funding is required for change activities including strategy development, communications, training, new processes and practices, restructuring, and testing innovations.

- If the aforementioned factors are not institutionalized then significant risk exists that the organization will revert to past behaviors. Change leaders must modify formal structures, procedures, and practices, employ rights, deploy innovation through trial projects, track progress through data collection, and coach employees through learning by doing (Armenakis and Bedeian, 1999).

- Transformation of large organizations requires congruence in subsystems and understanding of the interactions of these subsystems. Change in only a fraction of the subsystems can lead to failed transformation. Focusing on high-impact decision-making subsystems may be an effective way to prioritize (Amis, Slack and Hinings, 2004).

These eight factors provide useful guidance for infrastructure managers as they plan change. They point to the preconditions, opportunities for disruption, and sustained actions and resources necessary for transformation. However, change without visions or goals is often misdirected, and improved sensemaking is increasingly highlighted as the critical competency for the Anthropocene.

Sensemaking is the process of giving meaning to how the environment is changing, and of changing technologies, governance, and education at a fast enough pace to remain viable (Weick, 1995; Miller and Muñoz-Erickson, 2018). Sensemaking is the primary activity of knowledge organizations. While infrastructure organizations have been designed around goals of delivering services (often physical and characterized by public and not market value), they are likely to struggle to continue doing so under assumptions that service demand and conditions of delivery will remain predictable. As such they'll need to innovate how they architect knowledge. Knowledge organizations must be reflexive by observing, assessing, evaluating, reflecting on, and reforming their knowledge systems, i.e., how they make knowledge about their own systems and the environments they operate in (Miller and Muñoz-Erickson, 2018). They must establish a perspective of the organization and its environment, and question whether their knowledge systems are adequate. To do this, three key strategies are needed (Miller and Muñoz-Erickson, 2018). First, infrastructure organizations must re-align their knowledge and decision making. The alignment of knowledge and action is called co-production and involves developing a decision framework and then focusing on the knowledge necessary to implement the framework and monitoring the outcomes. Following the organizational change rules described by (Fernandez and Rainey, 2006), knowledge creation should occur among diverse stakeholders (producers and users) so as to avoid lock-in driven by a single stakeholder (e.g., the knowledge producer) assuming they know everything needed to act. Next, following the co-production process, knowledge systems innovation is needed (including monitoring, stress-testing, and upgrading) to ensure that processes are changing and functioning as required. Finally, a professional capacity for knowledge systems management must be built, including training the organization's workforce to create, analyze, and innovate knowledge systems and design. As per the adhocracy model of governance, the organization must ensure that there are people strategically located throughout who can reflect on the knowledge systems, how they behave and work, how they are organized, and how they generate, validate, communicate, and apply knowledge to critical decisions (Miller and Muñoz-Erickson, 2018).

3.1.6. CONCLUSION

For an infrastructure organization embracing the reality of the Anthropocene, this would mean a more flexible, adaptive, and reflexive mode of operation. Infrastructure organizations would analyze and explore current and future trends to not only understand threats to physical operations but also to gain information that might require organizational changes. This would entail sense-making activities across the organization that require an interdisciplinary approach that would enable staff and administration to analyze opportunities and risks across complex social, ecological, and technological domains at various spatial and temporal scales. This would be augmented by an embrace of a more collaborative governance model – working with other infrastructure organizations as well as other governmental, industry and community stakeholders. Knowledge systems design, then, is fundamental to infrastructure design in the Anthropocene. This would include boundary agents or units within and across infrastructure organizations to ensure crosscutting risks and concepts, such as resilience, are integrated. Now, climate or resilience expertise is often in a separate part of a bureau or in a different department altogether. More adaptive knowledge systems would break down those institutional barriers to integrate expertise on specific risks and across SETs domains. This knowledge integration would also be connected with design, implementation, and maintenance across infrastructure types. The management of green infrastructure by a parks department could enhance service delivery from stormwater bureaus as well as deliver thermal regulation benefits. Better coordination between infrastructure organizations as well as with a broader set of stakeholders would also support collaboration in how and what kind of services are delivered as whole (as opposed to siloing management and understanding).

Shifting from rigid governance models to those that emphasize agility and flexibility will not be without tradeoffs. The rigidity that is present has indeed created tremendous value and well-being for societies over the past century. The inertia inherent to the rigidity has likely helped projects to get built and to persist despite political and funding cycles that operate on much shorter timescales, allowed investors to risk capital and base their investment on the presumption of infrastructure permanence, and sent signals that there is a commitment to established development that new services should be built on. The foundational concern, however, is that when a system is unable to change appropriately or at the rate that its environment is changing, then it will lose viability. This could take many forms including the decentralization of power where new players control aspects of the system that previously were controlled by a single agency, or at the extreme where the agency or technologies are replaced. We argue that the tensions between legacy models that emphasize and perpetuate rigidity and new models that allow for flexibility and agility in the face of instability must be reconciled. If they are not, then the basic and critical systems that we rely on appear increasingly likely to be disrupted and irrelevant.

It is difficult to identify examples of major proactive shifts in infrastructure governance motivated by a recognition that environments and demands are likely to become unstable. There are many examples of agencies that have embraced sustainability or resilience (Feiock *et al.*, 2014; Carlos *et al.*, 2018), for example, but these efforts tend to focus on adding new considerations on top of existing priorities and the structures that have persisted. If indeed there are few to no cases where foundational transformation of infrastructure agencies for complexity exist, then this is quite telling. Despite growing evidence of accelerating, increasingly uncertain, and increasingly complex environments, infrastructure organizations do not appear to have the capabilities to recognize forthcoming change and/or meaningfully change governance processes to proactively respond. Radical change isn't needed overnight, and infrastructure services will remain relevant for some time. However, the increasing disconnect between what our infrastructure is governed and designed to do, and what we'll need them to do in the future can be expected to become increasingly apparent. And where existing services are unable to keep up with increasing complexity, new alternatives will emerge, adding more complexity to the delivery of basic and critical services. It's critical that agencies today invest in building the capacities for sensemaking and begin assessing how their organizations should be structured for the next century instead of the last.

3.1.7. REFERENCES

Abramson, M.A. and Lawrence, P.R. (2001) *Transforming Organizations*. Lanham, MD: Rowman & Littlefield.

ADOT (2019) *Organization Chart | Department of Transportation, Arizona Department of Transportation*. Available at: https://azdot.gov/about/inside-adot/organization-chart (Accessed: 30 October 2020).

Alford, J. and O'Flynn, J. (2012) *Rethinking Public Service Delivery: Managing with External Providers*. London, UK: Macmillan International Higher Education.

Amis, J., Slack, T. and Hinings, C.R. (2004) 'The Pace, Sequence, and Linearity of Radical Change', *Academy of Management Journal*, 47(1), pp. 15–39. Available at: https://doi.org/10.2307/20159558.

Armenakis, A.A. and Bedeian, A.G. (1999) 'Organizational Change: A Review of Theory and Research in the 1990s', *Journal of Management*, 25(3), pp. 293–315. Available at: https://doi.org/10.1177/014920639902500303.

Ausubel, J.H. and Herman, R. (eds) (1988) *Cities and Their Vital Systems: Infrastructure Past, Present, and Future*. Washington, DC: National Academies Press. Available at: https://doi.org/10.17226/1093.

Beniger, J.R. (1989) *The Control Revolution: Technological and Economic Origins of the Information Society*. Cambridge, MA: Harvard University Press.

Berkes, F., Colding, J. and Folke, C. (2008) *Navigating Social-Ecological Systems: Building Resilience for Complexity and Change*. Cambridge, UK: Cambridge University Press.

Boisot, M.H. (1998) *Knowledge Assets: Securing Competitive Advantage in the Information Economy*. Oxford, UK: OUP Oxford.

Bradbury, H. and Lichtenstein, B.M.B. (2000) 'Relationality in Organizational Research: Exploring The Space Between', *Organization Science*,

11(5), pp. 551–564. Available at:
https://doi.org/10.1287/orsc.11.5.551.15203.

van der Brugge, R., Rotmans, J. and Loorbach, D. (2005) 'The transition in Dutch water management', *Regional Environmental Change*, 5(4), pp. 164–176. Available at: https://doi.org/10.1007/s10113-004-0086-7.

Caltrans (2020) *Departmental Organizational Chart, California Department of Transportation*. Available at: https://dot.ca.gov/about-caltrans/departmental-organizational-chart (Accessed: 30 October 2020).

CAP-AZ (2010) *Organization Chart*. Phoenix, AZ: Central Arizona Project. Available at: https://web.archive.org/web/20150914162750/https://www.cap-az.com/documents/departments/finance/2010-11-Biennial-Budget-Organizational-Summaries.pdf (Accessed: 30 October 2020).

Carlos, M. *et al.* (2018) *Institutionalizing Urban Resilience: A Midterm Monitoring and Evaluation Report of 100 Resilient Cities*. New York, NY: Rockefeller Foundation. Available at: https://www.rockefellerfoundation.org/wp-content/uploads/Institutionalizing-Urban-Resilience-A-Midterm-Monitoring-and-Evaluation-Report-of-100-Resilient-Cities.pdf (Accessed: 31 October 2020).

Carnall, C.A. (2007) *Managing Change in Organizations*. Hoboken, NJ: Financial Times Prentice Hall.

Chandler, A.D. (1977) *The Visible Hand*. Cambridge, MA: Harvard University Press.

Chapman, J. (2004) *System Failure: Why Governments Must Learn to Think Differently*. New York, NY: Demos.

Chester, M., Markolf, S. and Allenby, B. (2019) 'Infrastructure and the environment in the Anthropocene', *Journal of Industrial Ecology*, 23(5), pp. 1006–1015. Available at: https://doi.org/10.1111/jiec.12848.

Chester, M.V. and Allenby, B. (2018) 'Toward adaptive infrastructure: flexibility and agility in a non-stationarity age', *Sustainable and Resilient Infrastructure*, 4(4), pp. 173–191. Available at: https://doi.org/10.1080/23789689.2017.1416846.

Chester, M.V. and Allenby, B. (2019) 'Infrastructure as a wicked complex process', *Elementa: Science of the Anthropocene*. Edited by A. Iles and M.E. Chang, 7(1), p. 21. Available at: https://doi.org/10.1525/elementa.360.

Cilliers, P. (2001) 'Boundaries, hierarchies and networks in complex systems', *International Journal of Innovation Management*, 05(02), pp. 135–147. Available at: https://doi.org/10.1142/S1363919601000312.

Coutard, O. (ed.) (2002) *The Governance of Large Technical Systems*. Oxfordshire, UK: Taylor & Francis. Available at: https://doi.org/10.4324/9780203016893.

Crosby, B.C. and Bryson, J.M. (2005) *Leadership for the Common Good: Tackling Public Problems in a Shared-Power World*. New York, NY: Wiley.

Desai, J.D. and Armanios, D.E. (2018) 'What Cannot Be Cured Must Be Endured: Understanding Bridge Systems as Institutional Relics', *Journal of Infrastructure Systems*, 24(4), p. 04018032. Available at: https://doi.org/10.1061/(ASCE)IS.1943-555X.0000451.

Diefenbach, T. (2007) 'The managerialistic ideology of organisational change management', *Journal of Organizational Change Management*, 20(1), pp. 126–144. Available at: https://doi.org/10.1108/09534810710715324.

Dooren, W.V., Bouckaert, G. and Halligan, J. (2015) *Performance Management in the Public Sector*. London, UK: Routledge.

Drath, W. (2001) *The Deep Blue Sea: Rethinking the Source of Leadership*. New York, NY: Wiley. Available at: https://www.amazon.com/Deep-Blue-Sea-Rethinking-Leadership/dp/0787949329 (Accessed: 22 October 2023).

Edwards, P. (2003) 'Infrastructure and Modernity: Scales of Force, Time, and Social Organization in the History of Sociotechnical Systems', in *Modernity and Technology*. Cambridge, MA: MIT Press, pp. 185–225.

Edwards, P.N. *et al.* (2007) 'Understanding Infrastructure: Dynamics, Tensions, and Design', in *Report of a Workshop on "History & Theory of Infrastructure: Lessons for New Scientific Cyberinfrastructures"*. Available at: https://deepblue.lib.umich.edu/bitstream/handle/2027.42/49353/UnderstandingInfrastructure2007.pdf (Accessed: 28 October 2020).

EPA (2020) *Organization Chart*. Washington, DC: Environmental Protection Agency. Available at: https://www.epa.gov/sites/production/files/2020-01/documents/usepa_orgchart_11x17.pdf (Accessed: 30 October 2020).

Feiock, R.C. *et al.* (2014) 'The Integrated City Sustainability Database', *Urban Affairs Review*, 50(4), pp. 577–589. Available at: https://doi.org/10.1177/1078087413515176.

Fernandez, S. and Rainey, H.G. (2006) 'Managing Successful Organizational Change in the Public Sector', *Public Administration Review*, 66(2), pp. 168–176. Available at: https://doi.org/10.1111/j.1540-6210.2006.00570.x.

Friedlander, A. (1995a) *Emerging Infrastructure: The Growth of Railroads*. Corporation for National Research Initiatives.

Friedlander, A. (1995b) *Natural Monopoly and Universal Service: Telephones and Telegraphs in the U.S. Communications Infrastructure, 1837-1940*. Corporation for National Research Initiatives.

Friedlander, A. (1996) *Power and Light: Electricity in the U.S. Energy Infrastructure, 1870-1940*. Corporation for National Research Initiatives.

GADOT (2020) *Organizational Chart*. Atlanta, GA: Georgia Department of Transportation. Available at: http://www.dot.ga.gov/AboutGeorgia/Documents/OrgChart.pdf (Accessed: 30 October 2020).

Geels, F.W. *et al.* (2016) 'The enactment of socio-technical transition pathways: A reformulated typology and a comparative multi-level analysis of the German and UK low-carbon electricity transitions (1990–2014)', *Research Policy*, 45(4), pp. 896–913. Available at: https://doi.org/10.1016/j.respol.2016.01.015.

Goldthau, A. (2014) 'Rethinking the governance of energy infrastructure: Scale, decentralization and polycentrism', *Energy Research & Social Science*, 1, pp. 134–140. Available at: https://doi.org/10.1016/j.erss.2014.02.009.

Graham, S. and Marvin, S. (2001) *Splintering Urbanism: Networked Infrastructures, Technological Mobilities and the Urban Condition*. London, UK: Routledge.

Grant, D. *et al.* (2005) 'Guest editorial: discourse and organizational change', *Journal of Organizational Change Management*, 18(1), pp. 6–15. Available at: https://doi.org/10.1108/09534810510579814.

Griffin, D. (2003) *The Emergence of Leadership: Linking Self-organization and Ethics*. London, UK: Routledge.

Habib, M., Brealey, R. and Cooper, I. (2000) 'The financing of large engineering projects', in R. Miller and D. Lessard (eds) *Habib, Michel; Brealey, Richard; Cooper, Ian (2000). The financing of large engineering projects. In: Miller, Roger; Lessard, Donald. The strategic management of large engineering projects: shaping institutions, risks, and governance. Cambridge, Massachusetts, 165-179*. Cambridge, MA: University of Zurich, pp. 165–179. Available at: https://www.zora.uzh.ch/id/eprint/189118/ (Accessed: 13 November 2023).

Head, B.W. and Alford, J. (2013) 'Wicked Problems: Implications for Public Policy and Management', *Administration & Society*, 47(6), pp. 711–739. Available at: https://doi.org/10.1177/0095399713481601.

Hines, P., Blumsack, S. and Schläpfer, M. (2015) 'Centralized versus Decentralized Infrastructure Networks', *ArXiv* [Preprint]. Available at: https://doi.org/10.48550/arXiv.1510.08792.

Hommels, A. (2005) 'Studying Obduracy in the City: Toward a Productive Fusion between Technology Studies and Urban Studies', *Science, Technology, & Human Values*, 30(3), pp. 323–351. Available at: https://doi.org/10.1177/0162243904271759.

Horan, S., McGrath, T. and Peterson, B. (2018) *The Modern Utility as a Portfolio Manager, Lek.* Available at: https://www.lek.com/insights/ei/modern-utility-portfolio-manager (Accessed: 19 February 2020).

Huxham, C. and Vangen, S.E. (2005) *Managing to Collaborate: The Theory and Practice of Collaborative Advantage.* London, UK: Psychology Press/Routledge.

Jaques, E. (1989) *Requisite organization: the CEO's guide to creative structure and leadership.* Arlington, Va.: Cason Hall.

Jordan, A. (2008) 'The Governance of Sustainable Development: Taking Stock and Looking Forwards', *Environment and Planning C: Government and Policy*, 26(1), pp. 17–33. Available at: https://doi.org/10.1068/cav6.

Judson, A.S. (1991) *Changing Behavior in Organizations: Minimizing Resistance to Change.* Oxford, UK: B. Blackwell.

Kaminsky, J.A. (2018) 'National Culture Shapes Private Investment in Transportation Infrastructure Projects around the Globe', *Journal of Construction Engineering and Management*, 144(2), p. 04017098. Available at: https://doi.org/10.1061/(ASCE)CO.1943-7862.0001416.

Karp, T. and Helgø, T.I.T. (2008) 'From Change Management to Change Leadership: Embracing Chaotic Change in Public Service Organizations', *Journal of Change Management*, 8(1), pp. 85–96. Available at: https://doi.org/10.1080/14697010801937648.

Kooiman, J. (1993) *Modern Governance: New Government-Society Interactions.* Thousand Oaks, CA: SAGE.

Kotter, J.P. (1995) 'Leading Change: Why Transformation Efforts Fail', *Harvard Business Review*, 1 May. Available at: https://hbr.org/1995/05/leading-change-why-transformation-efforts-fail-2 (Accessed: 9 May 2020).

La Porte, T.R. (1996) 'High Reliability Organizations: Unlikely, Demanding and At Risk', *Journal of Contingencies and Crisis Management*,

4(2), pp. 60–71. Available at: https://doi.org/10.1111/j.1468-5973.1996.tb00078.x.

Laurent, A. (2003) 'Entrepreneurial Government: Bureaucrats as Business People', in M.A. Abramson and A.M. Kieffaber (eds) *New Ways of Doing Business*. Lanham, MD: Rowman & Littlefield Publishers, pp. 13–47.

Leach, M., Scoones, I. and Stirling, A. (2010) 'Governing epidemics in an age of complexity: Narratives, politics and pathways to sustainability', *Global Environmental Change*, 20(3), pp. 369–377. Available at: https://doi.org/10.1016/j.gloenvcha.2009.11.008.

Lichtenstein, U.-B. *et al.* (2006) 'Complexity Leadership Theory: An Interactive Perspective on Leading in Complex Adaptive Systems', *Emergence*, 8(4), pp. 2–12.

Lockwood, T. and Papke, E. (2017) *Innovation by Design: How Any Organization Can Leverage Design Thinking to Produce Change, Drive New Ideas, and Deliver Meaningful Solutions*. Newburyport, Massachusetts: Red Wheel/Weiser.

Long Beach (2020) *Organizational Chart*. Long Beach, CA: City of Long Beach. Available at: https://lbwater.org/about-us/important-documents/org-chart/ (Accessed: 30 October 2020).

Loorbach, D. (2010) 'Transition Management for Sustainable Development: A Prescriptive, Complexity-Based Governance Framework', *Governance*, 23(1), pp. 161–183. Available at: https://doi.org/10.1111/j.1468-0491.2009.01471.x.

McKelvey, B. and Boisot, M. (2003) 'Transcendental organizational foresight in nonlinear contexts', in. *INSEAD Conference on expanding perspectives on strategy processes*, Fontainebleu, FR.

Miller, C.A. and Muñoz-Erickson, T.A. (2018) *The Rightful Place of Science: Designing Knowledge*. Tempe, AZ: Consortium for Science, Policy & Outcomes.

Mintzberg, H. (1979) *The Structuring of Organizations: A Synthesis of the Research*. Hoboken, NJ: Prentice-Hall.

Mintzberg, H. (1981) 'Organization Design: Fashion or Fit?', *Harvard Business Review*, 1 January. Available at: https://hbr.org/1981/01/organization-design-fashion-or-fit (Accessed: 22 October 2023).

Mumford, M.D., Bedell-Avers, K.E. and Hunter, S.T. (2008) 'Planning for innovation: A multi-level perspective', in M. D. Mumford, S. T. Hunter, and K. E. Bedell-Avers (eds) *Multi-Level Issues in Creativity and Innovation*. Bingley, UK: Emerald Group Publishing Limited (Research in

Multi-Level Issues), pp. 107–154. Available at: https://doi.org/10.1016/S1475-9144(07)00005-7.

Muñoz-Erickson, T.A. *et al.* (2016) 'Demystifying governance and its role for transitions in urban social–ecological systems', *Ecosphere*, 7(11), p. e01564. Available at: https://doi.org/10.1002/ecs2.1564.

Nadler, D.A. (1997) *Champions of Change: How CEOs and Their Companies are Mastering the Skills of Radical Change.* New York, NY: Wiley.

O'Brien, P. and Pike, A. (2015) '"The governance of local infrastructure funding and financing"', *Infrastructure Complexity*, 2(1), p. 3. Available at: https://doi.org/10.1186/s40551-015-0007-6.

Osborn, R.N. and Hunt, J.G. (Jerry) (2007) 'Leadership and the choice of order: Complexity and hierarchical perspectives near the edge of chaos', *The Leadership Quarterly*, 18(4), pp. 319–340. Available at: https://doi.org/10.1016/j.leaqua.2007.04.003.

Osborne, S.P. and Brown, K. (2005) *Managing Change and Innovation in Public Service Organizations.* London, UK: Psychology Press/Routledge. Available at: https://ebookcentral-proquest-com.ezproxy1.lib.asu.edu/lib/asulib-ebooks/reader.action?docID=200596.

Ostrom, E. (2008) 'Institutions and the Environment', *Economic Affairs*, 28(3), pp. 24–31. Available at: https://doi.org/10.1111/j.1468-0270.2008.00840.x.

Phoenix (2020) *Organization Chart.* Phoenix, AZ: City of Phoenix Street Transportation Department. Available at: https://www.phoenix.gov/streetssite/Documents/StreetsOrgChart.pdf (Accessed: 30 October 2020).

Schneider, M. (2002) 'A Stakeholder Model of Organizational Leadership', *Organization Science*, 13(2), pp. 209–220. Available at: https://doi.org/10.1287/orsc.13.2.209.531.

Schon, D.A. (1995) *Frame Reflection: Toward the Resolution of Intractable Policy Controversies.* New York, NY: Basic Books.

Seddon, J. (2008) *Systems Thinking in the Public Sector.* Dorset, UK: Triarchy Press.

Senge, P.M. (1990) *The Fifth Discipline: The Art and Practice of the Learning Organization.* 1st edn. New York, NY: Doubleday/Currency.

Serpa, C.M. and Ferreira, S. (2019) 'Rationalization and bureaucracy: Ideal-type bureaucracy by Max Weber' *Humanities & Social Sciences Reviews* 7(2), 187-195. Available at: https://doi.org/10.18510/hssr.2019.7220.

Simon, H.A. (1962) 'The Architecture of Complexity', *Proceedings of the American Philosophical Society*, 106(6), pp. 467–482.

Simon, H.A. (1996) *The Sciences of the Artificial*. Cambridge, MA: MIT Press.

Simon, H.A. (2002) 'Near decomposability and the speed of evolution', *Industrial and Corporate Change*, 11(3), pp. 587–599. Available at: https://doi.org/10.1093/icc/11.3.587.

Sovacool, B.K., Lovell, K. and Ting, M.B. (2018) 'Reconfiguration, Contestation, and Decline: Conceptualizing Mature Large Technical Systems', *Science, Technology, & Human Values*, 43(6), pp. 1066–1097. Available at: https://doi.org/10.1177/0162243918768074.

Spencer-Matthews, S. (2001) 'Enforced Cultural Change in Academe. A Practical Case Study: Implementing quality management systems in higher education', *Assessment & Evaluation in Higher Education*, 26(1), pp. 51–59. Available at: https://doi.org/10.1080/02602930020022282.

Stacey, R.D. (2007) *Strategic Management and Organisational Dynamics: The Challenge of Complexity to Ways of Thinking about Organisations*. Hoboken, NJ: Financial Times Prentice Hall.

Steffen, W. *et al.* (2015) 'The trajectory of the Anthropocene: The Great Acceleration', *The Anthropocene Review*, 2(1), pp. 81–98. Available at: https://doi.org/10.1177/2053019614564785.

Streatfield, P. (2001) *The Paradox of Control in Organizations*. 1st edn. London, UK: Psychology Press/Routledge. Available at: https://www.amazon.com/Paradox-Control-Organizations-Complexity-Emergence/dp/0415250323 (Accessed: 22 October 2023).

Teece, D.J. (2007) 'Explicating dynamic capabilities: the nature and microfoundations of (sustainable) enterprise performance', *Strategic Management Journal*, 28(13), pp. 1319–1350. Available at: https://doi.org/10.1002/smj.640.

Thompson, J.R. and Fulla, S.L. (2001) 'Effecting Change in a Reform Context', *Public Performance & Management Review*, 25(2), pp. 155–175. Available at: https://doi.org/10.1080/15309576.2001.11643652.

Uhl-Bien, M. and Arena, M. (2018) 'Leadership for organizational adaptability: A theoretical synthesis and integrative framework', *The*

Leadership Quarterly, 29(1), pp. 89–104. Available at: https://doi.org/10.1016/j.leaqua.2017.12.009.

Uhl-Bien, M., Marion, R. and McKelvey, B. (2007) 'Complexity Leadership Theory: Shifting leadership from the industrial age to the knowledge era', *The Leadership Quarterly*, 18(4), pp. 298–318. Available at: https://doi.org/10.1016/j.leaqua.2007.04.002.

USDOT (2017) *Organization Chart*. Washington, DC: U.S. Department of Transportation. Available at: https://www.transportation.gov/org-chart (Accessed: 30 October 2020).

USGS (2020) *Water Mission Area Organizational Chart*. Washington, DC: U.S. Geological Survey. Available at: https://www.usgs.gov/media/images/image-water-mission-area-organizational-chart (Accessed: 30 October 2020).

Weick, K.E. (1995) *Sensemaking in Organizations*. Thousand Oaks, CA: SAGE.

Wilson, J.Q. (1989) *Bureaucracy*. New York, NY: Basic Books.

Yukl, G.A. (1981) *Leadership in Organizations*. 1st edn. Englewood Cliffs, NJ: Prentice-Hall.

3.2

FLEXIBLE LEADERSHIP

3.2.1. INTRODUCTION

Leadership—comprised of formal and informal governance—and physical networks drive an infrastructure system's ability to respond to changing circumstances. The role of institutions is frequently overlooked in infrastructure practice and theory (Gim, Miller and Hirt, 2019; Helmrich *et al.*, 2021), but infrastructure organizations interpret the operating environments and establish how infrastructure function within them (Chester, Miller and Muñoz-Erickson, 2020). Neglecting infrastructure institutions and their organizations places infrastructure at risk of obsolescence in the Anthropocene. Institutions are defined here as knowledge, rules, and norms created by society that influence infrastructure systems, while organizations are a structured collection of people working toward a common goal (e.g., accessible, potable water within the drinking water sector) (North, 1990; Chester, Miller and Muñoz-Erickson, 2020). At the dawn of the Anthropocene social, ecological, and technological conditions have seen rapid growth and subsequent massive disruptions to Earth systems, indicating a new era founded in increasing instability (Steffen *et al.*, 2015). This is exhibited in the relationships between built infrastructure and changing climatic conditions, where built infrastructure are deteriorating, and even failing, sooner than expected (Burillo *et al.*, 2017; Underwood *et al.*, 2017; Ayyub, 2018; Bondank, Chester and Ruddell, 2018; Nasr *et al.*, 2019).

In the Anthropocene, infrastructure managers (i.e., individuals who design, build, maintain, and decommission infrastructure) can no longer rely on relatively stationary conditions, i.e., the assumption that the past may predict the future, which has been the foundational model of modern

---

This chapter was adapted from the following article with publisher and lead author permissions: Alysha Helmrich and Mikhail Chester, 2022, Navigating Exploitative and Explorative Leadership in Support of Infrastructure Resilience, *Frontiers in Sustainable Cities*, 17(791474), doi: 10.3389/frsc.2022.791474.

infrastructure (Olsen, 2015; Chester and Allenby, 2018; Markolf *et al.*, 2021). For instance, it is not unreasonable to expect gradual climate change to become increasingly significant, and if this transition were to happen rapidly, infrastructure institutions would need to respond within a reasonable timeframe and at a scale of uncertainty that is largely unfamiliar, marking a radical change in how they operate (Wilbanks and Fernandez, 2014; Chester, Markolf and Allenby, 2019; Chester, Miller and Muñoz-Erickson, 2020; Chester, Underwood and Samaras, 2020; Helmrich and Chester, 2020). Similar challenges have persisted beyond infrastructure management; the technology sector experiences a competitive environment with fast-paced technology evolution and demand changes that leads to frequent destabilization, highlighted by the stories of Xerox (Teece, 2007; Uhl-Bien and Marion, 2009) and Kodak (Courtney, Kirkland and Viguerie, 1997). The tensions between exploitation (i.e., efficiency within the status quo) and exploration (i.e., pursuing innovations and associated risks) of these technological organizations parallel tensions experienced by infrastructure managers navigating efficiency and adaptation for resilience (March, 1991; Papachroni, Heracleous and Paroutis, 2016). The consequences of designing infrastructure systems for efficiency are becoming increasingly evident with failures across a range of disturbances even beyond climate change (Chester, Underwood and Samaras, 2020; Underwood *et al.*, 2020), such as aging infrastructure and emerging technology (Arbesman, 2017; Chester and Allenby, 2018), terrorist attacks, cyber warfare (Ogie, 2017; Paté-Cornell *et al.*, 2018), and pandemics (Carvalhaes *et al.*, 2020).

Leadership within infrastructure organizations must be able to react quickly and effectively to changing environments to maintain longevity, making it imperative to study and question how infrastructure is governed today relative to disturbances. While infrastructure literature is in the nascent stages of studying governance in the context of resilience, leadership and organizational change literature has developed considerable insights around governance in the context of uncertainty and complexity. Governance is a system of rules, values, and norms that balances the responsibility, authority, and power of management and individuals to establish a cooperative behavior (Mintzberg, 1979; Kooiman, 1993; Faguet, 2014; Chester, Miller and Muñoz-Erickson, 2020). Formal governance (i.e., rule-based) and informal governance (i.e., relation-based) provide opportunities and challenges toward institutional resilience. Commonly, formal governance is characterized by power granted through hierarchical mechanisms and/or formal rules that regulate the autonomy of individuals within an organization (Boesen, 2007; Uhl-Bien and Marion, 2009). Organizational structures and contractual agreements exemplify hierarchical mechanisms, and these governance tools create an expectation of an individual's role within the organization. Concurrently, whether in tension or congruence, informal governance manifests within organizations, where power is established through social relationships amongst individuals through mutual trust, appreciation, and respect (Boesen, 2007; Uhl-Bien and Marion, 2009). In terms of formal governance, many infrastructure organizations rely heavily on hierarchical bureaucracies, this vertical dispersion of power (e.g., direct oversight) has managed conditions of stability with some adaptive capacity; however, this organizational structure appears problematic moving forward, ultimately restricting flexibility by perpetuating standardization, reducing collaboration, and diminishing the value of exploration (Mintzberg, 1979; Adler, 2001; Zhou, 2013; Chaffin, Gosnell and Cosens, 2014; Martela, 2019; Chester, Miller and Muñoz-Erickson, 2020) as explored in the following section.

Infrastructure organizations will need to better manage cooperative pursuits of efficiency and adaptation across periods of stability and instability. The ability of leaders to navigate the complexity between exploitation and exploration is referred to as 'enabling leadership' (Uhl-Bien, Marion and McKelvey, 2007; Uhl-Bien and Arena, 2017, 2018), which may be embedded in existing governance systems through acknowledging and bridging conflict as well as promoting and connecting innovation within the existing processes (Uhl-Bien and Arena, 2018). In order to achieve this, infrastructure organizations must assess and modify their processes of leadership when reconciling disturbances (Chester, Miller and Muñoz-Erickson, 2020). It has been difficult to replicate resilient organizations because, oftentimes, organizations that display adaptive capacity are not fully aware of what behaviors enabled them to do so (Uhl-Bien and Arena, 2018). As such, this chapter addresses:

1)   What are the leadership capabilities that organizations need to respond to rapidly changing conditions (i.e., shifts between stability and instability)?

2)   How do the identified leadership capabilities translate to infrastructure systems?

3.2.2. Governance of Infrastructure Systems

Infrastructure institutions were established when external conditions were more stable than experienced today (Chester, Miller and Muñoz-Erickson, 2020), and it is critical to evaluate the agility of infrastructure institutions, organizations, and governance, in order to maintain services in increasing instability (Little, 2004; Salet, Bertolini and Giezen, 2013; Omer, Mostashari and Lindemann, 2014; Chester and Allenby, 2018; Sovacool, Lovell and Ting, 2018). However, engineered infrastructure resilience literature tends to prioritize physical infrastructure (Omer, Mostashari and Lindemann, 2014; Gim, Miller and Hirt, 2019; Chester, Miller and Muñoz-Erickson, 2020), undermining the importance of decision-making towards emphasizing technological reliability (La Porte and Consolini, 1991; Schulman and Roe, 2016). This is further problematic since infrastructure systems are sociotechnical systems, and thus both the social (e.g., governance) and technological components need to operate, adapt, and transform in conjunction — or at least in parallel — to navigate the enterprise in the face of change (Hughes, 1983; Sovacool, Lovell and Ting, 2018; Gim, Miller and Hirt, 2019; Chester, Miller and Muñoz-Erickson, 2020). Public infrastructure systems were organized as centralized structures because movements toward industrialization and urbanization during the 19th and 20th century provided an opportunity to utilize economies of scale to meet increasing demand for services (Faguet, 2014; Ansell and Lindvall, 2020). A hierarchical and departmentalized organizational structure was established across infrastructure systems to simplify complicated problems by capitalizing on specialized expertise, while maintaining cost-effective coordination through a chain of command (Chandler, 1977; Friedlander, 1995b, 1995a, 1996; La Porte, 1996; Zhou, 2013; Chester, Miller and Muñoz-Erickson, 2020). These organizations are expected to maintain services without failure as the environments in which they operate become increasingly complex — a dynamic explored in high reliability organizations (Grabowski & Roberts, 2019; Roberts, 1990). As infrastructure systems mature, they are encountering lock-in, an inability or resistance to change due to past decisions (Corvellec, Zapata Campos and Zapata, 2013; Markolf *et al.*, 2018; Chester and Allenby, 2019); lock-in partially occurs because infrastructure systems become highly specialized to their environments.

Formal governance influences an organization's ability (or inability) to adapt by creating rules that control roles, responsibilities, and relationships (re: informal governance) amongst employees. This structure determines who holds decision-making power and the process of workflow for an organization. While few studies have classified organizational structures of infrastructure systems, the divisionalized form (Figure 3.2.1) is seemingly prevalent (Friedlander, 1995b, 1995a, 1996; Chester, Miller and Muñoz-Erickson, 2020). There are five components within this structure (Mintzberg, 1979):

- Operating Core – employees performing the routine tasks.

- Strategic Apex – employees aligning the organization toward a mission.

- Middle Line – employees navigating communications between the operating core and strategic apex.

- Technostructure – employees standardizing workflows.

- Staff – employees providing indirect services.

Figure 3.2.1: Conceptual Representation of Divisionalized Form

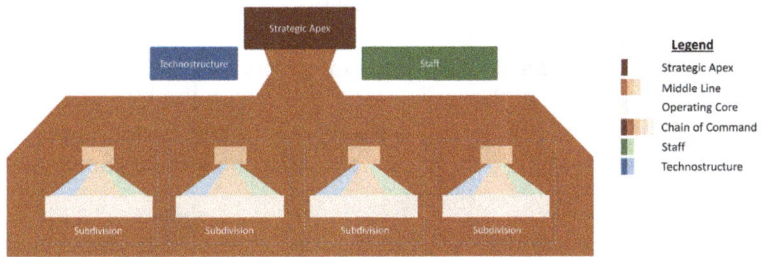

Adapted from Mintzberg (1979).

The divisionalized form relies on quality control of standardized outputs rather than direct supervision to monitor the operating core (Mintzberg, 1979); in terms of infrastructure systems, the goal of a standardized output is providing individuals with the same access to reliable, high-quality services (e.g., potable water, reliable electricity, accessible and safe travel) with socially-accepted controlled and uncontrolled failures (e.g., combined sewer overflows, rolling blackouts, traffic congestion during rush hours). A divisionalized organizational structure is most appropriate when organizations diversify, where the organization can manage additional products or address emerging needs through individual departments. This configuration permits some adaptive capacity since divisions can be added or subtracted within the organization as needed (Mintzberg, 1979; Zhou, 2013). However, departmentalization is not synonymous with governance decentralization, which is oftentimes promoted for infrastructure resilience (Helmrich *et al.*, 2021). This is because the divisionalized form still maintains a vertical hierarchy — or chain of command — with little dispersal of decision-making power since the divisions are closely managed by 'headquarters.' If a division were unable to maintain services, the headquarters could intervene. Additional critiques of divisionalized form have surfaced, including the prominence of risk-adversity and uneven power dynamics that derail innovation and limit holistic thinking (Adler, 2001; Cleaver and Whaley, 2018; Martela, 2019; Chester, Miller and Muñoz-Erickson, 2020). Highlighting a failed example of such hierarchical oversight, Flint, Michigan had been experiencing an economic crisis that caused emergency managers (state-level employees, re: headquarters) to intervene. The appointed, unelected emergency managers were not familiar with local contexts but were tasked with autocratic decision-making across a wide array of city affairs to balance budgets (Pauli, 2019). To stabilize water expenditures and obtain partial authority of the resource, Flint's water supply would be shifted from the Detroit Water and Sewerage Department (DWSD) to Karegnondi Water Authority, both sourced from Lake Huron. DWSD announced they would be terminating service prior to the completion of the new pipeline project, and the Flint River was chosen as an interim water source despite concerns. Ultimately, financial constraints, political pressure, failure of bureaucracy, and environmental injustice led to a contamination of the city's drinking water (Butler, Scammell and Benson, 2016; Masten, Davies and Mcelmurry, 2016; Pieper, Tang and Edwards, 2017; Pauli, 2019). Ineffective leadership by the emergency managers restricted Flint's ability to respond to the emergent disturbance, accentuating how institutional lock-in limits adaptive capacity.

Institutional lock-in contributes to the slow pace of change seen within infrastructure systems. First, there are consistent processes of change within organizations that occur within formal and informal governance, but lack of support for explorative behaviors can make it difficult for ongoing change to establish itself (Tsoukas and Chia, 2002; Deslatte and Swann, 2019). The establishment of formal governance may perpetuate lock-in through legacy rules and regulations, performance measures, and norms if the processes are inflexible. An adherence to exploitative behaviors helps organizations avoid overextending resources by ignoring noise in the environment; therefore, organizations must balance stability through exploitative behaviors and responsiveness with explorative behaviors while pursing ongoing change. Second, financing is crucial and infrastructure growth has been constrained in the past by federal financial decisions such as ending large grant programs, changing funding priorities, and managing increasing debt (Miller, 2000; Deslatte and Swann, 2019). However, the impacts of financial stress are not always predictable in the processes of change or subsequent outcomes (Deslatte and Swann, 2019; Garcia *et al.*, 2019). Third, and intertwined with funding opportunities, infrastructure systems are beholden to political and public interests, which may prioritize different outcomes or processes (Miller, Lessard and Group, 2000; Marshall and Alexandra, 2016; Garcia *et al.*, 2019). The factors presented here highlight a few examples of institutional lock-in an organization faces, which keep infrastructure on one path and restrict their ability to change at pace with their environments. Public service organizations are encountering increasing pressures to meet ambiguous objectives with high levels of risk aversion and unclear metrics of success (Deslatte and Swann, 2019).

3.2.3. LEADERSHIP CAPABILITIES FOR NAVIGATING STABILITY AND INSTABILITY

3.2.3.1. COLLABORATING: CREATING A VARIETY OF KNOWLEDGE THROUGH FORMAL AND INFORMAL NETWORKING

Organizations operating in complexity must develop a variety of knowledge to respond to changes in the environment. Collaboration creates a space for a heterogeneous organizational culture, and opportunities for collaboration are increased when formal and informal relationship building (i.e., networking) are abundant (Teece, 2007; Havermans *et al.*, 2015; Rosenhead *et al.*, 2019). There are two main components of this leadership capability. The first is developing a variety in stakeholders, anyone with influence on the infrastructure system, within or beyond the organization. Variety can be demonstrated in a number of ways, including function, authority, experience, location, etc. For instance, veteran employees can often make decisions without close oversight but are also prone to responding in familiar ways, creating informal institutional lock-in (Carley and Lee, 1998; Grote, 2019). Conversely, while new employees may need more guidance (e.g., training and/or mentorship), they are more likely to think creatively as their knowledge is unlikely redundant to the organization's (March, 1991). One way to build diversity, and avoid groupthink, is to not only sponsor internal personnel, but also external personnel, recruited from organizations such as research institutions, consumers, competitors, industry partners (Hooijberg, (Jerry) Hunt and Dodge, 1997; Teece, 2007; Pitelis and Wagner, 2019). These acts of inclusive community building enhance leadership's ability to think holistically and avoid groupthink (Carley and Lee, 1998; Uhl-Bien and Arena, 2017; Rosenhead *et al.*, 2019), assuming viable communication channels exist such as boundary spanning (Schneider, 2002; Taylor and Helfat, 2009). Notably, while organizational hierarchy is often considered the formal line of communication, informal relationships (e.g., trust, routine) are also significant, and formal and informal communication influence one another (Zhou, 2013).

Secondly, resilient organizations should encourage non-linear and interdependent relationships to promote transdisciplinary teamwork across all levels (Schneider, 2002; Lichtenstein and Plowman, 2009; Omer, Mostashari and Lindemann, 2014). Non-linearity allows any person to assume the role of a leader when contextually relevant (i.e., horizontal decentralization), and the interdependence creates a pressure for action that may diverge from routine conformity (Uhl-Bien, Marion and McKelvey, 2007; Mäkinen, 2018; Rosenhead *et al.*, 2019). A decentralized structure blurs the line between formal and informal communication and encourages co-current, overlapping—or even conflicting—endeavors, establishing 'messy institutions' that are arguably more adept to address complexity (Mintzberg, 1979; Teece, 2007; Lichtenstein and Plowman, 2009; Jansen, Simsek and Cao, 2012; Daviter, 2017). However, higher interdependence does come at an increased coordination cost (e.g., scheduling more meetings, hiring brokering personnel), which is necessary to ensure information is shared widely across the organization; this highlights the benefit of centralized structures which have proven their ability to be highly efficient at decomposing complicated problems — a person can trust that following protocol will achieve the goal (Teece, 2007; Taylor and Helfat, 2009; Zhou, 2013; Daviter, 2017; Uhl-Bien and Arena, 2017; Mäkinen, 2018; Martela, 2019; Nederveen Pieterse *et al.*, 2019). The takeaway from this leadership capability is that an organization must utilize their diversity by promoting multi-level collaboration which provides an opportunity for adaptive leadership to occur (Uhl-Bien, Marion and McKelvey, 2007; Mäkinen, 2018).

3.2.3.2. LEARNING: PERCEIVING AND EXPLORING THE ENVIRONMENT TO GENERATE INNOVATION

Learning is "both a product of knowledge and [the knowledge] source," where knowledge is not only shared but created and stored (Daviter, 2017; Serrat, 2017), and the ability to learn establishes a pathway for organizational change. Organizations learn from their experiences and those of their community; and they must store this information as to inform future decision-making (Serrat, 2017). Both personnel and organizations have the capacity to learn, and mutual learning promotes convergence; however, personnel must avoid conforming to formal governance or else adaptive capacity decreases (March, 1991; Uhl-Bien and Arena, 2017; Martela, 2019). To encourage learning, an organization and/or leaders should invest in individual and collective development (e.g., supporting ideas, sponsoring cross-training) and keep knowledge accessible (e.g., transparency) (Omer, Mostashari and Lindemann, 2014; Havermans *et al.*, 2015; Serrat, 2017; Bäcklander, 2019). A key component of learning is reflection, where leaders are not simply reacting but dissecting the effectiveness of their actions to inform future actions (Serrat, 2017; Rosenhead *et al.*, 2019). As organizations learn, they should act upon the information gathered by aligning or dealigning from existing formal and informal leadership strategies (e.g., structures, routines) to form a new, temporary equilibrium.

Leaders who are perceptive and explorative are in optimal positions to learn. Perceptive leaders are able to identify opportunities and challenges in advance of a disturbance and actively understand the implications and prepare a coordinated response (Teece, 2007; Jansen, Simsek and Cao, 2012; Omer, Mostashari and Lindemann, 2014; Grote, 2019; Pitelis and

Wagner, 2019). Leaders become increasingly valuable when they are able to recognize opportunities or threats outside of their expertise (Brown, 2004; Turner, Swart and Maylor, 2013; Papachroni, Heracleous and Paroutis, 2016). In order to develop independence and perceptiveness within employees, an organization should create a strong organizational culture around resilience and motivate employees through supporting organizational values (e.g., transparency, safety, sustainability) and incentives for explorative behaviors rather than focus on production output (La Porte, 1996; Gibson and Birkinshaw, 2004; Omer, Mostashari and Lindemann, 2014; Uhl-Bien and Arena, 2017; Martela, 2019; Nederveen Pieterse *et al.*, 2019; Pitelis and Wagner, 2019). While perception may be partially accredited to experience, it also relies on an organization's support of experimentation, providing a safe space to explore planned or spontaneous ideas (Gibson and Birkinshaw, 2004; Teece, 2007; Lichtenstein and Plowman, 2009; Havermans *et al.*, 2015; Arbesman, 2017; Serrat, 2017; Rosenhead *et al.*, 2019). A safe space can be established through supportive relationships and systems that provide resources to assist creative and explorative ideas, allow mistakes, provide independence, support risk-taking rather than monitoring milestones, penalizing errors, or establishing routines (Rosing, Frese and Bausch, 2011; Uhl-Bien and Arena, 2018). The cycle of creativity and experimentation allows an organization to continuously reinvent itself, not only through successes but failures (Lichtenstein and Plowman, 2009; Rosenhead *et al.*, 2019).

Innovation occurs when efforts of perception and exploration present an opportunity for a value-add to the organization, and if innovation is successful, allows an organization to evolve. The integration of an innovation into an organization is risky as the concept must be sponsored and adopted within the organization—institutionally, and possibly physically—before there is a return on investment (Galbraith, 1982; March, 1991; Rosing, Frese and Bausch, 2011; Cantarello, Martini and Nosella, 2012; Uhl-Bien and Arena, 2017, 2018; Mäkinen, 2018). There are two types of innovation an organization can pursue at either incremental or transformational scales: exploitative, seeking to increase efficiency, and explorative, focusing on new processes (March, 1991; Diesel and Scheepers, 2019). Explorative innovations are less frequently sponsored due to the emphasis on exploitative behaviors in established organizations and the long return on investment for explorative innovation (Galbraith, 1982; Teece, 2007; Rosing, Frese and Bausch, 2011). Innovation increases organization longevity by allowing organizations to not only meet current, but future, demands (Cantarello, Martini and Nosella, 2012). However, like collaboration, innovation is only one driver of an organization's adaptive capacity (Teece, 2007; Uhl-Bien and Arena, 2018).

3.2.3.3. LEADERSHIP: ENHANCING ADAPTIVE CAPACITY THROUGH FLEXIBLE FORMAL AND INFORMAL GOVERNANCE

The environments in which an organization operates are dynamic; therefore, leadership must be flexible to new ideas, responsive to shifts, and embrace uncertainty. Formal governance within organizations often-times seeks to routinize tasks for efficiency, but this creates a vulnerable position in the Anthropocene (March, 1991; Boisot and Mckelvey, 2011; Havermans *et al.*, 2015; Daviter, 2017). If an organization overemphasizes exploitative behaviors, it will likely be unable to adjust therefore collaps-ing due to a lack of flexibility, innovation, and adaptive capacity (Lichten-stein and Plowman, 2009; Papachroni, Heracleous and Paroutis, 2016; Sovacool, Lovell and Ting, 2018; Martin, 2019). Exploitative behaviors have been able to persist at scale due to the relatively stable environment, but the increasing uncertainty faced today demands organizations to change more rapidly, requiring organizational repositioning (Siggelkow and Levinthal, 2003; Teece, 2007, 2007; Sovacool, Lovell and Ting, 2018; Uhl-Bien and Arena, 2018; Martela, 2019). Sovacool et al. (2018) character-izes repositioning as transformation, technological substitution, reconfig-uration, dealignment and realignment. For example, one proposal of de-alignment and realignment is to implement temporary decentralization, where an organization shifts to a decentralized form when it encounters a disturbance and reverts to a centralized form when a temporary equilib-rium is found (Siggelkow and Levinthal, 2003). Temporary decentraliza-tion is already used within infrastructure systems during disaster recov-ery when immediate action is necessary to prevent cascading failures.

Leadership capabilities for complexity emphasize the capabilities to shift between exploitative and explorative behaviors (Hooijberg, (Jerry) Hunt and Dodge, 1997; Gibson and Birkinshaw, 2004; Havermans *et al.*, 2015; Uhl-Bien and Arena, 2018), highlighting the importance of respon-siveness and preparedness during uncertainty (Teece, 2007; Pitelis and Wagner, 2019). The organization should empower personnel (of varying levels of authority) to take initiative by providing them support (e.g., men-torship, motivation) and the necessary tools (e.g., resources, training) (Schneider, 2002; Cantarello, Martini and Nosella, 2012; Bäcklander, 2019; Martela, 2019). When an opportunity or threat is perceived, leaders must decide whether or not they will act upon the information (March, 1991; Zahra and George, 2002; Teece, 2007; Uhl-Bien and Arena, 2017, 2018; Bäcklander, 2019; Pitelis and Wagner, 2019). If an organization is too be-holden to exploitative behaviors, their bias may steer them toward inac-tion (Courtney, Kirkland and Viguerie, 1997; Teece, 2007; Pitelis and Wag-ner, 2019; Rosenhead *et al.*, 2019). To safeguard organizational longevity, leaders should recognize, accept, and endorse external uncertainty and rapid change (Lichtenstein and Plowman, 2009; Uhl-Bien and Arena, 2018;

Tourish, 2019). This is a difficult demand of leaders who will need to accept that they do not have control over the unpredictable system (Tsoukas and Chia, 2002; Rosenhead *et al.*, 2019; Tourish, 2019).

3.2.3.4. ENABLING LEADERSHIP: NAVIGATING CONFLICT AND CONTESTATION OF DRIFT AND CRISIS

Organizations can embrace the growing complexity and deep uncertainty of their external environments to leverage productive organizational change. An organization can embrace the potential of conflict, contestation, and controversy (i.e., tension) of complex systems to catalyze leadership capabilities including collaboration, learning, and flexibility toward creating enabling leadership (Lichtenstein and Plowman, 2009; Uhl-Bien and Arena, 2018; Rosenhead *et al.*, 2019; Tourish, 2019). Facilitating tension creates space for varying perspectives, constructive criticism, and innovation. In leadership and organizational change literature, the tension between exploitative and explorative behaviors is frequently discussed. This tension, however, should not always be viewed as a tradeoff where one must be disadvantaged to pursue the other. Instead, the tension can be seen as complementary, i.e., "continuous improvement," or interrelated, i.e., "distinct but equally necessary" (Gibson and Birkinshaw, 2004; Papachroni, Heracleous and Paroutis, 2016). An organization cannot seek to eliminate tension—as seen with hierarchical structures—as these disruptions, and eventual reconciliations, are what drive organizations to be successful over time (Gibson and Birkinshaw, 2004; Cantarello, Martini and Nosella, 2012; Uhl-Bien and Arena, 2017).

Given the emphasis on exploitative practices within organizations writ large, and particularly in hierarchical organizational structures, leadership should focus on integrating adaptive behaviors at multiple scales such as at the person, team, and organization (Gibson and Birkinshaw, 2004; Turner, Swart and Maylor, 2013). Ambidexterity literature proposes structural separation (i.e., providing separate divisions for exploitation and exploration), parallel structures (i.e., switching to an alternative organizational structure to address exploitation or exploration), and temporal balancing (i.e., approaching either exploitation or exploration exclusively as needed) as methods to manage tension; however, management of tension in practice is not always well-defined (Gibson and Birkinshaw, 2004; Turner, Swart and Maylor, 2013; Papachroni, Heracleous and Paroutis, 2016). Individuals have agency and make their own decisions, and bottom-up approaches have shown to inspire increased creativity and productivity within an organization (Marion and Uhl-Bien, 2001). Gibson and Birkinshaw (2004) proposed contextual ambidexterity as simultaneously pursuing exploitation and exploration. Similarly, Uhl-Bien and

Arena (2018) define enabling leadership as an emergent leadership style that is "creating, engaging and protecting [the] "adaptive space" needed to nurture and sustain the adaptability process in organizations" and which actively injects tension to support innovation (Uhl-Bien, Marion and McKelvey, 2007).

3.2.3.5. INTEGRATIVE LEADERSHIP FRAMEWORK

The following integrative leadership framework (Figure 3.2.2) proposes capabilities that may prepare organizations to navigate periods of stability and instability. Tension is an underlying characteristic of complex systems and drives each of the presented capabilities. Tension can form from internal or external stress, where organizations may experience drift as they continue to operate in routines created for a different environment, or shocks, where an organization must act immediately to a crisis. Leadership must acknowledge tension as an opportunity of productivity rather than an obstacle. Enabling leadership emerges from tension and the identified leadership capabilities, but also provides a reinforcing feedback that continues to strengthen an organization's ability to respond to stability and instability across scales (e.g., operating core to the strategic apex, individuals to the organization) (Uhl-Bien and Arena, 2017). It is crucial to acknowledge that leaders cannot control the outcomes but can merely guide an organization through tension, and leadership's influences may not always be predictable (Marion and Uhl-Bien, 2001; Uhl-Bien and Marion, 2009; Rosenhead *et al.*, 2019; Tourish, 2019).

Enabling leadership may also be fostered through the identified capabilities, providing organizations a degree of permanence not found when enabling leadership is driven by tension. Starting with collaboration, this leadership capability identifies the individuals in the working environment and how they interact. The inclusion of diverse stakeholders within and beyond the organization increases the number of opportunities and threats an organization will be able to perceive. If relationships between these stakeholders are non-linear and interdependent, more individuals have the opportunity to become leaders (re: flexible informal governance), further enhancing the ability of an organization to learn. This bottom-up approach also demonstrates the interdependence of informal and formal governance. This interdependence has not always been recognized, but it is critical in developing an organization's leadership (Weber and Khademian, 2008; Uhl-Bien and Marion, 2009). The formal governance structure must be supportive of emergent leaders, including personnel not necessarily assigned a leadership role. Leadership should nurture a culture of learning — including perception and exploration capabilities — to enable innovation. Once an opportunity is identified, an innovation still has a tumultuous path toward implementation within the formal governance structures. Innovation provides impetus for organizations to adapt. As discussed, generally organizations are prone to follow a routine and repositioning is a difficult and costly endeavor (Galbraith, 1982; Teece, 2007). This pathway between innovation and flexible formal governance also shows the value in complementary exploitative behaviors, so that an organization may incorporate new ideas into operations (Zahra and George, 2002). The combination of these leadership capabilities provides the opportunity to foster enabling leadership and, in turn, resilient organizations.

Figure 3.2.2: Integrative Leadership Framework

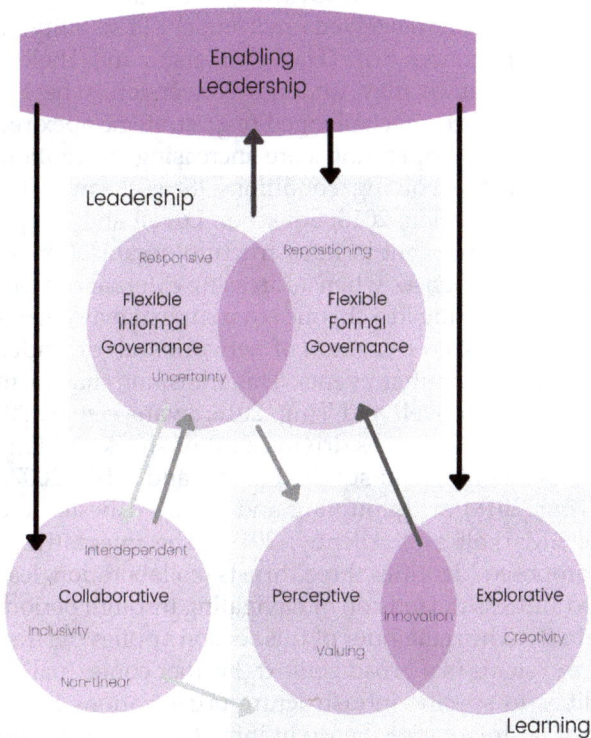

Organizations face tension from stresses and shocks (i.e., tension) and must be able to respond quickly and effectively to the changing dynamics (i.e., enabling leadership). The purple circles represent emerging capabilities in leadership and organizational change literature, with the fine print words indicating characteristics of the capability. The darkening arrows indicate growing momentum toward a virtuous cycle, where all arrows would eventually be saturated. Notably, it is the efforts of integrating all the themes that enables conflict navigation and, ultimately, resilience.

3.2.4. CONTEXTUALIZATION OF LEADERSHIP CAPABILITIES TO INFRASTRUCTURE

Infrastructure managers must reevaluate the assumptions of the operating environment, acknowledge the internal and external complexity of infrastructure systems, and actively navigate between tensions of efficiency and resilience; else, infrastructure systems may become obdurate and, potentially obsolete (Lemer, 1996; Chester and Allenby, 2018; Iwaniec *et al.*, 2019). The study of 'large technical systems' explores contestation, a state of nonconcurrence within an organization where authority may be challenged (Sovacool, Lovell and Ting, 2018). Contestation, oftentimes driven by drift and crisis, creates vulnerability through tension within an organization, leading to an opportunity to enhance adaptive capacity (Hughes, 1983; Summerton, 1994; Sovacool, Lovell and Ting, 2018). A system operating in a falsely perceived environment of stability, rather than a complex one, experiences drift (Hacker, Pierson and Thelen, 2015). In drift, existing governance may undergo conversion, where leadership may be directly or indirectly challenged (e.g., strategic apex reevaluating organizational goals or operating core increasing precautions, respectively) in response to evolving conditions (Streeck and Thelen, 2005; Hacker, Pierson and Thelen, 2015; Sovacool, Lovell and Ting, 2018). The divisionalized form prevalent in infrastructure organizational structures creates fragmented responses when addressing external complexity. Conversely, a system operating in a chaotic environment may encounter crisis, a rapid and/or publicized transition of network and/or leadership such as seen with extreme weather events. This transition may or may not be intentional (Sovacool, Lovell and Ting, 2018; Iwaniec *et al.*, 2019). These instances of instability — stresses driving drift or shocks initiating crisis — may lead to repositioning and survival (Geels and Schot, 2007; Sovacool, Lovell and Ting, 2018) or decoupling and failure (Chester and Allenby, 2018; Chester, Markolf and Allenby, 2019). The integrative leadership framework proposed identifies three thrusts (collaboration, learning, and leadership) to address the tension of navigating through periods of stability and instability. The remainder of this section applies the framework to infrastructure systems in a broad context. Further contextualizing leadership capabilities to specific infrastructure organizations is a critical endeavor toward addressing resilience in infrastructure institutions, which will ideally lead to more resilient infrastructure networks.

Leadership and organizational change literature reveals capabilities that boost longevity of organizations. While infrastructure systems exist in dynamic environments, they do not face the same pressures as the private organizations examined (Rashman, Withers and Hartley, 2009). Instead, infrastructure systems exist as public service organizations providing services that need to meet public expectations (Rashman, Withers and

Hartley, 2009, Rashman et al., 2009), with approximately 85% of (non-military) infrastructure assets being managed at a state or local level (Miller, 2000; Edwards, 2017; Saha and Ibrahima, 2020). In other words, the goods and services that infrastructure provide are intended to be accessible to everyone, and most of the time, the government will oversee the development and management of the systems. Water and transportation sectors provide conventional examples, but even the power sector, while often privately-owned (e.g., renewable energy), must abide by regulations to ensure the organization keeps public interest (e.g., social inclusion, sustainability (Osborne and Brown, 2005)) at the forefront of their mission. Public service organizations face higher pressures from political environments than private organizations, making this a fundamental tension to their operation (Osborne and Brown, 2005). Infrastructure systems experience significantly less pressure to change due to their natural monopoly model (i.e., lack of competition), which partially results from governmental oversight, economies of scale, and the challenge of pricing the value of public services (Miller, 2000). This drift can make infrastructure systems more prone to indulge in exploitative behaviors. Further, the high pressure to maintain services combined with a lack of resources and/or ability to experiment (Miller, 2000; Osborne and Brown, 2005), suppresses explorative behaviors.

Collaboration, the first thrust in the integrative leadership framework, is a skill practiced by infrastructure organizations (e.g., co-production, benchmarking, private-public partnerships) and has been recognized as a critical tool for resilience in infrastructure and social-ecological systems literature (Folke *et al.*, 2005; Biggs *et al.*, 2012; Park *et al.*, 2013; Chaffin, Gosnell and Cosens, 2014; Deslatte and Swann, 2019; Gim, Miller and Hirt, 2019). Infrastructure organizations collaborate with numerous diverse stakeholders, including other infrastructure managers within their sector, infrastructure managers of other sectors, state and federal agencies, private industries, academic institutions, non-profits (Rashman, Withers and Hartley, 2009; Muñoz-Erickson, 2014; Gim, Miller and Hirt, 2019). These stakeholders are identified based on their relevant knowledge and investment in the issue at hand, indicating that collaborators should change to address different disturbances (Weber and Khademian, 2008; Head and Alford, 2013). These relationships may be interconnected, dependent, or interdependent between individuals of varying authority or entire organizations (Weber and Khademian, 2008; Muñoz-Erickson, 2014). Resilience theories in social-ecological systems (SES) encourage collaboration through polycentricity, or governance through decentralization of power (Folke *et al.*, 2005; Biggs *et al.*, 2012; Chaffin, Gosnell and Cosens, 2014). SES theory is valuable in the framing of infrastructure resilience (Markolf *et al.*, 2018; Helmrich *et al.*, 2021). Further, relationships may be formal or informal. Informal relationships are widely prevalent in infrastructure projects, and they create mutual dependencies and trust amongst personnel (van Gestel *et al.*, 2008; Rashman, Withers and Hartley, 2009; Head and Alford, 2013). Collaboration, therefore, highlights the importance of teams and teamwork within organizations. Addressing instability requires teams that are collectively oriented through shared mental models and leadership, mutually supportive, monitoring performance, communicative and trusting, and adaptive (Baker, Day and Salas, 2006). The more increasingly non-linear relationships become, the more complex the system. While complexity is in part synonymous with unpredictability, complex organizations are necessary to help navigate increasingly complex environments (Folke *et al.*, 2005; Boisot and Mckelvey, 2011; Head and Alford, 2013). When there are numerous collaborators sharing (and creating) knowledge, learning becomes increasingly fruitful (Rashman, Withers and Hartley, 2009; Park *et al.*, 2013).

Learning, the second major thrust of the integrative leadership framework, requires infrastructure managers to gather new knowledge from theory and practice and integrate the knowledge into practice. Learning, within public service organizations, has been defined by Rashman et al. as 'a process of individual and shared thought and action, involving cognitive, social, behavioral, and technical elements' (2009). They further identify four organizational learning processes: individual perspective, shared understanding, diffusion, and embedding in organization. Learning requires a level of risk-taking to identify opportunities and threats (i.e., perception) and motivate exploration and experimentation. Infrastructure managers must assess risk of potential developments. There are a number of decision-making methods available to infrastructure managers: conventional (e.g., cost-benefit analysis, cost effectiveness analysis, risk assessment), environmental (e.g., life cycle analysis, environmental impact assessment), social (e.g., social impact assessment), deep uncertainty (e.g., real option analysis, robust decision making, info-gap analysis, adaptation pathways) (Helmrich and Chester, 2020). Notably, most of these are not required for development. It is also important for infrastructure organizations to understand the institutional pressures that reinforce particular actions (e.g., vested interests, knowledge asymmetry, mental models, cultural perceptions). By learning of the forces influencing the system as well as who and who does not hold power over those forces, infrastructure managers may be able to identify trigger points where the cycle may be broken (Marshall and Alexandra, 2016). Ideally, this would allow infrastructure managers to encourage adaptation and transformation more frequently—rather than, primarily, after crisis events (Huitema and Meijerink, 2010; Marshall and Alexandra, 2016; Iwaniec *et al.*, 2019). This highlights the need for 'policy entrepreneurs'—individuals internal or external to the infrastructure system who develop new ideas, create a network of support, and identify windows of opportunity for implementation

(Huitema and Meijerink, 2010). To further identify trigger points for change, or windows of opportunity, infrastructure managers should monitor evolving risks (embedded in both stresses and shocks, known and unknown) throughout the infrastructure life cycle (Park *et al.*, 2013; Woods, 2015).

Another challenge of risk analysis is the integration of qualitative knowledge into decision-making processes, as infrastructure assessments tend to focus on quantitative metrics (Park *et al.*, 2013). Leadership and organizational change literature acknowledges that unifying organizational mission and values (qualitative metrics) may guide leaders in their decision-making as well as establish motivation for learning (Little, 2004; Rashman, Withers and Hartley, 2009; Taylor and Helfat, 2009; Uhl-Bien and Arena, 2017). In regards to infrastructure organizations, value is oftentimes placed on efficient use of public resources, transparency, and democratic legitimacy, but there are competing values of sustainability, economic development, and environmental quality (van Gestel *et al.*, 2008; Rashman, Withers and Hartley, 2009). Values can change, or have fluctuating priority, over the lifetime of infrastructure; they can also change based on political environments (van Gestel *et al.*, 2008). This volatility highlights the importance of flexible leadership and a change-oriented organizational culture that embraces uncertainty (Folke *et al.*, 2005; van Gestel *et al.*, 2008; Rashman, Withers and Hartley, 2009; Chester and Allenby, 2018). Risk analysis is an important tool that allows infrastructure managers to perceive opportunities and threats and contributes toward learning.

When learning reveals a potential value-add for an organization, the opportunity to innovate arises; however, infrastructure systems struggle to deviate from the status quo (i.e., lock-in) and experiment with new ideas (Head and Alford, 2013; Chester, Markolf and Allenby, 2019; Yu *et al.*, 2020). Infrastructure managers can promote experimentation by investing in research and development, which requires an organizational culture shift and reallocation of resources (La Porte and Consolini, 1991; Chester, Markolf and Allenby, 2019; Gilrein *et al.*, 2019; Chester, Miller and Muñoz-Erickson, 2020). Errors can be embraced as tools for learning and decision-making within infrastructure systems, as long as managers are communicative and transparent with stakeholders and the surrounding community about the experimentation (La Porte and Consolini, 1991; La Porte, 1996). Experimentation—typically defined as small and deliberate changes to infrastructure technologies—requires 'learning-focused management approaches' and a degree of failure acceptance from infrastructure organizations, financers and insurers, and the public (Biggs *et al.*, 2012; Chester, Markolf and Allenby, 2019; Deslatte and Swann, 2019; Yu *et al.*, 2020). Safe-to-fail infrastructure provides an opportunity to embrace experimentation, with the approach accounting for various types of failure (social, ecological, technological) during the design process (Park *et al.*, 2013; Kim *et al.*, 2019; Yu *et al.*, 2020). The risks are then prioritized to minimize harm, and stakeholders are educated on expected failure pathways and the potential of unexpected failures (Kim *et al.*, 2019). Another approached proposed in SES literature is adaptive governance, where the interactions across multiple levels of power (individual to organization to institution) act collectively toward achieving a desired state (Folke *et al.*, 2005; Chaffin, Gosnell and Cosens, 2014). The choice of state can be revisited to ensure the system is responding to emerging conditions (Chaffin, Gosnell and Cosens, 2014; Cleaver and Whaley, 2018). Working across social, ecological, and technological dimensions highlights the advantage of collaboration in experimentation (Garud, Gehman and Kumaraswamy, 2011; Head and Alford, 2013; Gilrein *et al.*, 2019); the boundary spanning, presented in both of these approaches, challenges infrastructure organizations to invest in products and services that are tangentially related to their mission (e.g., a transportation department funding research on roadside photovoltaics (Aupperlee, 2018; Clines, 2018; Gilrein *et al.*, 2019)). Ideally, learning leads to sustained innovation, where value-adds are continuously identified and incorporated into the organization's formal governance (Garud, Gehman and Kumaraswamy, 2011). Innovation challenges the formal governance (i.e., bureaucratic nature) of public service organizations (Newell *et al.*, 2003; Osborne and Brown, 2005; Uhl-Bien and Arena, 2018).

Leadership, the final major thrust of the integrative leadership framework, is comprised of formal and informal governance, both of which hold valuable roles in agile infrastructure organizations. To reiterate, formal governance is defined as structured, explicit rules and regulations that require ample governance to circulate and enforce, but maintain low barriers to entry due to standardization (Boesen, 2007; Uhl-Bien and Arena, 2018). Meanwhile, informal governance consists of implicit norms based on relationships and trust; informal governance presents high entry barriers due to the contextualized nature (Boesen, 2007; Uhl-Bien and Arena, 2018). Infrastructure leadership is not currently established in a manner that is favorable toward advancing resilience within its organizations. Specifically, divisionalized form restricts the ability of personnel to collaborate and learn due to overemphasis on exploitative behaviors (Rashman, Withers and Hartley, 2009; Zhou, 2013; Chester *et al.*, 2021). Upon review of the five components of divisionalized form (operating core, strategic apex, middle line, technostructure, and staff), each component description exemplifies an exploitative behavior. No position is explicitly responsible for exploratory initiatives, and only the strategic apex holds explorative power with the ability to add and remove divisions (Mintzberg, 1979). Literature on public service and infrastructure organizations identifies organizational decentralization as a tool to promote collaboration, learning, and, importantly, adaptability for resilience (Finger and Brand, 1999; Rashman, Withers and Hartley, 2009; Head and Alford, 2013; Zhou, 2013; Chester, Miller and Muñoz-Erickson, 2020).

Decentralized organizations are capable of meeting complexity with complexity because they encourage non-linear relationships which emerge unpredictable endeavors (Lichtenstein and Plowman, 2009; Boisot and Mckelvey, 2011). Decentralized organizational structures emphasize the significance of informal relationships. It is crucial that personnel are included in the 'process of change' (i.e., a bottom-up approach) as their commitment can increase the success of infrastructure repositioning (Perry and Wise, 1990; Osborne and Brown, 2005), highlighting another key point: infrastructure organizations will need to continuously reposition to adjust to changing environments (Siggelkow and Levinthal, 2003; Head and Alford, 2013). Yet, there is a tradeoff. Organizations must also eliminate noise from decision-making to avoid overextending resources. Recalling temporary decentralization, this approach to governance may provide guidance, particularly within increasingly distributed infrastructure networks which are characterized by networks with coupled producer and consumer relationships that lead to coordinated responses (Helmrich *et al.*, 2021). This configuration has the potential to respond with more variety than centralized systems, while utilizing beneficial qualities of each. For instance, emergency response protocols will allow infrastructure managers to make more autonomous decisions during disasters. For example, a water treatment plant operator prevented a cyberattack where aggressors were attempting to tamper with sodium hydroxide concentrations (Chappell, 2021). This does, notably, add complexity to the governance system (La Porte and Consolini, 1991; Boisot and Mckelvey, 2011). Despite this, fluidity in organizational structure has been advocated as a tool within high reliability organizations, as well as resilient to meet demands in periods of stability and instability (Grabowski and Roberts, 2019).

3.2.5. DISCUSSION

Infrastructure institutions within the United States are often considered too critical to fail; however, this framing is preventing organizations from adapting and transforming to meet the complexity and deep uncertainty of the Anthropocene. While the leadership and organizational change literature assessed largely focused on private organizations, the emergent leadership capabilities have shown applicability and alignment with infrastructure. The lack of market competition may make public service organizations more open to collaboration and, subsequently, collective learning (Hartley and Benington, 2006; Rashman, Withers and Hartley, 2009); however, the need to provide specific services may cause risk adversity and an avoidance of failure, restricting flexibility and innovation (Newman and Chaharbaghi, 2000; Rashman, Withers and Hartley, 2009). Infrastructure systems must change at pace with the environments in which they operate (Hollnagel, Woods and Leveson, 2006). Unintentional and reductionistic approaches toward flexible leadership will not ensure reliable infrastructure services into the future (Lemer, 1996; Chester and Allenby, 2018; Iwaniec *et al.*, 2019). Operational and strategic planning of resilient infrastructure must confront the influence of formal and informal governance, and they must be inclusive when considering stakeholders who hold power over infrastructure networks. For instance, referring back to distributed systems, this configuration has created space for external stakeholders to exert unregulated authority over infrastructure. Individuals are routed through transportation networks via navigation applications hosted by Google, Waze, Apple, etc. These rerouting options, while providing real-time alternatives (re: flexibility and agility), channel vehicles through roads that were potentially not designed for high levels of traffic (e.g., shortcuts through residential neighborhoods). Network distribution is leading to disjointed governance of infrastructure systems when authority is not explicitly considered in the design consequences or the stakeholders are not operating in coordination. The integrative leadership framework highlights opportunities and challenges within infrastructure organizations and related stakeholders for more intentional fostering of collaboration and explorative behaviors while providing insights to how an organization may invest in its formal and informal governance to produce enabling leaders and institutional resilience.

To increase capacity for explorative behaviors in infrastructure organizations, there needs to be a cultural shift from exploitative behaviors (Chester, Miller, et al., 2020). It is clear that organizations need to change, and there several capabilities that support longevity in instable environments. The challenge is identifying pathways to transition organizations between exploitative and explorative behaviors, as seen with enabling leadership. These pathways must be dynamic to navigate future uncertainty. Infrastructure managers must realize the 'change' in organizational change will never be complete, meaning that all decisions and projects will be revisited. This acknowledgement is crucial to avoid inaction attributed to uncertainty, a topic readily explored in the field of deep uncertainty (Courtney, Kirkland and Viguerie, 1997; Walker, Lempert and Kwakkel, 2013). Instead, infrastructure managers, organizations, and institutions must learn to lead through the flux of instability. The proposed integrative leadership framework provides insight on leadership capabilities that organizations can invest in now to begin this transition (i.e., adaptation). Infrastructure organizations will also need to undergo transformational change to avoid path dependencies and reverting to a baseline of exploitative behavior. There are tools to help organizations address complexity and deep uncertainty such as horizon scanning and scenario planning. Infrastructure organizations may also find guidance from high reliability organizations, which promote strategies to minimize failures: learning, decentralization of power, incentivizing employee behavior, functional redundancy, and balancing tradeoffs of reliability and efficiency (Roberts, 1990; Roberts and Bea, 2001). Infrastructure organizations will need to educate and empower their personnel by creating environments in which their employees can partake in explorative behaviors. The environments in which infrastructure operate are changing rapidly. Technologies surrounding infrastructure are rapidly changing. Infrastructure organizations and institutions must also be able to respond rapidly change to advance infrastructure resilience.

3.2.6. CONCLUSION

The institutions that govern infrastructure—including their bureaucratic structures, leadership culture, administrative structure, rules, and norms—play a critical role in creating the capacities necessary to adapt and transform to known and unknown disturbances at pace with changing conditions (Hollnagel, Woods and Leveson, 2006; Pelling, O'Brien and Matyas, 2015; Woods, 2015). It is essential that infrastructure organizations are able to adjust at pace with the changes of the environments in which they operate (Hollnagel, Woods and Leveson, 2006). The current institutional practices surrounding infrastructure (e.g., lack of competition, risk-adversity) do not place pressure on infrastructure systems to change, but organizations must adjust to avoid becoming obsolete. The integrative leadership framework presented demonstrates an immediately applicable deployment of leadership capabilities that emerge enabling leadership; however, in alignment with (Uhl-Bien and Marion, 2009), organizational and institutional transformations still need to occur to ensure longevity. In order to enhance resilience, organizations must acknowledge they are operating in complex—rather than complicated—environments so that they may properly assess the tradeoffs for transitioning their leadership structures. Therefore, the takeaway should not be that investment in leadership capabilities will independently create resilient infrastructure organizations, but that they may provide a pathway towards retooling institutions to operate effectively in flux between states of stability and instability.

3.2.7. REFERENCES

Adler, P.S. (2001) 'Market, Hierarchy, and Trust: The Knowledge Economy and the Future of Capitalism', *Organization Science*, 12(2), pp. 215–234. Available at: https://doi.org/10.1287/orsc.12.2.215.10117.

Ansell, B.W. and Lindvall, J. (2020) *Inward Conquest: The Political Origins of Modern Public Services*. Cambridge: Cambridge University Press (Cambridge Studies in Comparative Politics). Available at: https://doi.org/10.1017/9781108178440.

Arbesman, S. (2017) *Overcomplicated: Technology at the Limits of Comprehension*. New York, NY: Penguin Publishing Group.

Aupperlee, A. (2018) 'Pennsylvania to build test track, training facility for next-gen vehicles', *TribLIVE.com*, 10 April. Available at: https://archive.triblive.com/news/pennsylvania/pennsylvania-to-build-test-track-training-facility-for-next-gen-vehicles/ (Accessed: 4 December 2023).

Ayyub, B.M. (2018) *Climate-Resilient Infrastructure*. Edited by Committee on Adaptation to a Changing Climate. Reston, VA: ASCE Press. Available at: https://doi.org/10.1061/9780784415191.

Bäcklander, G. (2019) 'Doing complexity leadership theory: How agile coaches at Spotify practise enabling leadership', *Creativity and Innovation Management*, 28(1), pp. 42–60. Available at: https://doi.org/10.1111/caim.12303.

Baker, D., Day, R. and Salas, E. (2006) 'Teamwork as an Essential Component of High-Reliability Organizations', *Health services research*, 41, pp. 1576–98. Available at: https://doi.org/10.1111/j.1475-6773.2006.00566.x.

Biggs, R. *et al.* (2012) 'Toward Principles for Enhancing the Resilience of Ecosystem Services', *Annual Review of Environment and Resources*, 37(1), pp. 421–448. Available at: https://doi.org/10.1146/annurev-environ-051211-123836.

Boesen, N. (2007) 'Governance and Accountability: How Do the Forma and Informal Interplay and Change?', in J. Jutting et al. (eds) *Informal Institutions How Social Norms Help or Hinder Development: How Social Norms Help or Hinder Development*. Danvers, MA: OECD Publishing.

Boisot, M. and Mckelvey, B. (2011) 'Complexity and organization–environment relations: Revisiting Ashby's law of requisite variety', in, pp. 279–298. Available at: https://doi.org/10.4135/9781446201084.n16.

Bondank, E.N., Chester, M.V. and Ruddell, B.L. (2018) 'Water Distribution System Failure Risks with Increasing Temperatures', *Environmental Science & Technology*, 52(17), pp. 9605–9614. Available at: https://doi.org/10.1021/acs.est.7b01591.

Brown, R. (2004) 'Consideration of the origin of Herbert Simon's theory of "satisficing" (1933-1947)', *Management Decision*, 42(10), pp. 1240–1256. Available at: https://doi.org/10.1108/00251740410568944.

Burillo, D. *et al.* (2017) 'Electricity demand planning forecasts should consider climate non-stationarity to maintain reserve margins during heat waves', *Applied Energy*, 206(June), pp. 267–277. Available at: https://doi.org/10.1016/j.apenergy.2017.08.141.

Butler, L.J., Scammell, M.K. and Benson, E.B. (2016) 'The Flint, Michigan, Water Crisis: A Case Study in Regulatory Failure and Environmental Injustice', *Environmental Justice*, 9(4), pp. 93–97. Available at: https://doi.org/10.1089/env.2016.0014.

Cantarello, S., Martini, A. and Nosella, A. (2012) 'A Multi-Level Model for Organizational Ambidexterity in the Search Phase of the Innovation Process', *Creativity and Innovation Management*, 21(1), pp. 28–48. Available at: https://doi.org/10.1111/j.1467-8691.2012.00624.x.

Carley, K.M. and Lee, J.S. (1998) 'Dynamic Organizations: Organizational Adaptation in a Changing Environment', *In Advances in Strategic Management*, 15, pp. 269–297.

Carvalhaes, T. *et al.* (2020) 'COVID-19 as a Harbinger of Transforming Infrastructure Resilience', *Frontiers in Built Environment*, 6. Available at: https://doi.org/10.3389/fbuil.2020.00148 (Accessed: 21 October 2023).

Chaffin, B., Gosnell, H. and Cosens, B. (2014) 'A decade of adaptive governance scholarship: synthesis and future directions', *Ecology and Society*, 19(3). Available at: https://doi.org/10.5751/ES-06824-190356.

Chandler, A.D. (1977) *The Visible Hand*. Cambridge, MA: Harvard University Press.

Chappell, B. (2021) 'FBI Called In After Hacker Tries To Poison Tampa-Area City's Water With Lye', *NPR*, 9 February. Available at: https://www.npr.org/2021/02/09/965791252/fbi-called-in-after-hacker-tries-to-poison-tampa-area-citys-water-with-lye (Accessed: 23 October 2023).

Chester, M. *et al.* (2021) 'Infrastructure resilience to navigate increasingly uncertain and complex conditions in the Anthropocene', *npj Urban Sustainability*, 1(1), pp. 1–6. Available at: https://doi.org/10.1038/s42949-021-00016-y.

Chester, M., Markolf, S. and Allenby, B. (2019) 'Infrastructure and the environment in the Anthropocene', *Journal of Industrial Ecology*, 23(5), pp. 1006–1015. Available at: https://doi.org/10.1111/jiec.12848.

Chester, M.V. and Allenby, B. (2018) 'Toward adaptive infrastructure: flexibility and agility in a non-stationarity age', *Sustainable and Resilient Infrastructure*, 4, pp. 1–19. Available at: https://doi.org/10.1080/23789689.2017.1416846.

Chester, M.V. and Allenby, B. (2019) 'Infrastructure as a wicked complex process', *Elementa: Science of the Anthropocene*. Edited by A. Iles and M.E. Chang, 7(1), p. 21. Available at: https://doi.org/10.1525/elementa.360.

Chester, M.V., Miller, T. and Muñoz-Erickson, T.A. (2020) 'Infrastructure governance for the Anthropocene', *Elementa: Science of the Anthropocene*, 8(1), p. 78. Available at: https://doi.org/10.1525/elementa.2020.078.

Chester, M.V., Underwood, B.S. and Samaras, C. (2020) 'Keeping infrastructure reliable under climate uncertainty', *Nature Climate Change*, 10(6), pp. 488–490. Available at: https://doi.org/10.1038/s41558-020-0741-0.

Cleaver, F. and Whaley, L. (2018) 'Understanding process, power, and meaning in adaptive governance: a critical institutional reading', *Ecology and Society*, 23(2). Available at: https://doi.org/10.5751/ES-10212-230249.

Clines, K. (2018) *"The Ray" is Ga.'s 18-mile proving ground for technology and ideas on I-85*, *Equipment World*. Available at: https://www.equipmentworld.com/roadbuilding/article/14969126/the-ray-is-gas-18-mile-proving-ground-for-technology-and-ideas-on-i-85 (Accessed: 4 December 2023).

Corvellec, H., Zapata Campos, M.J. and Zapata, P. (2013) 'Infrastructures, lock-in, and sustainable urban development: the case of waste incineration in the Göteborg Metropolitan Area', *Journal of Cleaner Production*, 50, pp. 32–39. Available at: https://doi.org/10.1016/j.jclepro.2012.12.009.

Courtney, H., Kirkland, J. and Viguerie, P. (1997) 'Strategy Under Uncertainty', *Harvard Business Review*, 1 November. Available at: https://hbr.org/1997/11/strategy-under-uncertainty (Accessed: 23 October 2023).

Daviter, F. (2017) 'Coping, taming or solving: alternative approaches to the governance of wicked problems', *Policy Studies*, 38(6), pp. 571–588. Available at: https://doi.org/10.1080/01442872.2017.1384543.

Deslatte, A. and Swann, W.L. (2019) 'Elucidating the Linkages Between Entrepreneurial Orientation and Local Government Sustainability Performance', *The American Review of Public Administration*, 50(1), pp. 92–109. Available at: https://doi.org/10.1177/0275074019869376.

Diesel, R. and Scheepers, C.B. (2019) 'Innovation climate mediating complexity leadership and ambidexterity', *Personnel Review*, 48(7), pp. 1782–1808. Available at: https://doi.org/10.1108/PR-11-2018-0445.

Edwards, P.N. (2017) 'Knowledge infrastructures for the Anthropocene', *The Anthropocene Review*, 4(1), pp. 34–43. Available at: https://doi.org/10.1177/2053019616679854.

Faguet, J.-P. (2014) 'Decentralization and Governance', *World Development*, 53, pp. 2–13. Available at: https://doi.org/10.1016/j.worlddev.2013.01.002.

Finger, M. and Brand, S.B. (1999) 'Organizational Learning and the Learning Organization: Developments in Theory and Practice', in *Organizational Learning and the Learning Organization: Developments in Theory and Practice*. London, UK: SAGE Publications Ltd, pp. 130–156. Available at: https://doi.org/10.4135/9781446218297.

Folke, C. *et al.* (2005) 'Adaptive Governance of Social-Ecological Systems', *Annu. Rev. Environ. Resour*, 15, pp. 441–73. Available at: https://doi.org/10.1146/annurev.energy.30.050504.144511.

Friedlander, A. (1995a) *Emerging Infrastructure: The Growth of Railroads*. Corporation for National Research Initiatives.

Friedlander, A. (1995b) *Natural Monopoly and Universal Service: Telephones and Telegraphs in the U.S. Communications Infrastructure, 1837-1940*. Corporation for National Research Initiatives.

Friedlander, A. (1996) *Power and Light: Electricity in the U.S. Energy Infrastructure, 1870-1940*. Corporation for National Research Initiatives.

Galbraith, J.R. (1982) 'Designing the innovating organization', *Organizational Dynamics*, 10(3), pp. 5–25. Available at: https://doi.org/10.1016/0090-2616(82)90033-X.

Garcia, M. *et al.* (eds) (2019) 'Towards urban water sustainability: Analyzing management transitions in Miami, Las Vegas, and Los Angeles', *Global Environmental Change*, 58, p. 101967. Available at: https://doi.org/10.1016/j.gloenvcha.2019.101967.

Garud, R., Gehman, J. and Kumaraswamy, A. (2011) 'Complexity Arrangements for Sustained Innovation: Lessons From 3M Corporation', *Organization Studies*, 32(6), pp. 737–767. Available at: https://doi.org/10.1177/0170840611410810.

Geels, F.W. and Schot, J. (2007) 'Typology of sociotechnical transition pathways', *Research Policy*, 36(3), pp. 399–417. Available at: https://doi.org/10.1016/j.respol.2007.01.003.

van Gestel, N. *et al.* (2008) 'Managing Public Values in Public-Private Networks: A Comparative Study of Innovative Public Infrastructure Projects', *Public Money & Management*, 28(3), pp. 139–145. Available at: https://doi.org/10.1111/j.1467-9302.2008.00635.x.

Gibson, C. and Birkinshaw, J. (2004) 'The Antecedents, Consequences, and Mediating Role of Organizational Ambidexterity', *Academy of Management Journal*, 47, pp. 209–226. Available at: https://doi.org/10.2307/20159573.

Gilrein, E.J. *et al.* (2019) 'Concepts and practices for transforming infrastructure from rigid to adaptable', *Sustainable and Resilient Infrastructure*, 6(3–4), pp. 213–234. Available at: https://doi.org/10.1080/23789689.2019.1599608.

Gim, C., Miller, C.A. and Hirt, P.W. (2019) 'The resilience work of institutions', *Environmental Science & Policy*, 97, pp. 36–43. Available at: https://doi.org/10.1016/j.envsci.2019.03.004.

Grabowski, M. and Roberts, K.H. (2019) 'Reliability seeking virtual organizations: Challenges for high reliability organizations and resilience engineering', *Safety Science*, 117, pp. 512–522. Available at: https://doi.org/10.1016/j.ssci.2016.02.016.

Grote, G. (2019) 'Leadership in Resilient Organizations', in S. Wiig and B. Fahlbruch (eds) *Exploring Resilience: A Scientific Journey from Practice to Theory*. Cham: Springer International Publishing (SpringerBriefs in Applied Sciences and Technology), pp. 59–67. Available at: https://doi.org/10.1007/978-3-030-03189-3_8.

Hacker, J.S., Pierson, P. and Thelen, K. (2015) 'Drift and conversion: Hidden faces of institutional change', in *Advances in comparative-historical analysis*. Cambridge, MA: Cambridge University Press, pp. 180–208.

Hartley, J. and Benington, J. (2006) 'Copy and Paste, or Graft and Transplant? Knowledge Sharing Through Inter-Organizational Networks', *Public Money & Management*, 26(2), pp. 101–108. Available at: https://doi.org/10.1111/j.1467-9302.2006.00508.x.

Havermans, L.A. *et al.* (2015) 'Exploring the Role of Leadership in Enabling Contextual Ambidexterity', *Human Resource Management*, 54(S1), pp. s179–s200. Available at: https://doi.org/10.1002/hrm.21764.

Head, B.W. and Alford, J. (2013) 'Wicked Problems: Implications for Public Policy and Management', *Administration & Society*, 47(6), pp. 711–739. Available at: https://doi.org/10.1177/0095399713481601.

Helmrich, A. *et al.* (2021) 'Centralization and decentralization for resilient infrastructure and complexity', *Environmental Research: Infrastructure and Sustainability*, 1(2), p. 021001. Available at: https://doi.org/10.1088/2634-4505/ac0a4f.

Helmrich, A.M. and Chester, M.V. (2020) 'Reconciling complexity and deep uncertainty in infrastructure design for climate adaptation', *Sustainable and Resilient Infrastructure*, 7(2), pp. 83–99. Available at: https://doi.org/10.1080/23789689.2019.1708179.

Hollnagel, E., Woods, D.D. and Leveson, N. (2006) *Resilience Engineering: Concepts and Precepts*. Aldershot, UK: Ashgate.

Hooijberg, R., (Jerry) Hunt, J.G. and Dodge, G.E. (1997) 'Leadership Complexity and Development of the leaderplex Model', *Journal of Management*, 23(3), pp. 375–408. Available at: https://doi.org/10.1177/014920639702300305.

Hughes, T.P. (1983) *Networks of Power: Electrification in Western Society, 1880-1930*. Baltimore, MD: Johns Hopkins University Press.

Huitema, D. and Meijerink, S. (2010) 'Realizing water transitions: the role of policy entrepreneurs in water policy change', *Ecology and Society*, 15(2). Available at: https://doi.org/10.5751/ES-03488-150226.

Iwaniec, D.M. *et al.* (2019) 'The Framing of Urban Sustainability Transformations', *Sustainability*, 11(3), p. 573. Available at: https://doi.org/10.3390/su11030573.

Jansen, J.J.P., Simsek, Z. and Cao, Q. (2012) 'Ambidexterity and performance in multiunit contexts: Cross-level moderating effects of structural and resource attributes', *Strategic Management Journal*, 33(11), pp. 1286–1303. Available at: https://doi.org/10.1002/smj.1977.

Kim, Y. *et al.* (2019) 'The Infrastructure Trolley Problem: Positioning Safe-to-fail Infrastructure for Climate Change Adaptation', *Earth's Future*, 7(7), pp. 704–717. Available at: https://doi.org/10.1029/2019EF001208.

Kooiman, J. (1993) *Modern Governance: New Government-Society Interactions*. Thousand Oaks, CA: SAGE.

La Porte, T. and Consolini, P. (1991) 'Working in Practice But Not in Theory: Theoretical Challenges of "High-Reliability Organizations"', *Journal of Public Administration Research and Theory*, 1(1), pp. 19–48. Available at: https://doi.org/10.1093/oxfordjournals.jpart.a037070.

La Porte, T.R. (1996) 'High Reliability Organizations: Unlikely, Demanding and At Risk', *Journal of Contingencies and Crisis Management*, 4(2), pp. 60–71. Available at: https://doi.org/10.1111/j.1468-5973.1996.tb00078.x.

Lemer, A.C. (1996) 'Infrastructure Obsolescence and Design Service Life', *Journal of Infrastructure Systems*, 2(4), pp. 153–161. Available at: https://doi.org/10.1061/(ASCE)1076-0342(1996)2:4(153).

Lichtenstein, B.B. and Plowman, D.A. (2009) 'The leadership of emergence: A complex systems leadership theory of emergence at successive organizational levels', *The Leadership Quarterly*, 20(4), pp. 617–630. Available at: https://doi.org/10.1016/j.leaqua.2009.04.006.

Little, R.G. (2004) 'The role of organizational culture and values in the performance of critical infrastructure systems', in *2004 IEEE International Conference on Systems, Man and Cybernetics (IEEE Cat. No.04CH37583)*. *2004 IEEE International Conference on Systems, Man and Cybernetics (IEEE Cat. No.04CH37583)*, pp. 4047–4052 vol.5. Available at: https://doi.org/10.1109/ICSMC.2004.1401164.

Mäkinen, E.I. (2018) 'Complexity Leadership Theory and the Leaders of Transdisciplinary Science', *Informing Science: The International Journal of an Emerging Transdiscipline*, 21, pp. 133–155. Available at: https://doi.org/10.28945/4009.

March, J.G. (1991) 'Exploration and Exploitation in Organizational Learning', *Organization Science*, 2(1), pp. 71–87. Available at: https://doi.org/10.1287/orsc.2.1.71.

Marion, R. and Uhl-Bien, M. (2001) 'Leadership in Complex Organizations Leadership in Complex Organizations Part of the Management Sciences and Quantitative Methods Commons Leadership in complex organizations', *The Leadership Quarterly*, 12(4), pp. 389–418. Available at: https://doi.org/10.1016/S1048- 9843(01)00092-3.

Markolf, S.A. *et al.* (2018) 'Interdependent Infrastructure as Linked Social, Ecological, and Technological Systems (SETSs) to Address Lock-in and Enhance Resilience', *Earth's Future*, 6(12), pp. 1638–1659. Available at: https://doi.org/10.1029/2018EF000926.

Markolf, S.A. *et al.* (2021) 'Re-imagining design storm criteria for the challenges of the 21st century', *Cities*, 109, p. 102981. Available at: https://doi.org/10.1016/j.cities.2020.102981.

Marshall, G.R. and Alexandra, J. (2016) 'Institutional Path Dependence and Environmental Water Recovery in Australia's Murray-Darling Basin', *Water Alternatives*, 9(3), pp. 679–703.

Martela, F. (2019) 'What makes self-managing organizations novel? Comparing how Weberian bureaucracy, Mintzberg's adhocracy, and self-organizing solve six fundamental problems of organizing', *Journal of Organization Design*, 8(1), p. 23. Available at: https://doi.org/10.1186/s41469-019-0062-9.

Martin, R.L. (2019) 'The High Price of Efficiency', *Harvard Business Review*, 1 January. Available at: https://hbr.org/2019/01/the-high-price-of-efficiency (Accessed: 19 October 2023).

Masten, S.J., Davies, S.H. and Mcelmurry, S.P. (2016) 'Flint Water Crisis: What Happened and Why?', *Journal AWWA*, 108(12), pp. 22–34. Available at: https://doi.org/10.5942/jawwa.2016.108.0195.

Miller, J.B. (2000) *Principles of Public and Private Infrastructure Delivery*. Boston, MA: Springer US. Available at: https://doi.org/10.1007/978-1-4757-6278-5.

Miller, R., Lessard, D.R. and Group, I.R. (2000) *The Strategic Management of Large Engineering Projects: Shaping Institutions, Risks, and Governance*. Cambridge, MA: MIT Press.

Mintzberg, H. (1979) *The Structuring of Organizations: A Synthesis of the Research*. Hoboken, NJ: Prentice-Hall.

Muñoz-Erickson, T.A. (2014) 'Co-production of knowledge–action systems in urban sustainable governance: The KASA approach', *Environmental Science & Policy*, 37, pp. 182–191. Available at: https://doi.org/10.1016/j.envsci.2013.09.014.

Nasr, A. *et al.* (2019) 'A review of the potential impacts of climate change on the safety and performance of bridges', *Sustainable and Resilient Infrastructure*, 6(3–4), pp. 192–212. Available at: https://doi.org/10.1080/23789689.2019.1593003.

Nederveen Pieterse, A. *et al.* (2019) 'Hierarchical leadership versus self-management in teams: Goal orientation diversity as moderator of their relative effectiveness', *The Leadership Quarterly*, 30(6), p. 101343. Available at: https://doi.org/10.1016/j.leaqua.2019.101343.

Newell, S. *et al.* (2003) '"Best practice" development and transfer in the NHS: the importance of process as well as product knowledge', *Health Services Management Research*, 16(1), pp. 1–12. Available at: https://doi.org/10.1258/095148403762539095.

Newman, V. and Chaharbaghi, K. (2000) 'The study and practice of leadership', *Journal of Knowledge Management*, 4(1), pp. 64–74. Available at: https://doi.org/10.1108/13673270010315966.

North, D.C. (1990) *Institutions, Institutional Change and Economic Performance.* Cambridge University Press. Available at: https://www.google.com/books/edition/Institutions_Institutional_Change_and_Ec/oFnWbTqgNPYC?hl=en&gbpv=0.

Ogie, R.I. (2017) 'Cyber Security Incidents on Critical Infrastructure and Industrial Networks', in *Proceedings of the 9th International Conference on Computer and Automation Engineering.* New York, NY: Association for Computing Machinery (ICCAE '17), pp. 254–258. Available at: https://doi.org/10.1145/3057039.3057076.

Olsen, J.R. (2015) *Adapting Infrastructure and Civil Engineering Practice to a Changing Climate.* (Books). Available at: https://doi.org/10.1061/9780784479193.fm (Accessed: 23 October 2023).

Omer, M., Mostashari, A. and Lindemann, U. (2014) 'Resilience Analysis of Soft Infrastructure Systems', *Procedia Computer Science,* 28, pp. 873–882. Available at: https://doi.org/10.1016/j.procs.2014.03.104.

Osborne, S.P. and Brown, K. (2005) *Managing Change and Innovation in Public Service Organizations.* London, UK: Psychology Press/Routledge. Available at: https://ebookcentral-proquest-com.ezproxy1.lib.asu.edu/lib/asulib-ebooks/reader.action?docID=200596.

Papachroni, A., Heracleous, L. and Paroutis, S. (2016) 'In pursuit of ambidexterity: Managerial reactions to innovation–efficiency tensions', *Human Relations,* 69(9), pp. 1791–1822. Available at: https://doi.org/10.1177/0018726715625343.

Park, J. *et al.* (2013) 'Integrating Risk and Resilience Approaches to Catastrophe Management in Engineering Systems', *Risk Analysis,* 33(3), pp. 356–367. Available at: https://doi.org/10.1111/j.1539-6924.2012.01885.x.

Paté-Cornell, M.-E. *et al.* (2018) 'Cyber Risk Management for Critical Infrastructure: A Risk Analysis Model and Three Case Studies', *Risk Analysis,* 38(2), pp. 226–241. Available at: https://doi.org/10.1111/risa.12844.

Pauli, B.J. (2019) *Flint Fights Back: Environmental Justice and Democracy in the Flint Water Crisis.* Cambridge, MA: The MIT Press. Available at: https://doi.org/10.7551/mitpress/11363.001.0001.

Pelling, M., O'Brien, K. and Matyas, D. (2015) 'Adaptation and transformation', *Climatic Change,* 133(1), pp. 113–127. Available at: https://doi.org/10.1007/s10584-014-1303-0.

Perry, J.L. and Wise, L.R. (1990) 'The Motivational Bases of Public Service', *Public Administration Review*, 50(3), pp. 367–373. Available at: https://doi.org/10.2307/976618.

Pieper, K.J., Tang, M. and Edwards, M.A. (2017) 'Flint Water Crisis Caused By Interrupted Corrosion Control: Investigating "Ground Zero" Home', *Environmental Science & Technology*, 51(4), pp. 2007–2014. Available at: https://doi.org/10.1021/acs.est.6b04034.

Pitelis, C.N. and Wagner, J.D. (2019) 'Strategic Shared Leadership and Organizational Dynamic Capabilities', *The Leadership Quarterly*, 30(2), pp. 233–242. Available at: https://doi.org/10.1016/j.leaqua.2018.08.002.

Rashman, L., Withers, E. and Hartley, J. (2009) 'Organizational learning and knowledge in public service organizations: A systematic review of the literature', *International Journal of Management Reviews*, 11(4), pp. 463–494. Available at: https://doi.org/10.1111/j.1468-2370.2009.00257.x.

Roberts, K. and Bea, R. (2001) 'Must accidents happen? Lessons from high-reliability organizations', *Academy of Management Executive*, 15. Available at: https://doi.org/10.5465/AME.2001.5229613.

Roberts, K.H. (1990) 'Managing High Reliability Organizations', *California Management Review*, 32(4), pp. 101–113. Available at: https://doi.org/10.2307/41166631.

Rosenhead, J. *et al.* (2019) 'Complexity theory and leadership practice: A review, a critique, and some recommendations', *The Leadership Quarterly*, 30(5), p. 101304. Available at: https://doi.org/10.1016/j.leaqua.2019.07.002.

Rosing, K., Frese, M. and Bausch, A. (2011) 'Explaining the heterogeneity of the leadership-innovation relationship: Ambidextrous leadership', *The Leadership Quarterly*, 22(5), pp. 956–974. Available at: https://doi.org/10.1016/j.leaqua.2011.07.014.

Saha, D. and Ibrahima, F.T. (2020) 'Who finances infrastructure, really? Disentangling public and private contributions', 15 January. Available at: https://blogs.worldbank.org/ppps/who-finances-infrastructure-really-disentangling-public-and-private-contributions (Accessed: 4 December 2023).

Salet, W., Bertolini, L. and Giezen, M. (2013) 'Complexity and Uncertainty: Problem or Asset in Decision Making of Mega Infrastructure Projects?', *International Journal of Urban and Regional Research*, 37(6), pp. 1984–2000. Available at: https://doi.org/10.1111/j.1468-2427.2012.01133.x.

Schneider, M. (2002) 'A Stakeholder Model of Organizational Leadership', *Organization Science*, 13(2), pp. 209–220. Available at: https://doi.org/10.1287/orsc.13.2.209.531.

Schulman, P. and Roe, E. (2016) *Reliability and Risk: The Challenge of Managing Interconnected Infrastructures.* Stanford, CA: Stanford University Press.

Serrat, O. (2017) 'Building a Learning Organization', in O. Serrat (ed.) *Knowledge Solutions: Tools, Methods, and Approaches to Drive Organizational Performance.* Singapore: Springer, pp. 57–67. Available at: https://doi.org/10.1007/978-981-10-0983-9_11.

Siggelkow, N. and Levinthal, D.A. (2003) 'Temporarily Divide to Conquer: Centralized, Decentralized, and Reintegrated Organizational Approaches to Exploration and Adaptation', *Organization Science*, 14(6), pp. 650–669. Available at: https://doi.org/10.1287/orsc.14.6.650.24840.

Sovacool, B.K., Lovell, K. and Ting, M.B. (2018) 'Reconfiguration, Contestation, and Decline: Conceptualizing Mature Large Technical Systems', *Science, Technology, & Human Values*, 43(6), pp. 1066–1097. Available at: https://doi.org/10.1177/0162243918768074.

Steffen, W. *et al.* (2015) 'The trajectory of the Anthropocene: The Great Acceleration', *The Anthropocene Review*, 2(1), pp. 81–98. Available at: https://doi.org/10.1177/2053019614564785.

Streeck, W. and Thelen, K. (2005) *Beyond Continuity: Institutional Change in Advanced Political Economies.* Oxford, UK: Oxford University Press.

Summerton, J. (1994) *Changing Large Technical Systems.* 1st edn. Boulder, CO: Westview Press.

Taylor, A. and Helfat, C.E. (2009) 'Organizational Linkages for Surviving Technological Change: Complementary Assets, Middle Management, and Ambidexterity', *Organization Science*, 20(4), pp. 718–739. Available at: https://doi.org/10.1287/orsc.1090.0429.

Teece, D.J. (2007) 'Explicating dynamic capabilities: the nature and microfoundations of (sustainable) enterprise performance', *Strategic Management Journal*, 28(13), pp. 1319–1350. Available at: https://doi.org/10.1002/smj.640.

Tourish, D. (2019) 'Is Complexity Leadership Theory Complex Enough? A critical appraisal, some modifications and suggestions for further research', *Organization Studies*, 40(2), pp. 219–238. Available at: https://doi.org/10.1177/0170840618789207.

Tsoukas, H. and Chia, R. (2002) 'On Organizational Becoming: Rethinking Organizational Change', *Organization Science*, 13(5), pp. 567–582. Available at: https://doi.org/10.1287/orsc.13.5.567.7810.

Turner, N., Swart, J. and Maylor, H. (2013) 'Mechanisms for Managing Ambidexterity: A Review and Research Agenda', *International Journal of Management Reviews*, 15(3), pp. 317–332. Available at: https://doi.org/10.1111/j.1468-2370.2012.00343.x.

Uhl-Bien, M. and Arena, M. (2017) 'Complexity leadership: Enabling people and organizations for adaptability', *Organizational Dynamics*, 46(1), pp. 9–20. Available at: https://doi.org/10.1016/j.orgdyn.2016.12.001.

Uhl-Bien, M. and Arena, M. (2018) 'Leadership for organizational adaptability: A theoretical synthesis and integrative framework', *The Leadership Quarterly*, 29(1), pp. 89–104. Available at: https://doi.org/10.1016/j.leaqua.2017.12.009.

Uhl-Bien, M. and Marion, R. (2009) 'Complexity leadership in bureaucratic forms of organizing: A meso model', *The Leadership Quarterly*, 20(4), pp. 631–650. Available at: https://doi.org/10.1016/j.leaqua.2009.04.007.

Uhl-Bien, M., Marion, R. and McKelvey, B. (2007) 'Complexity Leadership Theory: Shifting leadership from the industrial age to the knowledge era', *The Leadership Quarterly*, 18(4), pp. 298–318. Available at: https://doi.org/10.1016/j.leaqua.2007.04.002.

Underwood, B.S. *et al.* (2017) 'Increased costs to US pavement infrastructure from future temperature rise', *Nature Climate Change*, 7(10), pp. 704–707. Available at: https://doi.org/10.1038/nclimate3390.

Underwood, S. *et al.* (2020) 'Past and Present Design Practices and Uncertainty in Climate Projections are Challenges for Designing Infrastructure to Future Conditions', *Journal of Infrastructure Systems*, 26, p. 04020026. Available at: https://doi.org/10.1061/(ASCE)IS.1943-555X.0000567.

Walker, W.E., Lempert, R.J. and Kwakkel, J.H. (2013) 'Deep Uncertainty', in *Encyclopedia of Operations Research and Management Science*. Boston, MA: Springer, pp. 395–402. Available at: https://doi.org/10.1007/978-1-4419-1153-7_1140.

Weber, E.P. and Khademian, A.M. (2008) 'Wicked Problems, Knowledge Challenges, and Collaborative Capacity Builders in Network Settings', *Public Administration Review*, 68(2), pp. 334–349. Available at: https://doi.org/10.1111/j.1540-6210.2007.00866.x.

Wilbanks, T.J. and Fernandez, S. (2014) 'Framing Climate Change Implications for Infrastructures and Urban Systems', in T.J. Wilbanks and S. Fernandez (eds) *Climate Change and Infrastructure, Urban Systems, and Vulnerabilities*. Washington, DC: Island Press/Center for Resource Economics (NCA Regional Input Reports), pp. 17–40. Available at: https://doi.org/10.5822/978-1-61091-556-4_3.

Woods, D.D. (2015) 'Four concepts for resilience and the implications for the future of resilience engineering', *Reliability Engineering & System Safety*, 141, pp. 5–9. Available at: https://doi.org/10.1016/j.ress.2015.03.018.

Yu, D.J. *et al.* (2020) 'Toward General Principles for Resilience Engineering', *Risk Analysis*, 40(8), pp. 1509–1537. Available at: https://doi.org/10.1111/risa.13494.

Zahra, S.A. and George, G. (2002) 'Absorptive Capacity: A Review, Reconceptualization, and Extension', *The Academy of Management Review*, 27(2), pp. 185–203. Available at: https://doi.org/10.2307/4134351.

Zhou, Y.M. (2013) 'Designing for Complexity: Using Divisions and Hierarchy to Manage Complex Tasks', *Organization Science*, 24(2), pp. 339–355. Available at: https://doi.org/10.1287/orsc.1120.0744.

3.3

BALANCING EFFICIENCY AND RESILIENCE OBJECTIVES

3.3.1. INTRODUCTION AND BACKGROUND

Events like COVID-19 and the 2021 Winter Storm in Texas highlight a fundamental tension between efficiency (i.e. optimized system performance and use of resources) and resilience (i.e. capacity to identify, anticipate, prepare for, mitigate, and adapt to potentially disruptive changes and hazards) (Carvalhaes *et al.*, 2020). For instance, past recommendations for higher electricity generation reserve levels and increased weatherization of system components (i.e. resilience efforts) in the Texas power grid (FERC and NERC, 2011) were not heeded prior to the widespread outages that occurred during February 2021. The lack of resilience efforts was likely due, in part, to the perception that these actions were unnecessary or unjustifiably costly (i.e. not aligned with efficiency objectives) (Ball, 2021). This tension is intrinsic to the contradictory natures of efficiency and resilience. Efficiency strives to minimize waste (in the form of time, money, effort, resources, and other inputs) and maximize outputs/outcomes (Martin, 2019). Conversely, resilience is characterized by traits such as robustness, redundancy, diversity, flexibility, agility, and learning that appear to be antithetical to efficiency objectives (Biggs *et al.*, 2012; Park *et al.*, 2013; Woods, 2015; Chester and Allenby, 2018; Gilrein *et al.*, 2019). A description of each of these traits is provided in Table 3.3.1. Efficiency is particularly well suited for stable operating conditions and environments, while resilience is conducive to conditions of instability, complexity, and chaos (Helmrich and Chester, 2020). Likewise, efficiency is bolstered by

---

This chapter was adapted from the following article with publisher and lead author permissions: Samuel Markolf, Alysha Helmrich, Yeowon Kim, Ryan Hoff, and Mikhail Chester, 2022, Balancing Efficiency and Resilience Objectives in Pursuit of Sustainable Infrastructure Transformations, *Current Opinion in Environmental Sustainability*, 56(101181), doi: 10.1016/j.cosust.2022.101181.

processes of mechanization and standardization (Stanley, 2020), while re-silience is often bolstered by factors like creativity, improvisation, and ex-tensibility (Woods, 2015, 2018).

Table 3.3.1. Description of various traits associated with resilience.

Resilience Traits	Description
Robustness	System's ability to absorb disturbances, often via strengthening and hardening of system components.
Redundancy	The capacity or functionality of system components to compensate for each other.
Diversity	The variety, balance, and disparity of elements within a system.
Flexibility	System's ability to respond to both regular and irregular (non-incremental) changes.
Agility	System's ability to transform in response to unexpected changes or opportunities.

Adapted from (Biggs *et al.*, 2012; Anderies *et al.*, 2013; Park *et al.*, 2013; Woods, 2015; Chester and Allenby, 2018; Gilrein *et al.*, 2019).

Infrastructure systems are often built and managed according to pre-determined codes and practices (i.e. standardization). Additionally, they are often designed and built with the intent of lasting several decades — partly due to assumptions of system and environmental stability (Block-ley, Agarwal and Godfrey, 2012; Chester *et al.*, 2021; Markolf *et al.*, 2021). As a result, many of these systems appear to (implicitly or explicitly) em-phasize efficiency in their design and implementation — potentially at the expense of resilience. Contrary to other disciplines (e.g. ecology, leader-ship, and organizational change), the body of knowledge/practice related to infrastructure systems does not appear to contain much exploration of the efficiency–resilience tension. Although outside the scope of our anal-ysis, we acknowledge that there is associated work in many areas of liter-ature including (but not limited to) reliability engineering, robust control, risk management, multicriteria decision making, and decision making un-der deep uncertainty. Nonetheless, this chapter places particular empha-sis on applying knowledge from the ecological and social sciences to the engineering/infrastructure domain. Considering the increasingly com-plex, uncertain, and unstable conditions our infrastructure systems are likely to experience (Chester and Allenby, 2019), we strive to stimulate more explicit consideration and management of the efficiency–resilience tension within infrastructure systems — and ultimately help strike a dy-namic balance between the two. As detailed below, we explore and syn-thesize key bodies of knowledge on this topic and posit how they can be applied more directly to engineering and infrastructure systems.

This chapter is organized as follows. The next section synthesizes some of the key literature from ecology and ecological economics related to ef-ficiency and resilience. This discussion is particularly applicable to the physical components of infrastructure systems. Subsequently, we synthesize some of the key literature from business, management, and organizational theory related to efficiency and resilience. This discussion is particularly applicable to the institutional components of infrastructure systems. The final section posits how some of the key themes from these diffuse bodies of knowledge can be applied to help instill more balance between efficiency and resilience in engineering and infrastructure systems.

3.3.2. ECOLOGICAL SCIENCES AND THE 'WINDOW OF VITALITY' AS A BASIS FOR EFFICIENCY AND RESILIENCE ACROSS THE PHYSICAL ELEMENTS OF INFRASTRUCTURE

Regarding physical systems and networks, there is an established body of knowledge rooted in ecology and ecological economics that espouses the importance of both efficiency and resilience for the longevity of species and ecosystems (Ulanowicz, 2002, 2018; Goerner, Lietaer and Ulanowicz, 2009; Ulanowicz *et al.*, 2009; Homer-Dixon *et al.*, 2015; Stanley, 2020). Efficiency enhances the speed and amount of matter, energy, and information that species and ecosystems can process, while resilience enables species and ecosystems to persevere (and possibly transform) in the face of hazards, stressors, and extreme events. Traits linked to efficiency include centralization, streamlining, and specialization, while resilience is facilitated by traits like dispersity and redundancy (Ruth, 2006; Hopkins, 2014; Stanley, 2020). Notably, diversity and connectivity (i.e. higher transmission speed, capacity, and density among system components) appear to be two key features linking efficiency and resilience (Goerner, Lietaer and Ulanowicz, 2009; Biggs *et al.*, 2012; Homer-Dixon *et al.*, 2015). In general, higher connectivity and homogeneity (i.e. decreased diversity) contribute to increased system efficiency and decreased system resilience (Homer-Dixon *et al.*, 2015). Conversely, higher diversity and decreased connectivity translate to systems that are less efficient under stable conditions, but more adaptive to environmental shifts, crashes, shocks, or stressors (Bodin and Norberg, 2005; Scheffer *et al.*, 2012; Lever *et al.*, 2014; Homer-Dixon *et al.*, 2015). In sum, efficiency and resilience are complementary but often at odds with one another — greater resilience may result in less efficiency, and vice versa (Goerner, Lietaer and Ulanowicz, 2009).

Given the opposing directions in which efficiency and resilience can pull with respect to diversity and connectivity, tensions and trade-offs emerge between the two. However, these tensions may not always play out in straightforward manners. For example, greater connectivity can sometimes facilitate the flow of resources and assistance after a disruptive event — thereby contributing to system resilience (Biggs *et al.*, 2012; Scheffer *et al.*, 2012; Homer-Dixon *et al.*, 2015). Similarly, species/ecosystem resilience can sometimes lead to undesirable outcomes if it contributes to the preservation of deleterious system dynamics (Stanley, 2020; Eriksen *et al.*, 2021). Overall, the ecological literature posits that systems benefit from both efficiency and resilience. In an unconstrained world, the maximization of both would be advantageous. However, in reality systems must typically strike a dynamic balance between efficiency and resilience. Systems that exhibit sufficient (and balanced) efficiency and resilience have been described as functioning within the 'window of vitality' (Ulanowicz, 2002, 2018; Goerner, Lietaer and Ulanowicz, 2009; Ulanowicz *et al.*, 2009). Empirical examinations of the 'window of vitality' indicate that the efficiency–resilience spectrum in natural systems tends to lean slightly toward resilience (Ulanowicz, 2002, 2018; Goerner, Lietaer and Ulanowicz, 2009; Ulanowicz *et al.*, 2009). Given the evolutionary pathway that has led to the 'window of vitality', human/engineered systems may benefit from mimicking natural systems by placing additional emphasis on the resilience end of the spectrum.

In a similar vein, the concept of 'safe operating spaces (SOS)' has emerged as an approach for actively monitoring and navigating multiple misaligned objectives under dynamic and uncertain conditions (Dearing *et al.*, 2014; Carpenter *et al.*, 2015; Anderies, Mathias and Janssen, 2019). The idea of safe operating spaces has traditionally been applied to coupled social-ecological systems (e.g. fisheries, watersheds), and centers on supporting human well-being (e.g. equitable access to food, water, shelter, energy, education, economic opportunity) while staying within biophysical planetary boundaries (e.g. land use change, loss of biodiversity, ocean acidification, climate change, nitrogen and phosphorous cycles). Moving forward, there appear to be opportunities to apply elements of the safe operating spaces concept to the efficiency–resilience spectrum. For instance, one could envision a safe operating space bound by a 'floor' comprising performance and efficiency objectives and a 'ceiling' related to physical, sociotechnical, and environmental constraints. Similarly, properties of resilience and adaptive capacity would be crucial to remaining within (and possibly expanding or shifting) this safe operating space under variable or extreme conditions.

3.3.3. SOCIAL SCIENCES AND 'ORGANIZATIONAL AMBIDEXTERITY' AS A BASIS FOR EFFICIENCY AND RESILIENCE WITHIN THE INSTITUTIONAL ASPECTS OF INFRASTRUCTURE

Paralleling the translation of ecologically based concepts to the physical components of infrastructure, leadership and organizational theory appear well positioned to examine efficiency and resilience within the institutional context of infrastructure. These bodies of knowledge introduce and explore the tension between exploitation (i.e. risk-averse decisions) and exploration (i.e. risk-seeking decisions) as a space to ensure organizational longevity (Duncan, 1976; March, 1991; Tushman and O'Reilly, 1996; Uhl-Bien and Arena, 2018). Exploitative behaviors resemble aspects of efficiency and include rule enforcement, conformity through routines, rapid decision-making, and disciplinary approaches. Conversely, explorative behaviors include variability in the process, acceptance of failures, and diverse community building — resembling aspects of resilience (Rosing, Frese and Bausch, 2011; Havermans *et al.*, 2015). The effective management of the tension between exploitation and exploration is known as ambidexterity. The crux of ambidexterity is to 1) establish formal (e.g. organizational structures, rules, and regulations) and informal governances (e.g. leadership, trust) that sponsor explorational pursuits while simultaneously maintaining services, and 2) integrate successful explorative endeavors through institutional repositioning (Turner, Swart and Maylor, 2013; Sovacool, Lovell and Ting, 2018; Uhl-Bien and Arena, 2018).

Organizational ambidexterity is supported by the Law of Requisite Complexity (Boisot and McKelvey, 2011), and its predecessor, the Law of Requisite Variety (Ashby, 1956). The Law of Requisite Variety states that a system (re: organization) can appropriately adapt if the organization's range of responses is equivalent to — or greater than — the states in which it must operate (Ashby, 1956). Subsequently, the Law of Requisite Complexity states that, to be adaptable, an organization's internal complexity must match or surpass external complexity (Boisot and McKelvey, 2011). Achieving the requisite complexity (or variety) relies on an organization's ability to sense, learn, and react to the demands of its environment (Ashby, 1956; Hamel and Välikangas, 2003) — capacities that can align with both efficiency and resilience. Organizations can strive to reduce the range and variety of stimuli to which they are exposed via processes that align with exploitation such as routinization, streamlining, and simplification (March, 1991; Boisot and McKelvey, 2011). Similarly, organizations can strive to expand internal capacities to respond to a wider range and variety of stimuli via processes that align with exploration, such as increasing system diversity, variety, and complexity (March, 1991; Gell-Mann, 2002).

Due to the relative stability of the past, most infrastructure organizations and institutions appear to emphasize exploitative actions and outcomes (Uhl-Bien and Arena, 2018; Chester, Miller and Muñoz-Erickson, 2020), which in turn can accelerate and exacerbate organizational deficiencies in today's increasingly turbulent environments (Hamel and Välikangas, 2003). Ambidexterity and the Law of Requisite Complexity emphasize the importance of striking a dynamic balance between exploitative (efficiency-oriented) processes and explorative (resilience-oriented) processes. Too much emphasis on exploitive processes can result in an oversimplistic perception of external conditions, and hinder an organization's ability to effectively respond to changes, shocks, and surprises (Boisot and McKelvey, 2011; Carpenter *et al.*, 2015; Anderies, Mathias and Janssen, 2019). Conversely, too much emphasis on exploratory processes can be physically and cognitively expensive, and result in an overly responsive organization (i.e. responding to all stimuli regardless of their relevance; inability to distinguish the signal from the noise) (Boisot and McKelvey, 2011; Carpenter *et al.*, 2015). Papachroni et al. (Papachroni, Heracleous and Paroutis, 2016) posit that exploitation–exploration are not mutually exclusive, but instead are complementary and interrelated — further emphasizing that organizations should pursue both behaviors to remain relevant. Similarly, Anderies et al. (Anderies, Mathias and Janssen, 2019) suggest that different combinations of knowledge systems and policy types are needed to move between multiple safe operating spaces and avoid 'dead operating spaces' (Anderies, Mathias and Janssen, 2019). These perspectives parallel the ecological concept of the 'window of vitality' (discussed above), where a system's long-term persistence requires a balance between efficiency and resilience.

3.3.4. TOWARD A DYNAMIC BALANCE BETWEEN EFFICIENCY AND RESILIENCE IN INFRASTRUCTURE SYSTEMS

Ultimately, the diverse suite of literature reviewed in this chapter converges on the idea that both efficiency and resilience are vital for the long-term viability of systems — especially as they navigate recurrent fluctuations between conditions of stability and instability. Additionally, the importance of a dynamic balance between efficiency and resilience applies to both the physical and institutional components of (infrastructure) systems. For instance, relating the 'window of vitality' to the physical aspects of infrastructure systems reveals that a shift toward the resilience end of the spectrum is perhaps warranted — aligning infrastructure systems with the observed tendencies of natural systems. This shift would diverge from the current (implicit or explicit) emphasis on efficiency within infrastructure systems, which is likely due to several factors. For one, many infrastructure systems were designed under the assumption of long-term stability and rigidity. Considering factors like climate change, technological change, and population shifts, these assumptions appear to be increasingly at odds with the environments in which infrastructure must function (Moallemi, Kwakkel, *et al.*, 2020; Chester *et al.*, 2021). Said differently, hidden fragilities tend to emerge in systems that become well adapted to a particular set of inputs/forcing (i.e. the Law of Conservation of Fragility) (Hendrik, 1945; Csete and Doyle, 2002; Anderies *et al.*, 2013; Carpenter *et al.*, 2015; Anderies, Mathias and Janssen, 2019). Second, resilience is a system property that is often not readily observable until a disturbance occurs, whereas efficiency is typically easier to quantify (and operationalize)

(Kaika, 2017; Bruun, 2018; Stanley, 2020; Eriksen *et al.*, 2021). Similarly, emphasis on near-term conditions and outcomes can reinforce a proclivity toward established governance structures and operational practices (Dearing *et al.*, 2014; Lever *et al.*, 2014; Carpenter *et al.*, 2015; Mosberg, Nyukuri and Naess, 2017; Nightingale, 2017; Anderies, Mathias and Janssen, 2019; Wyborn *et al.*, 2019; Eriksen *et al.*, 2021). As a result, incentives and inertia emerge that tend to align with efficiency and depart from resilience. One potential response would be to place additional emphasis on metrics of variability such as shifts in the magnitude, frequency, duration, and direction of system performance and exogenous factors (e.g. temperature and precipitation). Significant movement in these so-called 'early warning signs' has been posited as an indication of declining system stability and resilience, as well as the possibility of an impending threshold (Dearing *et al.*, 2014).

The above factors can also be catalyzed and exacerbated (either consciously or unconsciously) by motivational effects (i.e. stakeholder motives that result in the consideration of certain alternatives and the ignorance or misjudgment of others), focused thinking (i.e. deliberate attention to specific issues and perspectives at the expense of others), and narrow thinking (i.e. unintentional or deliberate disregard for potential alternatives) (Montibeller and von Winterfeldt, 2015; Lahtinen, Guillaume and Hämäläinen, 2017; Zare *et al.*, 2017; Moallemi, Zare, *et al.*, 2020). For example, the costs (e.g. time, resources, conflict) of pluralistic governance and decision-making can sometimes be perceived as outweighing the benefits (e.g. increased capacity, creativity, and reflexivity) (Wyborn *et al.*, 2019). In turn, this perception can lead to a closing down of problem/solution spaces and a propensity toward existing incentive structures and models of analysis (Wyborn *et al.*, 2019). Finally, misalignments between incentives and impacts can arise from a variety of scale (geographic, temporal, and network) issues. For both efficiency and resilience, what is favorable for one actor or firm may not be favorable for the broader system(s)– and vice versa. Similarly, what is favorable in the near-term may not align with what is favorable in the long-term (Anderies *et al.*, 2013; Eriksen *et al.*, 2021). Efficiency-oriented efforts like standardization align with goals of reducing system variability and increasing predictability. On timescales conforming to things like terms of office, funding cycles, and immediate human needs, reduced system variability is appealing. However, reduced system variability in the near-term can lead to increased variability and risk of crossing critical thresholds in the long-term (Carpenter *et al.*, 2015). Therefore, managing variability appears to be inextricably linked to managing efficiency in the short run and resilience in the long run. Regardless of the impetus for efficiency-focused design and operation, there are a number of potentially dubious outcomes: 1) incomplete assessment or consideration of system context, dynamics, uncertainties, and trade-offs;

2) missed information and learning opportunities, system lock-in, diminished hardiness to shocks and stressors, and reduced safe operating spaces; 3) inequitable participation in the planning and implementation of interventions; 4) defining 'success' from an overly narrow or exclusionary perspective that primarily aligns with dominant agendas and powerful stakeholders; and 5) closing off potential solution pathways and outcomes in favor of 'traditional'/established approaches (Dearing *et al.*, 2014; Carpenter *et al.*, 2015; Anderies, Mathias and Janssen, 2019; Eriksen *et al.*, 2021). We conclude our discussion by outlining some potential approaches for assuaging these outcomes.

Within individual organizations and systems, infrastructure managers can enact the Complex Leadership Theory (CLT) framework to enhance organizational ambidexterity and navigate efficiency–resilience tensions (Uhl-Bien and Arena, 2017, 2018). Although exploitative and explorative behaviors are both practiced, infrastructure institutions tend to favor administrative leadership (i.e. exploitative behavior) that reduces complexity (Havermans *et al.*, 2015). CLT can advance resilience efforts by prompting infrastructure managers to consider the long-term consequences of decision making. In particular, CLT can facilitate the emergence of enabling leaders, who embrace both administrative and entrepreneurial leadership as operating conditions swing between stable and unstable (Havermans *et al.*, 2015; Uhl-Bien and Arena, 2017, 2018). An enabling leader is not simply someone who can partake in either behavior. Instead, they are pathfinders who can identify productive tensions and integrate knowledge toward continuous shifts in formal and informal governance under dynamic operating environments (Uhl-Bien and Arena, 2018). These continuous shifts and responses can be facilitated by exploratory modeling and analyses. Example approaches include design of experiments, stress-testing, worst-case scenario discovery, multi-objective decision making, and robust decision making (Moallemi, Kwakkel, *et al.*, 2020; Moallemi, Zare, *et al.*, 2020). These exploratory approaches can enable careful examination of rival decision paths, elucidation of system sensitivities and key decision criteria, identification of decisions and actions that produce (un)satisfactory trade-offs between multiple objectives, determination of scenarios that produce key performance thresholds (either positive or negative) and enhance overall confidence in decisions and methodological choices (Moallemi, Kwakkel, *et al.*, 2020; Moallemi, Zare, *et al.*, 2020).

Exploratory modeling can be helpful (but not necessarily prerequisite) for infrastructure managers to recognize and embrace the importance of boundary setting, boundary thresholds, and boundary spanning (when necessary) (Dearing *et al.*, 2014; Marchau *et al.*, 2019). Whether pursuing efficiency, resilience, or any other objectives, there can be merit in recognizing that infrastructure systems impact, and are impacted by, surrounding social, ecological, and technological systems (SETS) [61]. Considering infrastructure systems as coupled SETS builds upon work within socioecological systems (Walker *et al.*, 2006; Ostrom, 2009; Fischer *et al.*, 2015; Muneepeerakul and Anderies, 2017; Béné and Doyen, 2018) and provides an opportunity for further exploring concepts of enabling leadership within complex systems. Effectively identifying and operating within SETS boundaries can be aided by the practice of coproduction — "processes that iteratively bring together diverse groups and their ways of knowing and acting to create new knowledge and practices to transform societal outcomes (Wyborn *et al.*, 2019)". Hallmarks of coproduction include (but are not limited to) culturally appropriate engagement with all relevant stakeholders, open and flexible processes, frequent feedback from participants, acknowledging and addressing power dynamics, establishing pertinent boundary objects, clear and frequent communication, and sufficient resources to support sustained coordination and collaboration (Wyborn *et al.*, 2019). Many of these practices (e.g. involving others, group discussion, connectedness, diversity, boundary spanning, etc.) are also linked with cultivating enabling leadership (Havermans *et al.*, 2015). Although the science and practice of coproduction continue to evolve, potential outcomes of this approach include increased equity, improved processes and capacities, enhanced creativity and reflexivity among stakeholders, creation of new knowledge, deepened awareness of various issues, and broader understanding (Wyborn *et al.*, 2019). Coproduction may also facilitate the cultivation of new knowledge systems and policy mechanisms needed to navigate from one safe operating space to another (Anderies, Mathias and Janssen, 2019). Ultimately, these processes and their outcomes can potentially help various stakeholders navigate and establish a dynamic balance between efficiency and resilience across multiple SETS and safe operating spaces under varying conditions.

Although CLT, exploratory modeling, and coproduction can be undertaken in an ad-hoc and 'organic' manner, one possible approach for catalyzing these transformations within infrastructure systems is the formal consideration (and possible regulation) of resilience as a public/common good, and the lack of resilience as a negative externality (Homer-Dixon, 2006; Farley and Voinov, 2016; Wyborn *et al.*, 2019; Stanley, 2020). The steady improvements in air and water quality achieved via the Clean Air Act and Clean Water Act (and related policies) could serve as aspirational templates for establishing standards and policies for more explicitly addressing and reducing the negative externalities associated with a dearth of resilience. Doing so can complement, and be complemented by, exploratory modeling and coproduction. Ultimately, establishing resilience as a common good can help create incentives for implementing and coproducing attributes like diversity, redundancy, and robustness within infrastructure systems–resilience enhancing traits that can complement efficiency-oriented practices already in place.

The review and synthesis of diverse bodies of knowledge conducted in this analysis underscore the importance of striving to achieve a dynamic balance between efficiency and resilience within infrastructure systems. Furthermore, approaches such as CLT, coproduction, exploratory modeling techniques, and the establishment of resilience as a public good appear to be well positioned to help navigate tensions between efficiency and resilience. However, none of these concepts are a silver bullet. There will be systems and situations where other factors (in addition to or instead of efficiency and resilience) will take precedence. There will also be systems and situations where exploratory modeling and/or coproduction may not be necessary or appropriate. Finally, we acknowledge that the challenges, opportunities, and shortcomings of infrastructure systems cannot be fully distilled down to the tension between efficiency and resilience — especially given the complex, multi-objective, and varied nature of infrastructure systems. Nevertheless, we posit that efficiency and resilience are two of the most crucial 'levers' at our disposal for achieving system longevity, and perhaps more importantly, desirable outcomes for as many people as possible under as many conditions as possible. We are optimistic that the topics and discussions in this chapter can catalyze continued research and practice aimed at further exploring and critically examining the appropriate balance between efficiency and resilience (among other objectives) — as well as the tools, frameworks, and approaches for doing so. Collectively, these efforts can empower infrastructure institutions and systems to adapt to a wide range of stresses, shocks, and surprises, while helping them thrive under conditions of both stability and instability.

3.3.5. REFERENCES

Anderies, J. *et al.* (2013) 'Aligning Key Concepts for Global Change Policy: Robustness, Resilience, and Sustainability', *Ecology and Society*, 18(2). Available at: https://doi.org/10.5751/ES-05178-180208.

Anderies, J.M., Mathias, J.-D. and Janssen, M.A. (2019) 'Knowledge infrastructure and safe operating spaces in social–ecological systems', *Proceedings of the National Academy of Sciences*, 116(12), pp. 5277–5284. Available at: https://doi.org/10.1073/pnas.1802885115.

Ashby, W.R. (1956) *An Introduction to Cybernetics*. Hoboken, NJ: J. Wiley. Available at: https://doi.org/10.5962/bhl.title.5851.

Ball, J. (2021) 'The Texas Blackout Is the Story of a Disaster Foretold', *Texas Monthly*, 19 February. Available at: https://www.tex-asmonthly.com/news-politics/texas-blackout-preventable/ (Accessed: 20 October 2023).

Béné, C. and Doyen, L. (2018) 'From Resistance to Transformation: A Generic Metric of Resilience Through Viability', *Earth's Future*, 6(7), pp. 979–996. Available at: https://doi.org/10.1002/2017EF000660.

Biggs, R. *et al.* (2012) 'Toward Principles for Enhancing the Resilience of Ecosystem Services', *Annual Review of Environment and Resources*, 37(1), pp. 421–448. Available at: https://doi.org/10.1146/annurev-environ-051211-123836.

Blockley, D., Agarwal, J. and Godfrey, P. (2012) 'Infrastructure resilience for high-impact low-chance risks', *Proceedings of the Institution of Civil Engineers - Civil Engineering*, 165(6), pp. 13–19. Available at: https://doi.org/10.1680/cien.11.00046.

Bodin, Ö. and Norberg, J. (2005) 'Information Network Topologies for Enhanced Local Adaptive Management', *Environmental Management*, 35(2), pp. 175–193. Available at: https://doi.org/10.1007/s00267-004-0036-7.

Boisot, M. and McKelvey, B. (2011) 'Connectivity, Extremes, and Adaptation: A Power-Law Perspective of Organizational Effectiveness', *Journal of Management Inquiry*, 20(2), pp. 119–133. Available at: https://doi.org/10.1177/1056492610385564.

Bruun, J.A. (2018) 'Climate changing civil society: the role of value and knowledge in designing the Green Climate Fund', in S.B. Woodhouse Aurora Fredriksen, Sian Sullivan, Philip et al. (eds) *Valuing Development, Environment and Conservation: Creating Values that Matter*. London, UK: Routledge. Available at: https://doi.org/10.4324/9781315113463.

Carpenter, S.R. *et al.* (2015) 'Allowing variance may enlarge the safe operating space for exploited ecosystems', *Proceedings of the National Academy of Sciences*, 112(46), pp. 14384–14389. Available at: https://doi.org/10.1073/pnas.1511804112.

Carvalhaes, T. *et al.* (2020) 'COVID-19 as a Harbinger of Transforming Infrastructure Resilience', *Frontiers in Built Environment*, 6. Available at: https://doi.org/10.3389/fbuil.2020.00148 (Accessed: 21 October 2023).

Chester, M. *et al.* (2021) 'Infrastructure resilience to navigate increasingly uncertain and complex conditions in the Anthropocene', *npj Urban Sustainability*, 1(1), pp. 1–6. Available at: https://doi.org/10.1038/s42949-021-00016-y.

Chester, M.V. and Allenby, B. (2018) 'Toward adaptive infrastructure: flexibility and agility in a non-stationarity age', *Sustainable and Resilient Infrastructure*, 4, pp. 1–19. Available at: https://doi.org/10.1080/23789689.2017.1416846.

Chester, M.V. and Allenby, B. (2019) 'Infrastructure as a wicked complex process', *Elementa: Science of the Anthropocene*. Edited by A. Iles and M.E. Chang, 7(1), p. 21. Available at: https://doi.org/10.1525/elementa.360.

Chester, M.V., Miller, T. and Muñoz-Erickson, T.A. (2020) 'Infrastructure governance for the Anthropocene', *Elementa: Science of the Anthropocene*, 8(1), p. 78. Available at: https://doi.org/10.1525/elementa.2020.078.

Csete, M.E. and Doyle, J.C. (2002) 'Reverse Engineering of Biological Complexity', *Science*, 295(5560), pp. 1664–1669. Available at: https://doi.org/10.1126/science.1069981.

Dearing, J.A. *et al.* (2014) 'Safe and just operating spaces for regional social-ecological systems', *Global Environmental Change*, 28, pp. 227–238. Available at: https://doi.org/10.1016/j.gloenvcha.2014.06.012.

Duncan, R.B. (1976) 'The ambidextrous organization, designing dual structures for innovation', in *Strategies and implementation*. New York, NY: North-Holland Publishing, pp. 167–188.

Eriksen, S. *et al.* (2021) 'Adaptation interventions and their effect on vulnerability in developing countries: Help, hindrance or irrelevance?', *World Development*, 141, p. 105383. Available at: https://doi.org/10.1016/j.worlddev.2020.105383.

Farley, J. and Voinov, A. (2016) 'Economics, socio-ecological resilience and ecosystem services', *Journal of Environmental Management*, 183, pp. 389–398. Available at: https://doi.org/10.1016/j.jenvman.2016.07.065.

FERC and NERC (2011) *Report on Outages and Curtailments During the Southwest Cold Weather Event of February 1-5, 2011*. Federal Energy Regulatory Commission and North American Electric Reliability Corporation, p. 357. Available at: https://www.ferc.gov/sites/default/files/2020-04/08-16-11-report.pdf.

Fischer, J. *et al.* (2015) 'Advancing sustainability through mainstreaming a social–ecological systems perspective', *Current Opinion in Environmental Sustainability*, 14, pp. 144–149. Available at: https://doi.org/10.1016/j.cosust.2015.06.002.

Gell-Mann, M. (2002) 'What Is Complexity?', in A.Q. Curzio and M. Fortis (eds). Heidelberg, DE: Physica-Verlag HD, pp. 13–24. Available at: https://doi.org/10.1007/978-3-642-50007-7_2.

Gilrein, E.J. *et al.* (2019) 'Concepts and practices for transforming infrastructure from rigid to adaptable', *Sustainable and Resilient Infrastructure*, 6(3–4), pp. 213–234. Available at: https://doi.org/10.1080/23789689.2019.1599608.

Goerner, S.J., Lietaer, B. and Ulanowicz, R.E. (2009) 'Quantifying economic sustainability: Implications for free-enterprise theory, policy and practice', *Ecological Economics*, 69(1), pp. 76–81. Available at: https://doi.org/10.1016/j.ecolecon.2009.07.018.

Hamel, G. and Välikangas, L. (2003) 'The Quest for Resilience', *Harvard Business Review*, 1 September. Available at: https://hbr.org/2003/09/the-quest-for-resilience (Accessed: 20 October 2023).

Havermans, L.A. *et al.* (2015) 'Exploring the Role of Leadership in Enabling Contextual Ambidexterity', *Human Resource Management*, 54(S1), pp. s179–s200. Available at: https://doi.org/10.1002/hrm.21764.

Helmrich, A. *et al.* (2021) 'Centralization and decentralization for resilient infrastructure and complexity', *Environmental Research: Infrastructure and Sustainability*, 1(2), p. 021001. Available at: https://doi.org/10.1088/2634-4505/ac0a4f.

Helmrich, A.M. and Chester, M.V. (2020) 'Reconciling complexity and deep uncertainty in infrastructure design for climate adaptation', *Sustainable and Resilient Infrastructure*, 7(2), pp. 83–99. Available at: https://doi.org/10.1080/23789689.2019.1708179.

Hendrik, B.W. (1945) *Network Analysis and Feedback Amplifier Design*. New York, NY: D. Van Nostrand Company, Inc. Available at: http://archive.org/details/dli.ernet.15701 (Accessed: 20 October 2023).

Homer-Dixon, T. (2006) *The Upside of Down: Catastrophe, Creativity, and the Renewal of Civilization*. 1st edn. Washington, DC: Island Press.

Homer-Dixon, T. *et al.* (2015) 'Synchronous failure: the emerging causal architecture of global crisis', *Ecology and Society*, 20(3). Available at: https://doi.org/10.5751/ES-07681-200306.

Hopkins, R. (2014) *The Transition Handbook: From Oil Dependency to Local Resilience*. London, UK: Bloomsbury.

Kaika, M. (2017) '"Don't call me resilient again!": the New Urban Agenda as immunology … or … what happens when communities refuse to be vaccinated with "smart cities" and indicators', *Environment and*

Urbanization, 29(1), pp. 89–102. Available at:
https://doi.org/10.1177/0956247816684763.

Lahtinen, T.J., Guillaume, J.H.A. and Hämäläinen, R.P. (2017) 'Why pay attention to paths in the practice of environmental modelling?', *Environmental Modelling & Software*, 92, pp. 74–81. Available at:
https://doi.org/10.1016/j.envsoft.2017.02.019.

Lever, J.J. *et al.* (2014) 'The sudden collapse of pollinator communities', *Ecology Letters*, 17(3), pp. 350–359. Available at:
https://doi.org/10.1111/ele.12236.

March, J.G. (1991) 'Exploration and Exploitation in Organizational Learning', *Organization Science*, 2(1), pp. 71–87. Available at:
https://doi.org/10.1287/orsc.2.1.71.

Marchau, V.A.W.J. *et al.* (eds) (2019) *Decision Making under Deep Uncertainty: From Theory to Practice*. Cham: Springer International Publishing. Available at: https://doi.org/10.1007/978-3-030-05252-2.

Markolf, S.A. *et al.* (2021) 'Re-imagining design storm criteria for the challenges of the 21st century', *Cities*, 109, p. 102981. Available at:
https://doi.org/10.1016/j.cities.2020.102981.

Martin, R.L. (2019) 'The High Price of Efficiency', *Harvard Business Review*, 1 January. Available at: https://hbr.org/2019/01/the-high-price-of-efficiency (Accessed: 19 October 2023).

Moallemi, E.A., Kwakkel, J., *et al.* (2020) 'Exploratory modeling for analyzing coupled human-natural systems under uncertainty', *Global Environmental Change*, 65, p. 102186. Available at:
https://doi.org/10.1016/j.gloenvcha.2020.102186.

Moallemi, E.A., Zare, F., *et al.* (2020) 'Structuring and evaluating decision support processes to enhance the robustness of complex human–natural systems', *Environmental Modelling & Software*, 123, p. 104551. Available at: https://doi.org/10.1016/j.envsoft.2019.104551.

Montibeller, G. and von Winterfeldt, D. (2015) 'Cognitive and Motivational Biases in Decision and Risk Analysis', *Risk Analysis*, 35(7), pp. 1230–1251. Available at: https://doi.org/10.1111/risa.12360.

Mosberg, M., Nyukuri, E. and Naess, L.O. (2017) 'The Power of "Know-Who": Adaptation to Climate Change in a Changing Humanitarian Landscape in Isiolo, Kenya', *IDS Bulletin*, 48(4), pp. 79–92. Available at: https://doi.org/10.19088/1968-2017.154.

Muneepeerakul, R. and Anderies, J.M. (2017) 'Strategic behaviors and governance challenges in social-ecological systems', *Earth's Future*, 5(8), pp. 865–876. Available at: https://doi.org/10.1002/2017EF000562.

Nightingale, A.J. (2017) 'Power and politics in climate change adaptation efforts: Struggles over authority and recognition in the context of political instability', *Geoforum*, 84, pp. 11–20. Available at: https://doi.org/10.1016/j.geoforum.2017.05.011.

Ostrom, E. (2009) 'A General Framework for Analyzing Sustainability of Social-Ecological Systems', *Science*, 325(5939), pp. 419–422. Available at: https://doi.org/10.1126/science.1172133.

Papachroni, A., Heracleous, L. and Paroutis, S. (2016) 'In pursuit of ambidexterity: Managerial reactions to innovation–efficiency tensions', *Human Relations*, 69(9), pp. 1791–1822. Available at: https://doi.org/10.1177/0018726715625343.

Park, J. *et al.* (2013) 'Integrating Risk and Resilience Approaches to Catastrophe Management in Engineering Systems', *Risk Analysis*, 33(3), pp. 356–367. Available at: https://doi.org/10.1111/j.1539-6924.2012.01885.x.

Rosing, K., Frese, M. and Bausch, A. (2011) 'Explaining the heterogeneity of the leadership-innovation relationship: Ambidextrous leadership', *The Leadership Quarterly*, 22(5), pp. 956–974. Available at: https://doi.org/10.1016/j.leaqua.2011.07.014.

Ruth, M. (2006) 'A quest for the economics of sustainability and the sustainability of economics', *Ecological Economics*, 56(3), pp. 332–342. Available at: https://doi.org/10.1016/j.ecolecon.2005.09.012.

Scheffer, M. *et al.* (2012) 'Anticipating Critical Transitions', *Science*, 338(6105), pp. 344–348. Available at: https://doi.org/10.1126/science.1225244.

Sovacool, B.K., Lovell, K. and Ting, M.B. (2018) 'Reconfiguration, Contestation, and Decline: Conceptualizing Mature Large Technical Systems', *Science, Technology, & Human Values*, 43(6), pp. 1066–1097. Available at: https://doi.org/10.1177/0162243918768074.

Stanley, C. (2020) 'Living to Spend Another Day: Exploring Resilience as a New Fourth Goal of Ecological Economics', *Ecological Economics*, 178, p. 106805. Available at: https://doi.org/10.1016/j.ecolecon.2020.106805.

Turner, N., Swart, J. and Maylor, H. (2013) 'Mechanisms for Managing Ambidexterity: A Review and Research Agenda', *International Journal of Management Reviews*, 15(3), pp. 317–332. Available at: https://doi.org/10.1111/j.1468-2370.2012.00343.x.

Tushman, M.L. and O'Reilly, C.A. (1996) 'Ambidextrous Organizations: Managing Evolutionary and Revolutionary Change', *California*

Management Review, 38(4), pp. 8–29. Available at: https://doi.org/10.2307/41165852.

Uhl-Bien, M. and Arena, M. (2017) 'Complexity leadership: Enabling people and organizations for adaptability', *Organizational Dynamics*, 46(1), pp. 9–20. Available at: https://doi.org/10.1016/j.orgdyn.2016.12.001.

Uhl-Bien, M. and Arena, M. (2018) 'Leadership for organizational adaptability: A theoretical synthesis and integrative framework', *The Leadership Quarterly*, 29(1), pp. 89–104. Available at: https://doi.org/10.1016/j.leaqua.2017.12.009.

Ulanowicz, R.E. (2002) 'The balance between adaptability and adaptation', *Biosystems*, 64(1), pp. 13–22. Available at: https://doi.org/10.1016/S0303-2647(01)00170-8.

Ulanowicz, R.E. *et al.* (2009) 'Quantifying sustainability: Resilience, efficiency and the return of information theory', *Ecological Complexity*, 6(1), pp. 27–36. Available at: https://doi.org/10.1016/j.ecocom.2008.10.005.

Ulanowicz, R.E. (2018) 'Biodiversity, functional redundancy and system stability: subtle connections', *Journal of The Royal Society Interface*, 15(147), p. 20180367. Available at: https://doi.org/10.1098/rsif.2018.0367.

Walker, B. *et al.* (2006) 'A Handful of Heuristics and Some Propositions for Understanding Resilience in Social-Ecological Systems', *Ecology and Society*, 11(1). Available at: https://doi.org/10.5751/ES-01530-110113.

Woods, D.D. (2015) 'Four concepts for resilience and the implications for the future of resilience engineering', *Reliability Engineering & System Safety*, 141, pp. 5–9. Available at: https://doi.org/10.1016/j.ress.2015.03.018.

Woods, D.D. (2018) 'The theory of graceful extensibility: basic rules that govern adaptive systems', *Environment Systems and Decisions*, 38(4), pp. 433–457. Available at: https://doi.org/10.1007/s10669-018-9708-3.

Wyborn, C. *et al.* (2019) 'Co-Producing Sustainability: Reordering the Governance of Science, Policy, and Practice', *Annual Review of Environment and Resources*, 44(1), pp. 319–346. Available at: https://doi.org/10.1146/annurev-environ-101718-033103.

Zare, F. *et al.* (2017) 'Integrated water assessment and modelling: A bibliometric analysis of trends in the water resource sector', *Journal of Hydrology*, 552, pp. 765–778. Available at: https://doi.org/10.1016/j.jhydrol.2017.07.031.

3.4

DYNAMIC CRITICALITY

3.4.1. INTRODUCTION

The diversity of novel hazard dynamics raises questions about whether static framings of critical infrastructure (CI) are appropriate (Carlson and Doyle, 2002; Chester and Allenby, 2018; Gilrein *et al.*, 2019; Markolf *et al.*, 2022). For example, should infrastructure managers prepare for extreme weather-related events in the same way as a pandemic? Do current framings of criticality provide the flexibility to reprioritize resources across various hazards? Throughout this chapter the term *environment* will refer to the many external forces that affect infrastructure, including the natural environment, politics, cyber warfare, disruptive technologies, economic pressures, etc.

Static approaches have long characterized prioritization and resilience strategies for CI (Humphreys, 2019). Since 9/11, many governmental actions, such as presidential directives, congressional acts, and federal department policies, have attempted to inspire greater awareness for critical infrastructure protection and prioritization (Humphreys, 2019). The seminal definition of critical infrastructure came from the Patriot Act and is still used by the Department of Homeland Security (DHS) (CISA, 2019). Multiple lists of prioritized national CIs have been created and contain a mix of traditional civil infrastructure (i.e., those systems thought of as "utilities") and some social and ecological systems. There does not appear to be a concerted effort to support rapid transitions of resources to different infrastructure sensitive to the hazard. DHS and CISA use a two-tiered priority system for CI but do not have a dynamic prioritization process for when disturbances change (Moteff, 2015). Static framings continue to be

---

This chapter was adapted from the following article with publisher and lead author permissions: Ryan Hoff, Alysha Helmrich, Abbie Dirks, Yeowon Kim, Rui Li, and Mikhail Chester, 2023, Dynamic Criticality for Infrastructure Prioritization in Complex Environments, *Environmental Research Infrastructure and Sustainability*, 3(1), 015011, doi: 10.1088/2634-4505/acbe15.

standard practice for infrastructure organizations (Moteff, 2015; Clark, Seager and Chester, 2018; CISA, 2019).

Infrastructure organizations often lack the competencies to dynamically prioritize critical systems with quickly changing environments (Helmrich and Chester, 2022). As disasters unfold, managers need the competencies to make sense of the impacts and the most vulnerable services. COVID-19 is a valuable case study. Whereas energy, water, and other lifeline systems were largely uncompromised, parks (to house and socially distance the homeless) and digital communications became critical to health and well-being ((Isaacs and Chan, 2020); Criticality and prioritization for infrastructure may change conditionally (Clark, Seager and Chester, 2018); Montgomery *et al.*, 2021). Infrastructure managers need insight into how their organizations should prepare to morph and bend to chaotic events, identify changing environmental conditions, and rapidly pivot priorities. We refer to this as dynamic criticality, where a system can contextually adjust to environmental disturbances, dynamically prioritize resources, and balance robustness and adaptability (Roli *et al.*, 2018).

Many CI sectors have diverse operational requirements, so a framework for dynamic criticality must be broadly applicable and focus on infrastructure organizational management and not specific engineered systems. Toward this end, we start by reviewing the competencies of other sectors that appear to be able to pivot how they focus as hazards change, cross-compare these competencies, and then apply them to engineered infrastructure.

3.4.2. THEMATIC ANALYSIS

Cross-industry sectors that appear to have dynamic criticality capabilities were reviewed to improve the capabilities to dynamically define critical infrastructure and pivot resources depending on specific hazard contexts. Five sectors were selected and analyzed: 1) Leadership and organizational change; 2) Military and defense; 3) Medical emergency and triage; 4) Manufacturing; and 5) Disaster response. Several commonalities emerged across the sectors. A thematic analysis revealed four generalizable themes. First, many sectors showed methods for describing goals when dynamically shifting priorities. Second, several sectors exhibited capabilities towards configuring organizational structures to implement the goals. Third, a common theme of sensemaking appeared across sectors: making sense of an environment to open up decision-making (Weick, 1995). Fourth, organizations developed specific strategies for implementing flexibility amidst changing conditions. These four themes and their competencies are shown in Figure 3.4.1 and are discussed at length in this section.

268

Figure 3.4.1 – Dynamic Criticality Themes and Competencies

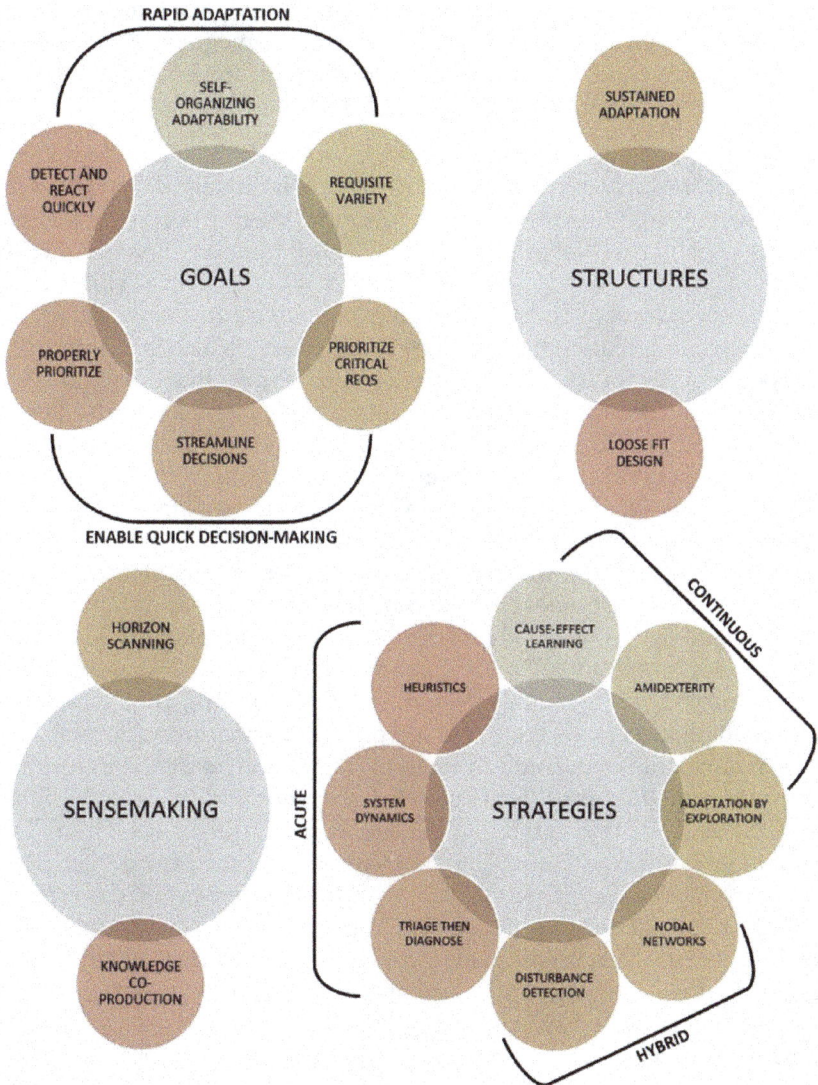

3.4.2.1. THEME 1: GOALS

Establishing goals was pivotal for sectors to implement dynamic criticality. Goals guide organizations toward responding to disturbances or chaos, which leads organizations to change structures, sensemaking, and strategies accordingly. Goals appear foundational for strategy development. The six competencies that emerged from the goals fell into two primary categories as shown in Figure 3.4.1. The first was a rapid adaptation to changing environments. Rapid adaptation includes self-organizing adaptability, requisite variety, and quick detection and reaction to disturbances. The first category of goals focused on enabling quick decision-making, including prioritization of resources during emergencies, identifying critical requirements for mission accomplishment, and building organizational relationships to facilitate dynamic decision-making.

When organizations set goals for rapid adaptation, this nudges the organization toward dynamic criticality, often indirectly. First, organizations with goals toward dynamic criticality select priorities more efficiently than others (Manville and Ober, 2003). Similarly, dynamic environments alternate unpredictably between stability and instability. In response, organizations should develop exploitative efficiency and explorative innovation. Exploitation focuses on efficiency, which is effective during periods of stability, and exploration is more effective during instability. This ambidexterity makes an organization efficient, agile, and flexible (March, 1991; Uhl-Bien and Arena, 2018). Second, requisite variety commonly appeared in both military and manufacturing goals. Requisite variety describes how systems in changing environments must have a repertoire of responses sufficient for their environment complexity (Naughton, 2017; Chester and Allenby, 2022). The military changed its operations to incorporate randomness for the timing and movement of forces which simultaneously confuses adversaries and builds adaptability and readiness for deployment (DOD, 2019). In manufacturing, RMSs match market needs and the pace of change (Koren, Gu and Guo, 2018; Tliba *et al.*, 2020). Rapid adaptation goals have also led manufacturing to use sensors, process monitoring, and analysis tools to detect and react to process and equipment disturbances (Frankowiak, Grosvenor and Prickett, 2005). For disaster response, the variability of disaster outcomes makes it unrealistic to standardize prioritization methods (such as in medical triage). Thus, disaster planners set general goals toward quick contextual discernment of criticality and speed of response (Applied Technology Council, 2016; FEMA, 2019). Although the goals found within the thematic analysis were different, they were generally oriented toward rapid adaptation. Ultimately, this appeared to inform organizational priorities, preventing reflexive decision-making. Goals that supported making faster decisions also improved dynamic criticality. The sectors showed many of the same principles for decision-making that Brehmer (1992) cites for the theory of DDM, such as decisiveness, delegation, taking responsibility, and avoiding fixation. In medical triage frameworks, the goal of appropriately prioritizing patients during emergencies is paramount to establishing decision criteria so medical staff can dynamically sort and prioritize patients specific to the scenario without time-consuming analysis, testing, or judgment (Aacharya, Gastmans and Denier, 2011; Storm-Versloot *et al.*, 2011). Secondly, several sectors set a goal to clarify the requirements needed to meet specific objectives. Military COG analysis uses critical requirements, vulnerabilities, and assets to select priorities for mission accomplishment (Perez, 2012; Schnaubelt, Larson and Boyer, 2014). Similar to how military planners need to identify various critical attributes, disaster response planners also must prioritize specific assets and resources during a response (O'Sullivan *et al.*, 2013; CISA, 2019). Disaster planning seeks to isolate the most critical assets and then shift those priorities dynamically. Also, the military realized that micro-management and lack of trust slowed the decision-making process. So, senior military commanders set goals to streamline the decision-making process. They built organizational relationships that empowered local commanders to prioritize and make decisions swiftly by creating a culture of trust, communication, and deep mutual understanding (Deployable Training Division, 2020).

3.4.2.2. THEME 2: STRUCTURES

The ability of organizations to change their governance models and processes to respond to changing conditions emerged across the sectors. Novel methods for transitioning governing structures appear to enable organizations to see game-changing disruptions and pivot resources more clearly in response. Two competencies emerged, as shown in Figure 3.4.1: 1) a commitment to sustained adaptation where the organization recognizes that its environment is in flux and structures itself to adjust course as needed, and 2) instituting processes that enable dynamic organizational structures and adaptive planning, referred to as "loose fit design." In CLT, organizations can pivot between efficiency and innovation governance models, the latter suitable for periods of instability (Uhl-Bien and Arena, 2018). RMSs are more flexible at handling demand and disruption shocks, adjusting the systems' orientation in response. RMSs achieve this flexible state through convertibility (capable of adaptation to new products), diagnosability (design quality assurance with the system, and not as an afterthought), customizability (designed around a family of products, and not just one), and scalability (cost-effective adaptation to future market demand). The loose-fit design has several associated properties. First, horizontal governance – shifting resources and decision-making authority to front-line workers who can coordinate and better sense change – creates organizational capabilities to diagnose and respond appropriately to chaos and change quickly. Formally, dynamic planning involves avoiding fixation – remaining focused on a set of increasingly obsolete challenges – and committing to a continuous cycle of reassessment of environmental conditions relative to organizational goals and processes (Brehmer, 1992; FEMA, 2016).

As goals inform decisions, organizational structures provide a foundation for sound decision-making. Organizations that confront frequent dynamism have streamlined processes and aligned their formal and informal structures to be more flexible as the environment changes. To maintain readiness, they must suppress natural apathy within structures during periods of equilibrium and maintain energy toward adaptability (Pascal and Henry, 2006).

3.4.2.3. THEME 3: SENSEMAKING

Dynamic environments forced organizations to develop new ways of understanding and interpreting the environment. In doing so, they are exercising sensemaking: taking in new knowledge, structuring it using novel techniques, and ultimately opening up decision-making opportunities (Weick, Sutcliffe and Obstfeld, 2005). In the thematic analysis, sensemaking presented two distinct competencies, shown in Figure 3.4.1: 1) the search for weak signals that may indicate changing environmental conditions in a process called horizon scanning; and 2) focusing on organizational co-production of knowledge. For dynamic force employment the military collects and interprets data to understand the operational environment, enabling it to alter its force structure dynamically (DOD, 2019). Similarly, disaster response planners for communities spend significant time understanding the dynamic environment within their area of responsibility to anticipate how different disturbances may affect the community (O'Sullivan *et al.*, 2013; DHS, 2021). Additionally, manufacturing systems constantly scan within their system to detect weak signs of equipment/process failure (Frankowiak, Grosvenor and Prickett, 2005) and also scan outside their systems (i.e., markets) to see hints of market changes that may trigger shifts in production or design (Tliba *et al.*, 2020). Organizational co-production of knowledge supports dynamic criticality primarily through network and collaboration. CLT creates informal social networks in organizations, allowing for a freer flow of ideas and collaboration, thus increasing innovation during disorder when old priorities suddenly become irrelevant and new ones must be identified (Uhl-Bien and Arena, 2018).

Similarly, managers of knowledge workers have shifted focus from task oversight towards knowing the capabilities of subordinates and building networked teams, creating more effective organizational knowledge toward shifting priorities during disturbances (Davenport, 2001). Indeed, the military has also identified this need for knowledge co-production with the concept of mission command. Senior commanders have a greater understanding of the strategic environment, while subordinate commanders have a better contextual awareness. Thus, mission command also shifts focus from oversight. The focus on building trust between the higher and lower ranks empowers and supports. The continuous dialogue toward shared understanding establishes trust and liberates senior commanders to focus on giving clear guidance and intent. Subordinate commanders are then empowered to swiftly decide and prioritize without asking for additional guidance (Deployable Training Division, 2020). Collectively, horizon scanning and co-production of knowledge during disturbances seek to combine information collection and synthesis with robust abilities to quickly and efficiently convert that information into relevant priorities.

3.4.2.4. THEME 4: STRATEGIES

The goals, structures, and sensemaking culminated in the creation of strategies. Eight strategies emerged from the review, describing acute (short-term), continuous (long-term), and hybrid decision-making timeframes. Acute strategies address disturbances with an apparent beginning and end (e.g., medical triage). Other sectors used continuous strategies oriented toward cyclical and ongoing problems (e.g., global military competition). Some sectors use hybrid strategies for scenarios with a clear beginning and end but require reassessment (e.g., lifeline disaster response). The three types are grouped and labeled accordingly in Figure 3.4.1.

The three acute strategies came from the triage and disaster response sectors. Time constraints of medical emergencies and disaster scenarios focus on rapid decision-making using pre-established frameworks, which requires deep systems knowledge to do dynamic criticality. In medical triage, this situational urgency requires prioritization via predetermined critical information heuristics. Action and priority-based thresholds are predetermined for efficiency, so a paramedic or triage nurse is not responsible for analyzing the patient's condition. They are trained for condition determination and prioritization via prescribed metrics, charts, data, and sensors (Aacharya, Gastmans and Denier, 2011), with some flexibility for tacit knowledge and experience to account for framework simplicity (van Pijkeren, Wallenburg and Bal, 2021). Sometimes, medical triage encounters situations where professionals must do initial sorting & emergency interventions and then detailed evaluation and determinations – such as in a mass casualty situation. Medical professionals begin with simple visual heuristics for prioritization: unresponsive, responsive, or walking. While simple, it is the most expedient strategy in the results. It displays how organizations can simplify a chaotic environment for tiered criticality prioritization. Emergency managers also found that in-depth knowledge of the system, connections, capabilities, and dependencies was an effective strategy to cut through the complexity and chaos during disasters. This knowledge enables decision-makers to make quick but contextualized decisions for prioritization (O'Sullivan *et al.*, 2013; DHS, 2021). While building and maintaining this knowledge is a continuous process, this strategy is acute when the knowledge is applied toward rapid prioritization, reducing waiting time and ambiguity.

Next, the continuous strategies shifted priorities during disturbances via a mixture of structures and sensemaking practices. Paradoxically, organizations that use continuous strategies appear to exist in a constant state of change where disturbances become a form of equilibrium. Business organizations have found that ambidexterity & organic, networked adaptation are necessary strategies for survival. Organizations intentionally vacillate between exploitation and exploration, building adaptability and dynamic decision-making and eliminating dependence on fragile and inaccurate market forecasts. Moreover, pursuing informal networks in organizations creates organic adaptation, which is more desirable when constantly engaging disturbances (Papachroni, Heracleous and Paroutis, 2016; Uhl-Bien and Arena, 2018). Also, toward continuous dynamic decision-making, people unconsciously fixate during stress and chaos. Intentional decision-makers must measure, analyze, and compare marginal gains for each action, allowing for iterative priority adjustments (Brehmer, 1992). This cause-effect learning towards improved decision-making helps decision-makers maintain a state of dynamic criticality. Similarly, the U.S. military continuously adapts to a rapidly changing global environment. Dynamic force employment and competition continuum doctrine emphasize adaptation by exploration through continuous change, inserting temporal and physical randomness in force movement (DOD, 2018), and dynamic engagement levels (i.e., peacetime, cooperation, and combat operations) (DOD, 2019). This proactive strategy seeks to constantly develop new "forms" that the organization can adopt to outpace competitors, focusing on attaining an adaptive state rather than seeking specific outcomes. These strategies all target the deep development of organizational adaptive capacity, the ability to reform and reshape when faced with new challenges.

Last, the military COG analysis, RMSs, and FEMA's community lifeline support frameworks were hybrids of acute and continuous strategies. They used continuous processes to achieve dynamic criticality, but disturbances also had a clear beginning and end. The military uses COG to derive priorities from critical attributes of the environment (e.g., financial systems, physical targets) linked to the mission's desired outcome. These nodal networks are used for single and continuous objectives (e.g., a long campaign or operation), constantly identifying new priorities and updating old ones. COG uses a network that maps capabilities, vulnerabilities, assets, and the COGs they orbit around. The top priorities are the network attributes connected to the desired end state (i.e., the target COG). Critical priorities shift if the end state shifts (Perez, 2012; Schnaubelt, Larson and

Boyer, 2014; Kornatz, 2016). Next, disturbance detection, adaptation identification, monitoring, and remembering is a cyclical process that RMS and
FEMA use for individual disturbances. The RMS process detects market
disturbances, develops manufacturing system adaptations, and monitors
market conditions' relevance. Key to this adaptation process is a rich archive of past disturbances and adaptations and the ability to recall them
for reuse – simplifying future adaptation development (Tliba *et al.*, 2020).
Similarly, emergency support management applies a cyclical process for
restoring essential CIs (e.g., water, electricity, shelter) after a disruption.
When an incident occurs, this triggers assessments, prioritizations, logistics, and responses, a process that loops until CIs are stabilized. Emergency managers also apply their archive, remember, and recall process
while updating plans so that emergency response goals are relevant to the
environment (FEMA, 2019).

3.4.3. INFRASTRUCTURE DYNAMIC CRITICALITY

Having described the themes and their competencies in the results and
in Figure 3.4.1, we turn to how infrastructure systems can use these capabilities to practice dynamic criticality. These themes can support infrastructure systems to detect disturbances early, pivot priorities, and balance robustness and adaptability. The thematic analysis showed that organizations that successfully confront chaos tend to engage disturbances
in three phases: prior, during, and post-disturbance (FEMA, 2016). The
(Park *et al.*, 2013) framework for Sensing, Anticipating, Adapting, and
Learning (SAAL) closely aligns with this process. Before chaos, infrastructure managers should probe, sense, and respond to the environment
(Chester and Allenby, 2019). The thematic analysis shows that most work
toward dynamic criticality happens before chaos. In alignment with the
four main themes, goals are set for adaptability and quick decision-making. Dynamic and flexible structures are formed. Organizations will practice sensemaking for past, present, and future environments and develop
adaptable strategies to engage disturbances. Then, organizations must
transition during disturbances to more acute and hybrid strategies. During disturbances, they will test sensemaking capabilities, execute plans,
and rapidly innovate. After the disturbance is over and stable conditions
return, organizations should shift towards expanding resilience for the future. It is time for organizations to learn, produce knowledge from the experiences, archive and remember, and change adaptations and plans for
future chaos cycles. This learning component is a looped cycle that links
all the other components of the adaptation process and thus deserves additional attention (Thomas *et al.*, 2019). This final section will contextualize
the themes and competencies for infrastructure, discussing them relative

to the prior, during, and post phases of disturbance engagement. Henceforth, italicized terms refer to the framework of themes and competencies shown in Figure 3.4.1.

The primary takeaway from the thematic analysis is that goals toward rapid adaptability and quick decision-making are essential to building capacity for dynamic criticality. In the thematic analysis, adaptable goals focused on capabilities that enabled quick shifts in priorities. Without dynamic criticality as a goal, it is unlikely to permeate the structures and operations of the organization. Goals bring inspiration to changes in organizational structures. For example, goals to exhibit requisite complexity will inspire an organization to look for more forms that an organization can take to fit the increasing forms of the environment (Brose, 2020; Brown, 2020; Chester and Allenby, 2022). After all, addressing complexity is about flux and unpredictability. The environment will always overcome more robust or efficient systems. So, these organizational goals, determined by leadership, will be a product of new governance that has embraced wicked complexity and uncertainty as the new normal (Chester and Allenby, 2021).

The thematic analysis indicated that, for more complex organizations, pre-established priority lists are less critical than building the capacity to engage chaos. Overemphasis on efficiency and optimization has led to rigidity and catastrophic failure in infrastructure. Leading up to the 2021 Winter Storm in Texas, electrical utility companies had neglected to upgrade system capacities and improve weatherization, resulting in an unprecedented cascading power outage and highlighting numerous areas where community resilience had been neglected (Markolf *et al.*, 2022). Dynamic criticality thinking would have encouraged utilities to invest time in developing loose-fit structures and build horizon scanning capacity for weather-related cascading failure scenarios. With these tools, they may have had the ability to pivot priorities and develop strategies for quick reactions to extreme storms. While prioritizing assets is necessary for developing readiness and strategies to engage disturbances, static priority lists have often been mistaken as a good plan for disturbances (Clark, Seager and Chester, 2018). This static thinking causes shortfalls when responding to novel or extreme disturbances that exceed historical precedent (Clark, Seager and Chester, 2018). These shortfalls exemplify how goals that focus on foundational requirements (e.g., requisite variety, detecting/reacting quickly) enable adaptative strategies and specific competencies such as ambidexterity, disturbance detection, and adaptation via exploration. This adaptive capacity gives organizations more tools to confront chaos when it comes (Lichtenstein *et al.*, 2007; Chester, Miller and Muñoz-Erickson, 2020).

Capacity development for infrastructure must happen in the pre-chaos space. Adaptive capacity is not expanded during chaos as much as used (Woods, 2015). Successful organizations spend considerable effort building organizational relationships toward cultures of sustained adaptation and practicing reactions to chaos from the cause-effect learning that exercises foster. These efforts may differ depending on the type of disturbance (i.e., practicing reactions for a hurricane will look much different than practicing for reactions to seasonal monsoon flooding). Hurricane Katrina showed how neglect of pre-chaos adaptations could hamper responses. Overdependence on robustness for resilience causes infrastructure organizations to undervalue the knowledge co-production that comes with intentional cooperation and collaboration. There was no consensus within and between agencies about pivoting priorities when critical infrastructure failed. Organizational relationships and cooperation quickly deteriorated without firm goals and consensus methods to triage and diagnose priorities (Leavitt and Kiefer, 2006). This lesson demonstrates how prior-to-disturbance efforts to build adaptive capacity for infrastructure should focus on leadership to enable ambidexterity, flexible structures, and knowledge co-production necessary to bolster innovation and build capacity (Helmrich *et al.*, 2021; Helmrich and Chester, 2022).

Reimagining infrastructure as knowledge enterprises and shifting to flexible loose-fit governance structures will grow the capacity to adapt by exploration much faster than traditional governance structures (Uhl-Bien and Arena, 2018). Infrastructure governance structures historically manifest as divisional bureaucracies, characterized by isolated divisions that often lack coordination and collaboration skillsets that may hinder many of the dynamic criticality competencies cited in this framework (Chester, Miller and Muñoz-Erickson, 2020). During the Northeast Blackout of 2003, time-crucial coordination and sensemaking between two personnel who worked across the hallway could have prevented the cascading failure in the initial minutes of the disaster (NERC, 2004; Pescaroli and Alexander, 2016). Thus, two organizational transformations are necessary to shift toward a more adaptable paradigm. The first is to transition to a knowledge enterprise, which focuses less on developing a product (i.e., infrastructure assets) and more on developing knowledge workers (i.e., technicians, operators, and engineers) who are responsible for systems (Chester, Miller and Muñoz-Erickson, 2020; Chester and Allenby, 2021). This transformation deemphasizes the importance of supervision and oversight and emphasizes leadership, empowerment, and sharing of knowledge (Davenport, 2001; Deployable Training Division, 2020), all pieces that bolster sensemaking and more adaptable governance structures. Therefore, shifting towards these principles may improve dynamic criticality via communication and coordination. Communication and coordination, in turn, increase idea syndication and expand sensemaking (Uhl-Bien and Arena,

2018). The second transition is to develop ambidexterity, switching between hierarchical and decentralized, ad-hoc structures during equilibrium and chaos, respectively (Siggelkow and Levinthal, 2003; Chester,
Miller and Muñoz-Erickson, 2020; Helmrich and Chester, 2020). These relationships also display the interconnected relationship between organizational structures and sensemaking. To this end, infrastructure organizations should practice the discomfort of shifting to emergency response
teams, diverse in expertise and empowered to take quick action to triage
and diagnose disturbances. Infrastructure organizations will familiarize
themselves with scenarios where structure shifts are necessary, diminishing lethargic responses that may hinder dynamic criticality (Alderson *et
al.*, 2022). Furthermore, an infrastructure organization that knows when to
shift between efficient and resilient structures gains the requisite variety
to match its environment, which also aids the dynamic prioritization process (Markolf *et al.*, 2022).

The nature of the disturbance and the outputs of sensemaking should
guide strategy selection and development. Infrastructure needs to practice
and exercise disturbance responses, not to be predictive, but to develop
familiarity with the discomfort of surprise and intimate knowledge of the
system dynamics. This practice expands the SAAL skillsets toward the
sensemaking competencies of horizon scanning and knowledge co-production (Miller and Muñoz-Erickson, 2018; Ancona, Williams and Gerlach, 2020; Alderson *et al.*, 2022; Chester and Allenby, 2022). Disturbances
manifested differently across the sectors of this study, and the diversity of
hazards battering infrastructure appears to be doing the same. Practically,
low-chaos disturbances may allow for node-networked responses with
multiple considerations for shifting priorities – much like COG analysis,
which uses critical capabilities to determine priorities dynamically. For
high-chaos disturbances, reflexive reactions may be more realistic, such as
triaging and diagnosing – much like how medical professionals sort patients into general categories during mass casualty events. Additionally,
multiple strategies could be nested within each other to increase flexibility. An infrastructure control center may develop a COG-like nodal network based on triage-like assessments from multiple teams transmitting
information, pushing back against the degradation of rationality that often
occurs during dynamic decision-making (Brehmer, 1992). When chaos is
so high, some organizations have no choice but to simplify the environment – as discussed in requisite variety (Boisot and Mckelvey, 2011). But
this simplification must also be balanced with proper sensemaking, lest
infrastructure managers misdiagnose problems (Chester and Allenby,
2022). So, the nesting of strategies may be a reasonable compromise to
these problems. Additionally, strategy selection may present an opportunity for human-supervised artificial intelligence systems to assist with

sensemaking, reducing confusion and subjective bias while bolstering speed and agility (Markolf, Chester and Allenby, 2021).

Although most sensemaking competencies should be built pre-chaos, they are tested and exercised more intensely during chaos. Horizon scanning and knowledge co-production remain essential to leading through chaos (Ancona, Williams and Gerlach, 2020) and analyzing how the chaos will affect the infrastructure system. The COVID-19 pandemic revealed that infrastructure organizations often neglect sensemaking to anticipate hazards (Carvalhaes *et al.*, 2020). For example, in the summer of 2021, hospitals began to rapidly consume the available supply of liquid oxygen due to the surge in severe COVID patients. Consequently, there was concern that water utilities would run out of the resource – commonly used as a critical water treatment component (Rosen, 1973). Until the realization of resource constraints, no one had considered the interdependencies that might have caused liquid oxygen to become the critical priority for the water utility. Managers were forced to revert to simplified decision-making thresholds for water consumption and conservation (Lusk, Krimsky and Taylor, 2021). But nodal networked thinking, horizon scanning as a discipline, and disturbance detection could have identified this vulnerability before it became a crisis. Making sense of a system requires an in-depth analysis of connections, interdependencies, and stakeholders (O'Sullivan *et al.*, 2013). It is necessary to keep up with real-time shifts in criticality (Clark, Seager and Chester, 2018).

Finally, cause-effect learning for the future is a best practice for dynamic criticality – although it appears to be among the hardest of competencies to retain (Westrum, 2006; Thomas *et al.*, 2019). When comparing the different sectors of this study, manufacturers appeared to do this more competently. They intentionally archive and recall previous strategies when a new market disturbance is detected. It saves time and effort in reinventing new strategies and helps an organization remain familiar with other competencies for adaptation to disturbances (Tliba *et al.*, 2020). Additionally, newly developed strategies contribute to an ever-growing "snowball" of remembered potential responses (Sweet *et al.*, 2014), which continue to grow requisite variety and contribute to a streamlined decision-making process. Therefore, remembering for infrastructure is foundational to requisite variety because of the interactive feedback loops between cause-effect learning and other aspects of sensemaking (Clark *et al.*, 2018). Moreover, remembering is an essential component of organizational cognition (Cooke *et al.*, 2013), and cognition links to knowledge co-production concerning systems and how responses should be tailored accordingly (Miller and Muñoz-Erickson, 2018). So, infrastructure organizations must practice remembering to practice cognition, which is ultimately necessary for sensemaking and strategy development for dynamic criticality.

3.4.4. CONCLUSION

Infrastructure organizations must implement practices towards dynamic criticality during times of chaos to remain viable in rapidly changing and increasingly unpredictable environments. Other sectors provide insights into the competencies that enable rapid pivots to reprioritize knowledge and resources. Chaos is not predictable or comprehendible (Chester and Allenby, 2019). Static priorities to engage chaos will remain unknowable, much like an "event horizon of chaos" for infrastructure. Thus, the results of this study show that if infrastructure organizations wish to approach dynamic criticality amidst disturbances, they should focus on maximizing adaptive capacity. Specifically, during periods of equilibrium, they should set goals for rapid adaptation and quick decision-making. They should alter their formal structures in ways that are friendly to sustained adaptation, which can be dynamic, flexible, and shiftable when disturbances occur. These goals and structures will then enable sensemaking competencies, allowing the organizations to scan the horizon for threats and make sense of increasing information flow (before, during, and after disturbances). These efforts will give way to the final sought-after product: practical strategies for dynamic criticality. Beneficial future research may be the historical analysis of disturbances and how dynamic criticality was or was not achieved by infrastructure organizations. But we urge caution regarding developing specific decision-making frameworks, as they may lead to strategy entrenchment and a decrease in adaptive capacity. The primary lesson from this work is that strategies are also dynamic and unique to disturbances. Thus, focusing on adaptive capacity will benefit infrastructure organizations more than a rigid list of priorities.

3.4.5. REFERENCES

Aacharya, R.P., Gastmans, C. and Denier, Y. (2011) 'Emergency department triage: an ethical analysis', *BMC Emergency Medicine*, 11(1), p. 16. Available at: https://doi.org/10.1186/1471-227X-11-16.

Alderson, D.L. *et al.* (2022) 'Surprise is inevitable: How do we train and prepare to make our critical infrastructure more resilient?', *International Journal of Disaster Risk Reduction*, 72, p. 102800. Available at: https://doi.org/10.1016/j.ijdrr.2022.102800.

Allenby, B. and Chester, M. (2018) 'Infrastructure in the Anthropocene', *Issues in Science and Technology*, 1, pp. 58–64.

Ancona, D., Williams, M. and Gerlach, G. (2020) 'The Overlooked Key to Leading Through Chaos', *MIT Sloan Management Review*, 62(1). Available at: https://sloanreview.mit.edu/article/the-overlooked-key-to-leading-through-chaos/ (Accessed: 21 October 2023).

Applied Technology Council (2016) *Critical assessment of lifeline system performance: Understanding societal needs in disaster recovery | Prevention-Web*. 16–917–39. Gaithersburg, MD: U.S. Department of Commerce National Institute of Standards and Technology, Engineering Laboratory. Available at: https://doi.org/10.6028/NIST.GCR.16-917-39 (Accessed: 21 October 2023).

Boisot, M. and Mckelvey, B. (2011) 'Complexity and organization–environment relations: Revisiting Ashby's law of requisite variety', in, pp. 279–298. Available at: https://doi.org/10.4135/9781446201084.n16.

Brehmer, B. (1992) 'Dynamic decision making: Human control of complex systems', *Acta Psychologica*, 81(3), pp. 211–241. Available at: https://doi.org/10.1016/0001-6918(92)90019-A.

Brose, C. (2020) *The Kill Chain: Defending America in the Future of High-Tech Warfare*. New York, NY: Hachette Books.

Brown, C.Q. (2020) *Accelerate change or lose*. Chief of Staff, United States Air Force. Available at: https://www.af.mil/Portals/1/documents/csaf/CSAF_22/CSAF_22_Strategic_Approach_Accelerate_Change_or_Lose_31_Aug_2020.pdf.

Carlson, J.M. and Doyle, J. (2002) 'Complexity and robustness', *Proceedings of the National Academy of Sciences*, 99(suppl_1), pp. 2538–2545. Available at: https://doi.org/10.1073/pnas.012582499.

Carvalhaes, T. *et al.* (2020) 'COVID-19 as a Harbinger of Transforming Infrastructure Resilience', *Frontiers in Built Environment*, 6. Available at: https://doi.org/10.3389/fbuil.2020.00148 (Accessed: 21 October 2023).

Chester, M.V. and Allenby, B. (2018) 'Toward adaptive infrastructure: flexibility and agility in a non-stationarity age', *Sustainable and Resilient Infrastructure*, 4(4), pp. 173–191. Available at: https://doi.org/10.1080/23789689.2017.1416846.

Chester, M.V. and Allenby, B. (2019) 'Infrastructure as a wicked complex process', *Elementa: Science of the Anthropocene*. Edited by A. Iles and M.E. Chang, 7(1), p. 21. Available at: https://doi.org/10.1525/elementa.360.

Chester, M.V. and Allenby, B. (2021) 'Toward adaptive infrastructure: the Fifth Discipline', *Sustainable and Resilient Infrastructure*, 6(5), pp. 334–338. Available at: https://doi.org/10.1080/23789689.2020.1762045.

Chester, M.V. and Allenby, B. (2022) 'Infrastructure autopoiesis: requisite variety to engage complexity', *Environmental Research: Infrastructure and Sustainability*, 2(1), p. 012001. Available at: https://doi.org/10.1088/2634-4505/ac4b48.

Chester, M.V. and Allenby, B.R. (2020) 'Perspective: The Cyber Frontier and Infrastructure', *IEEE Access*, 8, pp. 28301–28310. Available at: https://doi.org/10.1109/ACCESS.2020.2971960.

Chester, M.V., Miller, T. and Muñoz-Erickson, T.A. (2020) 'Infrastructure governance for the Anthropocene', *Elementa: Science of the Anthropocene*, 8(1), p. 78. Available at: https://doi.org/10.1525/elementa.2020.078.

CISA (2019) *A Guide to Critical Infrastructure Security and Resilience*. DHS. Available at: https://www.cisa.gov/sites/default/files/publications/Guide-Critical-Infrastructure-Security-Resilience-110819-508v2.pdf (Accessed: 21 October 2023).

Clark, S.S. *et al.* (2018) 'The vulnerability of interdependent urban infrastructure systems to climate change: could Phoenix experience a Katrina of extreme heat?', *Sustainable and Resilient Infrastructure*, 4(1), pp. 21–35. Available at: https://doi.org/10.1080/23789689.2018.1448668.

Clark, S.S., Seager, T.P. and Chester, M.V. (2018) 'A capabilities approach to the prioritization of critical infrastructure', *Environment Systems and Decisions*, 38(3), pp. 339–352. Available at: https://doi.org/10.1007/s10669-018-9691-8.

Cooke, N.J. *et al.* (2013) 'Interactive Team Cognition', *Cognitive Science*, 37(2), pp. 255–285. Available at: https://doi.org/10.1111/cogs.12009.

Davenport, T.H. (2001) 'Knowledge work and the future of management', in W.G. Bennis, T.G. Cummings, and G.M. Spreitzer (eds) *The Future of Leadership: Today's Top Leadership Thinkers Speak to Tomorrow's Leaders*. New York, NY: Jossey-Bass/Wiley, pp. 41–58.

Deployable Training Division (2020) *Insights and Best Practices Focus Paper: Mission Command*. Joint Staff J7. Available at: https://www.jcs.mil/Portals/36/Documents/Doctrine/fp/mission-command_fp_2nd_ed.pdf?ver=2020-01-13-083451-207#.

DHS (2021) *Developing and Maintaining Emergency Operations Plans*. Department of Homeland Security. Available at:

https://www.fema.gov/sites/default/files/documents/fema_cpg-101-v3-developing-maintaining-eops.pdf.

DOD (2018) *Summary of National Defense Strategy*. Department of Defense. Available at: https://dod.defense.gov/Portals/1/Documents/pubs/2018-National-Defense-Strategy-Summary.pdf.

DOD (2019) *Joint Doctrine Note 1-19: Competition Continuum*. Department of Defense. Available at: https://www.jcs.mil/Portals/36/Documents/Doctrine/jdn_jg/jdn1_19.pdf?ver=2019-06-10-113311-233.

FEMA (2016) *National Disaster Recovery Framework*. Department of Homeland Security. Available at: https://www.fema.gov/sites/default/files/2020-06/national_disaster_recovery_framework_2nd.pdf.

FEMA (2019) *National Response Framework - Fourth Edition*. Washington, DC: Department of Homeland Security.

Frankowiak, M., Grosvenor, R. and Prickett, P. (2005) 'A review of the evolution of microcontroller-based machine and process monitoring', *International Journal of Machine Tools and Manufacture*, 45(4), pp. 573–582. Available at: https://doi.org/10.1016/j.ijmachtools.2004.08.018.

Gilrein, E.J. *et al.* (2019) 'Concepts and practices for transforming infrastructure from rigid to adaptable', *Sustainable and Resilient Infrastructure*, 6(3–4), pp. 213–234. Available at: https://doi.org/10.1080/23789689.2019.1599608.

Helmrich, A. *et al.* (2021) 'Centralization and decentralization for resilient infrastructure and complexity', *Environmental Research: Infrastructure and Sustainability*, 1(2), p. 021001. Available at: https://doi.org/10.1088/2634-4505/ac0a4f.

Helmrich, A. and Chester, M. (2022) 'Navigating Exploitative and Explorative Leadership in Support of Infrastructure Resilience', *Frontiers in Sustainable Cities*, 4. Available at: https://doi.org/10.3389/frsc.2022.791474 (Accessed: 21 October 2023).

Helmrich, A.M. and Chester, M.V. (2020) 'Reconciling complexity and deep uncertainty in infrastructure design for climate adaptation', *Sustainable and Resilient Infrastructure*, 7(2), pp. 83–99. Available at: https://doi.org/10.1080/23789689.2019.1708179.

Humphreys, B.E. (2019) *Critical infrastructure: Emerging trends and policy considerations for congress*. R45809. Congressional Research Service. Available at: https://crsreports.congress.gov/product/pdf/R/R45809.

Isaacs, A.M. and Chan, B. (2020) *Zoom: The Challenge of Scaling with COVID-19 on the Horizon*. Berkeley, CA: Berkeley-Haas Case Series. University of California, Berkeley. Haas School of Business.

Koren, Y., Gu, X. and Guo, W. (2018) 'Reconfigurable manufacturing systems: Principles, design, and future trends', *Frontiers of Mechanical Engineering*, 13(2), pp. 121–136. Available at: https://doi.org/10.1007/s11465-018-0483-0.

Kornatz, S.D. (2016) 'The Primacy of COG Planning: Getting Back to Basics', *Joint Forces Quarterly 82*, 24(3), pp. 91–97.

Leavitt, W.M. and Kiefer, J.J. (2006) 'Infrastructure Interdependency and the Creation of a Normal Disaster: The Case of Hurricane Katrina and the City of New Orleans', *Public Works Management & Policy*, 10(4), pp. 306–314. Available at: https://doi.org/10.1177/1087724X06289055.

Lichtenstein, B.B. *et al.* (2007) 'Complexity Dynamics of Nascent Entrepreneurship', *Journal of Business Venturing*, 22(2), pp. 236–261. Available at: https://doi.org/10.1016/j.jbusvent.2006.06.001.

Lusk, M.G., Krimsky, L.S. and Taylor, N. (2021) 'How COVID-19 Exposed Water Supply Fragility in Florida, USA', *Urban Science*, 5(4), p. 90. Available at: https://doi.org/10.3390/urbansci5040090.

Manville, B. and Ober, J. (2003) 'Beyond Empowerment: Building a Company of Citizens [5]', *Harvard Business Review*, 1 January. Available at: https://hbr.org/2003/01/beyond-empowerment-building-a-company-of-citizens (Accessed: 22 October 2023).

March, J.G. (1991) 'Exploration and Exploitation in Organizational Learning', *Organization Science*, 2(1), pp. 71–87. Available at: https://doi.org/10.1287/orsc.2.1.71.

Markolf, S.A. *et al.* (2022) 'Balancing efficiency and resilience objectives in pursuit of sustainable infrastructure transformations', *Current Opinion in Environmental Sustainability*, 56, p. 101181. Available at: https://doi.org/10.1016/j.cosust.2022.101181.

Markolf, S.A., Chester, M.V. and Allenby, B. (2021) 'Opportunities and Challenges for Artificial Intelligence Applications in Infrastructure Management During the Anthropocene', *Frontiers in Water*, 2. Available at: https://doi.org/10.3389/frwa.2020.551598.

Miller, C.A. and Muñoz-Erickson, T.A. (2018) *The Rightful Place of Science: Designing Knowledge*. Tempe, AZ: Consortium for Science, Policy & Outcomes.

Montgomery, M.P. *et al.* (2021) 'Hand hygiene during the COVID-19 pandemic among people experiencing homelessness—Atlanta, Georgia, 2020', *Journal of Community Psychology*, 49(7), pp. 2441–2453. Available at: https://doi.org/10.1002/jcop.22583.

Moteff, J.D. (2015) *Critical Infrastructures: Background, Policy, and Implementation*. RL30153. Congressional Research Service. Available at: www.crs.gov.

Naughton, J. (2017) *Ashby's Law of Requisite Variety, Edge.org*. Available at: https://www.edge.org/response-detail/27150 (Accessed: 8 July 2021).

NERC (2004) *A Review of System Operations Leading up to the Blackout of August 14, 2003*. Available at: https://www.nerc.com/pa/rrm/ea/August 14 2003 Blackout Investigation DL/Operations_Report_FINAL.pdf.

O'Sullivan, T.L. *et al.* (2013) 'Unraveling the complexities of disaster management: A framework for critical social infrastructure to promote population health and resilience', *Social Science & Medicine*, 93, pp. 238–246. Available at: https://doi.org/10.1016/j.socscimed.2012.07.040.

Papachroni, A., Heracleous, L. and Paroutis, S. (2016) 'In pursuit of ambidexterity: Managerial reactions to innovation–efficiency tensions', *Human Relations*, 69(9), pp. 1791–1822. Available at: https://doi.org/10.1177/0018726715625343.

Park, J. *et al.* (2013) 'Integrating Risk and Resilience Approaches to Catastrophe Management in Engineering Systems', *Risk Analysis*, 33(3), pp. 356–367. Available at: https://doi.org/10.1111/j.1539-6924.2012.01885.x.

Pascal, R.T. and Henry, J. (2006) 'Surfing the Edge of Chaos', in *Creative Management and Development*. Thousand Oaks, CA: SAGE.

Perez, C. (2012) *Addressing the Fog of COG: Perspectives on the Center of Gravity in US Military Doctrine*. Fort Leavenworth, KS: Combat Studies Institute Press.

Pescaroli, G. and Alexander, D. (2016) 'Critical infrastructure, panarchies and the vulnerability paths of cascading disasters', *Natural Hazards*, 82(1), pp. 175–192. Available at: https://doi.org/10.1007/s11069-016-2186-3.

van Pijkeren, N., Wallenburg, I. and Bal, R. (2021) 'Triage as an infrastructure of care: The intimate work of redistributing medical care in nursing homes', *Sociology of Health & Illness*, 43(7), pp. 1682–1699. Available at: https://doi.org/10.1111/1467-9566.13353.

Roli, A. *et al.* (2018) 'Dynamical Criticality: Overview and Open Questions', *Journal of Systems Science and Complexity*, 31(3), pp. 647–663. Available at: https://doi.org/10.1007/s11424-017-6117-5.

Rosen, H.M. (1973) 'Use of Ozone and Oxygen in Advanced Wastewater Treatment', *Journal (Water Pollution Control Federation)*, 45(12), pp. 2521–2536.

Schnaubelt, C.M., Larson, E.V. and Boyer, M.E. (2014) *Vulnerability Assessment Method Pocket Guide: A Tool for Center of Gravity Analysis*. RAND Corporation, Arroyo Center. Available at: https://www.rand.org/pubs/tools/TL129.html (Accessed: 22 October 2023).

Siggelkow, N. and Levinthal, D.A. (2003) 'Temporarily Divide to Conquer: Centralized, Decentralized, and Reintegrated Organizational Approaches to Explorationand Adaptation', *Organization Science*, 14(6), pp. 650–669. Available at: https://doi.org/10.1287/orsc.14.6.650.24840.

Storm-Versloot, M.N. *et al.* (2011) 'Comparison of an Informally Structured Triage System, the Emergency Severity Index, and the Manchester Triage System to Distinguish Patient Priority in the Emergency Department', *Academic Emergency Medicine*, 18(8), pp. 822–829. Available at: https://doi.org/10.1111/j.1553-2712.2011.01122.x.

Sweet, D.S. *et al.* (2014) 'Sustainability Awareness and Expertise: Structuring the Cognitive Processes for Solving WickedProblems and Achieving an Adaptive-State', in Igor Linkov (ed.) *Sustainable Cities and Military Installations*. Dordrecht: Springer Netherlands (NATO Science for Peace and Security Series C: Environmental Security), pp. 79–129. Available at: https://doi.org/10.1007/978-94-007-7161-1_5.

Thomas, J.E. *et al.* (2019) 'A resilience engineering approach to integrating human and socio-technical system capacities and processes for national infrastructure resilience', *Journal of Homeland Security and Emergency Management*, 16(2). Available at: https://doi.org/10.1515/jhsem-2017-0019.

Tliba, K. *et al.* (2020) 'Model Based Systems Engineering approach for the improvement of manufacturing system flexibility', in *2020 21st International Conference on Research and Education in Mechatronics (REM)*. 2020 21st International Conference on Research and Education in Mechatronics (REM), pp. 1–6. Available at: https://doi.org/10.1109/REM49740.2020.9313871.

Uhl-Bien, M. and Arena, M. (2018) 'Leadership for organizational adaptability: A theoretical synthesis and integrative framework', *The Leadership Quarterly*, 29(1), pp. 89–104. Available at: https://doi.org/10.1016/j.leaqua.2017.12.009.

Weick, K.E. (1995) *Sensemaking in Organizations*. Thousand Oaks, CA: SAGE.

Weick, K.E., Sutcliffe, K.M. and Obstfeld, D. (2005) 'Organizing and the Process of Sensemaking', *Organization Science*, 16(4), pp. 409–421. Available at: https://doi.org/10.1287/orsc.1050.0133.

Westrum, R. (2006) 'A Typology of Resilience Situations', in *Resilience Engineering*. 1st edn. CRC Press, pp. 55–65. Available at: https://doi.org/https://doi.org/10.1201/9781315605685-8.

Woods, D.D. (2015) 'Four concepts for resilience and the implications for the future of resilience engineering', *Reliability Engineering & System Safety*, 141, pp. 5–9. Available at: https://doi.org/10.1016/j.ress.2015.03.018.

Section 4

ENVIRONMENT AND CLIMATE

In Section 4 we focus on the changing relationship between engineered infrastructure and natural environments. In the previous sections we referred to environments in the broadest of senses. Here we describe several challenges and approaches to navigate the changing boundaries between the built and natural worlds.

Of particular focus is the unraveling of mental models that assume environment predictability (stationarity) towards futures where non-stationarity and deep uncertainty in how the environment will interact with infrastructure necessitates new models. These new models will require a pivot in how we plan for infrastructure, recognizing that robustness-centric thinking has limits and when those limits are reached extensibility and sustained adaptation are needed.

The section focuses heavily on climate change as a key driver for environmental non-stationarity and deep uncertainty. Novel models -- including extensibility, sustained adaptation, safe-to-fail, and biomimicry -- while nascent are potentially critical for managing infrastructure in the face of environmental surprise.

4.1

INFRASTRUCTURE AND THE ENVIRONMENT

4.1.1. INTRODUCTION

Infrastructure has in many ways reflected human social and demographic structures. Infrastructure have for much of human history been the designed and built set of technological systems that mediate between humans, their communities, and their broader environment. Early hunter-gatherer humans had minimal-to-no shared infrastructure. The Neolithic revolution brought agriculture and its associated technologies as infrastructure, as well as town systems. During the classical era infrastructure became quite sophisticated but mostly remained within empires. An exception being transportation, where marine fleets and caravans (e.g., Roman, Islamic, Chinese empires) created trade infrastructure that extended far beyond political boundaries. The Romans brought advancements in hydraulics (including aqueducts), pavements (for roads), communications (mail), and energy use (petroleum), technologies that would not see widespread use again for centuries after the empire's collapse (Walski, 2006). Agricultural technologies including irrigation (e.g., dikes and dams) and farming tools were developed throughout this period by Chinese dynasties. The modern globalized era and now the Anthropocene has accelerated the pace and increased the scale of technology, where the distinctions between nature and infrastructure are blurring.

The infrastructure that evolved largely since the Industrial Revolution were viewed as separate systems from these larger natural systems. This was especially true in cities where a large number of people and infrastructure coalesced, and there was an increasing need to keep environmental hazards at bay while extracting large quantities of resources from the environment. The interfaces of infrastructure and the environment were largely defined as designing 1) with resources from the environment;

This chapter was adapted from the following article with publisher permission: Mikhail Chester, Samuel Markolf, and Braden Allenby, 2019, Infrastructure and the Environment in the Anthropocene, *Journal of Industrial Ecology*, 23(5), pp. 1006-1015, doi: 10.1111/jiec.12848.

2) to remove and transform resources from the environment; and 3) to withstand environmental perturbations. These interfaces largely determined (and continue to determine) the location and design of infrastructure. Unexpected changes in natural systems make slow-changing (i.e., obdurate) infrastructure obsolete, as evidenced by a third intake pipe at Lake Mead completed in 2015 for Las Vegas—the first two becoming obsolete as water levels have dropped (we acknowledge that there are deep and unique complexities surrounding water policy in the Western United States, like for most large infrastructure).

Furthermore, infrastructure are at risk to changing environmental hazards such as climate change (U.S. National Climate Assessment, 2014). Symptoms of the problem are that 1) the resources that we use to construct infrastructure are becoming more and more global (and less local), escalating their costs and reducing ease of access; 2) the infrastructure that move resources are largely inflexible and any changes in resource supply and/or external conditions means there is a mismatch in utilization (and we often have difficulty predicting how much we'll use and by when); 3) the cycle time of changes in human, built, and natural systems has accelerated, resulting in legacy infrastructure which is still in place but increasingly obsolete, if not dysfunctional; and 4) climate and other earth systems changes threaten the model that we use that infrastructure must be able to continue to provide reliable service despite perturbations. The crux of the problem is not change itself but rather the volatility, scale, and cycle time associated with change.

The scale and scope of human activities has changed dramatically in the past century and show signs of accelerating into the future—particularly with regard to trends of increasing urbanization (UNESCO, 2010; Romero-Lankao and Dodman, 2011; *Urbanization, Biodiversity and Ecosystem Services: Challenges and Opportunities: A Global Assessment*, 2013; UN-DESA, 2014). Economic development and other aspects of global history were certainly very different around the world (e.g., Europe, the United States, and Asia), but the current convergence reflects the competitive global environment that we see today.

Prior to the Anthropocene, human activities and their impacts were small relative to earth systems. Separating infrastructure and the environment made sense since human caused environmental impacts were within the capacity of the environment to absorb quickly. As human society has progressed, the significance of our role as hyperkeystone species has emerged (Worm and Paine, 2016). Keystone species exhibit disproportionate and often unexpected effects on natural systems, and are found in every major habitat. The term hyperkeystone has been coined to describe the influence that humans have over not just other keystone species but also natural systems (Worm and Paine, 2016). Across many earth systems — climate, water, nitrogen, phosphorous, precious metals, etc. — evidence accumulates that human activities are creating global and unpredictable impacts that create unintended consequences (Syvitski, 2012). The reach of human activities has grown so large that infrastructure, and the activities and technologies that they support, are directly and at scale managing natural systems.

Examples of infrastructure and the environment converging are found across regional and global scales. Consider urban water supply. Arizona is a state with diverse climates and geography but whose population largely exists in semi-arid desert regions of the state. Water supply to this population, contrary to popular belief, is diverse with 17% from in-state rivers and associated reservoirs, 40% from groundwater, 3% reclaimed, and 40% imported from the Colorado River (ADWR, 2016). A large portion of imported water is banked. Wastewater is sent to either power plants or used for recharging groundwater aquifers.

Similarly, land use laws and regulations designed to protect New York City's drinking water sources have far-reaching implications elsewhere in the state (City of New York, 2010). In other words, New York's nature has been designed as part of water supply infrastructure. Water in California is also tightly designed: the mountains store snow, which becomes runoff, which is heavily managed as it moves toward the ocean to optimize its use by people.

Like many other regions, natural flows from precipitation are now highly managed and non-stationarity — i.e., historical flows are becoming less and less relevant for predicting future flows — is the norm. Watersheds — particularly for urban water supply — are intensely managed. Thinking of the environment as water infrastructure is necessary for maintaining supply. In many ways, the dichotomy between infrastructure and the "environment" no longer exists; they are one and the same.

Wildlife in some contexts is becoming entirely managed by humans through technology. The upstream river migration of salmon in the Northwest United States is controlled by management techniques including ladders, a need that emerged after dams were constructed (Kareiva, Marvier and McClure, 2000). Aquaculture (ocean fish farming) represents the development of marine infrastructure and associated technologies to directly manage populations (Goldburg and Naylor, 2005).

Anthropogenic carbon dioxide, driven increasingly by urban activity (Dhakal, 2009, 2010; Kennedy *et al.*, 2009; Parshall *et al.*, 2010), has resulted in the atmosphere becoming a design space for technology. Efforts to manage atmospheric carbon dioxide concentrations involve the deployment of new transportation and energy technologies to reduce additions of green-house gases, as well as technologies to remove carbon dioxide (i.e., carbon capture and storage). There are even discussions on reducing incoming solar radiation to manage the energy balance of the planet (Ming *et al.*, 2014). When it comes to managing the energy balance of the planet, infrastructure and the atmosphere are becoming one. The atmosphere is becoming testbed of emergent and planned human activities. Going forward we may not be able to afford the inefficiency (from a human perspective) and unexpected transients that such a sloppy approach to infrastructure implies.

Fundamentally, the models that we have used to build and operate infrastructure have separated infrastructure and the environment as independent systems. This framing was a useful simplification when natural systems were not affected in significant ways by human activity. However, as human activities have grown, their reach into the environment has increased to the point where some natural systems are now largely managed by humans. And as human activities continue to grow and the need to secure reliable natural resources and services increases, it is likely that we'll see increased management of environmental systems. As such, in the long term, major earth systems will continue to become part of our infrastructure designs. In the Anthropocene, infrastructure will become how we design nature and the planet, integrating relevant human, natural, and built systems in design through technologies that support local, regional, and even global activities.

4.1.2. INFRASTRUCTURE OBDURACY

Our core infrastructure systems have developed a relatively high amount of obduracy over time. Broadly speaking, obduracy refers to rigidity and difficulty or resistance to change (Hommels, 2000). Perhaps nowhere is obduracy more prevalent or impactful than in cities. "Despite the fact that cities are considered to be dynamic and flexible spaces...it is very difficult to radically alter a city's design: once in place, urban structures become fixed, obdurate" (Hommels, 2005). Obduracy arises for several reasons: 1) changes in demand for the services provided by our infrastructure systems have historically been relatively gradual and predictable (we demand water, power, and vehicle travel much like we did decades, even a century ago); 2) from a financial perspective, the high upfront capital cost associated with many infrastructure projects often necessitates a longer lifespan; 3) infrastructure systems often operate in economic settings with minimal competition (and impetus for innovation) — many infrastructure systems are either publicly funded or natural monopolies; 4) many infrastructure systems are installed under the assumption of stable environmental and societal conditions (i.e., stationarity is assumed); and 5) many infrastructure systems are constructed of long-lasting materials and legal structures (such as right-of-ways) that have inherently long cycle times.

Historically, obdurate infrastructure has provided some level of benefit in terms of reliability, predictability, safety, and efficiency. But, as seen in the example of waste incineration in Goteborg, Sweden, the obduracy that contributed to and emerged from once desirable services (diverting waste from landfills and using it as a fuel source), can eventually inhibit the adoption of innovative and more preferable solutions (e.g., recycling, waste reduction programs, generating electricity from wind or solar power, etc.) (Corvellec, Zapata Campos and Zapata, 2013). In the context of the Anthropocene, obduracy may become problematic for a couple different reasons. First, it appears to be a major factor in the growing and complex relationship that infrastructure has with earth and social systems. For example, cities such as Detroit that have experienced large population decreases have had to grapple with infrastructure that is now underutilized and poorly funded. The inability of the infrastructure (and related institutions and practices) to adjust and adapt to the decrease in population and demand has created challenges related to funding, maintenance, consumer affordability, public health, and environmental degradation (Faust, Abraham and McElmurry, 2016). The obdurate nature of most infrastructure (and related institutions) can also conflict with sustainability-oriented goals such as increased urban density, transit-oriented development, and distributed solar power. Similarly, the obduracy of many infrastructure systems contributes to the ongoing expansion of resource extraction efforts (Krausmann *et al.*, 2009). For example, the fact that the transportation sector has been consistently and nearly exclusively powered by fossil fuels for over a century has contributed to the need for increasing efforts to secure fuel (e.g., offshore oil drilling, hydraulic fracturing, plant-based ethanol, etc.). Finally, the longer certain infrastructure systems are around, the more opportunity there is for social, cultural, and other technological systems to couple and co-evolve with the infrastructure systems — thereby increasing complexity and scale (Hughes, 1987; Bijker, Hughes and Pinch, 2012; B. R. Allenby, 2012).

The other concern with infrastructure obduracy in the context of the Anthropocene is that the cycles of infrastructure evolution (i.e., maintenance, renovation, and replacement schedules) may increasingly become out of sync with environmental (e.g., climate change, resource accessibility and availability) and cultural (e.g., societal demand, preferences, and needs) cycle times. Historically, infrastructure has had cycle times that align well with our social, economic, and technological systems. However, social, environmental, and technological changes appear to be accelerating at a faster rate than infrastructure systems (Kurzweil, 2005; Allenby and Sarewitz, 2011; Marchant, 2011). For example, the growing trend toward electric vehicles (Madrigal, 2017) means that the economic, politic, cultural, and environmental viability of a project like the Keystone XL pipeline may come into question over the coming decades. Similarly, with an increasing push toward online shopping, ride-share systems, and autonomous vehicles, how viable is it to continue to size much of our parking infrastructure — and all of the land-use and environmental issues associated with it (Shoup, 2011) — to handle high volumes that infrequently occur (e.g., the largest shopping days of the year, special events, etc.)? Aside from accelerating technological and social changes, fluctuations in nitrogen, phosphorus, water, carbon, climate, and natural resource cycles should increasingly challenge the assumption that infrastructure can be designed, installed, and operated under stable environmental conditions. For example, in response to sea level rise and tidal flooding events, the City of Miami Beach has spent roughly $100 million (as part of a larger $500 million effort) to raise street levels and install pumping stations throughout the island. At the same time, concerns have been raised about the impact that the effluent from the pumping stations is having on water quality and ecosystem health in Biscayne Bay (Flechas and Staletovich, 2015; Staletovich, 2015; Flechas, 2017). The issue of non-stationarity — the concept that historical conditions are increasingly a poor predictor of future conditions (Milly *et al.*, 2008) — raises important questions about the design, implementation, and obduracy of our infrastructure systems. Engineers, over the past century, have generally assumed stationarity in environmental systems in order to design for robustness (Park, Seager and Rao, 2011), an assumption that was non-problematic so long as the life time of infrastructure components and the larger environmental systems was roughly equivalent. The stationarity assumption was reasonable because the cycle times of infrastructure (i.e., how quickly we change infrastructure and its associated technologies) were roughly in sync with the environment. However, as environmental (and other) conditions begin to change more rapidly (e.g., climate change; "dead zones" in the Gulf of Mexico), such implicit assumptions no longer hold. Moving to a system of designing for future conditions that are difficult to anticipate seems to be in order. Decisions will need to be made about what infrastructure life span and future conditions should be assumed, especially given the uncertainty in financial and environmental conditions (Chester and Allenby, 2018). This challenge is not specific to climate but also other earth systems that have become profoundly affected by humans.

Ultimately, one of our primary challenges will be meshing infrastructure cycle times (through concepts like flexibility, agility, and adaptability) with the much more rapid cycle times of coupled social, economic, and environmental systems. Similarly, it will increasingly become a challenge for infrastructure managers and practitioners to grapple with the fact that infrastructure decisions are having a growing influence on earth systems. And as urban populations grow, cities will increasingly drive changes in the environment. For example, it has been shown that a relatively modest commitment to ethanol fuel in the transportation sector can have significant implications on global agricultural practices, land use, water and nutrient cycles, and the carbon cycle (von Blottnitz and Curran, 2007; de Fraiture, Giordano and Liao, 2008; Searchinger *et al.*, 2008; Gerbens-Leenes, Hoekstra and van der Meer, 2009; Simpson *et al.*, 2009; Hertel *et al.*, 2010; Lagi *et al.*, 2011). Similarly, urban areas typically have higher market shares of electric vehicles (EVs) than the national average (OECD, 2017). To the extent that the EV market continues to grow as expected and cities remain at the forefront of this growth, urban areas will likely be at the center of major alterations to the material flows and environmental impacts of the electricity sector, the petro-chemical sector, and global supply chains of earth elements like lithium, cobalt, and nickel (i.e., crucial components of batteries). This is not to say that cities are necessarily the cause of negative environmental outcomes, but instead should be viewed as a major driver of the consumption of the policies and complexities that have been introduced in the city and more broadly. Such systemic implications, which we know are there and will be manifested even if we cannot predict specifics, must be part of infrastructure design and management in the Anthropocene. There was a time when ignoring systems-level implications (e.g., the impact of U.S. ethanol use and policy on global food prices, land use patterns, and environmental quality) might have been appropriate to simplify analysis and management. However, that appears to no longer be the case. A major component of addressing these challenges will be overcoming lock-in — the idea that today's systems are constrained by past decisions, even if the context of the past choices is no longer relevant or if new alternatives have emerged that may be more effective (Corvellec, Zapata Campos and Zapata, 2013). It will be important to recognize that changes will not only need to be made to the physical configuration of infrastructure, but also to the institutional, financial, and cultural forces associated with the physical infrastructure. It will also be important to attempt to reconcile with the growing complexity that arises from obdurate and interconnected infrastructure and social systems operating at increasingly large scales.

4.1.3. INFRASTRUCTURE IN THE ANTHROPOCENE

Our conceptual models of how infrastructure interface with the environment will need to be shifted and we will need to move toward integrated management of technological and ecological systems in cities. We are already seeing, and can expect acceleration of, natural systems moving toward being increasingly human managed. This will require new kinds of engineering skills and professional niches, not to mention engineering education. These managed resources will increasingly fall under the purview of engineering. Engineering, which has traditionally treated the environment as a separate entity that either provides resources or creates hazards, will increasingly need the competencies to design and manage techno-environment (and even social) systems as one. Examples of this are already apparent. Apple hires engineers to ensure that they can design and manufacture given the availability of rare earth metals (Apple, 2017). Cargill, a major global food producer, has invested in cultured meat (Cargill, 2017).

The coming century will be defined by how we view and position infrastructure in the Anthropocene. As we increasingly manage natural systems, our infrastructure becomes our environment – particularly in urban settings. With this transition different principles will emerge. Recognizing the complexity (defined partly by an inability to predict emergent behaviors as a result of feedbacks and nonlinearities) of techno-environment systems, emphases will need to shift away from optimizing and toward satisficing (Simon, 1947, 1957). If we allow existing models of infrastructure to persist and a perpetuation of thinking that separates infrastructure from the environment then we can expect 1) cycle times of infrastructure that don't match the environment and our changing needs/technologies; and 2) greater failures (and costs) of infrastructure. Agile (physical structure and the rules, policies, norms, and actors who manage and operate it, will need to be able to maintain function in a non-stationarity future) and flexible (ability to meet changing demands (in the face of both predictable and unpredictable challenges) infrastructure will be key to adapting infrastructure in the short term (Chester and Allenby, 2018).

Thomas Kuhn in his classic book *The Structure of Scientific Revolutions* (Kuhn, 1962) made an important distinction between periods of "normal science," where incremental advances are made within accepted frameworks of theory and belief, and "paradigm shifts," where the frameworks themselves are overthrown, and entirely new ones introduced. Examples in science include the shift from Newtonian physics to quantum mechanics, and, in geology, the shift to plate tectonics from a belief in a static and fixed planetary structure. What we are asserting here is that engineering at the dawn of the Anthropocene must undergo a paradigm shift, in the Kuhnian sense: it is no longer adequate to view infrastructure as simply localized artifacts within an unchanging exogenous human, natural, and built context. Rather, engineering, and infrastructure, are now means by which that context is itself designed and shaped.

Design for adaptive capacity. Infrastructure that is able to quickly respond to changing environmental conditions in the future will be much more adept at adapting to changing demands (Chester and Allenby, 2018). The paradigms that define our current infrastructure, and the inherent obduracy that is created, are an artifact of the traditional paradigms that separate infrastructure from the environment. These existing paradigms are insufficient to address the growing complexities of our world; our infrastructure so far hasn't needed to be agile and flexible. But in the future, with rapid and unpredictable changes in earth systems, accelerated disruptive technologies, and the potential for conflict over increasingly scarce local resources, infrastructure will need to be reimagined. For example, the SMART Tunnel in Kuala Lumpur, Malaysia serves the dual purposes of improving mobility and diverting floodwaters away from city center during storms. Under normal conditions, the tunnel serves as a motorway. Under moderate storm conditions, half of the tunnel diverts floodwater while the other half remains operational to motorists. Under extreme storm conditions, automated systems activate water-tight gates and allow to whole tunnel to divert and temporarily storm floodwater (Darby and Wilson, 2006; Hing, Welch and Giap, 2006). Non-stationarity in many domains can be expected to define the Anthropocene. The competencies needed to produce agile and flexible infrastructure may include multifunctionality, roadmapping, focus on software over hardware, resilience-based thinking, compatibility, connectivity, and modularity of components, organic and change-oriented management, and transdisciplinary education (Chester and Allenby, 2018). These competencies exist in industries such as ICT and manufacturing which have shown an ability to quickly adapt to changing demands. Given the obdurate nature of infrastructure, the capacity to adapt (i.e., resilience) becomes critical for addressing some of the complexity of our changing environments, including earth systems such as climate change, and resource availability. Agile and flexible infrastructure might decrease the cycle times of infrastructure so they are on par or ahead of changes in the environment. They would allow adjustment in real time to unpredictable change, thus maintaining more efficient system integration and operation.

Design for complexity. While adaptation strategies can address environmental perturbations, the paradigms that we use to design and operate infrastructure will need to shift toward working with complex techno-environmental systems. Infrastructure by themselves is becoming more and more complex as they grow to encompass old and new technologies, become interdependent, and replace hardware with software. This is particularly true in cities where large numbers of technical and social infrastructure come together, often with decades or centuries old systems. Understanding how a small perturbation cascades through and across infrastructure is becoming increasingly challenging. But complexity is not just

299

the result of hardware, but of interconnected institutional and environmental systems too. As is the case of Arizona, designing water conveyance and distribution systems for urban water delivery is an exercise of hydrological management with direct implications for economic growth and food production. If we choose to deploy atmospheric carbon capture and storage technologies, we are explicitly designing the carbon dioxide content of the atmosphere, and implicitly designing atmospheric dynamics, climate patterns, and ocean circulation patterns. And we may not fully grasp the dynamics of these systems and resulting effects, which may impact other infrastructure and people. This reinforces the need for flexibility and agility, which are required to respond to unpredictable shifts in system state. Emerging technologies, such as artificial intelligence, may be increasingly available to help engineers deal with such complexity rapidly and effectively. However, few engineers are currently trained in using these tools effectively.

Complexity and uncertainty will need to be central to the training of those who design and operate infrastructure in the Anthropocene. Infrastructure encompass a rich diversity of components, organizations, and rules. They include the layering of technologies over decades, some of which despite their antiquated capabilities continue to be used today. For example, despite some difficulties associated with finding replacement parts, equipment dating back to the 1930s enables the reliable signaling and control in the New York City subway (Limer, 2015; Fitzsimmons, 2017). The confluence of these factors means that "we often are left with only the extremes of understanding: either a general notion of how the thing works, even if its innards are at best murky to us, or an examination of its bits and pieces, without an inkling of how it all fits together and how we can expect it to behave (Arbesman, 2016)."

Adding to this complexity is the non-stationarity in future environmental and other factors. Yet those who design and operate infrastructure are often taught based on stationary and reductionist approaches (B. Allenby, 2012). Going forward, these infrastructure managers will need to accept that they cannot fully predict the emergent behavior of perturbed systems, or how the systems will be utilized or needed in the future. Unpredictability will be the new normal. As such, competencies around design and management under deep uncertainty will be essential. Resilience-based thinking that emphasizes adaptive capacity, or the ability to move between different approaches for design and management, should be central to academic and job training.

Changing infrastructure takes time and deploying new paradigms will require us to address forces that lock in current practices. For many resources, the means to deliver adequate supply is technologically feasible but implementation is slowed by non-technical forces. For example, the technologies to deliver potable water over long distances are established (e.g., long-distance conveyance and desalination). But bringing them online on time as part of future water supply requires addressing planning, permitting, financing, rights, and other challenges. If not done properly or expeditiously, the transition period between technologies could be costly and dysfunctional.

The difficulty of change in complex adaptive systems can be viewed through dissipative structure theory. Complex adaptive systems are self-organizing, they drop less valuable elements (dissipation) to incorporate new elements, thereby contributing to the conditions for adapting and growing (Annila and Salthe, 2009). We can apply the dissipative structure theory to infrastructure growth and change. To change infrastructure on large scales quickly means that we in many ways need to work against the current momentum designed around technologies that consume abundant and finite resources. We need to accelerate the dissipation of elements that are reliant on a consumption model that focuses on these resources in what we anticipate will be a new model, while at the same time recognizing the complexity, that biofeedstocks and carbon capture and storage (for example) may reverse our views on particular technologies. While humans may have the technical means to institute these changes, the forces that prevent this dissipation from happening (financial, cultural, political, etc.) have high inertial resistance to change, and thus allow the current system to continue.

To design, manage, and work within complex techno-environmental systems, engineering domains will need broad competencies and a recognition that complexity is likely to increase. As we accept that infrastructure and the environment have in some ways and are in many ways becoming one in the same, it will be necessary for engineers to not only recognize this but to work within it. In addition to the usual disciplinary expertise, expertise around material flows, earth systems, impact analysis, and others will be needed to create a systemic perspective of technologies and identify the transition, scarcity, and other challenges that may affect the systems.

4.1.4. REFRAMING INFRASTRUCTURE

Infrastructure, particularly in engineering, has been synonymous with the physical implementation of structures that provide services to society, a definition that carries less and less meaning as technology progresses and we make new progress toward intentionally designing systems that in the past were not steered by humans. As this intentional design continues and even accelerates in the Anthropocene, our definition of infrastructure will need to change. Cyberspace, technological capability, health care, and other systems that we attempt to increasingly manage and interconnect with other systems will become design spaces. While our current gray infrastructure may indeed persist, albeit in possibly very different forms, it may exist as part of complex systems that provide services in very different ways than what we've experienced in the past.

Currently, engineers are taught to design in response to a set of objectives and constraints which, even in complex domains, can be quantified. But what happens when we know we're doing design that is intended to extend well into a future that is fundamentally unpredictable, and where the design objectives and constraints we have now are at best contingent? How do we design infrastructure that is in some meaningful sense contingent on future conditions we might not even be able to conceive (e.g., urban infrastructure for the AI city? Or, less speculatively, urban transportation infrastructure, designed for decades, as we shift to autonomous vehicles providing a service rather than owned vehicles parked for 95% of the time?).

The management of infrastructure in the Anthropocene and the societal goals it enables is not a challenge for the engineering or infrastructure professions exclusively. Society as a whole must recognize that the scale and scope of human activities are becoming so large, and often locked-in, that fast and large-scale change is becoming difficult, if not impossible. Designing for adaptive capacity and complexity is not simply a job for engineers. Collective action and broad changes in worldviews are needed, tasks that have proven difficult for other professions, such as climate scientists. The emerging field of cultural evolution (Henrich and McElreath, 2003) suggests that worldviews co-evolve with the societies they support and that, absent catastrophic collapse, they are hard to change.

As the tendrils of human activity reach further and our natural systems become increasingly managed systems, casting infrastructure as a coupled human-environmental system will become increasingly important. No longer should we view infrastructure as separate from the environment; they are becoming inexorably linked in ways that will require us to manage them as single (and complex) systems.

4.1.5. REFERENCES

ADWR (2016) *Annual Report 2016*. Phoenix, AZ: Arizona Department of Water Resources, p. 30. Available at: https://www.azwater.gov/about-us/adwr-strategic-plan-and-annual-reports.

Allenby, B. (2012) *The Theory and Practice of Sustainable Engineering*. Pearson Prentice Hall.

Allenby, B.R. (2012) *The Theory and Practice of Sustainable Engineering*. Hoboken, NJ: Upper Saddle River: Pearson Prentice Hall.

Allenby, B.R. and Sarewitz, D. (2011) *The Techno-Human Condition*. Cambridge, MA: The MIT Press. Available at: https://doi.org/10.7551/mitpress/8714.001.0001.

Annila, A. and Salthe, S. (2009) 'Economies Evolve by Energy Dispersal', *Entropy*, 11(4), pp. 606–633. Available at: https://doi.org/10.3390/e11040606.

Apple (2017) *Environmental Responsibility Report*. Cupertino, CA: Apple, p. 58. Available at: https://www.apple.com/environment/pdf/Apple_Environmental_Responsibility_Report_2017.pdf.

Arbesman, S. (2016) *Overcomplicated: Technology at the Limits of Comprehension*. New York, NY: Penguin Publishing Group.

Bijker, W.E., Hughes, T.P. and Pinch, T. (2012) *The social construction of technological systems: new directions in the sociology and history of technology*. Anniversary ed. Cambridge, MA: MIT Press (UPCC book collections on Project MUSE).

von Blottnitz, H. and Curran, M.A. (2007) 'A review of assessments conducted on bio-ethanol as a transportation fuel from a net energy, greenhouse gas, and environmental life cycle perspective', *Journal of Cleaner Production*, 15(7), pp. 607–619. Available at: https://doi.org/10.1016/j.jclepro.2006.03.002.

Cargill (2017) *Protein innovation: Cargill invests in cultured protein | Cargill*. Available at: https://www.cargill.com/story/protein-innovation-cargill-invests-in-cultured-meats (Accessed: 5 November 2023).

Chester, M. and Allenby, B. (2018) 'Toward Adaptive Infrastructure: Flexibility and Agility in a Non-Stationarity Age', *Sustainable and Resilient Infrastructure*, 3, pp. 1–15. Available at: https://doi.org/10.1080/23789689.2017.1416846.

City of New York (2010) *Rules and Regulations for the Protection from Contamination, Degradation and Pollution of the New York City Water Supply and its Sources*. New York City, NY.

Corvellec, H., Zapata Campos, M.J. and Zapata, P. (2013) 'Infrastructures, lock-in, and sustainable urban development: the case of waste incineration in the Göteborg Metropolitan Area', *Journal of Cleaner Production*, 50, pp. 32–39. Available at: https://doi.org/10.1016/j.jclepro.2012.12.009.

Darby, A. and Wilson, R. (2006) 'Design of the smart project, Kuala Lumpur, Malaysia', in. *International Conference and Exhibition on Tunnelling and Trenchless Technology*, pp. 7–9.

Dhakal, S. (2009) 'Urban energy use and carbon emissions from cities in China and policy implications', *Energy Policy*, 37(11), pp. 4208–4219. Available at: https://doi.org/10.1016/j.enpol.2009.05.020.

Dhakal, S. (2010) 'GHG emissions from urbanization and opportunities for urban carbon mitigation', *Current Opinion in Environmental Sustainability*, 2(4), pp. 277–283. Available at: https://doi.org/10.1016/j.cosust.2010.05.007.

Faust, K.M., Abraham, D.M. and McElmurry, S.P. (2016) 'Water and Wastewater Infrastructure Management in Shrinking Cities', *Public Works Management & Policy*, 21(2), pp. 128–156. Available at: https://doi.org/10.1177/1087724X15606737.

Fitzsimmons, E.G. (2017) 'Key to Improving Subway Service in New York? Modern Signals', *The New York Times*, 1 May. Available at: https://www.nytimes.com/2017/05/01/nyregion/new-york-subway-signals.html (Accessed: 5 November 2023).

Flechas, J. (2017) 'Miami Beach to begin new $100 million flood prevention project in face of sea level rise', *Miami Herald*, 28 January. Available at: https://www.miamiherald.com/news/local/community/miami-dade/miami-beach/article129284119.html (Accessed: 5 November 2023).

Flechas, J. and Staletovich, J. (2015) 'Miami Beach's battle to stem rising tides', *Miami Herald*, 23 October. Available at: https://www.miamiherald.com/news/local/community/miami-dade/miami-beach/article41141856.html (Accessed: 5 November 2023).

de Fraiture, C., Giordano, M. and Liao, Y. (2008) 'Biofuels and implications for agricultural water use: blue impacts of green energy', *Water Policy*, 10(S1), pp. 67–81. Available at: https://doi.org/10.2166/wp.2008.054.

Gerbens-Leenes, P.W., Hoekstra, A.Y. and van der Meer, Th. (2009) 'The water footprint of energy from biomass: A quantitative assessment and consequences of an increasing share of bio-energy in energy supply', *Ecological Economics*, 68(4), pp. 1052–1060. Available at: https://doi.org/10.1016/j.ecolecon.2008.07.013.

Goldburg, R. and Naylor, R. (2005) 'Future seascapes, fishing, and fish farming', *Frontiers in Ecology and the Environment*, 3(1), pp. 21–28. Available at: https://doi.org/10.1890/1540-9295(2005)003[0021:FSFAFF]2.0.CO;2.

Henrich, J. and McElreath, R. (2003) 'The evolution of cultural evolution', *Evolutionary Anthropology: Issues, News, and Reviews*, 12(3), pp. 123–135. Available at: https://doi.org/10.1002/evan.10110.

Hertel, T.W. *et al.* (2010) 'Effects of US Maize Ethanol on Global Land Use and Greenhouse Gas Emissions: Estimating Market-mediated Responses', *BioScience*, 60(3), pp. 223–231. Available at: https://doi.org/10.1525/bio.2010.60.3.8.

Hing, N.K., Welch, D.N. and Giap, T.S. (2006) 'Stormwater Management and Road Tunnel (SMART): a bypass solution to mitigate flooding in Kuala Lumpur city center', in. *International Conference and Exhibition on Tunnelling and Trenchless Technology*, pp. 7–9.

Hommels, A. (2000) 'Obduracy and Urban Sociotechnical Change: Changing Plan Hoog Catharijne', *Urban Affairs Review*, 35(5), pp. 649–676. Available at: https://doi.org/10.1177/10780870022184589.

Hommels, A. (2005) 'Studying Obduracy in the City: Toward a Productive Fusion between Technology Studies and Urban Studies', *Science, Technology, & Human Values*, 30(3), pp. 323–351. Available at: https://doi.org/10.1177/0162243904271759.

Hughes, T.P. (1987) 'The Evolution of Large Technological Systems', in W.E. Bijker and T. Pinch (eds) *The Social construction of technological systems: new directions in the sociology and history of technology*. Cambridge, MA: MIT Press.

Kareiva, P., Marvier, M. and McClure, M. (2000) 'Recovery and Management Options for Spring/Summer Chinook Salmon in the Columbia River Basin', *Science*, 290(5493), pp. 977–979. Available at: https://doi.org/10.1126/science.290.5493.977.

Kennedy, C. *et al.* (2009) 'Greenhouse Gas Emissions from Global Cities', *Environmental Science & Technology*, 43(19), pp. 7297–7302. Available at: https://doi.org/10.1021/es900213p.

Krausmann, F. *et al.* (2009) 'Growth in global materials use, GDP and population during the 20th century', *Ecological Economics*, 68(10), pp. 2696–2705. Available at: https://doi.org/10.1016/j.ecolecon.2009.05.007.

Kuhn, T.S. (1962) *The Structure of Scientific Revolutions*. Chicago, IL: University of Chicago Press.

Kurzweil, R. (2005) *The Singularity Is Near: When Humans Transcend Biology*. New York, NY: Viking Books.

Lagi, M. *et al.* (2011) 'The Food Crises: A quantitative model of food prices including speculators and ethanol conversion'. arXiv. Available at: https://doi.org/10.48550/arXiv.1109.4859.

Limer, E. (2015) 'The Tech that Runs the NYC Subway Is Positively Ancient', *Popular Mechanics*, 30 July. Available at: https://www.popular-mechanics.com/technology/infrastructure/a16685/ancient-mta-tech-nyc-subway/ (Accessed: 5 November 2023).

Madrigal, A.C. (2017) 'All the Promises Automakers Have Made About the Future of Cars', *The Atlantic*, 7 July. Available at: https://www.theatlantic.com/technology/archive/2017/07/all-the-promises-automakers-have-made-about-the-future-of-cars/532806/ (Accessed: 5 November 2023).

Marchant, G.E. (2011) 'The Growing Gap Between Emerging Technologies and the Law', in G.E. Marchant, B.R. Allenby, and J.R. Herkert (eds) *The Growing Gap Between Emerging Technologies and Legal-Ethical Oversight: The Pacing Problem*. Dordrecht: Springer Netherlands (The International Library of Ethics, Law and Technology), pp. 19–33. Available at: https://doi.org/10.1007/978-94-007-1356-7_2.

Milly, P.C.D. *et al.* (2008) 'Stationarity Is Dead: Whither Water Management?', *Science*, 319(5863), pp. 573–574. Available at: https://doi.org/10.1126/science.1151915.

Ming, T. *et al.* (2014) 'Fighting global warming by climate engineering: Is the Earth radiation management and the solar radiation management any option for fighting climate change?', *Renewable and Sustainable Energy Reviews*, 31, pp. 792–834. Available at: https://doi.org/10.1016/j.rser.2013.12.032.

OECD (2017) *Global EV Outlook 2017: Two million and counting*. Paris: Organisation for Economic Co-operation and Development. Available at: https://doi.org/10.1787/9789264278882-en (Accessed: 5 November 2023).

Park, J., Seager, T.P. and Rao, P.S.C. (2011) 'Lessons in risk- versus resilience-based design and management', *Integrated Environmental Assessment and Management*, 7(3), pp. 396–399. Available at: https://doi.org/10.1002/ieam.228.

Parshall, L. *et al.* (2010) 'Modeling energy consumption and CO2 emissions at the urban scale: Methodological challenges and insights from the United States', *Energy Policy*, 38(9), pp. 4765–4782. Available at: https://doi.org/10.1016/j.enpol.2009.07.006.

Romero-Lankao, P. and Dodman, D. (2011) 'Cities in transition: transforming urban centers from hotbeds of GHG emissions and vulnerability to seedbeds of sustainability and resilience: Introduction and Editorial overview', *Current Opinion in Environmental Sustainability*, 3(3), pp. 113–120. Available at: https://doi.org/10.1016/j.cosust.2011.02.002.

Searchinger, T. *et al.* (2008) 'Use of U.S. Croplands for Biofuels Increases Greenhouse Gases Through Emissions from Land-Use Change', *Science*, 319(5867), pp. 1238–1240. Available at: https://doi.org/10.1126/science.1151861.

Shoup, D.C. (2011) *The high cost of free parking*. Updated. Chicago, IL: American Planning Association.

Simon, H.A. (1947) *Administrative Behavior*. Basingstoke, UK: Macmillan Company.

Simon, H.A. (1957) *Models of Man: Social and Rational; Mathematical Essays on Rational Human Behavior in Society Setting*. Berkeley, CA: Wiley.

Simpson, T.W. *et al.* (2009) 'Chapter 9: Impact of Ethanol Production on Nutrient Cycles and Water Quality: The United States and Brazil as Case Studies', in *Biofuels: Environmental Consequences and Interactions with Changing Land Use*. Ithaca, NY: Cornell University Library's Initiative in Publishing, pp. 153–167. Available at: https://hdl.handle.net/1813/46219 (Accessed: 5 November 2023).

Staletovich, J. (2015) 'Beyond the high tides, South Florida water is changing', *Miami Herald*, 26 October. Available at: https://www.miamiherald.com/news/local/environment/article41416653.html (Accessed: 5 November 2023).

Syvitski, J. (2012) 'Anthropocene: An epoch of our making - IGBP', *Global Change Magazine*, (78), p. 12.

UNDESA (2014) *World Urbanization Prospects*. Report ST/ESA/SER.A/366. United Nations Department of Economic and Social Affairs. Available at: https://www.un.org/en/development/desa/publications/2014-revision-world-urbanization-prospects.html (Accessed: 15 November 2023).

UNESCO (2010) *Global trend towards urbanisation, UNESCO*. Available at: https://web.archive.org/web/20151108173832/http://www.unesco.org/education/tlsf/mods/theme_c/popups/mod13t01s009.html.

Urbanization, Biodiversity and Ecosystem Services: Challenges and Opportunities: A Global Assessment (2013). Cham, NL: Springer Nature. Available at: https://library.oapen.org/bitstream/id/5fb8553f-807a-4220-bdde-57348158f19b/2013_Book_UrbanizationBiodiversityAndEco.pdf (Accessed: 5 November 2023).

U.S. National Climate Assessment (2014) *Climate Change Impacts in the United States*. Washington, DC: U.S. Global Change Research Program. Available at: https://doi.org/10.7930/J0Z31WJ2.

Walski, T.M. (2006) 'A history of Water distribution', *Journal AWWA*, 98(3), pp. 110–121. Available at: https://doi.org/10.1002/j.1551-8833.2006.tb07611.x.

Worm, B. and Paine, R.T. (2016) 'Humans as a Hyperkeystone Species', Trends in Ecology & Evolution, 31(8), pp. 600–607. Available at: https://doi.org/10.1016/j.tree.2016.05.008.

4.2

CLIMATE-READY INFRASTRUCTURE

The most recent international report on climate change paints a picture of disruption to society unless there are drastic and rapid cuts in greenhouse gas emissions.

Although it's early days, some cities and municipalities are starting to recognize that past conditions can no longer serve as reasonable proxies for the future. This is particularly true for the country's infrastructure. Highways, water treatment facilities and the power grid are at increasing risk to extreme weather events and other effects of a changing climate.

The problem is that most infrastructure projects have typically ignored the risks of climate change. In our work researching sustainability and infrastructure, we encourage and are starting to shift toward designing man-made infrastructure systems with adaptability in mind.

4.2.1. DESIGNING FOR THE PAST

Infrastructure systems are the front line of defense against flooding, heat, wildfires, hurricanes, and other disasters. City planners and citizens often assume that what is built today will continue to function in the face of these hazards, allowing services to continue and to protect us as they have done so in the past. But these systems are designed based on histories of extreme events.

Pumps, for example, are sized based on historical precipitation events. Transmission lines are designed within limits of how much power they can move while maintaining safe operating conditions relative to air temperatures. Bridges are designed to be able to withstand certain flow rates

This chapter was adapted from the following article with publisher permission: Mikhail Chester, Braden Allenby, and Samuel Markolf, 2018, What is climate-ready infrastructure? Some cities are starting to adapt. *The Conversation*.

in the rivers they cross. Infrastructure and the environment are intimately connected.

Now, however, the country is more frequently exceeding these historical conditions and is expected to see more frequent and intense extreme weather events. Said another way, because of climate change, natural systems are now changing faster than infrastructure.

How can infrastructure systems adapt? First let's consider the reasons infrastructure systems fail at extremes:

- The hazard exceeds design tolerances. This was the case of Interstate 10 flooding in Phoenix in fall 2014, where the intensity of the rainfall exceeded design conditions.

- During these times there is less extra capacity across the system: When something goes wrong there are fewer options for managing the stressor, such as rerouting flows, whether it's water, electricity or even traffic.

- We often demand the most from our infrastructure during extreme events, pushing systems at a time when there is little extra capacity.

Gradual change also presents serious problems, partly because there is no distinguishing event that spurs a call to action. This type of situation can be especially troublesome in the context of maintenance backlogs and budget shortfalls which currently plague many infrastructure systems. Will cities and towns be lulled into complacency only to find that their long-lifetime infrastructure systems are no longer operating like they should?

Currently the default seems to be securing funding to build more of what we've had for the past century. But infrastructure managers should take a step back and ask what our infrastructure systems need to do for us into the future.

4.2.2. AGILE AND FLEXIBLE BY DESIGN

Fundamentally new approaches are needed to meet the challenges not only of a changing climate, but also of disruptive technologies.

These include increasing integration of information and communication technologies, which raises the risk of cyberattacks. Other emerging technologies include autonomous vehicles and drones as well as intermittent renewable energy and battery storage in the place of conventional power systems. Also, digitally connected technologies fundamentally alter individuals' cognition of the world around us. Consider how our mobile devices can now reroute us in ways that we don't fully understand based on our own travel behavior and traffic across a region.

Yet our current infrastructure design paradigms emphasize large centralized systems intended to last for decades and that can withstand environmental hazards to a preselected level of risk. The problem is that the level of risk is now uncertain because the climate is changing, sometimes in ways that are not very well-understood. As such, extreme events forecasts may be a little or a lot worse.

Given this uncertainty, agility and flexibility should be central to our infrastructure design. In our research, we've seen how a number of cities have adopted principles to advance these goals already, and the benefits they provide.

In Kuala Lampur, traffic tunnels are able to transition to stormwater management during intense precipitation events, an example of multifunctionality.

Across the United States, citizen-based smartphone technologies are beginning to provide real-time insights. For instance, the CrowdHydrology project uses flooding data submitted by citizens that the limited conventional sensors cannot collect.

Infrastructure designers and managers in a number of U.S. locations, including New York, Portland, Miami and southeast Florida, and Chicago, are now required to plan for this uncertain future—a process called roadmapping. For example, Miami has developed a $500 million plan to upgrade infrastructure, including installing new pumping capacity and raising roads to protect at-risk oceanfront property.

These competencies align with resilience-based thinking and move the country away from our default approaches of simply building bigger, stronger, or more redundant.

4.2.3. PLANNING FOR UNCERTAINTY

Because there is now more uncertainty with regard to hazards, resilience instead of risk should be central to infrastructure design and operation in the future. Resilience means systems can withstand extreme weather events and come back into operation quickly.

This means infrastructure planners cannot simply change their design parameter—for example, building to withstand a 1,000-year event instead of a 100-year event. Even if we could accurately predict what these new risk levels should be for the coming century, is it technically, financially, or politically feasible to build these more robust systems?

This is why resilience-based approaches are needed that emphasize the capacity to adapt. Conventional approaches emphasize robustness, such as building a levee that is able to withstand a certain amount of sea level rise. These approaches are necessary but given the uncertainty in risk we need other strategies in our arsenal.

For example, providing infrastructure services through alternative means when our primary infrastructure fails, such as deploying microgrids ahead of hurricanes. Or put simply, planners can design infrastructure systems such that when they fail, the consequences to human life and the economy are minimized.

This is a practice recently implemented in the Netherlands, where the Rhine delta rivers are allowed to flood but people are not allowed to live in the flood plain and farmers are compensated when their crops are lost.

Uncertainty is the new normal, and reliability hinges on positioning infrastructure to operate in and adapt to this uncertainty. If the country continues to commit to building last century's infrastructure, we can continue to expect failures of these critical systems and the losses that come along with them.

4.3

COMPLEXITY AND DEEP UNCERTAINTY

4.3.1. INTRODUCTION

Managers must be able to adapt infrastructure to the emerging climate patterns that are changing more rapidly than the design life of infrastructure systems. The rigidity and long design lives of infrastructure may result in systems where the rate of change in climatic conditions that they must be robust against is exceeded.

Infrastructure managers often encounter lock-in from financial, political, technical, social, cultural, and technological barriers, preventing transformative reimagining of infrastructure (Chester and Allenby, 2018). This can be perpetuated since the individual consumer demand of services such as electricity and water have not changed drastically (i.e. individuals continue to expect instantaneous access); however, pressures of urbanization, population growth, and climate change are expected to increase demand and induce stress on existing systems (Ayyub, 2018). Infrastructure services are expected to become less reliable (beyond natural deterioration) due to gradually changing climate patterns while also becoming increasingly vulnerable to extreme weather events (e.g. tropic storms, winter storms)—even when considering modern design standards. This change introduces an emerging challenge to how managers design infrastructure. Typically, engineers design infrastructure parameters to historical climate patterns and extremes. Extreme storm and heat events can impact the integrity of asphalt roads and bridges, causing chronic degradation, acute damage, and higher demand and failure risk (Nasr *et al.*, 2019). For instance, in order to determine the thermal design conditions for roadways, managers refer to standards that encourage the use of 1964 to 1995 climate data to determine temperature extremes (Underwood *et al.*, 2017).

This chapter was adapted from the following article with publisher and lead author permissions: Alysha Helmrich and Mikhail Chester, 2022, Reconciling Complexity and Deep Uncertainty in Infrastructure Design for Climate Adaptation, *Sustainable and Resilient Infrastructure*, 7(2), pp. 83-99, doi: 10.1080/23789689.2019.1708179.

These historic extreme temperatures are now being surpassed on a regular occurrence, meaning roadway surfaces will fail more frequently (i.e. roadways will have a shorter design life than intended). Extreme temperatures also impact water and power infrastructure (Burillo *et al.*, 2017; Bondank, Chester and Ruddell, 2018).

Designing infrastructure to historical climate conditions poses a large risk. Investigating stormwater infrastructure design, storm events are integrated by utilizing historical intensity-duration-frequency (IDF) curves to calculate a design storm standard. Typically, for large hydraulic infrastructure, a 100-year design storm standard is chosen (Bauer, 2011). A 100-year design storm is a precipitation event that has a 1 in 100 (or 1%) chance of occurring in a given year based on the historically-derived IDF curves. The use of design storms has come under scrutiny given the reliance on historical data (Packman and Kidd, 1980; Adams and Howard, 1986; Hirabayashi *et al.*, 2013; Watt and Marsalek, 2013; Harvey and Connor, 2017; Koerth, 2017; Ayyub, 2018). Recent catastrophic floods, such as in Houston, TX (three 500-year floods in three years; Ingraham, 2017) and Ellicott City, MD (two 1,000-year floods in two years; Bacon, 2018), have resulted in review of employing the 100-year design storm standard by elevating the issue to mainstream media, motivating city officials to update flood mitigation plans, and encouraging further research (Swartz, 2018).

Climate change is also expected to increase extreme cold (Francis and Vavrus, 2012), droughts (Strzepek *et al.*, 2010), sea level rise (Hansen *et al.*, 2016), and wildfires (Westerling, 2016). These events may be amplified in regions that already experience these hazards, or they may migrate to regions with no past experience of dealing with these threats, increasing risk. Additionally, climate change is expected to increase the magnitude and frequency of extreme weather events (Cheng and AghaKouchak, 2014). Recent catastrophic events such as Hurricanes Katrina (2005), Sandy (2012), Harvey (2017), and Maria (2017) devastated aged and new infrastructure alike. This devastation is not a consequence of neglect by any single (or several) infrastructure manager(s), but partly a consequence of the fundamental decision-making and design approaches utilized in infrastructure that perpetuate lock-in, enforce planning based on historical conditions, and restrict transformative change (Chester, Markolf and Allenby, 2019). Infrastructure managers heavily rely upon economic analysis, such as cost-benefit analysis, when appraising alternative solutions. These approaches look to quantify the tradeoffs in monetary terms for evaluation, but not all tradeoffs are easily quantifiable—specifically those related to environmental and social outcomes (Atkinson, Bateman and Mourato, 2012), which can unintentionally simplify the problem when excluded. As infrastructure managers construct in a world with undefined environmental design parameters due to a rapidly changing climate that undermines the expected lifetime of infrastructure, they must understand the associated complexity and uncertainty (Chester and Allenby, 2019). Climate change is an issue of complexity because there is not a singular solution to address the elaborate interactions between economic, environmental, and social drivers that are causing emergent climatic behaviors. It is also a problem of deep uncertainty, partially related to modelling parameters and assumptions, but primarily driven by the inability to know how socioeconomic systems will respond. This uncertainty may be alleviated or further complexed by new scientific discoveries (Walker, Lempert and Kwakkel, 2013). A variety of decision-making methods have been created to account for these attributes (Table 4.3.1 highlights a few of these strategies); however, infrastructure design is still most commonly approached with conventional decision-making methods which do not inherently account for complexity and uncertainty (Walker, Haasnoot and Kwakkel, 2013; Sánchez-Silva, 2018).

Infrastructure managers must integrate these attributes — complexity and uncertainty — into their design approaches so that infrastructure may continue to provide the services the public has come to expect despite the pressures of climatic shifts and extreme events. For this study, the term 'infrastructure design approaches' refers to various strategies of designing, operating, and maintaining infrastructure for climate change — whether designing new or retrofitting old infrastructure. Infrastructure managers implementing new infrastructure will have increased flexibility, not needing to work around legacy components; however, they will still need to address lock-in such as institutional constraints. The impact of climate change on infrastructure is complex due to the interdependence of infrastructure systems, multiple technologies, competing stakeholders, and other factors that result in their emergent behaviors being systemically unpredictable (Chester and Allenby, 2019). These complex system dynamics with positive and negative feedback loops between numerous infrastructure systems (and individual components) that are not necessarily working in cohesion can cause cascading failures and social consequences beyond what is initially predicted (Rinaldi, Peerenboom and Kelly, 2001), emphasizing that infrastructure managers cannot only consider how climate change will impact infrastructure design but how failure may have cascading effects to other infrastructure sectors and services.

Table 4.3.1: Infrastructure Decision-making Methods

Decision-Making Methods	Description	Source
Conventional		
Cost-Benefit Analysis (CBA)	CBA is based upon the comparison of costs and benefits (monetized) across potential designs.	(Dittrich, Wreford and Moran, 2016)
Cost Effectiveness Analysis (CEA)	CEA is a comparison of infrastructure alternatives effectiveness evaluated by a single, non-monetized parameter.	(Dittrich, Wreford and Moran, 2016)
Risk Assessment	A risk assessment is a component of risk analysis, which seeks to quantify the probability and magnitude of a risk associated with an infrastructure project.	(Yoe, 2012)
Environmental		
Life Cycle Assessment	This approach considers the emissions and wastes of the infrastructure throughout its entire lifespan (raw material extraction to disposal).	(Baumann and Tillman, 2004)
Environmental Impact Assessment*	An environmental impact assessment considers the impact of development on the environment and further looks to assess avoidance or minimization of those impacts.	(Banhalmi-Zakar, 2012)
Social		
Social Impact Assessment	This qualitative approach assesses the social and cultural consequences of development.	(Burdge and Vanclay, 1996)
Deep Uncertainty		
Real Option Analysis	This method expands upon CBA by adding an adaptable learning component for uncertainty of a singular parameter.	(Swart, Raskin and Robinson, 2004)
Robust Decision Making	In this approach, a wide variety of scenarios are assessed to determine design parameters. This increases robustness but decreases optimization.	(Lempert, Popper and Bankes, 2003)
Info-Gap Analysis	Info-gap analysis focuses on quantifying the information the decision maker knows and does not know by considering uncertainty, risk, and robustness.	(Ben-Haim, 2006)
Adaptation Pathways	The primary focus of this method is to determine which decisions can be made now and which decisions can be made later at identified tipping points.	(Haasnoot *et al.*, 2012)

Adaptive infrastructure systems should be approached with flexible and agile designs in order to address future challenges such as climate change (Chester and Allenby, 2018). To achieve adaptive systems, Chester and Allenby (2018) propose ten competencies: roadmapping, designing for obsolescence, hardware-to-software focus, risk-to-resilience based, compatibility, connectivity, modularity, organic structures, a culture of change, and transdisciplinary education. While achieving completely flexible and agile infrastructure will likely take a transformative alteration of hard and soft infrastructure, where hard infrastructure consists of physical systems such as water, power, and transportation networks and soft infrastructure entails institutions such as politics and finance, emerging hard infrastructure design approaches, including safe-to-fail infrastructure and adaptive management, are being discussed as pathways forward (Dittrich, Wreford and Moran, 2016; Kim *et al.*, 2019).

Infrastructure managers need a methodology to navigate climate-related complexity and uncertainty when approaching infrastructure design. In Section 4.3.2, the chapter will address this gap by mapping existing frameworks for complexity and deep uncertainty together. In Section 4.3.3, there is exploration of how existing infrastructure design approaches manage these topics as related to climate change. The Cynefin Framework and levels of deep uncertainty (hereafter referred to as the Deep Uncertainty Framework) address complexity and uncertainty faced by decision-makers and help provide recommendations for making decisions in varying degrees of these attributes as shown in Section 4.3.4. In summary, this study applies these frameworks to infrastructure design and seeks to analyze how well existing hard infrastructure design approaches account for the concepts of climate-related complexity and uncertainty in which infrastructure managers operate. By implementing infrastructure design approaches that are increasingly flexible and agile, future infrastructure managers will be further prepared to adapt existing infrastructure to emerging climate patterns and brace for unknowable events as discussed in the final section.

4.3.2. FRAMEWORKS FOR COMPLEXITY AND DEEP UNCERTAINTY

Two existing frameworks that focus on complexity and deep uncertainty present opportunities for advancing infrastructure design and management. The Cynefin Framework has been proposed by Chester and Allenby (2019) as a way to conceptualize complexity in infrastructure design and highlights climate change as one of the contributing factors of this complexity. The Deep Uncertainty Framework is also applied since it is frequently identified in climate literature when exploring the uncertainty of climate change (Kandlikar, Risbey and Dessai, 2005; Walker, Haasnoot and Kwakkel, 2013; Olsen, 2015; USACE, 2015; Dittrich, Wreford and Moran, 2016; Döll and Romero-Lankao, 2016; Helgeson, 2018; Manocha and Babovic, 2018) The Cynefin and Deep Uncertainty Frameworks presented in the following subsections will 1) introduce frameworks for managers to evaluate and respond to complexity and uncertainty, 2) provide explicit examples of infrastructure and climate change within each context, and 3) build a tool to evaluate existing infrastructure design approaches' capacity to handle climate-related complexity and uncertainty. It is increasingly important to understand how existing infrastructure design approaches respond to climate change as public demand pushes for climate change to be addressed with new passing policy (Gustafson *et al.*, 2019).

4.3.2.1. CYNEFIN FRAMEWORK

The Cynefin Framework (Figure 1.4.2) is a leading management strategy for understanding and making decisions in domains of increasing complexity (Snowden and Boone, 2007). The framework describes four primary domains: obvious, complicated, complex, and chaotic. The first domain, obvious (previously known as simple), is the domain of known knowns, where there is a clear cause-and-effect relationship that reveals a solution. In this domain, all the information is known to make a decision; and, therefore, decision-makers need only to understand the situation, evaluate their options, and take action. The second domain is complicated, where there are known unknowns and cause-and-effect relationships are not clearly apparent. Projects classified as complicated are not straight forward and might have multiple solutions but can be solved with expertise. In this domain, a decision-maker will still sense and respond to a problem; but instead of categorizing the options, they will need to analyze them. This is the realm that infrastructure managers have been operating under by assuming a calculated environmental parameter exists (e.g. design storms for precipitation events). The third domain, complexity, is the domain of unknown unknowns, which emphasizes unpredictability and emerging behaviors. Decision-makers now need to research the problem and associated feedback loops through probing before sensing and responding; however, there will not be a 'right' solution since not all the information can be known. The fourth domain is chaos, where there is no ability to distinguish cause-and-effect relationships. In this domain, a decision-maker should act first to create order, and then sense and respond to the problem. Each of these four primary domains are represented within infrastructure design approaches and their ability to integrate climate complexity as seen in Table 4.3.2. There is a fifth domain, disorder, which occurs when decision-makers cannot identify which of the four primary domains they are operating in. In this situation, decision-makers must step back and evaluate the situation to determine which of the four primary domains they are operating.

Table 4.3.2. Cynefin framework domains in infrastructure design relative to climate.

Domain	Climate Scenario	Infrastructure Application
Obvious	Recognizing current weather	Managing day-to-day operations, e.g. not operating airplanes in extreme heat
Complicated	Extrapolating historical climate patterns	Determining environmental design parameters, e.g. design storm standards for water infrastructure
Complex	Analyzing a range of predicted climate scenarios	Implementing infrastructure that expects unpredictability, e.g. planning for failures
Chaos	Experiencing an extreme weather event	Responding to an event without all the information, e.g. immediate response to a level 5 hurricane

Infrastructure systems are now operating under the domain of complexity while infrastructure managers continue to design within the complicated domain (Chester and Allenby, 2019). This complexity is derived from the variety of dynamic interactions infrastructure systems have with the natural, built, and social environment. Climate non-stationarity innately moves infrastructure systems from the complicated to complex domain. Infrastructure managers should no longer calculate design parameters from stationary climate datasets but instead incorporate climate forecasts (Underwood *et al.*, 2017). Furthermore, the built environment itself is becoming more complex as new technologies are implemented on top of legacy components (Arbesman, 2017), reducing the need for massive infrastructure overhauls, but relying upon infrastructure managers passing down knowledge and production continuing to manufacture parts. Additionally, complexity has increased within the built environment from the interactions between infrastructure systems as seen with cascading failures (Rinaldi, Peerenboom and Kelly, 2001). Infrastructure managers must work under the assumption of complexity to manage these factors; however, it is largely unknown how infrastructure should be managed in this capacity — particularly regarding climate change.

4.3.2.2. DEEP UNCERTAINTY FRAMEWORK

The Deep Uncertainty Framework (Figure 4.3.1) was created to understand the uncertainty decision-makers face when making decisions. By recognizing uncertainty, a decision-maker can make more confident decisions — and avoid paralysis — by understanding the risk involved (Courtney, Kirkland and Viguerie, 1997).

Figure 4.3.1: Deep Uncertainty Framework

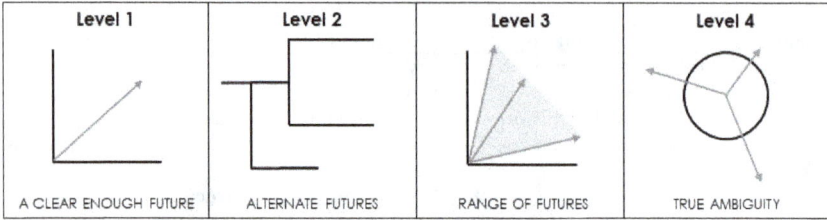

Level 1	Level 2	Level 3	Level 4
A CLEAR ENOUGH FUTURE	ALTERNATE FUTURES	RANGE OF FUTURES	TRUE AMBIGUITY

Adapted from Courtney et al. (1997).

There are four levels of deep uncertainty that fall between the extremes of complete certainty and total ignorance. The first level is a clear enough future (Level 1) where the decision-maker understands the outcome with small tolerances for uncertainty. In this stage, uncertainty is nearly negligible, and decision-makers do not need to consider the uncertainty-related risk involved. If a stationary climate is assumed, there is little concern regarding how the infrastructure will perform because the environmental design parameters are considered known. In this level, decision-makers may use conventional, environmental, or social decision-making methods as were seen in Table 4.3.1 to obtain an optimal solution. The second level, alternate futures or discrete scenarios (Level 2), describes situations where there are multiple potential outcomes with quantifiable probabilities of occurrence. Climate science has not yet reached this level of certainty due to the complexity of technological, climate, social, and environmental interactions. In this level, decision-makers can only make the best decision based on what occurs, which they can only know retrospectively. In order to make decisions in level two, decision-makers should evaluate each plausible scenario for tradeoffs and consider the probability of that event occurring to make an appropriate decision; therefore, at this point, decision-makers may continue to use conventional, environmental, or social decision-making methods or deep uncertainty decision-making methods listed in Table 4.3.1. These trade-offs between probability, risk, and consequence happen frequently in infrastructure management as it is expensive (money, time, resources, etc.) to build infrastructure to withstand the largest risk — especially if it is unlikely to occur. It has been acknowledged that making progress in applicability and understanding risk of deep uncertainty decision-making for infrastructure managers can greatly help address climate change in infrastructure design (Shortridge and Camp, 2019). The next two levels represent deep uncertainty, which climate change has long been attributed to, and these levels often provide the alleged basis for inaction (Kandlikar, Risbey and Dessai, 2005). Level 3, range of futures, describes where numerous outcomes are possible within a range predicted by key variables. The decision-making process for Level 3 is similar to that of Level 2, but the decision-maker must create their own unique scenarios within the range of predicted occurrences that have the highest likelihood of happening for evaluation. A direct solution cannot be computed, but a decision-maker can test for robustness within this level. Infrastructure managers can implement robust infrastructure, which is designed to handle a wide range of scenarios instead of optimizing for one potential outcome. Since there is not a clear, direct solution within a range of futures, infrastructure managers may also implement incremental design until there is a more distinct alternative. Modern climate change models fall into this category. They can predict temperature fluctuations for internally consistent, future scenarios that assign socioeconomic, land use, emission, and climate data; however, it is uncertain how any of these technological, political, and social factors will play out, embedding climate models with a range of uncertainty (van Vuuren *et al.*, 2011). The fourth level of uncertainty, true ambiguity (Level 4), is a future that cannot be predicted. In this level, decision-makers should break down what they know, what they can learn, and what they cannot learn. Decision-makers may then track the variables they do know to make incremental changes to their plans as the knowledge becomes available. As with the Cynefin Framework, each level of uncertainty is represented within infrastructure design approaches and their ability to integrate climate uncertainty into design as seen in Table 4.3.3. Courtney et. al. (Courtney, Kirkland and Viguerie, 1997) assert that most decision-makers will treat problems as Level 1 or 4 uncertainty and apply the same decision-making methods regardless; however, most problems are Level 2 or 3 uncertainty and should not be analyzed in the same manner.

Table 4.3.3. Deep uncertainty framework levels in infrastructure design approaches relative to climate.

Level	Climate Scenario	Infrastructure Application
Level 1	Utilizing a single extrapolation of climate behavior	Designing infrastructure to manage a fixed environmental design parameter, e.g. a threshold for extreme heat
Level 2	Applying probabilities of climate behavior extrapolations	Constructing infrastructure that designs to the most probable climate scenario, e.g. raising substations to likely flood-safe heights
Level 3	Testing a range of potential climate behaviors	Building infrastructure that may manage minimum and maximum environmental parameters, e.g. phase-change materials in pavements
Level 4	Realizing not all climate behaviors can be known in advance	Creating infrastructure that may be adapted to new information, e.g. adding diverse perspectives to a design team to encourage new ideas

4.3.2.3. UNDERSTANDING COMPLEXITY, UNCERTAINTY, AND INFRASTRUCTURE DESIGN WITHIN CLIMATE CHANGE

By operating within complexity and deep uncertainty, an infrastructure manager will have established the underlying assumption necessary to achieve flexible and agile infrastructure and, ultimately, adaptive infrastructure systems. Chester and Allenby (2019) connect flexible and agile infrastructure to the Cynefin Framework to understand how managers should make decisions under different domains of complexity. This is necessary as infrastructure must continue to deliver services reliably in an unpredictable environment. Climate science faces uncertainty due to the lack of confidence surrounding the location, timing, and magnitude of climatic change (Ayyub, 2018) and has been asserted a problem of deep uncertainty (Easterling and Fahey, 2018), operationalizing the Deep Uncertainty Framework principles will likely be necessary to make decisions in the future as confirmed by emerging decision-making methods such as real option analysis, robust decision making, info-gap analysis, and adaptation pathways. Complexity and deep uncertainty are a unified problem, where one cannot be addressed without addressing the other. Complexity is defined by unpredictability and emerging behaviors, which lend to uncertainty of how a system may evolve. Meanwhile, uncertainty is driven by the inability to determine an outcome with confidence, which is heightened by complexity. These concepts are intricately linked and can be visualized when mapping the Cynefin and Deep Uncertainty Frameworks. By navigating these frameworks, infrastructure managers may analyze their designs with a transdisciplinary perspective to create adaptive infrastructure systems. The mapping of these frameworks does not seek to provide an assessment for infrastructure managers to achieve a quantifiable goal, but, instead, looks to provide guidance so that managers may understand how complexity and deep uncertainty is embedded in the design parameters and react accordingly to the presence of these attributes.

By navigating the Cynefin and Deep Uncertainty Frameworks (Figure 4.3.2), infrastructure managers can assess the reality of underlying complexity and uncertainty within their design choices. A clear enough future (Level 1) from the Deep Uncertainty Framework maps to both the obvious and complicated domains from the Cynefin Framework, depending on the circumstance. When infrastructure design is approached with a fixed set of climatic parameters for the design condition that are either known (obvious) or can be easily calculated (complicated), it is assuming a Level 1 uncertainty as the climate pattern is not expected to change. This approach is commonly seen in avoidance and anticipatory pathways to climate change. Avoidance pathways ignore climate change and operate with the assumption of a stationary climate. Anticipatory pathways recognize climate change and seek to climate-proof development upon implementation (Asian Development Bank, 2015); however, this still assumes known parameters by accepting an extent of climate change for the environmental design parameter. There are instances where this assumption is valid in infrastructure design, most obviously when the design life is short or within a few months to a couple years, where the weather and climate respectively can be reasonably predicted or if the design is a smaller component of a larger system that may be easier to replace or has lower failure consequences. By assuming alternate futures (Level 2) uncertainty into an infrastructure design approach, a decision-maker has increased complexity by accepting that there is not a single optimal future but predictable, discrete scenarios. Now, infrastructure managers may need to introduce decision-making methods such as real options analysis or robust decision-making to best understand the knowable risks associated with a particular infrastructure design. An assumption of Level 2 uncertainty related to climate is best applied for infrastructure with a decadal design life such as road surfaces or public buses. This application is ideal for two reasons: 1) climate predictions are more precise due to the shorter time span, and 2) the short design life allows for reevaluation to address long-term uncertainty incrementally.

By a range of futures (Level 3), infrastructure managers are operating under complexity through evaluation of a continuous climate model. The representative concentration pathways (RCPs) adopted by the Intergovernmental Panel on Climate Change highlight select pathways from a range of climate futures that can be utilized in infrastructure design, which can be recalled as a decision strategy for Level 3 uncertainty. Here, infrastructure design must be able to account for the deep uncertainty of climate change and the complexity of interactions that further drives the uncertainty. There is often no singular solution to reasonably accommodate all RCPs, but infrastructure design may be approached in a manner where it is flexible and agile enough to adapt to new information. The inclusion of RCPs within an infrastructure design approach may be seen as an adaptation strategy to climate change, where infrastructure managers plan to adapt to future complexities and deep uncertainties; and, therefore, they approach decision-making and infrastructure design with intentions of adjusting or reiterating upon the first design implementation. This approach is optimal for infrastructure with a design life of decades to centuries as it allows planning for change throughout the design life. For instance, integrating green infrastructure into a stormwater management system helps reduce volume throughout the man-made pipe system. As rainfall intensity increases, infrastructure managers may choose to implement more green infrastructure to maintain the integrity of the pipes. Finally, true ambiguity (Level 4) of the Deep Uncertainty Framework can be approached with the same incremental design changes as Level 3 uncertainty but with recognition that it also associates with the chaotic domain in the Cynefin Framework since chaotic events cannot be predicted. While infrastructure cannot be designed specifically to handle unforeseeable events, infrastructure that is designed to manage increasing complexity and uncertainty will have an increased likelihood of withstanding these abnormalities.

Figure 4.3.2: Navigating Between the Cynefin and Deep Uncertainty Frameworks

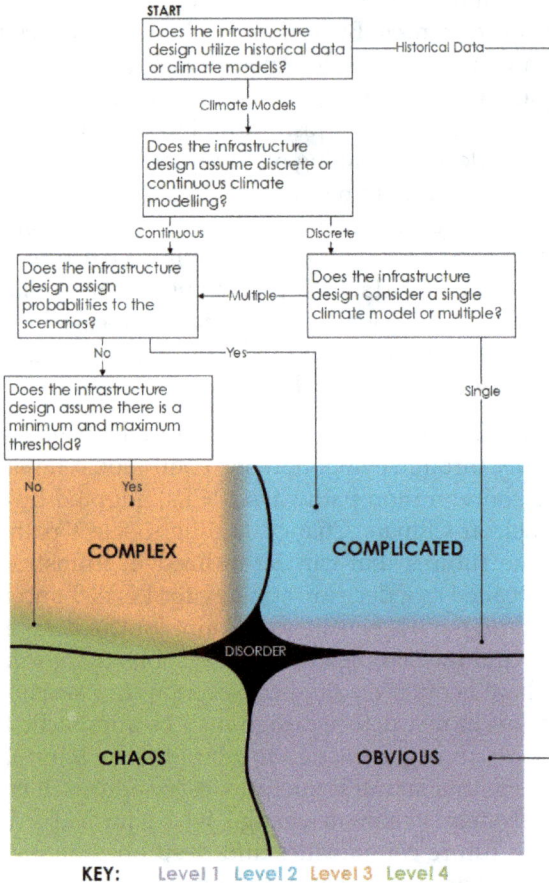

4.3.3. CAPACITY OF INFRASTRUCTURE DESIGN APPROACHES TO MANAGE CLIMATE COMPLEXITY AND UNCERTAINTY

Infrastructure managers utilize a variety of design approaches across the water, power, and transportation sectors, but there are no universal methodologies established for infrastructure design, management, and transformation (Hansman *et al.*, 2006; Shortridge and Camp, 2019). Typically, traditional fail-safe approaches are utilized, designing infrastructure to legal safety standards and the customer's satisfaction in the most efficient manner available (Ayyub, 2018). However, there have been proposed approaches for climate adaptation to address complexity and uncertainty (Olsen, 2015) such as armoring, low regret, safe-to-fail, and adaptive management (Table 4.3.4). Each of these infrastructure design approaches have varying capacities to manage climate-related complexity and deep uncertainty, and these capacities can be explored by navigating the Cynefin and Deep Uncertainty Frameworks to evaluate the flexibility and agility of existing infrastructure design approaches. Each infrastructure design approach is analyzed independently throughout this section to understand their origins and fundamental assumptions; however, it is important to realize that these practices may be used in conjunction throughout a system, which is encouraged in the discussion. For example, adaptive management can be applied to fail-safe approaches to increase infrastructure flexibility: the infrastructure may be strengthened after implementation or even rebuilt at pre-determined intervals. However, the infrastructure managers would no longer be operating with the fundamental assumptions of fail-safe design but those of adaptive management even though it may initially appear to be a fail-safe approach.

Table 4.3.4. Infrastructure Design Approaches

Infrastructure Design Approaches	Description	Examples
Traditional Fail-Safe	Infrastructure designed to withstand stress up to a pre-determined design parameter, and when these parameters are breached, the design fails in uncontrolled ways (Kim *et al.*, 2019).	Formulating material for a roadway based upon historical maximum temperatures.
Armoring Fail-Safe	Infrastructure that utilizes a fail-safe approach but uses more stringent environmental conditions for parameters.	Increasing a levee height based upon future climate predictions.
Low Regret	Infrastructure that will perform well across a range of futures without changing the function of the system and having co-benefits (Olsen, 2015).	Employing transmissions lines with capacity to handle low and high predicted peak demands.
Safe-to-Fail	Infrastructure designed to lose function in controlled ways, thus different types of failure consequence are experienced as expected based on prioritized decisions (Kim *et al.*, 2019).	Implementing a green space that retains water during storm events but otherwise serves as recreational space.
Adaptive Management	Infrastructure designed to for risk-adverse incremental adjustments, where this changeability increases the ability of infrastructure to react to known and unknown uncertainties (Allenby, 2012; Sánchez-Silva, 2019).	Constructing a building along a coastline that has an easily modifiable first floor to prepare for sea level rise.

4.3.3.1. TRADITIONAL FAIL-SAFE INFRASTRUCTURE

Traditional fail-safe methods use historical climate patterns to determine these parameters. This is a risky, avoidance approach to infrastructure design due to non-stationarity; therefore, fail-safe infrastructure may be stretched beyond its designed capacity, resulting in cascading, and potentially catastrophic, failures. When considering the complexity and uncertainty frameworks, it is evident that the underlying assumption of traditional fail-safe design approaches align with Level 1 uncertainty and the complicated domain. By operating under these assumptions, traditional fail-safe infrastructure design is inflexible and perpetuating lock-in because the design process does not consider any deviations from the expected climate pattern. The challenge of lock-in with traditional fail-safe design is further fortified by social norms, which perpetuate the application of this infrastructure design approach. For instance, within the stormwater field, infrastructure managers (e.g. developers, regulators, engineers, practitioners, etc.) are accustomed to the traditional fail-safe design approach and the risks associated with it (Roy *et al.*, 2008); consequently, their institutions are optimized to handle these expectations.

4.3.3.2. ARMORING INFRASTRUCTURE

Armoring (also referred to as hardening or strengthening) infrastructure utilizes the same approach as traditional fail-safe infrastructure; however, climate forecasts are used to determine design parameters. Typically, this increase of robustness occurs after a failure. For instance, if winds during an extreme event wreck power lines, standards may be adjusted to protect infrastructure from this higher wind speed. This approach remains risky as infrastructure managers deliberate between the probabilities of alternative scenarios and trade-offs to determine the design parameters. There is no confirmation that building to a particular climate scenario – or even the worst-case climate scenario – will protect the infrastructure in the future since there is uncertainty in climate modelling. Therefore, if the new environmental parameters are exceeded, the infrastructure will fail even more catastrophically than traditional fail-safe due to designing to a stronger magnitude of event. This can be seen in the levee effect and safe development paradox where hazardous areas are made safer by the government such as levees designed to withstand a larger flooding event, and developers feel more protected and continue to build within the flood zone (Burby, 2006).

Armoring infrastructure design continues to operate under the complicated domain because the approach does not consider the emerging behaviors of climate and remains an inflexible approach that identifies a singular climate outcome as a design parameter for the infrastructure. The armoring approach may assume Level 2 uncertainty if infrastructure managers compare alternative futures to determine the parameters, or it may assume Level 1 uncertainty if managers simply design to the worst-case event that has occurred. Notably, if Level 2 uncertainty is involved, the infrastructure will be simplified to Level 1 uncertainty upon implementation since the design will target a single scenario. This infrastructure design approach still perpetuates lock-in, ignoring climate-related complexity and deep uncertainty. It is important to recognize that this anticipatory approach is not adaptive toward climate change and exhibits the same concerns as traditional fail-safe infrastructure despite being utilized in practice as a response to climate change.

4.3.3.3. LOW REGRET INFRASTRUCTURE

This strategy differs from armoring in that the infrastructure is designed to manage more than one climate scenario. This strategy is oftentimes inflexible; however, the approach steps away from optimization and moves toward robustness. A robust approach minimizes risk over the lifetime of the system as long as the infrastructure's design life is shorter than the occurrence of climate patterns altering beyond the minimum and maximum design parameters. The robust strategy of low regret infrastructure inherently assumes complexity and Level 3 uncertainty. First, the design approach considers the complex domain by acknowledging there are unknowable environmental design parameters due to emergent behaviors in climate patterns. Second, the approach accommodates a range of potential futures (Level 3) uncertainty; it does not yet accept true ambiguity (Level 4) since it is an inflexible approach to design. If adaptive management approaches were combined with the low regret strategy as explored by Olsen (2015), the system would have the potential to address Level 4 uncertainty. The integration of iteration has not been universally adopted into low regret infrastructure literature (Preston, Mustelin and Maloney, 2013; Dittrich, Wreford and Moran, 2016). By the definition adopted here, low regret infrastructure still perpetuates lock-in due to the inflexibility to adapt the infrastructure over time, but it does take an anticipatory approach to climate change recognizing deep uncertainty.

4.3.3.4. SAFE-TO-FAIL INFRASTRUCTURE

Safe-to-fail design methodology embraces the happenstance of extreme events by expecting and containing such occasions (Park *et al.*, 2013). Desired outcomes of safe-to-fail infrastructure include maintaining services; minimizing consequences; promoting social and ecosystem services; designing decentralized, autonomous infrastructure; and encouraging transdisciplinary perspectives (Kim *et al.*, 2017) through the use of design strategies that follow the competencies of flexible and agile infrastructure (Ahern, 2011; Park *et al.*, 2013; Chester and Allenby, 2018). Safe-to-fail infrastructure ultimately assumes complexity and Level 4 uncertainty by recognizing that the interactions between infrastructure and the natural environment are not predictable. Therefore, safe-to-fail designs infrastructure to handle this unpredictability (i.e. uncertainty) by controlling failure and managing both the working and failing operational states of the system. This infrastructure design approach embodies an adaptable strategy that embraces modularity and learning so that the design may be adjusted to emergent climate patterns. If complexity and deep uncertainty simplifies, safe-to-fail infrastructure would still be operational, allowing this infrastructure design approach to also operate within the other domains. Safe-to-fail infrastructure has the capacity to manage complexity and uncertainty of climate change and provides a valuable adaptive strategy for infrastructure managers.

4.3.3.5. ADAPTIVE MANAGEMENT

This flexible approach to infrastructure assumes complexity and Level 4 uncertainty, but has the capacity to address complicatedness and all levels of uncertainty although that would potentially bean overexertion of resources. Adaptive management must consider financial, political, environmental, technical, social, cultural, and technological inputs to determine a best course of action forward. These inputs are web of dependencies and interdependencies, which results in emergent behaviors and exemplify complexity. At this time, it is difficult for infrastructure managers to assess the benefits of flexible infrastructure, but researchers are working toward developing long-term evaluation methods (Špačková and Straub, 2017). Adaptive management should be approached as a transdisciplinary problem due to the complexities involved, and infrastructure managers should look to integrate multiple perspectives into the design process (Chester and Allenby, 2019).

Concerning uncertainty, adaptive management addresses Level 4, or true ambiguity, as there is no determination of the design conditions expected near the end of lifetime of the infrastructure. Instead, the approach looks to make incremental and experimental adaptations as new information is available. This means that the infrastructure managers can address climatic deep uncertainty while making less risky decisions. In order for this to work properly, infrastructure managers cannot indefinitely resign from the next incremental design change but must accept a threshold of uncertainty or a frequency of adaptation. Recent literature by the Committee on Adaptation to a Changing Climate (Ayyub, 2018) promotes the observational method, a form of adaptive risk management to address climate change but also recognizes limitations of available knowledge exploring deep uncertainty in practice. Altogether, this approach embraces the concept of flexible and agile infrastructure, alleviating lock-in and providing an adaptation strategy to climate change.

4.3.4. DISCUSSION

Climate change is a wickedly complex problem surrounded by deep uncertainty, and infrastructure managers are in the nascent stages of integrating measures within their designs to protect against known and unknown hazards. Existing infrastructure design approaches are positioned to address a range of complexity and uncertainty challenges (Figure 4.3.3). However, these approaches are not necessarily designed to directly address these attributes, but provide promising qualities that can be employed to support resilient infrastructure in a future marked by these challenges.

Figure 4.3.3: Capacity of Design Approaches to Handle Climate-related Complexity and Uncertainty

		Cynefin Framework				Legend
		Obvious	Complicated	Complex	Chaotic	Fail-Safe: Traditional
Deep Uncertainty Framework	Level 1		✓✓ ✓ ✓✓	✓✓✓		Fail-Safe: Armoring
	Level 2		✓✓ ✓✓	✓✓✓		Low Regret
	Level 3		✓✓✓	✓✓✓		Safe-to-Fail
	Level 4		✓✓	✓✓		Adaptive Management

Current state-of-practice remains largely focused on fail-safe approaches (Kim *et al.*, 2019), which operate in the complicated domain. This simplification of complex problems as complicated problems are likely to perpetuate lock-in. As this chapter has shown, by only considering the effects of climate non-stationarity, infrastructure design can be considered a problem of complexity and deep uncertainty. Yet, there are numerous other factors — interactions and interdependencies between natural, built, and social environments, accretion, cascading failures — that increase the complexity and uncertainty of infrastructure design. The reconciliation of the Cynefin and Deep Uncertainty Frameworks is not exclusive to climate non-stationarity but may be extended to consider these other factors.

Furthermore, as seen in Figure 4.3.4, none of the infrastructure design approaches were mapped to the obvious or chaotic domains of the Cynefin Framework. There are no obvious approaches because infrastructure design fundamentally requires expertise and are not problems of categorization as seen in the obvious domain. Additionally, no approaches are explicitly classified as chaotic since this is the domain of unknowables that cannot be foreshadowed and, therefore, cannot be planned for; however, as designs become more flexible and agile, they reduce the potential magnitude of impact of chaotic situations. In the examination of the Deep Uncertainty Framework, all levels of uncertainty are accounted for across the six identified design approaches. This indicates that a specific infrastructure design approach is neither a valid or invalid approach for designing infrastructure to address climate change, but it may be better adept to addressing a particular scale of complexity and uncertainty. Infrastructure systems assuming complexity and deep uncertainty will be more equipped to handle chaotic and truly ambiguous scenarios by implementing systems that are capable of adaptation (and proactively investing in that competence).

4.3.4.1. HOW TO APPROACH INFRASTRUCTURE DESIGN FOR CLIMATE-RELATED COMPLEXITY AND UNCERTAINTY?

Not every component of an infrastructure system will need to be designed exclusively for complexity and deep uncertainty to improve resilience. Resilience is the ability of a system to rebound from a disruption to intended functionality, increase robustness to maintain a state of functionality in increasing complexity, an ability to dampen the impacts of a disruption, and produce sustained adaptability (Woods, 2015). This indicates that infrastructure managers must be able to determine when an inflexible strategy (traditional fail-safe, armoring, and low regret) should be chosen in lieu of a flexible strategy (safe-to-fail and adaptive management) to address the varying degrees of climate-related complexity and uncertainty within an infrastructure system. One way to place the appropriateness of an approach is to consider scale. Overall, an infrastructure system (e.g. stormwater management of a watershed) should be flexible and agile, but not every individual component (e.g. a pump) must be designed in this mindset to achieve a resilient system. Therefore, infrastructure managers should consider the goals and characteristics of their design to determine an appropriate approach.

Figure 4.3.4 shows characteristics of infrastructure design approaches with their capacity to address climate-related complexity and uncertainty and when they are best applied. The key difference in the characteristics of infrastructure addressing complicatedness and low uncertainty compared to complexity and deep uncertainty is fundamentally the focus on optimizing versus satisficing. Optimization is traditionally defined in engineering as maximizing the performance or efficiency of the primary function while minimizing costs. Satisficing (an agglomeration of the words satisfy and suffice) is settling on a course of action that may not be optimal, but is good enough (Chester, Markolf and Allenby, 2019). Therefore, there are times when one side of the spectrum may be more appropriate for designing than the other. There are instances in infrastructure design where there is reduced complexity and uncertainty. This can arise when an infrastructure has a planned obsolescence within a short time frame. It can also occur when a component or sub-system has minor failure consequences (spatially and temporally) within a system. Lastly, some components are fragile to environmental parameters and must operate in a fail-safe manner. Components and subsystems with short design lives and few failure consequences will benefit most from a fail-safe design choice since it will not be a question of 'if' the infrastructure will fail but 'when' as the environmental design conditions are exceeded by gradual climate change or extreme events and a short design life forces adaptation.

Climate science is not yet definitive enough to make accurate long-term assessments of the environmental conditions in which infrastructure will need to operate due to uncertainty of socioeconomic responses, meaning projects with longer design lives or larger failure consequences should be approached in a way that considers complexity and deep uncertainty (Easterling and Fahey, 2018). This finding aligns with the recommendations of the Committee on Adaptation to a Changing Climate, which states that higher levels of uncertainty analysis should be utilized for critical infrastructure with design lives greater than thirty years (Ayyub, 2018). In order to make the tradeoff between fail-safe and safe-to-fail infrastructure, infrastructure managers must consider their project as part of a larger system that interacts with the given infrastructure sector, other infrastructure sectors, society (including their institutions and governances), and the environment so that they may understand the potential failure consequences and/or cascading effects when assessing the design. Furthermore, infrastructure managers cannot assume a previous system is adaptive or resilient purely because it has exhibited these competencies in the past, since the lack of failure does not necessarily indicate resilience (Hollnagel, Woods and Leveson, 2006). Infrastructure managers must continue to question their design processes and evaluate them to emerging behaviors and information. It is important to recognize that the ability of an infrastructure manager to make decisions is constrained by their institution and resources. Some of these constraints are obvious such as funding and time, but others may be deeply embedded in the system and not yet recognized. To effectively address complexity and deep uncertainty in infrastructure design, there needs to be a change in culture within the establishment as well as increased education across disciplines—two competencies identified by Chester and Allenby (2018) for flexible and agile infrastructure. Institutions that design infrastructure are large hierarchies with many subdivisions that must all be educated on the long-term goals of address complexity and uncertainty to achieve resilience because a system must be consider holistically to achieve resilience. Bastidas-Arteage and Stewart (2019) recommend for institutions to have high engagement with stakeholders, identify decision-relevant information, integrate from national to local levels, employ decision-making tools based in science, and secure funding to achieve resilient infrastructure. To achieve the goals outlined in this discussion, institutions should promote multidisciplinary teams to expand perspective. This challenges the current institutional structure, which struggles to integrate multidisciplinary perspectives due to constraints such as funding and time. However, by reforming institutions and investing in diverse perspectives and acknowledging opportunities for flexibility, infrastructure managers will have the capacity to expand the design life of infrastructure in today's changing climate.

Figure 4.3.4: Characteristics and recommendations for application of infrastructure design approaches considering capacity to ad-dress climate-related complexity and uncertainty

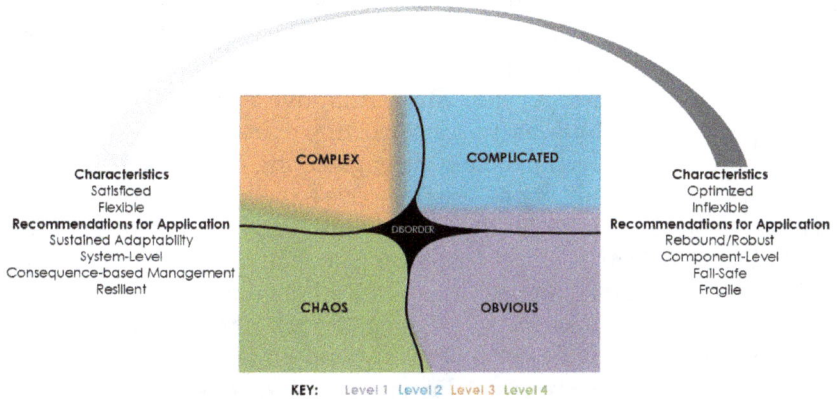

	COMPLEX	COMPLICATED	
Characteristics			**Characteristics**
Satisficed			Optimized
Flexible			Inflexible
Recommendations for Application	DISORDER		**Recommendations for Application**
Sustained Adaptability			Rebound/Robust
System-Level			Component-Level
Consequence-based Management			Fail-Safe
Resilient	CHAOS	OBVIOUS	Fragile

KEY:　Level 1　Level 2　Level 3　Level 4

At present, infrastructure managers generally seek immediate solutions (i.e. fail-safe infrastructure) rather than proposing systems that will need to be continuously maintained and managed (Park *et al.*, 2013), which leaves the majority of infrastructure incapable of addressing deep uncertainty. This requires a significant change in mindset from current engineering practice from one of optimization to one of satisficing—and ultimately flexibility—as advocated in deep uncertainty decision-making literature to 'monitor and adapt' rather than 'predict and act' (Walker, Haasnoot and Kwakkel, 2013; Chester and Allenby, 2019). A low regret strategy operates within complexity and Level 3 deep uncertainty, and this approach is best applied to projects that have a design life within a couple generations or integrated into systems that need to increase flexibility. Low regret can address water, power, and transportation infrastructure, and this leaves low regret infrastructure being the most common infrastructure design approach promoted to address Level 3 uncertainty in the power and transportation sector since safe-to-fail and adaptive management has not penetrated these sectors as it has in the water sector. Low regret infrastructure can be a costly way forward if everything is designed and implemented to operate for all climate scenarios; therefore, it needs to be determined in what situations this infrastructure design approach would be preferred in a system. Infrastructure managers can implement safe-to-fail and adaptive management strategies, which address upwards of Level 4 uncertainty, to fill this void of flexible infrastructure. Safe-to-fail and adaptive management infrastructure are best equipped to manage climate-related complexity and deep uncertainty due to their flexible and agile nature. Therefore, these design approaches should be utilized for large-scale infrastructure systems that have long design lives, larger consequences upon failure, and flexibility toward environmental design parameters (refer to Figure 4.3.5). However, adaptive management, while seen as an inherent component of socio-ecological systems (SES) (Cote and Nightingale, 2012), is infrequently applied beyond conceptual theory to water, power, and transportation infrastructure design (Chester and Allenby, 2019) although promoted by the new manual of practice, Climate-Resilient Infrastructure (Ayyub, 2018). Adaptive management accounts for deep uncertainty by enabling infrastructure managers to evaluate their options over time, which parallels the engineering design process where engineers are taught to reiterate — or improve — their design as more information becomes available. Infrastructure managers should explore how to conduct a reiterative adaptive management approach within large systems, which has significant barriers of lock-in previously explored. In order to be successful, infrastructure managers must clearly define and communicate objectives; work collaboratively; and monitor, learn, and adjust strategies to adapt to emerging climate science. As new information becomes available, infrastructure managers can either decide between maintaining current practices and implementing new strategies. Both approaches — safe-to-fail and adaptive management — need further exploration to become influential within design. Particularly, regarding adaptive management,

The examination of adaptive capacity for infrastructure to manage deep uncertainty reveals two important paths forward. First, infrastructure design addressing Level 3 deep uncertainty needs to be flexible and agile. As aforementioned, Chester and Allenby (2018) identified ten competencies for flexible and agile infrastructure: roadmapping, designing for obsolescence, hardware-to-software focus, risk-to-resilience based, compatibility, connectivity, modularity, organic structures, a culture of change, and transdisciplinary education. Infrastructure managers should not design an inflexible structure when there are not clear or likely environmental design conditions because that infrastructure is subjected to become antiquated and risk failure. With no learning or management in place, the infrastructure will need to be rebuilt to meet new design conditions (if it is under-designed), which is costly and time-consuming. Gilrein et. al. (2019) identify fifty infrastructure practices that demonstrate adaptability. For example, one way to increase the flexibility of infrastructure by utilizing the hardware-to-software, compatibility, connectivity, and modularity competencies is to incorporate information and communication technology (ICT), making smart infrastructure with a feedback loop between design, operation, and maintenance (Stephens *et al.*, 2013; Trindade *et al.*, 2017). While ICT innately increases complexity and uncertainty through the integration of an additional technological layer to infrastructure, infrastructure managers can leverage monitored changes between the built infrastructure and surrounding environment with emerging climate science to adapt the system to specific feedback. Another pathway, utilizing competencies such as risk-to-resilience thinking, is to strategically integrate green infrastructure into system designs. Green infrastructure (GI) is broadly defined as "environmental or sustainability goals that cities are trying to achieve through a mix of natural approaches" (Foster, Lowe and Winkelman, 2011). GI may be fail-safe or safe-to-fail depending upon implementation, but a clear benefit of green infrastructure (versus grey infrastructure) is the ability to provide ecosystem services in addition to the intended services.

4.3.5. CONCLUSION

The climate is changing faster than the design life of infrastructure, leaving infrastructure vulnerable as it must operate in conditions it was not designed to withstand. This is primarily credited to infrastructure lock-in — structural and institutional — which is associated with standards and incentives that encourage infrastructure managers to continue implementing inflexible fail-safe infrastructure. These standards and incentives have been implemented where 'traditional risk analysis is used to determine the acceptable likelihood and magnitude of an event to which infrastructure is expected to withstand' (Markolf *et al.*, 2018), leaving infrastructure vulnerable to changing conditions. Climate change can be an overwhelming concept to evaluate in infrastructure design, but it is important to not let emergent behaviors paralyze efforts to create resilient infrastructure. By embracing the Cynefin and Deep Uncertainty Frameworks, infrastructure managers have tools to assess the contexts of complexity and deep uncertainty and may respond accordingly to address those contexts within design. The preceding review of existing infrastructure design approaches shows that those being employed today are capable of addressing varying contexts of climate-related complexity and deep uncertainty, and even increasing a buffer for truly ambiguous events. However, the majority of infrastructure approaches applied in practice operate in the context of complicatedness and uncertainty. Infrastructure managers should pursue approaches that assume complexity and deep uncertainty in system design while also looking to expand the capacity of less adaptive approaches to become more resilient by integrating opportunities for flexibility and agility. This navigation of the Cynefin and Deep Uncertainty Frameworks improves comprehensibility of integrating conceptual attributes of complexity and deep uncertainty into design practice so that infrastructure managers may understand the contexts in which they are operating and implement appropriate design for the situation.

4.3.6. REFERENCES

Adams, B.J. and Howard, C.D.D. (1986) 'Design Storm Pathology', *Canadian Water Resources Journal / Revue canadienne des ressources hydriques*, 11(3), pp. 49–55. Available at: https://doi.org/10.4296/cwrj1103049.

Ahern, J. (2011) 'From fail-safe to safe-to-fail: Sustainability and resilience in the new urban world', *Landscape and Urban Planning*, 100(4), pp. 341–343. Available at: https://doi.org/10.1016/j.landurbplan.2011.02.021.

Allenby, B.R. (2012) *The Theory and Practice of Sustainable Engineering*. Hoboken, NJ: Upper Saddle River: Pearson Prentice Hall.

Arbesman, S. (2017) *Overcomplicated: Technology at the Limits of Comprehension*. New York, NY: Penguin Publishing Group.

Asian Development Bank (2015) *Guidelines for Climate Proofing Investment in the Energy Sector*. Mandaluyong City, Philippines: Asian Development Bank.

Atkinson, G., Bateman, I. and Mourato, S. (2012) 'Recent advances in the valuation of ecosystem services and biodiversity', *Oxford Review of Economic Policy*, 28(1), pp. 22–47. Available at: https://doi.org/10.1093/oxrep/grs007.

Ayyub, B.M. (2018) *Climate-Resilient Infrastructure*. Edited by Committee on Adaptation to a Changing Climate. Reston, VA: ASCE Press. Available at: https://doi.org/10.1061/9780784415191.

Bacon, J. (2018) *Ellicott City flooding: Why a 1-in-1,000-year rain event happened again*. Available at: https://www.usatoday.com/story/weather/2018/05/28/ellicott-city-flooding-why-1-000-year-rain-event-happened-again/649502002/ (Accessed: 24 October 2023).

Banhalmi-Zakar, Z. (2012) 'Introduction to environmental impact assessment', *Australian Planner*, 50(4), pp. 365–366. Available at: https://doi.org/10.1080/07293682.2012.747551.

Bastidas-Arteaga, E. and Stewart, M.G. (2019) 'Chapter Twelve - Recommendations for Infrastructure Decision-Makers Under a Changing

Climate', in E. Bastidas-Arteaga and M.G. Stewar (eds) *Climate Adaptation Engineering*. Oxford, UK: Butterworth-Heinemann, pp. 353–360. Available at: https://doi.org/10.1016/B978-0-12-816782-3.00012-7.

Bauer, J. (2011) *Summary of state stormwater standards*. Environmental Protection Agency, p. 144. Available at: https://www3.epa.gov/npdes/pubs/sw_state_summary_standards.pdf.

Baumann, H. and Tillman, A.-M. (2004) *The Hitch Hiker's Guide to LCA: An orientation in life cycle assessment methodology and application*. Lund, SE: Studentlitteratur AB.

Ben-Haim, Y. (2006) *Info-Gap Decision Theory: Decisions Under Severe Uncertainty*. Amsterdam, NL: Elsevier. Available at: https://books.google.com/books?id=yR9H_WbkIHkC&dq=info+gap+analysis&lr=&source=gbs_navlinks_s.

Bondank, E.N., Chester, M.V. and Ruddell, B.L. (2018) 'Water Distribution System Failure Risks with Increasing Temperatures', *Environmental Science & Technology*, 52(17), pp. 9605–9614. Available at: https://doi.org/10.1021/acs.est.7b01591.

Burby, R. (2006) 'Hurricane Katrina and the Paradoxes of Government Disaster Policy: Bringing About Wise Governmental Decisions for Hazardous Areas', *The ANNALS of the American Academy of Political and Social Science*, 604. Available at: https://doi.org/10.1177/0002716205284676.

Burdge, R.J. and Vanclay, F. (1996) 'Social Impact Assessment: A Contribution to the State of the Art Series', *Impact Assessment*, 14(1), pp. 59–86. Available at: https://doi.org/10.1080/07349165.1996.9725886.

Burillo, D. *et al.* (2017) 'Electricity demand planning forecasts should consider climate non-stationarity to maintain reserve margins during heat waves', *Applied Energy*, 206(June), pp. 267–277. Available at: https://doi.org/10.1016/j.apenergy.2017.08.141.

Cheng, L. and AghaKouchak, A. (2014) 'Nonstationary Precipitation Intensity-Duration-Frequency Curves for Infrastructure Design in a Changing Climate', *Scientific Reports*, 4(1), p. 7093. Available at: https://doi.org/10.1038/srep07093.

Chester, M., Markolf, S. and Allenby, B. (2019) 'Infrastructure and the environment in the Anthropocene', *Journal of Industrial Ecology*, 23(5), pp. 1006–1015. Available at: https://doi.org/10.1111/jiec.12848.

Chester, M.V. and Allenby, B. (2018) 'Toward adaptive infrastructure: flexibility and agility in a non-stationary age', *Sustainable and Resilient*

Infrastructure, 4, pp. 1–19. Available at:
https://doi.org/10.1080/23789689.2017.1416846.

Chester, M.V. and Allenby, B. (2019) 'Infrastructure as a wicked complex process', *Elementa: Science of the Anthropocene*. Edited by A. Iles and M.E. Chang, 7(1), p. 21. Available at: https://doi.org/10.1525/elementa.360.

Cote, M. and Nightingale, A.J. (2012) 'Resilience thinking meets social theory: Situating social change in socio-ecological systems (SES) research', *Progress in Human Geography*, 36(4), pp. 475–489. Available at: https://doi.org/10.1177/0309132511425708.

Courtney, H., Kirkland, J. and Viguerie, P. (1997) 'Strategy Under Uncertainty', *Harvard Business Review*, 1 November. Available at: https://hbr.org/1997/11/strategy-under-uncertainty (Accessed: 23 October 2023).

Dittrich, R., Wreford, A. and Moran, D. (2016) 'A survey of decision-making approaches for climate change adaptation: Are robust methods the way forward?', *Ecological Economics*, 122, pp. 79–89. Available at: https://doi.org/10.1016/j.ecolecon.2015.12.006.

Döll, P. and Romero-Lankao, P. (2016) 'How to embrace uncertainty in participatory climate change risk management—A roadmap', *Earth's Future*, 5(1), pp. 18–36. Available at: https://doi.org/10.1002/2016EF000411.

Easterling, D.R. and Fahey, D.W. (2018) *Impacts, Risks, and Adaptation in the United States: Fourth National Climate Assessment*. Washington, DC: U.S. Global Change Research Program, pp. 1–470. Available at: https://doi.org/10.7930/NCA4.2018.CH8 (Accessed: 24 October 2023).

Foster, J., Lowe, A. and Winkelman, S. (2011) *The Value Of Green Infrastructure For Urban Climate Adaptation*. Washington, DC: The Center for Clear Air Policy, p. 52. Available at: https://www.ca-ilg.org/sites/main/files/file-attachments/the-value-of-green-infrastructure-for-urban-climate-adaptation_ccap-february-2011.pdf?1376354679.

Francis, J.A. and Vavrus, S.J. (2012) 'Evidence linking Arctic amplification to extreme weather in mid-latitudes', *Geophysical Research Letters*, 39(6). Available at: https://doi.org/10.1029/2012GL051000.

Gilrein, E.J. *et al.* (2019) 'Concepts and practices for transforming infrastructure from rigid to adaptable', *Sustainable and Resilient Infrastructure*, 6(3–4), pp. 213–234. Available at: https://doi.org/10.1080/23789689.2019.1599608.

Gustafson, A. *et al.* (2019) 'Americans' worry about global warming is increasing.', *Yale Program on Climate Change Communication*, 21 February. Available at: https://climatecommunication.yale.edu/publications/a-growing-majority-of-americans-think-global-warming-is-happening-and-are-worried/ (Accessed: 24 October 2023).

Haasnoot, M. *et al.* (2012) 'Exploring pathways for sustainable water management in river deltas in a changing environment', *Climatic Change*, 115(3), pp. 795–819. Available at: https://doi.org/10.1007/s10584-012-0444-2.

Hansen, J. *et al.* (2016) 'Ice melt, sea level rise and superstorms: evidence from paleoclimate data, climate modeling, and modern observations that 2 °C global warming could be dangerous', *Atmospheric Chemistry and Physics*, 16(6), pp. 3761–3812. Available at: https://doi.org/10.5194/acp-16-3761-2016.

Hansman, R.J. *et al.* (2006) 'Research agenda for an integrated approach to infrastructure planning, design and management', *International Journal of Critical Infrastructures*, 2(2/3), p. 146. Available at: https://doi.org/10.1504/IJCIS.2006.009434.

Harvey, C. and Connor, J. (2017) 'History of the Application of Design Storm Frequency and Intensity', in *World Environmental and Water Resources Congress 2017. World Environmental and Water Resources Congress 2017*, Sacramento, California: American Society of Civil Engineers, pp. 1–9. Available at: https://doi.org/10.1061/9780784480595.001.

Helgeson, C. (2018) 'Structuring Decisions Under Deep Uncertainty', *Topoi*, 39(2), pp. 257–269. Available at: https://doi.org/10.1007/s11245-018-9584-y.

Hirabayashi, Y. *et al.* (2013) 'Global flood risk under climate change', *Nature Climate Change*, 3(9), pp. 816–821. Available at: https://doi.org/10.1038/nclimate1911.

Hollnagel, E., Woods, D.D. and Leveson, N. (2006) *Resilience Engineering: Concepts and Precepts*. Aldershot, UK: Ashgate.

Ingraham, C. (2017) 'Houston is experiencing its third "500-year" flood in 3 years. How is that possible?', *Washington Post*, 29 August. Available at: https://www.washingtonpost.com/news/wonk/wp/2017/08/29/houston-is-experiencing-its-third-500-year-flood-in-3-years-how-is-that-possible/?utm_term=.0b262e09c0f1 (Accessed: 24 October 2023).

Kandlikar, M., Risbey, J. and Dessai, S. (2005) 'Representing and communicating deep uncertainty in climate-change assessments', *Comptes*

Rendus Geoscience, 337(4), pp. 443–455. Available at: https://doi.org/10.1016/j.crte.2004.10.010.

Kim, Y. *et al.* (2017) 'Fail-safe and safe-to-fail adaptation: decision-making for urban flooding under climate change', *Climatic Change*, 145(3), pp. 397–412. Available at: https://doi.org/10.1007/s10584-017-2090-1.

Kim, Y. *et al.* (2019) 'The Infrastructure Trolley Problem: Positioning Safe-to-fail Infrastructure for Climate Change Adaptation', *Earth's Future*, 7(7), pp. 704–717. Available at: https://doi.org/10.1029/2019EF001208.

Koerth, M. (2017) 'It's Time To Ditch The Concept Of "100-Year Floods"', *FiveThirtyEight*, 30 August. Available at: https://fivethirtyeight.com/features/its-time-to-ditch-the-concept-of-100-year-floods/ (Accessed: 24 October 2023).

Lempert, R.J., Popper, S.W. and Bankes, S.C. (2003) *Shaping the Next One Hundred Years: New Methods for Quantitative, Long-Term Policy Analysis*. Santa Monica, CA: RAND, p. 210. Available at: https://doi.org/10.7249/MR1626 (Accessed: 24 October 2023).

Manocha, N. and Babovic, V. (2018) 'Sequencing Infrastructure Investments under Deep Uncertainty Using Real Options Analysis', *Water*, 10(2), p. 229. Available at: https://doi.org/10.3390/w10020229.

Markolf, S.A. *et al.* (2018) 'Interdependent Infrastructure as Linked Social, Ecological, and Technological Systems (SETSs) to Address Lock-in and Enhance Resilience', *Earth's Future*, 6(12), pp. 1638–1659. Available at: https://doi.org/10.1029/2018EF000926.

Nasr, A. *et al.* (2019) 'A review of the potential impacts of climate change on the safety and performance of bridges', *Sustainable and Resilient Infrastructure*, 6(3–4), pp. 192–212. Available at: https://doi.org/10.1080/23789689.2019.1593003.

Olsen, J.R. (2015) *Adapting Infrastructure and Civil Engineering Practice to a Changing Climate*. (Books). Available at: https://doi.org/10.1061/9780784479193.fm (Accessed: 23 October 2023).

Packman, J.C. and Kidd, C.H.R. (1980) 'A logical approach to the design storm concept', *Water Resources Research*, 16(6), pp. 994–1000. Available at: https://doi.org/10.1029/WR016i006p00994.

Park, J. *et al.* (2013) 'Integrating Risk and Resilience Approaches to Catastrophe Management in Engineering Systems', *Risk Analysis*, 33(3), pp. 356–367. Available at: https://doi.org/10.1111/j.1539-6924.2012.01885.x.

Preston, B.L., Mustelin, J. and Maloney, M.C. (2013) 'Climate adaptation heuristics and the science/policy divide', *Mitigation and Adaptation Strategies for Global Change*, 20(3), pp. 467–497. Available at: https://doi.org/10.1007/s11027-013-9503-x.

Rinaldi, S.M., Peerenboom, J.P. and Kelly, T.K. (2001) 'Identifying, understanding, and analyzing critical infrastructure interdependencies', *IEEE Control Systems Magazine*, 21(6), pp. 11–25. Available at: https://doi.org/10.1109/37.969131.

Roy, A.H. *et al.* (2008) 'Impediments and Solutions to Sustainable, Watershed-Scale Urban Stormwater Management: Lessons from Australia and the United States', *Environmental Management*, 42(2), pp. 344–359. Available at: https://doi.org/10.1007/s00267-008-9119-1.

Sánchez-Silva, M. (2018) 'Managing Infrastructure Systems through Changeability', *Journal of Infrastructure Systems*, 25(1), p. 04018040. Available at: https://doi.org/10.1061/(ASCE)IS.1943-555X.0000467.

Sánchez-Silva, M. (2019) 'Flexibility of infrastructure management decisions: the case of a project expansion', *Structure and Infrastructure Engineering*, 15(1), pp. 72–81. Available at: https://doi.org/10.1080/15732479.2018.1486439.

Shortridge, J. and Camp, J.S. (2019) 'Addressing Climate Change as an Emerging Risk to Infrastructure Systems', *Risk Analysis*, 39(5), pp. 959–967. Available at: https://doi.org/10.1111/risa.13234.

Snowden, D.J. and Boone, M.E. (2007) 'A Leader's Framework for Decision Making', *Harvard Business Review*, 1 November, pp. 68–76, 149.

Špačková, O. and Straub, D. (2017) 'Long-term adaption decisions via fully and partially observable Markov decision processes', *Sustainable and Resilient Infrastructure*, 2(1), pp. 37–58. Available at: https://doi.org/10.1080/23789689.2017.1278995.

Stephens, J.C. *et al.* (2013) 'Getting Smart? Climate Change and the Electric Grid', *Challenges*, 4(2), pp. 201–216. Available at: https://doi.org/10.3390/challe4020201.

Strzepek, K. *et al.* (2010) 'Characterizing changes in drought risk for the United States from climate change', *Environmental Research Letters*, 5(4), p. 044012. Available at: https://doi.org/10.1088/1748-9326/5/4/044012.

Swart, R.J., Raskin, P. and Robinson, J. (2004) 'The problem of the future: sustainability science and scenario analysis', *Global Environmental Change*, 14(2), pp. 137–146. Available at: https://doi.org/10.1016/j.gloenvcha.2003.10.002.

Swartz, M. (2018) 'Troubled Waters: A Year After Harvey, Has Houston Learned Anything?', *Texas Monthly*, 22 August. Available at: https://www.texasmonthly.com/news-politics/harvey-anniversary-houston-preparing-next-big-storm/ (Accessed: 24 October 2023).

Trindade, E.P. *et al.* (2017) 'Sustainable development of smart cities: a systematic review of the literature', *Journal of Open Innovation: Technology, Market, and Complexity*, 3(3), pp. 1–14. Available at: https://doi.org/10.1186/s40852-017-0063-2.

Underwood, B.S. *et al.* (2017) 'Increased costs to US pavement infrastructure from future temperature rise', *Nature Climate Change*, 7(10), pp. 704–707. Available at: https://doi.org/10.1038/nclimate3390.

USACE (2015) *Climate Change Adaptation Plan and Report*. Norfolk, VA: US Army Corps of Engineers. Available at: https://cdm16021.contentdm.oclc.org/utils/getfile/collection/p266001coll1/id/5265.

van Vuuren, D.P. *et al.* (2011) 'The representative concentration pathways: an overview', *Climatic Change*, 109(1), p. 5. Available at: https://doi.org/10.1007/s10584-011-0148-z.

Walker, W.E., Haasnoot, M. and Kwakkel, J.H. (2013) 'Adapt or Perish: A Review of Planning Approaches for Adaptation under Deep Uncertainty', *Sustainability*, 5(3), pp. 955–979. Available at: https://doi.org/10.3390/su5030955.

Walker, W.E., Lempert, R.J. and Kwakkel, J.H. (2013) 'Deep Uncertainty', in *Encyclopedia of Operations Research and Management Science*. Boston, MA: Springer, pp. 395–402. Available at: https://doi.org/10.1007/978-1-4419-1153-7_1140.

Watt, E. and Marsalek, J. (2013) 'Critical review of the evolution of the design storm event concept', *Canadian Journal of Civil Engineering*, 40(2), pp. 105–113. Available at: https://doi.org/10.1139/cjce-2011-0594.

Westerling, A.L. (2016) 'Increasing western US forest wildfire activity: sensitivity to changes in the timing of spring', *Philosophical Transactions of the Royal Society B: Biological Sciences* [Preprint]. Available at: https://doi.org/10.1098/rstb.2015.0178.

Woods, D.D. (2015) 'Four concepts for resilience and the implications for the future of resilience engineering', *Reliability Engineering & System Safety*, 141, pp. 5–9. Available at: https://doi.org/10.1016/j.ress.2015.03.018.

Yoe, C. (2012) *Principles of Risk Analysis: Decision Making Under Uncertainty*. Boca Raton, FL: CRC Press. Available at: https://doi.org/10.1201/b11256.

4.4

SAFE-TO-FAIL

4.4.1. INTRODUCTION

The evolving role of infrastructure coupled with changing environmental conditions raises the question: is it possible to create an infrastructure system that will not catastrophically fail? Given the increasing frequency of extreme weather events (Guerreiro *et al.*, 2018) and the challenges for infrastructure to withstand these events, there is a growing need to consider infrastructure failures explicitly in the development process. Thinking about infrastructure failures in system development may at first sound inappropriate since infrastructure design practices focus on achieving an optimum functional capacity by balancing system cost and performance through technical models. Infrastructure failure is assumed preventable by adding safety margins following familiar engineering design criteria. Current infrastructure development practice often lacks considerations for disaster management. Of growing concern is the non-stationary climate risks which challenge the robustness afforded by traditional infrastructure development practices, and raise questions of whether catastrophic infrastructure failures are inevitable (Boin and McConnell, 2007). If infrastructure systems are bound to fail, then decision-makers face an "infrastructure trolley problem", i.e., they must make decisions about how to incur the consequences when systems are eventually compromised. The trolley problem is a popular ethical experiment: should you pull a lever to divert a runaway trolley from its current path where it will hit multiple people, to another path where it will hit one? This choice juxtaposes various moral viewpoints including the intentionality of causing harm (Thomson, 1985). The infrastructure trolley problem means that the trade-offs of

This chapter was adapted from the following journal articles with publisher and lead author permissions: Yeowon Kim, Mikhail Chester, Daniel Eisenberg, and Charles Redman, 2019, The Infrastructure Trolley Problem: Positioning Safe-to-fail Infrastructure for Climate Change Adaptation, *Earth's Future*, 7(7), pp. 704-717, doi: 10.1029/2019EF001208. Yeowon Kim, Thomaz Carvalhaes, Alysha Helmrich, Samuel Markolf, Ryan Hoff, Mikhail Chester, Rui Li, and Nasir Ahmad, 2022, Leveraging SETS Resilience Capabilities for Safe-to-Fail Infrastructure under Climate Change, *Current Opinion in Environmental Sustainability*, 51(101153), doi: 10.1016/j.cosust.2022.101153.

costs experienced from future infrastructure failure must be managed prior to construction. This perspective is a stark change from previous approaches to urban planning and development, but is rooted in the emerging issues that infrastructure systems face. Building upon historical perspectives of the role infrastructure performs in urban development, this chapter presents an overview of climate and infrastructure challenges, suggests a new perspective for defining infrastructure failures, demonstrates dilemmas in the development process, and provides initial guidance for developing infrastructure systems that are safe-to-fail.

4.4.1.1. EMERGENCE OF FAIL-SAFE DESIGN

While the meaning of the word infrastructure has changed significantly in the last century, planning and development practices, embedded technologies, and the services provided by infrastructure remain largely unchanged in the same time period. Built infrastructure are planned, designed, constructed, operated, and maintained to ensure systems remain functional, safe, and sturdy for long service lives, i.e., fail-safe, typically 50 years to sometimes more than 100 years. Many stakeholders like politicians, city authorities, safety officers, engineers and designers are involved in the development of infrastructure by means of codes, regulations, capital, laws, policies, and institutions that guide infrastructure performance against hazards (Rasmussen, 1997). The standardization of development practices is codified and intended to produce functional, long-lasting systems with acceptably low risks of failures (Olsen, 2015). Although contemporary design standards provide greater consistency and reliability than in the 1930s, development practices themselves have remained stagnant over time and have yet to match the dynamism of modern society. For example, approaches to managing infrastructure risks by calculating possible hazards, and basing designs on acceptable tolerances have not changed much since their initial inception in the 1960s (Olsen, 2015).

Traditionally, engineers design for probable conditions to ensure a fail-safe system and incorporate safety margins to account for unknowns beyond the predictable risks. Risk predictions are often based on historical observations and statistical analysis, which then are translated into frequency terms or annual exceedance probability (AEP) of specific events (Kennedy and Paretti, 2014). While these historical development practices appear effective for constructing reliable infrastructure, the breadth of hard, soft, and critical systems are not often considered. Associated infrastructure developed to reduce predicted risks may have the unintended consequence of increasing risk to unpredicted events. For example, elevated levees give people the confidence of being protected against flood in a low-lying area so as to build houses, when, in fact, floods may surpass the predicted intensity and the risk of unintended levee breach may cause extensive damages to the area. Advancing new development practices that consider the breadth of complexity in contemporary use of the word infrastructure is necessary to face future challenges.

4.4.1.2. INFRASTRUCTURE IN A FUTURE OF NON-STATIONARY CLIMATE

A key limitation of infrastructure development practices is what appears to be their inability to adapt to recent volatility in climate. During the development process, weather-related hazards are often expressed as prepackaged datasets and charts showing the statistics on temperature, precipitation, wind speed. For instance, the "design storm" is a common operational threshold for designing drainage systems represented by return year frequencies (e.g., 100-year event). "100-year return period" is an insufficient probability representation of one in 100 chances of occurrence in any given one year (i.e., the 1% annual exceedance probability, AEP). It often leads to a misunderstanding of statistical weather hazard prediction that having experienced a 100-year event in a year anticipates the safety for the next 99 years. Given the recent variability in weather, these historical distributions may be becoming less useful in planning infrastructure performance for risks under a changing and unpredictable environment (Milly *et al.*, 2008). Future climate projections are not required in planning and strategic decision-making activities despite years of data and climate model development from the scientific community (Weaver *et al.*, 2013; Olsen, 2015; Lempert, 2016; Shortridge, Guikema and Zaitchik, 2017). Furthermore, infrastructure systems that are designed for 100-year events often fail to account for the design life. The probability of failure for 1% AEP system increases to 22%, 39% and 63% over a range of 25, 50, and 100 years, respectively (Read and Vogel, 2015). Infrastructure design standards and the infrastructure themselves remain difficult to change even when political, social, economic, and environmental systems change around them (Chester and Allenby, 2018). Despite studies that have characterized the impact of climate change on infrastructure (Pregnolato *et al.*, 2017; Dawson *et al.*, 2018), dams, pipelines, roadways, power plants, and other infrastructure continue to be managed without consideration of climate variability.

On-going debate in the scientific community on whether to use stationary or non-stationary statistical distributions to predict the frequency and/or intensity of future climate extremes highlights the need for infrastructure development practices that transcend the risk-based approach to future weather extremes (Warren *et al.*, 2018). High resolution climate simulations can improve the representation of extreme weather in certain regions (Mahajan *et al.*, 2015). Also, there have been advancements in modeling convective storms, which may help design infrastructure for future flash floods (Prein *et al.*, 2017). Models adopting stationary distributions are reliable and practical by enhancing the credibility of predicted extreme frequencies with uncertainty assessment (Serinaldi and Kilsby, 2015). Still, increasing attention has been devoted to using models that take non-stationarity into account for extreme frequency analysis by incorporating relevant covariates such as time and temperature (Milly *et al.*, 2008). Several studies have demonstrated that the use of non-stationary distributions in frequency analysis of rainfall data can improve a fit to the observed data than stationary models (Tramblay *et al.*, 2013; H. Kim *et al.*, 2017). Regardless, the general consensus across climate studies is that there are increasing uncertainties in predicting extremes due to urbanization and anthropogenic changes. Historical development practices that rely on statistical, frequency-based data cannot capture these unpredictable future events. Hence, infrastructure development practices meant to manage future disasters must embrace this unpredictability by strategic decisions that incorporate knowledge elicited by climate scientists, policy makers, as well as engineers for effective infrastructure risk management (Katz, 2010; Gilroy and McCuen, 2012; Lins, 2012).

4.4.1.3. RESILIENT INFRASTRUCTURE DEVELOPMENT AND CLIMATE CHANGE ADAPTATION

Here we build on theories of infrastructure resilience to advance a new development paradigm that is responsive to future weather extremes. Resilience has become a popular concept describing a system's ability to manage perturbations, challenges, or shocks since its first introduction in ecology (Holling, 1973). The concept is being used in various disciplines including business, psychology, ecology, engineering, and disaster risk management (Martin-Breen and Anderies, 2011; Alexander, 2013; Meerow and Newell, 2015; Rose, 2017). Especially in disaster risk management, resilience has been highlighted as a key attribute defined as "the ability to plan and prepare for, absorb, recover from, and adapt to adverse events" (NRC, 2012). In response, resilient infrastructure systems have been extensively recognized as an alternative to traditional infrastructure by managing unforeseen and unknown threats (Chang *et al.*, 2014; Chester and Allenby, 2018). Given that the notion of resilience has a malleable and multidisciplinary nature, there is no clear-cut standard that measures 'infrastructure resilience'. Thus, implementing resilience in practice entails unavoidable subjective representation of the concept by decision-makers in consideration of implementation context embodying social, ecological, and technological systems in the affected region.

Climate change adaptation studies suggest that resilience is a key attribute that societies must consider when implementing infrastructure systems confronted by a non-stationary climate (McDaniels *et al.*, 2008; IPCC, 2014; Linkov *et al.*, 2014; Chester and Allenby, 2018; Miller, Chester and Munoz-Erickson, 2018). However, there is often a gap when communicating resilience from research to practice (Chang *et al.*, 2014; Alduce *et al.*, 2015; Meerow and Stults, 2016). Few studies explore decision frameworks that promote resilient infrastructure, and many of those that do suggest that resilient infrastructure development requires the consideration not only of biophysical but also social and institutional factors such as institutional capacity, spatial variability, social vulnerability, and level of serviceability of existing infrastructure (McDaniels *et al.*, 2008; O'Brien, Hayward and Berkes, 2009; Chmutina *et al.*, 2014; Francis and Bekera, 2014; Y. Kim *et al.*, 2017). Decision-makers are in the position to understand these complex consideration factors when governing climate adaptation and infrastructure development strategies. However, there is a lack of understanding of how to depart from current infrastructure design standards and development practices. A novel resilient infrastructure development practice is needed that recognizes the various contexts in which infrastructure are deployed and guides decision makers in practice.

We propose the use of the recently introduced paradigm of "safe-to-fail" as a guiding decision approach for developing resilient infrastructure system under non-stationary climate. We further advance the paradigm into a descriptive decision theory of safe-of-fail, which encourages decision-makers to engage with dilemmas of infrastructure risk management through assessing social, institutional, and/or biophysical capacity responding to various types of failure consequences. Among the few strategies suggested for developing resilient infrastructure systems (Park *et al.*, 2013), a safe-to-fail approach is becoming increasingly attractive to communities vulnerable to natural disasters and non-stationary climate risks (Tye, Holland and Done, 2015; Y. Kim *et al.*, 2017). We suggest that safe-to-fail infrastructure development practice aims to guide infrastructure investment and design for unpredicted risk scenarios and building adaptive capacity for affected communities. The safe-to-fail infrastructure development can also support decision-makers to consider resilience in social, ecological, and technological dimensions by engaging with local governments, practitioners, community members, and utility owners, because they face "infrastructure trolley problem" situations where future infrastructure failures will affect stakeholders in unequal ways. Safe-to-fail theory can guide these decisions by anticipating infrastructure failures to ensure controlled aftermaths, and thus, help decision-makers be more strategic in infrastructure development process. Safe-to-fail decision consideration includes guidance on how much to invest in infrastructure development, what infrastructure functions to maintain, where to direct the impact of failures, which assets and values to prioritize for protection, how and when to recover from disruption, and which organization to react at emergency. We establish a guiding decision theory of safe-to-fail for infrastructure systems and discuss how to incorporate failure consequences in the development practice.

4.4.2. ADVANCING THE SAFE-TO-FAIL PARADIGM

The safe-to-fail paradigm originates from green infrastructure and safety science research (Lister, Hargreaves and Czerniak, 2007; Möller and Hansson, 2008). Both the green infrastructure and safety science perspectives are valuable since they accept that unexpected failures are inevitable. Yet, there is no consensus on what failure means or how this paradigm guides the development of resilient infrastructure to manage the consequences of system failure. In particular, green infrastructure and safety science literature emphasize different design objectives, namely: i) experimental design strategies that expect a failure, and ii) a system that fails while causing minimum harm. Here we overview the existing literature and redefine safe-to-fail for resilient infrastructure decision making.

4.4.2.1. THE EMERGENCE OF SAFE-TO-FAIL PARADIGM

Green infrastructure literature focuses on small-scale design innovations with expectation of innocuous failures (i.e., trial and errors and learning-by-doing) and strengthening the ecological value of infrastructure (Lister, Hargreaves and Czerniak, 2007; Novotny, Ahern and Brown, 2010; Ahern, 2011). Failure in this sense is an experience that can be useful in the future, so an adaptive approach is limited by reliable experiments to planning and design where failure impacts can be naturally absorbed in the ecosystem (Lister, Hargreaves and Czerniak, 2007; Novotny, Ahern and Brown, 2010). Green infrastructure studies suggest that science, professional practice, and stakeholder participation need to be integrated with urban development to achieve intended ecosystem services (Ahern, Cilliers and Niemelä, 2014). Specific examples that demonstrate this perspective are green infrastructure and low impact development practices such as permeable pavements, bioswales, and urban tree canopies that capture rainfall and attenuate drainage flows (Ahern, 2011).

The safety science perspective argues that reducing risks and adding safety barriers are necessary to contain the impact of infrastructure failure within designed system tolerances (González *et al.*, 1997; Möller and Hansson, 2008; Butler *et al.*, 2014; Mugume *et al.*, 2015). Failure here means service disruptions, and thus, these studies tend to focus on maintaining system function or system recovery practices. Particular examples that illustrate this perspective are underground nuclear waste repositories (Möller and Hansson, 2008) and storage tanks or parallel pipes in urban drainage systems (Mugume *et al.*, 2015). This characterization of safe-to-fail in-cludes risk analysis and critical infrastructure security studies which em-phasize awareness of unforeseen risks (Boin and McConnell, 2007; Block-ley, Agarwal and Godfrey, 2012). A study of critical infrastructure crisis management based on risk analysis underlines the adaptive behavior of infrastructure managers in an effective and rapid response to an aftermath of system breakdown (Boin and McConnell, 2007). Another study calls on engineers to recognize the 'low-chance but potentially high-impact' risks arising from interdependencies of complex infrastructure where the sys-tem behavior may not be fully understood, and to design the system as robust to unforeseen risks (Blockley, Agarwal and Godfrey, 2012). This safe-to-fail framing in safety science recognizes unpredicted disastrous risks that may cause a rare system break-down and a need of processes to ensure systems degrade in a way that allows some control of the safety of people. In the resilience engineering perspective, the risk-based approach is further questioned by advocating for a resilience-based approach that advances from a fail-safe overconfidence mentality of large and robust in-frastructure that leads to a lack of failure preparations (Park, Seager and Rao, 2011; Park *et al.*, 2013).

What is missing from current safe-to-fail literature is an operational definition of infrastructure failure. While the goal of safe-to-fail research has largely been to explore design strategies in the areas of green infra-structure and safety science to better manage infrastructure performance under risks, there is disagreement on what infrastructure failure or safe-to-fail means and how to manage infrastructure failure consequences in the infrastructure development process. Without a clear definition of fail-ure, the current literature is insufficient to address climate non-station-arity. For instance, novel green infrastructure practices provide additional ecosystem services such as multifunctionality (Ahern, Cilliers and Nie-melä, 2014), however, studies do not elucidate how the additional features control failure consequences in uncertain futures.

4.4.3. REFRAMING SAFE-TO-FAIL AND FAIL-SAFE INFRASTRUC-TURE

Whereas several authors discuss the paradigm of safe-to-fail, no stud-ies have systematically assessed the implications that the safe-to-fail par-adigm has on infrastructure development practices. We assert that this is due to both a lack of definition of infrastructure failure and a lack of ad-dressing how safe-to-fail infrastructure supports adaptation to changing and unforeseeable future conditions. Using our definition of failure, the important features delineating fail-safe and safe-to-fail development are the different decision approaches for incorporating failure consequences that infrastructure are designed for.

4.4.3.1. UNDERSTANDING INFRASTRUCTURE FAILURE

Currently in infrastructure planning, the word "failure" is almost exclusively considered in prevention activities. We refer to current infrastructure development practices as "fail-safe" because they focus on making failure a rare and preventable event as long as plans and designs are followed and maintained. We extend this notion to define infrastructure failure in two parts: (1) when infrastructure stop serving its intended service and/or function, and (2) when infrastructure disruption by a hazard causes social, economic, and environmental impact. Type-1 failures arise when infrastructure is overwhelmed by predicted risks or discontinues its intended function, e.g., failure to convey excessive rainfall runoff through the drainage structure due to limited pipe capacity. Type-2 failures arise when infrastructure is overwhelmed by consequences of Type-1 failure resulting in severe damage to the system itself, ecosystem services, physical assets, and livelihood. While fail-safe is focused on avoiding Type-1 failure, we argue that safe-to-fail requires us to consider Type-2 failures as well in the development process and to re-evaluate the risks, particularly in situations where Type-2 failures occur without knowing the cause or prognosis of Type-1 failures due to an unforeseeable threat.

Catastrophic failures occur in contemporary fail-safe infrastructure, not because of a lack of data on potential risks, but a lack of consideration for the consequences caused when infrastructure themselves fail, i.e., Type-1 failure and resulting Type-2 failure. Our definition of infrastructure failures responds to the uncertainty of future climate risk by focusing attention on understanding consequences when infrastructure services are lost rather than unpredictable causes of failure. A number of studies have demonstrated the significance of understanding conceivable impacts of infrastructure failure, highlighting the relationships of infrastructure service loss and its consequences (Bludau, Zirkelbach and Kuenzel, 2008; Oxenford and Williams, 2014; Revi *et al.*, 2014; Wilbanks and Fernandez, 2014; The Trust for Public Land, 2016). Table 4.4.1 summarizes infrastructure service examples and their Type-1 and Type-2 failures related to climate change.

Table 4.4.1. The relationship of infrastructure services and their failure including Type-1 and Type-2

Intended infrastructure services	Type-1 failure – infrastructure service loss from identified hazards	Type-2 failure – consequences of infrastructure service loss regardless of identified hazards
Control floods	• Drainage overflow • Flood levee breach • Stormwater pipe pump malfunction/breakdown	• Destruction of property and public infrastructure • Contamination of water sources • Water logging • Loss of business • Loss of livelihood options • Increase in water-borne and water-related diseases
Mitigate extreme heat exposure	• AC failure • Reflective roof malfunction (e.g., condensation) • Reflective or permeable pavements breakdown • Urban vegetation destruction	• Heat-related illness • Increased urban heat island • Damage to properties • Increased air pollution • Increase in energy demand • Business disruption • High maintenance cost (e.g., irrigation for vegetation) • Street storm sewer clogging
Secure water resources	• Water main break • Treatment deficiency • Source water quality and/or quantity control instability	• Increased water shortages • Electricity shortages (where hydropower is a source) • Water-related diseases (through use of contaminated water) • Food prices and food insecurity from reduced supplies
Protect coastal area from storm surge	• Storm surge gate malfunction • Flood barrier breakdown	• Loss of livelihood • Damage to properties • Effects on coastal vegetation and ecosystems • Threats to commerce, business, and tourism
Manage power and energy networks and provide electricity	• Transmission system malfunction • Power grid deterioration • Substation breakdown	• Electricity shortages • Propagating failure across multiple systems due to interdependency of the power grids with other infrastructure systems (e.g., water distribution system uses electric power to pump water, transit networks, electric power plants, ICT).
Support transportation	• Roadway/railway destruction	• Impact of mobility on livelihood (e.g. daily commute) and related services (e.g., freight and retail industry, fuel delivery for plants) • Loss of evacuation routes and emergency services

Understanding infrastructure failure in terms of Type-1 and Type-2 failures expands beyond existing safe-to-fail literature to guide infrastructure development practices. When infrastructure failure is discussed in existing literature, it often refers to structural disruption, component malfunction, operational error, and physical breakdown (Möller and Hansson, 2008; Blockley, Agarwal and Godfrey, 2012). These failures are derived from a static development perspective that understands infrastructure not as ubiquitous systems in a region, but as composed of multiple elements performing isolated tasks. We redefine failure to focus on the consequential disruptions caused by infrastructure service losses, and expand this perspective to include the intended or unintended consequences infrastructure may bring to affected populations and regions. We further argue that infrastructure failures can occur when infrastructure compromise or stop their functioning regardless of the cause, rather than a breakdown in a particular part of structure. We build upon resilience scholarship that examines infrastructure as systems rather than isolated parts. In this respect, there is less importance on calculating the exact probability of extreme climatic risks or component losses in infrastructure design, because the definition of Type-1 and Type-2 failures are not contingent on initiating events. Thus, both stationary and non-stationary models of future climate risks can be considered in infrastructure development practices through this definition.

4.4.3.2. REFRAMING FAIL-SAFE

We define fail-safe infrastructure as built systems that are designed to avoid failure and to be fully functional up to safety thresholds, but lose all function when thresholds are exceeded. Under the fail-safe approach, a given system is characterized in one of two states: functioning or failed. Fail-safe infrastructure maintains the functioning state at all costs, and failure is typically understood as losing system function, i.e., Type-1 failure and/or physical disruption. The stability of fail-safe infrastructure ensures their services (e.g., flood protection) available in the near-term against predicted hazards, yet in unpredictable future events like natural disasters may cause catastrophic service losses. This means fail-safe infrastructure is unable to manage unintended consequences because they are developed to be robust in the near-term (10-30 years) and are difficult to maintain and at greater risk of failure in the long-term (40-100 years). The consequences of fail-safe infrastructure failure is often catastrophic because these consequences do not inform design. Risks that transcend designed safety thresholds thereby cause significant damages to the infrastructure itself and other dependent systems. After-failure actions for fail-safe systems are usually rebuilding and restoration back to the previous functioning state.

Historical and current infrastructure development practices are fail-safe as the consequences of failure, i.e., Type-2 failure, are not considered during the development process. Current infrastructure focuses on optimizing the service delivery given financial constraints and safety thresholds. This development approach is incomplete as large infrastructure with low probability of failure, long-lifetimes, and oversized to handle unforeseen threats will inevitably fail. The fail-safe approach emphasizing near-term reliability and risk management may only increase future damages, as the larger and more permanent an infrastructure is, the greater the damages caused by its failure (Park, Seager and Rao, 2011). While incorporating failure consequences in risk analysis may seem feasible, even the best models cannot fully prescribe future non-stationarities including extreme weather, population growth, social demographics, urban form, and policies (Christensen *et al.*, 2007; Shortridge, Guikema and Zaitchik, 2017). Moreover, model results do not provide an understanding of system status when stresses exceed the functional range of system capacity. While the durability or safety of local elements can be improved based on climate model forecasts, the consequences of system failure are not computationally simulated. Hence, engineers and decision-makers who are involved in multiple stages of infrastructure development need to recognize the possible failures that are not captured in models.

Oversizing, a robustness strategy, has been the primary mechanism used to avoid failure (Olsen, 2015) in fail-safe infrastructure. Design standards are a key element to oversizing by setting minimum thresholds for robustness that are serviceable, safe, durable, and constructible. Design standards reflecting changing stresses to systems such as increased storm frequencies and intensities, high variability of available water sources, groundwater depletion, extreme heat, and environmental loads can also increase the robustness of infrastructure to future climates (Muller *et al.*, 2015; Slota and Bowker, 2017). However, oversizing is fail-safe because it is based on the assumption that failure is avoidable, and will not serve a future with unpredictable climate extremes. For example, oversizing is not efficient in non-stationary climate conditions where high uncertainties exist, because the analytics of prediction models may diverge from the range of design criteria.

The design standards and development practices used in New Orleans prior to Hurricane Katrina exemplify the limitations of oversizing and fail-safe infrastructure. Heavy precipitation and storm surge are frequently expected weather phenomenon in New Orleans due to its geography. The city enlisted planners and engineers to upgrade flooding protection measures such as levees and floodwalls. However, the plan was insufficient to take into account the inevitable complexity of interdependent infrastructure (Leavitt and Kiefer, 2006). Moreover, the infrastructure failures at the scale of what happened by Hurricane Katrina, i.e., the combination of tidal surge and fluvial flooding from the Mississippi river, were not adequately considered by engineering design standards in addition to the wind speeds which were very rare according to the statistical hazard prediction (Boin and McConnell, 2007; Wilbanks and Fernandez, 2014). Poor maintenance of levee system also exacerbated the problem causing subsidence failure and breaches. Several national and international reports on infrastructure systems highlight the significant need of maintenance and upgrade (Boesen, 2007; Morrison and Fay, 2007; NRC, 2010), but maintaining these systems only with stagnated standards means that they are limited in capacity to respond to the changing environment. Taken together, traditional, robust infrastructure development strategies which resist external shocks that disrupt their integrity and/or protect their local urban environment could not be expected to survive weather extremes like Hurricane Katrina. Climate change further brings to question the efficacy of traditional fail-safe development practices into the future.

4.4.3.3. REFRAMING SAFE-TO-FAIL

We define safe-to-fail infrastructure as built systems designed to lose function in controlled ways, thus different types of failure consequence are experienced as incurred by prioritized decisions, even when safety thresholds are exceeded in unpredicted risks. Under a safe-to-fail approach, a given system can fall into at least three different states: functioning, limited functioning/incurred failure, and full failure with chosen consequences. Here, the functioning state is a normal state where the system performs all of its intended function within the designed capacity against a predicted range of hazards. Limited functioning or incurred failure is when the system stops its service and causes Type-1 failure, but limits the impact of Type-2 failure within the system. Full failure and loss of system function still occurs in safe-to-fail systems, but the consequences of the Type-1 failure are controlled to ensure that the overall impact of Type-2 failure (i.e., loss of life, ecosystem, economy, physical assets and disruption of livelihood) are minimized and experienced based on development decisions. Thus, safe-to-fail development practice requires that the system remains adaptable to control the consequences at full failure by recovering lost function or transforming to serve new purposes. Safe-to-fail considers unpredicted risks caused by non-stationary climate and supports long-term climate adaptation. Creating safe-to-fail infrastructure systems helps climate adaptation by forcing cities to examine their institutional capacity to manage unpredictable risks and to develop more adaptive coping mechanisms to future risks. This is possible because frequent, controlled infrastructure failures help prevent risky development practices from becoming locked-in prior to unpredictable weather extremes and ensures that the calculated risks are re-assessed and consequential impacts are experienced as expected (Blockley, Agarwal and Godfrey, 2012). Moreover, loss of infrastructure services forces city planners and engineers to constantly reassess infrastructure service needs to help cope with changing climates.

In safe-to-fail infrastructure development, multi-stakeholder engagement is a key element for assessing the institutional capacity to respond to infrastructure failure consequences. Safe-to-fail infrastructure develop-ment combines design standards, supporting policy, and stakeholder en-gagement to create infrastructure where the consequences of failure are managed. In a safe-to-fail approach, it is important to plan and design for the consequences of both Type-1 and Type-2 failure scenarios. This re-quires incorporating knowledge of multiple stakeholders to take into ac-count the spatial context and complexity of interdependent infrastructure systems. It is more straightforward to manage consequences in Type-1 failure scenarios, because planning of limited system functions and recov-ery practices is already possible in fail-safe infrastructure development. For example, many infrastructure systems already have planned failure operations that limit system function to reduce damages, including drain-age pump shutdown to avoid overheating and load shedding in power systems. Safe-to-fail development practices are more intricate, since man-aging Type-2 failure requires consideration of the social, economic, and environmental attributes in the affected region. Damages caused by Type-2 failures require multiple stakeholders including practitioners, local gov-ernments, communities, and engineers to understand consequences and to decide how interdependent systems should fail to strategically prepare for anticipated damages.

The "Room for the River" strategy used in the Netherlands is a good example of safe-to-fail infrastructure development practice that uses a combination of design, policy and stakeholder engagement (Zevenbergen *et al.*, 2013). The city of Lent, where flooding has been a chronic problem and was becoming more intense, intentionally expanded flood-prone ar-eas into nearby farmland, which the Dutch decision-makers chose to transform into vegetated flood buffer during heavy rainfall and recrea-tional parks at other times. When heavy precipitation occurs, high vol-umes of water are diverted from the river to buffers and parks. While it compromises the economic and recreational values of the parks, it signif-icantly reduces the overall human loss, and economic loss from cata-strophic damages.

Indian Bend Wash, green infrastructure designed to manage heavy rainfall events, is also an example of safe-to-fail infrastructure design for Type-2 failures. Its greenbelt stretches 18 kilometers through Scottsdale, Arizona, one of the major cities in the Phoenix Metropolitan Area. At the time of its development, the primary function of the system was to accom-modate runoff, attenuate flows, and reduce flooding. Instead of adopting a design of a concrete channel structure suggested by the Army Corps, Scottsdale practitioners opted for a bioretention basin consisting of parks, golf courses, and other activities. When a record storm (~1 % of AEP) hit the area in 2014, the Indian Bend Wash accommodated excessive runoff in the designated wash, reducing the intensity of flooding in nearby areas.

When the rain stopped, the wash helped drain city streets and neighborhoods (The City of Scottsdale Communications and Public Affairs, 2004). Indian Bend Wash exemplifies safe-to-fail because it serves the same primary function as conventional storm drainage systems, but was developed with involvement of local practitioners and citizens and further designed to control the consequences when the wash stops accommodating excessive runoffs by considering the infrastructural recovery capacity of nearby area from flooding. Thus, greater investment in safe-to-fail infrastructure is one way to advance current infrastructure development practices to manage the unpredictable weather events that climate change brings.

4.4.4. SAFE-TO-FAIL AND THE INFRASTRUCTURE TROLLEY PROBLEM

The potential benefits of safe-to-fail development to adapt to unpredictable climate risks brings additional decision dilemmas to infrastructure development. The safe-to-fail approach urges stakeholders to make explicit decisions about failure consequences, meaning that decisions made today will have a direct connection to eventual undesirable futures. The addition of failure considerations in the development process further incorporates multiple stakeholders, their context-specific needs, and assumptions about failure consequences. This complicates already difficult decision-making processes and creates dilemmas for infrastructure development. We characterize the dilemma of designing for failure as an infrastructure trolley problem, i.e., prioritizing the consequences of infrastructure failures that may be experienced by different attributes and population in a city. These decisions raise societal and ethical questions regarding whom and what should be prioritized to remain safe when infrastructure fail (Cutter, 2016).

A practical way to demonstrate this dilemma is by considering oversizing and stakeholder engagement activities of infrastructure development in cost-benefit analysis (CBA). CBA is a major decision support framework used by governments and institutions to organize and calculate social and economic costs and benefits, inherent trade-offs, and economic efficiency of a policy, program, or project (Kull, Mechler and Hochrainer-Stigler, 2013). Currently in infrastructure development, CBA provides a quantitative way to prioritize risk reduction and service provision activities based on comparing benefits of an actual or planned investment with direct and indirect costs due to Type-1 failures (Table 4.4.2). In comparison, Table 4.4.2 also shows additional cost categories that are difficult to calculate, rarely included in infrastructure development CBAs, and generally associated with Type-2 failures, including social costs and losses due to business disruption, and intangible costs. Fail-safe infrastructure development practices like oversizing are not amenable to CBAs that consider additional costs caused by Type-2 failure presented in Table 4.4.2, because it is difficult to know how to set design thresholds for Type-2 failure consequences such as homelessness, loss of business revenue, interdependent service failures, loss of heritage, and psychological stress. In contrast, stakeholder engagement activities common in safe-to-fail development allow decision-makers to consult with a broader range of consequences, and may prioritize avoiding Type-2 social loss, business interruptions, and intangible costs before calculating Type-1 depending on their capacity to manage different failure types and associated costs.

Table 4.4.2. Costs resulting from Type-1 and Type-2 infrastructure failures.

Cost category	Associated impacts
Direct tangible costs Indirect costs	▪ Infrastructure damage ▪ Cost of emergency services and disaster assistance ▪ Cost of reconstruction and recovery ▪ Cost of planning and implementation of risk prevention measures
Direct social loss Losses due to business interruption Intangible costs	▪ Deaths ▪ People affected (e.g., missing, displaced, homeless, livelihood damaged) ▪ Property losses (residential and commercial) ▪ Agricultural loss ▪ Increase in government debt ▪ Loss of revenue ▪ Losses due to the absence of public services (e.g., telecommunication, transportation, gas, water, electricity) ▪ Negative impacts on stock market prices ▪ Increase in unemployment ▪ Environmental losses ▪ Health impacts ▪ Cultural heritage losses ▪ Psychological stress

Direct tangible costs and indirect costs are often considered in fail-safe infrastructure development by design standards. Direct social loss, losses due to business interruption, and intangible costs needs to be considered in safe-to-fail infrastructure development (adapted from OECD, 2015, 2016; IDDR, 2017).

The strength of the safe-to-fail approach is that it encourages decision makers to assess the different types of costs in their decision context and recognize the acceptable costs based on their institutional capacity to manage infrastructure failure, protect vulnerable populations and critical assets, identify affected regions, and recover from failure. The infrastructure trolley problem, based on utilitarian decision theory, suggests that the best decision is to prioritize the needs of many over the needs of few (Bennis, Medin and Bartels, 2010). Fail-safe infrastructure development practices like oversizing have implicit bias in how the needs of the many are defined, as they are only possible when costs are amenable to calculation.

Safe-to-fail offers a transformative utilitarian approach by considering a greater range of costs, but introduces the following dilemma: being explicit about institutional capacity and failure consequences (i.e., costs) is highly context dependent and limits the use of standard design protocols and precedent development practices. For example, choosing how to prioritize costs and how infrastructure manages Type-2 failures may introduce costs that we are not able to calculate, limiting the use of CBA for decision support.

Still in the near-term, safe-to-fail cost prioritizations may conflict with standard development practices for managing Type-1 failures indoctrinated in design manuals, and thus difficult to result in clear and actionable development plans. Safe-to-fail infrastructure development is a challenging process that needs long-term management strategies and operation considerations that require existing systems to adapt to rapidly changing climate. In addition, safe-to-fail requires more careful attention of decision-makers to embrace marginalized groups, who tend to be more vulnerable to unpredictable risks in the stakeholder engagement, to inform decisions of cost prioritization. The safe-to-fail approach aims to address these challenges by emphasizing stakeholder engagement to rank the relative importance and ramifications among different cost categories, and thus, the estimation of overall cost can be adjusted in order to focus on prioritizing between different types of costs and trade-offs.

Room for the River offers a good example of how to overcome this challenge. The Dutch decision-makers choose to divert the high volume of water from the river to a nearby vegetated area when heavy precipitation occurs in order to safely flood the high-risk areas. This decision compromises nearby vegetated areas during flooding and creates direct tangible costs and losses due to business interruption, but prioritizes reducing social loss, indirect costs, and intangible costs that might incur due to uncontrolled levee failure. Although the project took about 10 years to implement the new infrastructure development and risk management practices while consulting with local governments, practitioners, engineers, civic societies, and community members, the decisions have been well informed to affected regions and population. The Dutch project is considered a successful example of a long-lasting sustainable infrastructure solution to a chronic flooding problem.

4.4.5. TOWARDS SAFE-TO-FAIL INFRASTRUCTURE DEVELOPMENT UNDER A NON-STATIONARY CLIMATE

Safe-to-fail infrastructure development supports climate change adaptation strategies by considering the uncertainties inherent in climate models and/or risk analysis. While climate prediction has improved substantially, there remains significant uncertainty in these projections due to interrelationships of systems, nonlinearities in biophysical processes, adoption of greenhouse gas emitting technologies, and the adoption or lack of greenhouse gas mitigation policy (Chester *et al.*, 2014; Hulme, 2016). This reaffirms a need for a new infrastructure development paradigm that manages unforeseen risks by building adaptive capacity without compromising systems when infrastructure failure. Traditionally, infrastructure is designed as fail-safe – they are designed against infrequent weather events and such that they cannot fail. Yet when failure occurs the consequences to human life, economic loss and other infrastructure are enormous. Risk-based fail-safe approaches are often contingent on statistical analysis of identified risk, thus, often do not account for the uncertainties associated with future climate change, making them inadequate for resilience. Safe-to-fail is valuable for a climate impacted future by introducing uncertain and unidentified future risks in infrastructure development decisions.

Green infrastructure is often conflated with safe-to-fail; they may or may not be safe-to-fail. Green infrastructure is a valuable practice that enhances natural processes while delivering environmental, social, and economic benefits. However, green infrastructure designed without consideration of Type-2 failures are fail-safe. A common example is small-scale rain gardens that experience ponding leading to nearby flooding, possible health impacts and ecosystem disruptions. Green infrastructure systems designed with Type-2 failures in mind, like bioretention basins in the Room for the River project and the Indian Bend Wash, are safe-to-fail. These two examples provide further evidence for how to prioritize decisions with broad stakeholder engagement as a means to achieve safe-to-fail development.

Safe-to-fail development connects to the resilience of social, ecological, and technological systems, as infrastructure will influence future social, environmental, and economic costs incurred by extreme events. Infrastructure is built to adapt to climate change, but basing development decisions on probabilistic models and risk analyses may not serve resilience by focusing on fail-safe systems that are robust to Type-1 failures. Safe-to-fail not only considers these methods but expands upon them with a multidisciplinary perspective that considers failure consequences via identified Type-2 failures. Safe-to-fail development decisions require consideration of infrastructure's functional capacity (e.g., safety thresholds) and social capacities to respond to risks such as institutional capacity, spatial variability, social vulnerability, and serviceability.

Safe-to-fail infrastructure development is challenging because the risks and performance of long-lasting infrastructure are often difficult to predict. In addition, safe-to-fail decision dilemmas may be inimical to the use of standard engineering design practices. Particularly for hard infrastructure, it is difficult to make alterations post construction to adapt to changing stresses. Safe-to-fail infrastructure development requires a broader scope of knowledge and decision support than fail-safe to untangle the decision dilemma of the infrastructure trolley problem, and extra steps in the development process to consider context-specific information including geography, existing infrastructure services, social vulnerability, different types of failure cost, and institutional adaptation capacities, among others. One approach to achieve safe-to-fail is to use multi-stakeholder engagement to help decision-makers to determine the acceptable level of "failure" and its cost. Thus, the functions of safe-to-fail infrastructure may vary in different cities and regions depending on which assets and values are prioritized for protection and what their capacities are for undertaking different types of failure costs. Ideally, for every city, a safe-to-fail infrastructure system can be developed by deciding whom or what should remain safe during failed infrastructure states, with consequential trade-offs between different assets, values, locations, and people.

4.4.6. SAFE-TO-FAIL AS SETS

Urban systems are composed of intertwined SETS subsystems that collectively produce essential functions and resilience dynamics characteristic of a city. Technological systems (T-systems) — such as infrastructure and the built environment — are embedded in social (e.g. institutions and infrastructure management) and ecological systems (e.g. natural resource processes). Simultaneously, T-systems shape social systems (e.g. the distribution of public services to people and protection of communities from climate hazards) and ecological systems (e.g. modifying natural resource processes and enhancing ecosystem values via engineered solutions), such that we cannot understand cities' resilience capabilities (i.e. system capacity and behavior in responding to disturbances) to climate hazards without an understanding of the interactions within and between each SETS domain (Seto, Solecki and Griffith, 2015; McPhearson *et al.*, 2016; Kim, Mannetti, *et al.*, 2021).

In response to climate hazards, T-focused approaches for infrastructure resilience often emphasize recovery of physical components and mechanical processes to ensure the provision of critical services in cities (e.g. back-up electrical transmissions and redundant water supply mains). In this way, infrastructure's aftermath response emphasizes 'bouncing back' from a perturbation, where the disturbed object's inherent materiality is restored to provide critical functions like electricity and potable water (Hollnagel, Woods and Leveson, 2006; Francis and Bekera, 2014; Vale, 2014). Studies providing definitions and guidelines for infrastructure resilience abound in the engineering resilience literature, which supports the planning of T-systems that are robust to disturbances (Hosseini, Barker and Ramirez-Marquez, 2016; Yodo and Wang, 2016). Engineering resilience studies tend to focus on reinforcing the ability of infrastructure systems to withstand predetermined hazard envelopes or analyzing risks to infrastructure performance in terms of probability predictions (Yazdani, Otoo and Jeffrey, 2011; Chopra *et al.*, 2016; Donovan and Work, 2017; Huang *et al.*, 2018). Such technocentric approaches are often aligned with FS design in their view of infrastructure systems resilience. Risk management decisions made without considering the social or ecological context of infrastructure often affect the overall adaptive capacity to climate hazards in cities and overlook potential impacts on other systems. For instance, elevated levees cannot control the damages to homes and ecosystems if floods overflow the levee or if the levee itself breaches (Kroes *et al.*, 2006; Doorn, Gardoni and Murphy, 2019). In addition, social factors affecting T-systems, such as limited funding available for an infrastructure project and the socially acceptable safety in design, largely contribute to infrastructure performance and their capacity to reduce vulnerability from climate hazards (Grabowski *et al.*, 2017). Hence, a few studies have advocated for a need to evaluate infrastructure systems in consideration of the interactions with social systems such as political, financial (financing and affordability), governance, community engagement, equity, decision-making, public health, education, and so on (Grabowski *et al.*, 2017; Clark *et al.*, 2018; Wyborn *et al.*, 2019; Gardoni and Murphy, 2020). The SETS perspective adds explicit consideration of these additional systems, and their dynamics, which may have been overlooked or considered in isolation previously.

The SETS perspective builds upon social-ecological systems (SES) literature, which has critically framed resilience in terms of the sustainability of human-environment interactions (Folke, 2006; Walker *et al.*, 2006). With the rapidly growing number of cities experiencing extreme weather events, the importance of understanding urban systems as SES and their resilience to climatic hazards has followed, which may help contextualize infrastructure systems (Folke, 2006; Anderies, 2014). A few key studies have extended the SES perspective to include the role of built infrastructure as a means for delivering and managing ecosystem services for society (Folke *et al.*, 2010; Anderies, 2014; Yu *et al.*, 2015; Grimm *et al.*, 2017). However, a limitation of the SES perspective in addressing urban resilience is that it overlooks T-systems as a mediating actor in complex urban systems and underrepresents technology in SES sustainability dialogue (Reddy and Allenby, 2020). For example, SES-based institutional analysis and development framework only consider T-systems to be contextual factors defining biophysical conditions, rather than viewing technological, social, and ecological systems as commensurate in shaping the dynamics of cities (Ostrom, 2009, 2011). SES interactions with T-systems have often been marginalized in the design and management of infrastructure systems. In responding to Hurricane Maria, for example, Lugo (2020) outlines the lack of ecological monitoring and administrative capacities (e.g. emergency sensors, institutional information flows, decision autonomy) that led to insufficient anticipatory efforts, further failures, and repair delays for electrical systems in Puerto Rico (Lugo, 2020). At the same time, SES perspectives usually view T-systems as a subset of social systems (Anderies, 2014; Markolf *et al.*, 2018; Gim, Miller and Hirt, 2019). However, as components of infrastructure systems are entangled among SETS components, T-systems must be addressed alongside SES. The SETS view of urban systems is necessary to uncover the synergies and conflicts across SETS domains in addressing climate challenges through infrastructure systems.

Problem framings of infrastructure systems that are approached with narrow technocentric solutions are becoming increasingly insufficient under non-stationary climate. SETS resilience approaches challenge such framings by prioritizing agility (e.g. adaptive planning) to surprises over robustness and rigidity (Chester, Underwood and Samaras, 2020). Earlier in this chapter we described safe-to-fail (STF) to frame resilient infrastructure development as planning for system failure during design to elucidate new solution pathways that minimize the impacts when infrastructure fails (e.g. traffic service disruptions due to storm drainage overflows). The STF response to 'beyond-design' hazards focuses on comprehensive risk management across SETS. Thus, 'anticipated' structural or functional system failure may occur based on the SETS risk management decisions in order to minimize damages to people, the economy, or ecosystems. In this chapter, we argue that STF planning and design unveils the SETS view of urban systems resilience in responding to climate hazards as it requires decisions for prioritization of SETS capabilities and potential impact transfers from one domain to another upon system failure.

4.4.7. ADDRESSING SETS IRREDUCIBILITY THROUGH SAFE-TO-FAIL INFRASTRUCTURE

STF infrastructure planning and design incorporate resilience strategies with a consideration of how SETS and their subsystems interact with infrastructure (Y. Kim *et al.*, 2017; Kim *et al.*, 2019). We define this STF design process as 'leveraging SETS resilience capabilities', that is, identifying components and functions across SETS that can be substituted to deal with critical service loss or system failure impacts, to proactively plan for infrastructure failures for comprehensive risk management in urban areas. Unlike traditional infrastructure planning that follows a set of technical design specifications for safety management, STF planning and design requires an understanding of regional SETS capabilities (e.g. identification of socio-economic vulnerability to climate risks, institutional readiness to extreme weather events, financial capacity for recovery, adequate infrastructure system, ecosystem responses to hazards, emergency planning, etc.) and trade-offs (i.e. risk management decisions that compromise incompatible SETS resilience capabilities; vulnerability transfers among affected SETS) within the decision context for improved adaptive capacity and more comprehensive urban climate risk management. In comparison to FS approaches, STF urges stakeholders to critically examine trade-offs across SETS due to the unintended transfer of vulnerability from one domain to another or within components of each SETS domain. For instance, a dense city experiencing housing problems may allow developments close to floodplains and vegetated flood mitigation buffers to overflow but equip the area with advanced flood warning systems and flood insurance programs (Ranger *et al.*, 2011; Alias *et al.*, 2020). Thus, the risk of physical damage in the ecological and built environment domain is substituted by additional institutional capacities in the social domain. In Table 4.4.3, we summarize the design principles of traditional infrastructure (i.e. FS) and STF in responding to climate hazards.

Table 4.4.3. Design principles of fail-safe and safe-to-fail infrastructure and examples showing how these design approaches leverage social, ecological, and technological systems (SETS) capabilities for infrastructure resilience.

	Fail-safe	Safe-to-fail
Design principle	• Preservation of status quo • Failure prevention	• Adaptation to changing conditions • Failure impacts management
Design focus	• Advanced risk probability calculations and safety margins • System shutdown for a rare, catastrophic event	• Comprehensive risk impact assessments • Compromised system function for a rare, catastrophic event
Failure response	• Rebuild • Back to normal or decision limited by lock-in	• Recovery • Adapting to new normal
Example	Strengthen/back-up engineered system capabilities (T) to maintain the system function or avoid structural failures such as dam/levee spillways and oversized/backup storm drainage pipelines.	Lowered and reinforced road sections (T) in floodplains that are designed to allow the controlled overflow of stormwater drainage systems during the intense flooding and direct them to wetlands and recharges (E) despite traffic disruptions (S, trade-off).

STF approaches address the irreducibility of SETS through the Infrastructure Trolley Problem. The Infrastructure Trolley Problem, where there is a strategic choice between what and who is impacted by a failure, reveals the inherent moral dilemma of incorporating failure in design and planning. It also underscores the potential consequences of infrastructure failures that may be experienced differently by SETS attributes in a city. In other words, the consequences of STF infrastructure failure will have varying levels of impact and be judged by different values along SETS dimensions in cities. Infrastructure managers implementing a STF approach must identify potential disturbances and associated failure consequences, prioritize diverse values of stakeholders, and navigate the associated trade-offs to implement a design (Blockley, Agarwal and Godfrey, 2012). This navigation encourages infrastructure managers to prioritize impacts and identify trade-offs across SETS. However, infrastructure managers must also adhere to rules and regulations that lower risks, such as emphasizing public safety and reducing environmental impact (Olsen, 2015; Ayyub, 2018). Therefore, infrastructure failure is additionally defined by the consequences on the social and ecological domains – again highlighting the irreducibility of SETS systems. Failure management is not a simple task given the complex urban systems in which infrastructure operates, requiring STF approaches to be iterative with reassessments of prioritizations and trade-offs throughout the infrastructure systems life. Ultimately, for resilience efforts and objectives to be fully realized, SES frameworks should strive to more explicitly recognize and consider the influence and importance of technological systems (i.e. move from SES to SETS perspectives), while T-systems should strive to more explicitly anticipate, consider, and balance the social and ecological impacts that can arise from failure (i.e. move from FS to STF perspectives).

As a process of navigating tensions across SETS resilience capabilities, which remains largely unexplored, the STF approach provides a critical opportunity to incorporate SETS dynamics into the system design and planning. For instance, infrastructure has empowered humans to live in harsh environments (e.g. large-scale movement of water via canals and pipelines in dry areas, implementation of dams and levees in flood-prone areas, and adoption of refrigeration and air conditioning in hot areas), connect distant and remote locations (e.g. transportation of people and goods via ship, rail, road, and air), and create global economies (e.g. identification, extraction, and transformation of natural resources into products). Thus, underappreciation of T-systems can translate to an underappreciation of risks/vulnerabilities within the urban system, as well as mechanisms by which resilience can be enhanced. In addition, given the role of infrastructure as a key intermediary in connecting social and ecological systems, risk and resilience principles (or lack thereof) within T-systems are implicitly integrated into the broader SES dynamics. Therefore, SES approaches to resilience appear to be unwittingly underappreciating sources of catastrophic failure by underappreciating the influence of T-systems across SETS. Conversely, technocentric FS approaches overemphasize T-systems and account for social and ecological domains as external design conditions rather than embedded system characteristics. STF approaches directly challenge the reduction of complex urban systems as narrowly technological or as strictly socio-ecological systems, and necessitate a recognition that SETS domains are interconnected and interdependent (Rinaldi, Peerenboom and Kelly, 2001; Markolf *et al.*, 2018; Chester, Markolf and Allenby, 2019; Helmrich and Chester, 2020). Because of this level of complexity, it is necessary to anticipate that known and unknown hazards will occur, which highlights the irreducibility of SETS resilience considerations in STF infrastructure planning and design.

4.4.8. CHALLENGES AND OPPORTUNITIES OF SAFE-TO-FAIL INFRASTRUCTURE TRANSFORMATION

Several questions for constructing and operating STF, with a SETS lens, still need to be answered to address issues related to resilience governance (Meerow and Newell, 2019), including (but not limited to): i) who is responsible for navigating trade-offs of SETS resilience capabilities?; ii) how to engage with stakeholders for prioritizing decisions in addressing the Infrastructure Trolley Problem?; and iii) how might the role of institutions change to encourage STF approaches? With the necessity for considering failure consequences in STF infrastructure development, practitioners need to decide whom, where, and why people and infrastructure systems experience certain failure outcomes. In addition, these decisions must entail how resources across SETS will be provided and how the community will respond after the failure (e.g. emergency response plan). FS decisions allow decision-makers to transfer the responsibility of failing infrastructure systems to technological capabilities based on design manuals and climate prediction models or to those that own, operate, or use them. On the other hand, STF infrastructure development allocates the responsibility to domain experts and stakeholders across SETS dimensions. While this distribution of power allows for clearer understandings of dynamics between the domains and potential consequences of infrastructure failure, it also diffuses responsibility for that failure. In turn, this diffusion of responsibility can confound recovery efforts if the domain experts and stakeholders remain isolated from one another. Therefore, in order to provide space for effective STF planning and design — and acknowledge the irreducibility of SETS — infrastructure organizations should re-evaluate their organizational structures and relationships with stakeholders to support collaboration.

4.4.9. NAVIGATING SETS TRADE-OFFS

A STF approach asserts that stakeholders — willing to participate across SETS domains and from varying levels of authority — are responsible for the effective operations of infrastructure services. Therefore, stakeholder engagement (i.e. knowledge co-production) is critical for assessing SETS resilience capabilities and trade-offs within STF planning and design. For instance, when considering climate hazard impact profiles, tangible costs of infrastructure failure, like property loss, can be easily assumed in absolute economic terms, but additional impact categories considered in SETS capabilities are not easily captured without the inclusion of broad stakeholder opinion or valuing (Eakin *et al.*, 2007; Bessette *et al.*, 2017; Zuluaga, Karney and Saxe, 2021). Infrastructure failure consequences such as displacement, homelessness, livelihood damage, unemployment, environmental losses, and health impacts may be uniquely experienced depending on the affected stakeholders' capacity to respond and adjust to each disturbance (Romero-Lankao and Norton, 2018). Thus, two challenges emerge: i) ensuring social equity in risk mitigation (Eakin *et al.*, 2016) and ii) providing equitable opportunities for all stakeholders wishing to contribute to decision-making processes (Pescaroli and Alexander, 2016; Reddy and Allenby, 2020). Stakeholders affected by development decisions across SETS domains must be informed and consulted in the decision-making process, which will require active deconstruction of existing power dynamics regarding ownership of infrastructure systems (Grabowski *et al.*, 2017). For example, if stakeholder engagement is not effective at including vulnerable populations who have a lower capacity to respond to health issues or unemployment caused by infrastructure failures, then SETS trade-off decisions may make the same people more vulnerable to planned failures (Magnan *et al.*, 2016). Notably, complete stakeholder engagement is an inherent challenge, especially in cities with large, diverse populations (Chase, Siemer and Decker, 2002; Wyborn *et al.*, 2019; Eriksen *et al.*, 2021).

4.4.10. SETS STAKEHOLDER ENGAGEMENT

Several studies have demonstrated approaches for integrating diverse stakeholder views to help assess risk vulnerability, prioritize decisions with diverse objectives, and elucidate the SETS resilience capabilities for climate risk management, that is, addressing the Infrastructure Trolley Problem. Walpole et al. incorporated practitioners' mental models into ecological restoration decisions (Walpole *et al.*, 2020) and Kim et al. addressed the practitioners' shared/discrete views in implementing resilience strategies for infrastructure development (Kim, Grimm, *et al.*, 2021). Bessette et al. developed a values-informed mental model for understanding communities' climate risk management decisions (Bessette *et al.*, 2017) and York et al. demonstrated an inter-level feedback process for collective climate actions decision-making across individuals and organizations (York *et al.*, 2021). Particularly, Perrone et al. demonstrated the value of stakeholder engagement in evaluating the causes, consequences, and policies for flood management from both environmental and socio-economic perspectives through a participatory modeling approach for the Bradano River, Italy (Perrone *et al.*, 2020). In an effort to engage historically underrepresented communities and address social equity in urban adaptation planning, Amorim-Maia et al. proposed the adoption of place-based and place-making approaches, as well as the promotion of cross-identity climate action and community resilience building (Amorim-Maia *et al.*, 2022). Nonetheless, an exhaustive study for integrating SETS resilience capabilities, revealed through stakeholder engagement, into infrastructure decisions appears warranted for future STF planning.

4.4.11. INSTITUTIONAL CHANGE

Institutions that manage infrastructure systems will need to adapt to accommodate STF infrastructure transformation. Whereas current infrastructure regulations focus on refining design guidelines for system construction and maintenance, STF regulations may also require additional governance capabilities such as community-building (internal and external) and knowledge sharing so organizations may learn from one another (Chester and Allenby, 2019). For example, STF development may require sharing of data on infrastructure performance, decision criteria for prioritizing the SETS capabilities, protocols for emergency system operation, and compensation of failure impacts. One regulatory shift that promotes STF development is for city governments to require insurance companies to provide accumulated information on infrastructure risks and damages experienced in the region. This information may be shared with the city government and the affected stakeholders to assess the current SETS capabilities based on the empirical data. Shifts in one sector (e.g. design firms) will require shifts in other sectors, like governmental organizations, utilities, insurance companies, operation, and regulation (de Aguiar and Freire, 2017).

Transformation to infrastructure solutions that incorporate SETS resilience capabilities with STF design is steadily occurring. Incremental adaptation (organic but gradual system evolution that is tightly coupled to established paths, for example, strengthening infrastructure) and transformation (intentional deviations from the status quo during 'windows of opportunity' often found in the aftermath of extreme disturbances, for example, rapid adoption of an emergent technology) are two mediums for infrastructure transformation (Bolton and Foxon, 2015; Tongur and Engwall, 2017; Iwaniec *et al.*, 2019). Similarly, resilient infrastructure planning methods are being developed to incorporate SETS thinking into future solutions (Lin *et al.*, 2021). While STF infrastructure transformation is happening in the course of incremental adaptation, it is challenging because it requires design practices to be less path-dependent than previously established approaches. The most approachable window of opportunity for the rapid adoption of STF infrastructure would be when existing infrastructure systems reach design capacity and need to be upgraded or replaced, but technological solutions are not always ideal candidate solutions. While projects can focus myopically on efficient optimization for infrastructure planning, commonly featuring path dependency and business-as-usual solutions (Tongur and Engwall, 2017), SETS thinking uses a larger toolset of solution possibilities, which ought to increase the probability of reaching a sustainable solution. For example, as summer temperatures increase in Phoenix, Arizona, cooling and electrical demand loads increase, pushing the power grid closer to critical limits (Burillo *et al.*,

2017). Power failure during critical summer temperatures can have impacts that ultimately lead to human deaths. Technical solutions such as updating aging power lines and adding backup generators, while a necessary component of the solution, cannot be the only solution considering costs and technical thresholds for extreme temperatures (Bartos *et al.*, 2016). The city has been working with vulnerable communities to diversify responses to power system failures that leverage the various components of SETS capabilities. Social programs included educational programs to show children how to operate safely in the heat during summer break and providing funding for residents to improve insulation in homes. Ecological solutions included strategic green-space development to decrease ambient air temperatures and increase shade. Technological solutions included installing strategically placed drinking fountains and constructing splash pads. This example shows how SETS resilience capabilities are leveraged to provide safe-to-fail infrastructure responses to deal with extreme heat for the identified communities (Guardaro *et al.*, 2020).

4.4.12. CONCLUSIONS

Leveraging the resilience capabilities across SETS domains in STF approaches appears to support graceful extensibility in resilience engineering. Contemporary framings of infrastructure resilience describe strategies when systems are perturbed within and beyond their design conditions (Woods, 2015). Within their design conditions, rebound (bouncing back) and robustness (hardening) are appropriate. However, when perturbations exceed design conditions then extensibility becomes appropriate — extending adaptive capacity in the face of surprise. Extending the capacity of such large and extensive infrastructure systems is a monumental challenge. If the extension is viewed purely through a technological lens, then few options exist — for example, how do you provide water through an alternative technology to millions of city residents when the primary drinking water system has failed? Or how do you decide on the size of drainage pipes when the intensity of a 100-year storm keeps changing? STF leveraging SETS resilience capabilities offers pathways towards graceful extensibility by leveraging social and ecosystem capabilities in anticipating and planning for failure. For example, The Netherlands' Room for the River calls on social systems when rivers flood and infrastructure fail, to subsidize farmers for lost crops — far cheaper than elevating and maintaining levees (Roth and Warner, 2007). Arizona's Indian Bend Wash has initially leveraged ecosystem capabilities to attenuate flooding when monsoon rains overwhelm the stormwater system (The City of Scottsdale Communications and Public Affairs, 2004). And more recently, the City of Scottsdale is working on an updated master plan for infrastructure through multiple rounds of community feedback, which not only responds to the shifting hydrologic risks by updating aging infrastructure for flood management, but also asks the question of how social and ecological values of Indian Bend Wash as recreational parks and aquatic centers might affect the community when they are compromised by overflow (City of Scottsdale, 2020). In contrast to how other resilience frameworks incorporate capabilities of the three domains, STF appears to be better suited for leveraging SETS capabilities during the design phase to open up new adaptation strategies and infrastructure transformations — aligning it with traits of graceful extensibility upon surprises.

STF infrastructure planning and design offer transformational opportunities for infrastructure systems to evolve from a techno-centric or SES-centric solution space to an interactive system leveraging various SETS resilience capabilities — presenting new strategies for navigating uncertainties and disasters in the Anthropocene. External shocks such as extreme weather phenomena are not only disrupting the infrastructure system itself, but also the urban environment including people and property. Despite traditional infrastructure protection achieved by ensuring the robustness of built systems, climate change is altering the perspectives of cities to recognize infrastructure risks that are not predicted with climate models. Thus, there is a coupling between STF infrastructure planning, SETS climate adaptations, desired urban futures, and the likelihood of unprecedented non-stationary weather events. Major institutional and technological changes happening with national and international climate adaptation plans should give cities a chance to adapt to the change by transforming the processes that have contributed to vulnerability rather than focusing on reducing specific risks of climate change by a set of interventions (O'Brien, 2018). Hence, infrastructure transformations towards resilience in the face of climate uncertainties and non-stationarity must take what we know now and proceed to STF approaches that incorporate SETS capabilities.

4.4.13. REFERENCES

de Aguiar, T.R.S. and Freire, F. de S. (2017) 'Shifts in modes of governance and sustainable development in the Brazilian oil sector', *European Management Journal*, 35(5), pp. 701–710. Available at: https://doi.org/10.1016/j.emj.2017.05.001.

Ahern, J. (2011) 'From fail-safe to safe-to-fail: Sustainability and resilience in the new urban world', *Landscape and Urban Planning*, 100(4), pp. 341–343. Available at: https://doi.org/10.1016/j.landurbplan.2011.02.021.

Ahern, J., Cilliers, S. and Niemelä, J. (2014) 'The concept of ecosystem services in adaptive urban planning and design: A framework for supporting innovation', *Landscape and Urban Planning*, 125, pp. 254–259. Available at: https://doi.org/10.1016/j.landurbplan.2014.01.020.

Aldunce, P. *et al.* (2015) 'Resilience for disaster risk management in a changing climate: Practitioners' frames and practices', *Global Environmental Change*, 30, pp. 1–11. Available at: https://doi.org/10.1016/j.gloenvcha.2014.10.010.

Alexander, D.E. (2013) 'Resilience and disaster risk reduction: an etymological journey', *Natural Hazards and Earth System Sciences*, 13(11), pp. 2707–2716. Available at: https://doi.org/10.5194/nhess-13-2707-2013.

Alias, N.E. *et al.* (2020) 'Community responses on effective flood dissemination warnings — A case study of the December 2014 Kelantan Flood, Malaysia', 13, pp. 1–13. Available at: https://doi.org/10.1111/jfr3.12552.

Amorim-Maia, A.T. *et al.* (2022) 'Intersectional climate justice: A conceptual pathway for bridging adaptation planning, transformative action, and social equity', *Urban Climate*, 41, p. 101053. Available at: https://doi.org/10.1016/j.uclim.2021.101053.

Anderies, J.M. (2014) 'Embedding built environments in social–ecological systems: resilience-based design principles', *Building Research & Information*, 42(2), pp. 130–142. Available at: https://doi.org/10.1080/09613218.2013.857455.

Ayyub, B.M. (2018) *Climate-Resilient Infrastructure*. Edited by Committee on Adaptation to a Changing Climate. Reston, VA: ASCE Press. Available at: https://doi.org/10.1061/9780784415191.

Bartos, M. *et al.* (2016) 'Impacts of rising air temperatures on electric transmission ampacity and peak electricity load in the United States', *Environmental Research Letters*, 11(11), p. 114008. Available at: https://doi.org/10.1088/1748-9326/11/11/114008.

Bennis, W.M., Medin, D.L. and Bartels, D.M. (2010) 'The Costs and Benefits of Calculation and Moral Rules', *Perspectives on Psychological Science*, 5(2), pp. 187–202. Available at: https://doi.org/10.1177/1745691610362354.

Bessette, D.L. *et al.* (2017) 'Building a Values-Informed Mental Model for New Orleans Climate Risk Management', *Risk Analysis*, 37(10), pp. 1993–2004. Available at: https://doi.org/10.1111/risa.12743.

Blockley, D., Agarwal, J. and Godfrey, P. (2012) 'Infrastructure resilience for high-impact low-chance risks', *Proceedings of the Institution of*

Civil Engineers - Civil Engineering, 165(6), pp. 13–19. Available at: https://doi.org/10.1680/cien.11.00046.

Bludau, C., Zirkelbach, D. and Kuenzel, H. (2008) 'Condensation Problems in Cool Roofs', in. *11th International Conference on Durability of Building Materials and Components*, Istanbul, Turkey. Available at: https://www.researchgate.net/publication/355544827_Condensa-tion_Problems_in_Cool_Roofs.

Boesen, N. (2007) 'Governance and Accountability: How Do the Forma and Informal Interplay and Change?', in J. Jutting et al. (eds) *Informal Institutions How Social Norms Help or Hinder Development: How Social Norms Help or Hinder Development*. Danvers, MA: OECD Publishing.

Boin, A. and McConnell, A. (2007) 'Preparing for Critical Infrastructure Breakdowns: The Limits of Crisis Management and the Need for Resilience', *Journal of Contingencies and Crisis Management*, 15(1), pp. 50–59. Available at: https://doi.org/10.1111/j.1468-5973.2007.00504.x.

Bolton, R. and Foxon, T.J. (2015) 'Infrastructure transformation as a socio-technical process — Implications for the governance of energy distribution networks in the UK', *Technological Forecasting and Social Change*, 90, pp. 538–550. Available at: https://doi.org/10.1016/j.techfore.2014.02.017.

Burillo, D. *et al.* (2017) 'Electricity demand planning forecasts should consider climate non-stationarity to maintain reserve margins during heat waves', *Applied Energy*, 206(June), pp. 267–277. Available at: https://doi.org/10.1016/j.apenergy.2017.08.141.

Butler, D. *et al.* (2014) 'A New Approach to Urban Water Management: Safe and Sure', *Procedia Engineering*, 89, pp. 347–354. Available at: https://doi.org/10.1016/j.proeng.2014.11.198.

Carse, A. (2016) 'Keyword: infrastructure: How a humble French engineering term shaped the modern world', in A. Morita, C.B. Jensen, and P. Harvey (eds) *Infrastructures and Social Complexity*. Oxford, UK: Routledge, pp. 45–57. Available at: https://doi.org/10.4324/9781315622880-11 (Accessed: 15 October 2023).

Chang, S.E. *et al.* (2014) 'Toward Disaster-Resilient Cities: Characterizing Resilience of Infrastructure Systems with Expert Judgments', *Risk Analysis*, 34(3), pp. 416–434. Available at: https://doi.org/10.1111/risa.12133.

Chase, L., Siemer, W. and Decker, D. (2002) 'Designing stakeholder involvement strategies to resolve wildlife management controversies', *Wildlife Society Bulletin*, 30(3), pp. 937–950.

Chester, M., Markolf, S. and Allenby, B. (2019) 'Infrastructure and the environment in the Anthropocene', *Journal of Industrial Ecology*, 23(5), pp. 1006–1015. Available at: https://doi.org/10.1111/jiec.12848.

Chester, M.V. *et al.* (2014) 'Positioning infrastructure and technologies for low-carbon urbanization', *Earth's Future*, 2(10), pp. 533–547. Available at: https://doi.org/10.1002/2014EF000253.

Chester, M.V. and Allenby, B. (2018) 'Toward adaptive infrastructure: flexibility and agility in a non-stationarity age', *Sustainable and Resilient Infrastructure*, 4, pp. 1–19. Available at: https://doi.org/10.1080/23789689.2017.1416846.

Chester, M.V. and Allenby, B. (2019) 'Infrastructure as a wicked complex process', *Elementa: Science of the Anthropocene*. Edited by A. Iles and M.E. Chang, 7(1), p. 21. Available at: https://doi.org/10.1525/elementa.360.

Chester, M.V., Underwood, B.S. and Samaras, C. (2020) 'Keeping infrastructure reliable under climate uncertainty', *Nature Climate Change* 10(6), pp. 488–490. Available at: https://doi.org/10.1038/s41558-020-0741-0.

Chmutina, K. *et al.* (2014) 'Towards Integrated Security and Resilience Framework: A Tool for Decision-makers', *Procedia Economics and Finance*, 18, pp. 25–32. Available at: https://doi.org/10.1016/S2212-5671(14)00909-5.

Chopra, S.S. *et al.* (2016) 'A network-based framework for assessing infrastructure resilience: a case study of the London metro system', *Journal of The Royal Society Interface*, 13(118), p. 20160113. Available at: https://doi.org/10.1098/rsif.2016.0113.

Christensen, J.H. *et al.* (2007) 'Regional Climate Projections', in *Climate Change 2007: The Physical Science Basis, Contribution of Working Group I to the Fourth Assessment Re-port of the Intergovernmental Panel on Climate Change*. Cambridge, UK and New York, NY: Cambridge University Press, pp. 847–940. Available at: https://www.ipcc.ch/report/ar4/wg1/.

City of Scottsdale (2020) *Indian Bend Wash Preliminary Master Plan*. Scottsdale, AZ: The City of Scottsdale. Available at: https://www.scottsdaleaz.gov/Assets/ScottsdaleAZ/Construction/Indian+Bend+Wash+Master+Plan/Master-Plan-Download.pdf.

Clark, S.S. *et al.* (2018) 'The vulnerability of interdependent urban infrastructure systems to climate change: could Phoenix experience a Katrina of extreme heat?', *Sustainable and Resilient Infrastructure*, 4(1), pp. 21–35. Available at: https://doi.org/10.1080/23789689.2018.1448668.

Cutter, S.L. (2016) 'Resilience to What? Resilience for Whom?', *The Geographical Journal*, 182(2), pp. 110–113. Available at: https://doi.org/10.1111/geoj.12174.

Dawson, R.J. *et al.* (2018) 'A systems framework for national assessment of climate risks to infrastructure', *Philosophical Transactions of the Royal Society A: Mathematical, Physical and Engineering Sciences*, 376(2121), p. 20170298. Available at: https://doi.org/10.1098/rsta.2017.0298.

DHS (2013) *NIPP 2013: Partnering for Critical Infrastructure Security and Resilience*.

Donovan, B. and Work, D.B. (2017) 'Empirically quantifying city-scale transportation system resilience to extreme events', *Transportation Research Part C: Emerging Technologies*, 79, pp. 333–346. Available at: https://doi.org/10.1016/j.trc.2017.03.002.

Doorn, N., Gardoni, P. and Murphy, C. (2019) 'A multidisciplinary definition and evaluation of resilience: the role of social justice in defining resilience', *Sustainable and Resilient Infrastructure*, 4(3), pp. 112–123. Available at: https://doi.org/10.1080/23789689.2018.1428162.

Eakin, H. *et al.* (2007) 'A stakeholder driven process to reduce vulnerability to climate change in Hermosillo, Sonora, Mexico', *Mitigation and Adaptation Strategies for Global Change*, 12(5), pp. 935–955. Available at: https://doi.org/10.1007/s11027-007-9107-4.

Eakin, H. *et al.* (2016) 'Adapting to risk and perpetuating poverty: Household's strategies for managing flood risk and water scarcity in Mexico City', *Environmental Science & Policy*, 66, pp. 324–333. Available at: https://doi.org/10.1016/j.envsci.2016.06.006.

Eriksen, S. *et al.* (2021) 'Adaptation interventions and their effect on vulnerability in developing countries: Help, hindrance or irrelevance?', *World Development*, 141, p. 105383. Available at: https://doi.org/10.1016/j.worlddev.2020.105383.

Folke, C. (2006) 'Resilience: The emergence of a perspective for social-ecological systems analyses', *Global Environmental Change*, 16(3), pp. 253–267. Available at: https://doi.org/10.1016/j.gloenvcha.2006.04.002.

Folke, C. *et al.* (2010) 'Resilience Thinking: Integrating Resilience, Adaptability and Transformability', *Ecology and Society*, 15(4). Available at: https://www.jstor.org/stable/26268226 (Accessed: 23 October 2023).

Francis, R. and Bekera, B. (2014) 'A metric and frameworks for resilience analysis of engineered and infrastructure systems', *Reliability Engineering & System Safety*, 121, pp. 90–103. Available at: https://doi.org/10.1016/j.ress.2013.07.004.

Gardoni, P. and Murphy, C. (2020) 'Society-based design: promoting societal well-being by designing sustainable and resilient infrastructure', *Sustainable and Resilient Infrastructure*, 5(1–2), pp. 4–19. Available at: https://doi.org/10.1080/23789689.2018.1448667.

Gilroy, K.L. and McCuen, R.H. (2012) 'A nonstationary flood frequency analysis method to adjust for future climate change and urbanization', *Journal of Hydrology*, 414–415, pp. 40–48. Available at: https://doi.org/10.1016/j.jhydrol.2011.10.009.

Gim, C., Miller, C.A. and Hirt, P.W. (2019) 'The resilience of institutions', *Environmental Science & Policy*, 97, pp. 36–43. Available at: https://doi.org/10.1016/j.envsci.2019.03.004.

González, O. *et al.* (1997) 'Adaptive Fault Tolerance and Graceful Degradation Under Dynamic Hard Real-time Scheduling', *Computer Science Department Faculty Publication Series* [Preprint]. Available at: https://scholarworks.umass.edu/cs_faculty_pubs/188.

Grabowski, Z.J. *et al.* (2017) 'Infrastructures as Socio-Eco-Technical Systems: Five Considerations for Interdisciplinary Dialogue', *Journal of Infrastructure Systems*, 23(4), p. 02517002. Available at: https://doi.org/10.1061/(ASCE)IS.1943-555X.0000383.

Greenwald, D. (1965) 'Infrastructure (Social Overhead Capital)', in *McGraw-Hill Dictionary of Modern Economics*. New York, NY: McGraw-Hill.

Grimm, N.B. *et al.* (2017) 'Does the ecological concept of disturbance have utility in urban social–ecological–technological systems?', *Ecosystem Health and Sustainability*, 3(1), p. e01255. Available at: https://doi.org/10.1002/ehs2.1255.

Guardaro, M. *et al.* (2020) 'Building community heat action plans story by story: A three neighborhood case study', *Cities*, 107, p. 102886. Available at: https://doi.org/10.1016/j.cities.2020.102886.

Guerreiro, S.B. *et al.* (2018) 'Future heat-waves, droughts and floods in 571 European cities', *Environmental Research Letters*, 13(3), p. 034009. Available at: https://doi.org/10.1088/1748-9326/aaaad3.

Helmrich, A.M. and Chester, M.V. (2020) 'Reconciling complexity and deep uncertainty in infrastructure design for climate adaptation', *Sustainable and Resilient Infrastructure*, 7(2), pp. 83–99. Available at: https://doi.org/10.1080/23789689.2019.1708179.

Holling, C.S. (1973) 'Resilienceand Stability of Ecological Systems', *Annual Review of Ecology, Evolution, and Systematics*, 4(1), pp. 1–23. Available at: https://doi.org/10.1146/annurev.es.04.110173.000245.

Hollnagel, E., Woods, D.D. and Leveson, N. (2006) 'Prologue', in *Resilience Engineering: Concepts and Precepts*. Aldershot, UK: Ashgate.

Hosseini, S., Barker, K. and Ramirez-Marquez, J.E. (2016) 'A review of definitions and measures of system resilience', *Reliability Engineering & System Safety*, 145, pp. 47–61. Available at: https://doi.org/10.1016/j.ress.2015.08.006.

Huang, H. *et al.* (eds) (2018) *Resilience Engineering for Urban Tunnels*. Reston, VA: American Society of Civil Engineers.

Hulme, M. (2016) '1.5 °C and climate research after the Paris Agreement', *Nature Climate Change*, 6(3), pp. 222–224. Available at: https://doi.org/10.1038/nclimate2939.

IPCC (2014) *Climate Change 2014: Synthesis Report. Contribution of Working Groups I, II and III to the Fifth Assessment Report of the Intergovernmental Panel on Climate Change*. 9789291691432. Geneva, Switzerland: UN International Panel on Climate Change, p. 151.

IRDR (2017) *Disaster loss data: Raising the standard*, p. 10. Available at: https://www.preventionweb.net/publication/disaster-loss-data-raising-standard (Accessed: 24 October 2023).

Iwaniec, D.M. *et al.* (2019) 'The Framing of Urban Sustainability Transformations', *Sustainability*, 11(3), p. 573. Available at: https://doi.org/10.3390/su11030573.

Katz, R.W. (2010) 'Statistics of extremes in climate change', *Climatic Change*, 100(1), pp. 71–76. Available at: https://doi.org/10.1007/s10584-010-9834-5.

Kennedy, J.R. and Paretti, N.V. (2014) *Evaluation of the magnitude and frequency of floods in urban watersheds in Phoenix and Tucson, Arizona*. 2014–5121. U.S. Geological Survey. Available at: https://doi.org/10.3133/sir20145121.

Kim, H. *et al.* (2017) 'Appropriate model selection methods for nonstationary generalized extreme value models', *Journal of Hydrology*, 547, pp. 557–574. Available at: https://doi.org/10.1016/j.jhydrol.2017.02.005.

Kim, Y. *et al.* (2017) 'Fail-safe and safe-to-fail adaptation: decision-making for urban flooding under climate change', *Climatic Change*, 145(3), pp. 397–412. Available at: https://doi.org/10.1007/s10584-017-2090-1.

Kim, Y. *et al.* (2019) 'The Infrastructure Trolley Problem: Positioning Safe-to-fail Infrastructure for Climate Change Adaptation', *Earth's Future*, 7(7), pp. 704–717. Available at: https://doi.org/10.1029/2019EF001208.

Kim, Y., Grimm, N.B., *et al.* (2021) 'Capturing practitioner perspectives on infrastructure resilience using Q-methodology', *Environmental Research: Infrastructure and Sustainability*, 1(2), p. 025002. Available at: https://doi.org/10.1088/2634-4505/ac0f98.

Kim, Y., Mannetti, L.M., *et al.* (2021) 'Social, Ecological, and Technological Strategies for Climate Adaptation', in Z.A. Hamstead et al. (eds) *Resilient Urban Futures*. Berlin, DE: Springer International Publishing (The Urban Book Series), pp. 29–45. Available at: https://doi.org/10.1007/978-3-030-63131-4_3.

Kroes, P. *et al.* (2006) 'Treating socio-technical systems as engineering systems: some conceptual problems', *Systems Research and Behavioral Science*, 23(6), pp. 803–814. Available at: https://doi.org/10.1002/sres.703.

Kull, D., Mechler, R. and Hochrainer-Stigler, S. (2013) 'Probabilistic cost-benefit analysis of disaster risk management in a development context', *Disasters*, 37(3), pp. 374–400. Available at: https://doi.org/10.1111/disa.12002.

Leavitt, W.M. and Kiefer, J.J. (2006) 'Infrastructure Interdependency and the Creation of a Normal Disaster: The Case of Hurricane Katrina and the City of New Orleans', *Public Works Management & Policy*, 10(4), pp. 306–314. Available at: https://doi.org/10.1177/1087724X06289055.

Lempert, R.J. (2016) 'Infrastructure design must change with climate', *Orange County Register*, 12 August. Available at: https://www.ocregister.com/2016/08/12/infrastructure-design-must-change-with-climate/ (Accessed: 24 October 2023).

Lin, B.B. *et al.* (2021) 'Integrating solutions to adapt cities for climate change', *The Lancet Planetary Health*, 5(7), pp. e479–e486. Available at: https://doi.org/10.1016/S2542-5196(21)00135-2.

Linkov, I. *et al.* (2014) 'Changing the resilience paradigm', *Nature Climate Change*, 4(6), pp. 407–409. Available at: https://doi.org/10.1038/nclimate2227.

Lins, H.F. (2012) 'A Note on Stationarity and Nonstationarity', in *Commission for Hydrology, Advisory Working Group. Fourteenth Session of the Commission for Hydrology*, World Meteorological Organization.

Lister, N.-M., Hargreaves, G. and Czerniak, J. (2007) 'Sustainable Large Parks: Ecological design or designer ecology?', in *Large Parks*. Princeton, NJ: Princeton Architectural Press, pp. 31–51.

Lugo, A.E. (2020) 'Effects of Extreme Disturbance Events: From Ecesis to Social–Ecological–Technological Systems', *Ecosystems*, 23(8), pp. 1726–1747. Available at: https://doi.org/10.1007/s10021-020-00491-x.

Magnan, A.K. *et al.* (2016) 'Addressing the risk of maladaptation to climate change', *WIREs Climate Change*, 7(5), pp. 646–665. Available at: https://doi.org/10.1002/wcc.409.

Mahajan, S. *et al.* (2015) 'Fidelity of Precipitation Extremes in High Resolution Global Climate Simulations', *Procedia Computer Science*, 51(1), pp. 2178–2187. Available at: https://doi.org/10.1016/j.procs.2015.05.492.

Markolf, S.A. *et al.* (2018) 'Interdependent Infrastructure as Linked Social, Ecological, and Technological Systems (SETSs) to Address Lock-in and Enhance Resilience', *Earth's Future*, 6(12), pp. 1638–1659. Available at: https://doi.org/10.1029/2018EF000926.

Martin-Breen, P. and Anderies, J.M. (2011) 'Resilience: A Literature Review', *The Bellagio Initiative, The Future of Philanthropy and Development in the Pursuit of Human Wellbeing* [Preprint]. Available at: https://opendocs.ids.ac.uk/opendocs/handle/20.500.12413/3692.

McDaniels, T. *et al.* (2008) 'Fostering resilience to extreme events within infrastructure systems: Characterizing decision contexts for mitigation and adaptation', *Global Environmental Change*, 18(2), pp. 310–318. Available at: https://doi.org/10.1016/j.gloenvcha.2008.03.001.

McPhearson, T. *et al.* (2016) 'Advancing understanding of the complex nature of urban systems', *Ecological Indicators*, 70, pp. 566–573. Available at: https://doi.org/10.1016/j.ecolind.2016.03.054.

Meerow, S. and Newell, J.P. (2015) 'Resilience and Complexity: A Bibliometric Review and Prospects for Industrial Ecology', *Journal of Industrial Ecology*, 19(2), pp. 236–251. Available at: https://doi.org/10.1111/jiec.12252.

Meerow, S. and Newell, J.P. (2019) 'Urban resilience for whom, what, when, where, and why?', *Urban Geography*, 40(3), pp. 309–329. Available at: https://doi.org/10.1080/02723638.2016.1206395.

Meerow, S. and Stults, M. (2016) 'Comparing Conceptualizations of Urban Climate Resilience in Theory and Practice', *Sustainability*, 8(7), p. 701. Available at: https://doi.org/10.3390/su8070701.

Miller, T., Chester, M. and Munoz-Erickson, T. (2018) 'Rethinking Infrastructure in an Era of Unprecedented Weather Events', *Issues in Science and Technology*, 34(2), pp. 46–58.

Milly, P.C.D. *et al.* (2008) 'Stationarity Is Dead: Whither Water Management?', *Science*, 319(5863), pp. 573–574. Available at: https://doi.org/10.1126/science.1151915.

Möller, N. and Hansson, S.O. (2008) 'Principles of engineering safety: Risk and uncertainty reduction', *Reliability Engineering & System Safety*,

93(6), pp. 798–805. Available at:
https://doi.org/10.1016/j.ress.2007.03.031.

Morrison, M. and Fay, M. (2007) *Infrastructure in Latin America : Recent Developments and Key Challenges, Volume 1.* Washington, DC: World Bank. Available at: http://hdl.handle.net/10986/7179 (Accessed: 24 October 2023).

Mugume, S.N. *et al.* (2015) 'A global analysis approach for investigating structural resilience in urban drainage systems', *Water Research*, 81, pp. 15–26. Available at: https://doi.org/10.1016/j.watres.2015.05.030.

Muller, M. *et al.* (2015) 'Built infrastructure is essential', *Science*, 349(6248), pp. 585–586. Available at: https://doi.org/10.1126/science.aac7606.

Novotny, V., Ahern, J. and Brown, P. (2010) 'Planning, Retrofitting, and Building the Next Urban Environment', in. *Water Centric Sustainable Communities*, Wiley. Available at:
https://doi.org/10.1002/9780470949962.

NRC (2010) *Adapting to the Impacts of Climate Change.* Washington, DC: National Academies Press. Available at:
https://doi.org/10.17226/12783.

NRC (2012) *Disaster Resilience: A National Imperative.* Washington, DC: National Academies Press. Available at:
https://doi.org/10.17226/13457.

O'Brien, K. (2018) 'Is the 1.5°C target possible? Exploring the three spheres of transformation', *Current Opinion in Environmental Sustainability*, 31, pp. 153–160. Available at:
https://doi.org/10.1016/j.cosust.2018.04.010.

O'Brien, K., Hayward, B. and Berkes, F. (2009) 'Rethinking Social Contracts: Building Resilience in a Changing Climate', *Ecology and Society*, 14(2). Available at: https://doi.org/10.5751/ES-03027-140212.

OECD (2015) *Climate Change Risks and Adaptation: Linking Policy and Economics.* Paris, FR: Organisation for Economic Co-operation and Development. Available at: https://doi.org/10.1787/9789264234611-en (Accessed: 24 October 2023).

OECD (2016) *Joint Expert Meeting on Disaster Loss Data Improving the Evidence Base on the Costs of Disasters Key Findings from an OECD Survey.* Paris, FR: Organization for Economic Co-operation and Development, p. 39. Available at: https://www.oecd.org/gov/risk/Issues-Paper-Improving-Evidence-base-on-the-Costs-of-Disasters.pdf.

Olsen, J.R. (2015) *Adapting Infrastructure and Civil Engineering Practice to a Changing Climate.* (Books). Available at: https://doi.org/10.1061/9780784479193.fm (Accessed: 23 October 2023).

Ostrom, E. (2009) 'A General Framework for Analyzing Sustainability of Social-Ecological Systems', *Science*, 325(5939), pp. 419–422. Available at: https://doi.org/10.1126/science.1172133.

Ostrom, E. (2011) 'Background on the Institutional Analysis and Development Framework', *Policy Studies Journal*, 39(1), pp. 7–27. Available at: https://doi.org/10.1111/j.1541-0072.2010.00394.x.

Oxenford, J.L. and Williams, S.I. (2014) 'Understanding the causes for water system failure', *Journal AWWA*, 106(1), pp. E41–E54. Available at: https://doi.org/10.5942/jawwa.2014.106.0006.

Park, J. *et al.* (2013) 'Integrating Risk and Resilience Approaches to Catastrophe Management in Engineering Systems', *Risk Analysis*, 33(3), pp. 356–367. Available at: https://doi.org/10.1111/j.1539-6924.2012.01885.x.

Park, J., Seager, T.P. and Rao, P.S.C. (2011) 'Lessons in risk- versus resilience-based design and management', *Integrated Environmental Assessment and Management*, 7(3), pp. 396–399. Available at: https://doi.org/10.1002/ieam.228.

Perrone, A. *et al.* (2020) 'A participatory system dynamics modeling approach to facilitate collaborative flood risk management: A case study in the Bradano River (Italy)', *Journal of Hydrology*, 580, p. 124354. Available at: https://doi.org/10.1016/j.jhydrol.2019.124354.

Pescaroli, G. and Alexander, D. (2016) 'Critical infrastructure, panarchies and the vulnerability paths of cascading disasters', *Natural Hazards*, 82(1), pp. 175–192. Available at: https://doi.org/10.1007/s11069-016-2186-3.

Pregnolato, M. *et al.* (2017) 'Impact of Climate Change on Disruption to Urban Transport Networks from Pluvial Flooding', *Journal of Infrastructure Systems*, 23(4), p. 04017015. Available at: https://doi.org/10.1061/(ASCE)IS.1943-555X.0000372.

Prein, A.F. *et al.* (2017) 'Increased rainfall volume from future convective storms in the US', *Nature Climate Change*, 7(12), pp. 880–884. Available at: https://doi.org/10.1038/s41558-017-0007-7.

Ranger, N. *et al.* (2011) 'An assessment of the potential impact of climate change on flood risk in Mumbai', *Climatic Change*, 104(1), pp. 139–167. Available at: https://doi.org/10.1007/s10584-010-9979-2.

Rasmussen, J. (1997) 'Risk management in a dynamic society: a modelling problem', *Safety Science*, 27(2), pp. 183–213. Available at: https://doi.org/10.1016/S0925-7535(97)00052-0.

Read, L.K. and Vogel, R.M. (2015) 'Reliability, return periods, and risk under nonstationarity', *Water Resources Research*, 51(8), pp. 6381–6398. Available at: https://doi.org/10.1002/2015WR017089.

Reddy, A. and Allenby, B. (2020) 'Overlooked Role of Technology in the Sustainability Movement: A Pedagogical Framework for Engineering Education and Research', *ASME Journal of Engineering for Sustainable Buildings and Cities*, 1(021003). Available at: https://doi.org/10.1115/1.4046852.

Revi, A. *et al.* (2014) *Urban areas. In: Climate Change 2014: Impacts, Adaptation, and Vulnerability. Part A: Global and Sectoral Aspects. Contribution of Working Group II to the Fifth Assessment Report of the Intergovernmental Panel on Climate Change.* Cambridge, UK and New York, NY: Cambridge University Press.

Rinaldi, S.M., Peerenboom, J.P. and Kelly, T.K. (2001) 'Identifying, understanding, and analyzing critical infrastructure interdependencies', *IEEE Control Systems Magazine*, 21(6), pp. 11–25. Available at: https://doi.org/10.1109/37.969131.

Romero-Lankao, P. and Norton, R. (2018) 'Interdependencies and Risk to People and Critical Food, Energy, and Water Systems: 2013 Flood, Boulder, Colorado, USA', *Earth's Future*, 6(11), pp. 1616–1629. Available at: https://doi.org/10.1029/2018EF000984.

Rose, A. (2017) 'Defining Resilience Across Disciplines', in *Defining and Measuring Economic Resilience from a Societal, Environmental and Security Perspective.* Singapore: Springer, pp. 19–27. Available at: https://doi.org/10.1007/978-981-10-1533-5_3.

Roth, D. and Warner, J. (2007) 'Flood Risk, Uncertainty and Changing River Protection Policy in the Netherlands: The Case of "Calamity Polders"', *Tijdschrift voor Economische en Sociale Geografie*, 98(4), pp. 519–525. Available at: https://doi.org/10.1111/j.1467-9663.2007.00419.x.

Serinaldi, F. and Kilsby, C.G. (2015) 'Stationarity is undead: Uncertainty dominates the distribution of extremes', *Advances in Water Resources*, 77, pp. 17–36. Available at: https://doi.org/10.1016/j.advwatres.2014.12.013.

Seto, K.C., Solecki, W.D. and Griffith, C.A. (2015) *The Routledge Handbook of Urbanization and Global Environmental Change.* London, UK: Routledge.

Shortridge, J., Guikema, S. and Zaitchik, B. (2017) 'Robust decision making in data scarce contexts: addressing data and model limitations for infrastructure planning under transient climate change', *Climatic Change*, 140(2), pp. 323–337. Available at: https://doi.org/10.1007/s10584-016-1845-4.

Slota, S.C. and Bowker, G.C. (2017) 'Chapter 18: How Infrastructures Matter', in U. Felt et al. (eds) *The Handbook of Science and Technology Studies, Fourth Edition*. Cambridge, MA: MIT Press.

The City of Scottsdale Communications and Public Affairs (2004) *Indian Bend Wash*. Scottsdale, AZ: The City of Scottsdale Communications and Public Affairs.

The Trust for Public Land (2016) *The benefits of green infrastructure for heat mitigation and emissions reductions in cities*. Atlanta, GA: Urban Climate Lab at the Georgia Institute of Technology, p. 36. Available at: https://www.tpl.org/resource/benefits-green-infrastructure-heat-mitigation-and-emissions-reductions-cities (Accessed: 24 October 2023).

Thomson, J.J. (1985) 'The Trolley Problem', *The Yale Law Journal*, 94(6), pp. 1395–1415. Available at: https://doi.org/10.2307/796133.

Tongur, S. and Engwall, M. (2017) 'Exploring window of opportunity dynamics in infrastructure transformation', *Environmental Innovation and Societal Transitions*, 25, pp. 82–93. Available at: https://doi.org/10.1016/j.eist.2016.12.003.

Tramblay, Y. *et al.* (2013) 'Non-stationary frequency analysis of heavy rainfall events in southern France', *Hydrological Sciences Journal*, 58(2), pp. 280–294. Available at: https://doi.org/10.1080/02626667.2012.754988.

Tye, M.R., Holland, G.J. and Done, J.M. (2015) 'Rethinking failure: time for closer engineer–scientist collaborations on design', *Proceedings of the Institution of Civil Engineers - Forensic Engineering*, 168(2), pp. 49–57. Available at: https://doi.org/10.1680/feng.14.00004.

Vale, L.J. (2014) 'The politics of resilient cities: whose resilience and whose city?', *Building Research & Information*, 42(2), pp. 191–201. Available at: https://doi.org/10.1080/09613218.2014.850602.

Walker, B. *et al.* (2006) 'A Handful of Heuristics and Some Propositions for Understanding Resilience in Social-Ecological Systems', *Ecology and Society*, 11(1). Available at: https://doi.org/10.5751/ES-01530-110113.

Walpole, E.H. *et al.* (2020) 'The science and practice of ecological restoration: a mental models analysis of restoration practitioners', *Environment Systems and Decisions*, 40(4), pp. 588–604. Available at: https://doi.org/10.1007/s10669-020-09768-x.

Warren, R.F. *et al.* (2018) 'Advancing national climate change risk assessment to deliver national adaptation plans', *Philosophical Transactions of the Royal Society A: Mathematical, Physical and Engineering Sciences*, 376(2121), p. 20170295. Available at: https://doi.org/10.1098/rsta.2017.0295.

Weaver, C.P. *et al.* (2013) 'Improving the contribution of climate model information to decision making: the value and demands of robust decision frameworks', *WIREs Climate Change*, 4(1), pp. 39–60. Available at: https://doi.org/10.1002/wcc.202.

Wilbanks, T.J. and Fernandez, S. (eds) (2014) *Climate Change and Infrastructure, Urban Systems, and Vulnerabilities*. Washington, DC: Island Press/Center for Resource Economics. Available at: https://doi.org/10.5822/978-1-61091-556-4.

Woods, D.D. (2015) 'Four concepts for resilience and the implications for the future of resilience engineering', *Reliability Engineering & System Safety*, 141, pp. 5–9. Available at: https://doi.org/10.1016/j.ress.2015.03.018.

Wyborn, C. *et al.* (2019) 'Co-Producing Sustainability: Reordering the Governance of Science, Policy, and Practice', *Annual Review of Environment and Resources*, 44(1), pp. 319–346. Available at: https://doi.org/10.1146/annurev-environ-101718-033103.

Yazdani, A., Otoo, R.A. and Jeffrey, P. (2011) 'Resilience enhancing expansion strategies for water distribution systems: A network theory approach', *Environmental Modelling & Software*, 26(12), pp. 1574–1582. Available at: https://doi.org/10.1016/j.envsoft.2011.07.016.

Yodo, N. and Wang, P. (2016) 'Engineering Resilience Quantification and System Design Implications: A Literature Survey', *Journal of Mechanical Design*, 138(111408). Available at: https://doi.org/10.1115/1.4034223.

York, A.M. *et al.* (2021) 'Integrating institutional approaches and decision science to address climate change: a multi-level collective action research agenda', *Current Opinion in Environmental Sustainability*, 52, pp. 19–26. Available at: https://doi.org/10.1016/j.cosust.2021.06.001.

Yu, D.J. *et al.* (2015) 'Effect of infrastructure design on commons dilemmas in social–ecological system dynamics', *Proceedings of the National*

Academy of Sciences, 112(43), pp. 13207–13212. Available at: https://doi.org/10.1073/pnas.1410688112.

Zevenbergen, C. *et al.* (2013) *Herk, S. van, Douma, J., & Paap, L. van R. Tailor made collaboration: A clever combination of process and content.* Utrecht, NL: Rijkswaterstaat Room for the River.

Zuluaga, S., Karney, B.W. and Saxe, S. (2021) 'The concept of value in sustainable infrastructure systems: a literature review', *Environmental Research: Infrastructure and Sustainability*, 1(2), p. 022001. Available at: https://doi.org/10.1088/2634-4505/ac0f32.

4.5

BIOMIMICRY

4.5.1. INTRODUCTION

The field of ecology has a long and rich history of studying and theorizing resilience. Since natural systems have shown capacity to respond to disturbances within their tolerances and restructure for disturbances beyond what they have previously experienced, it may be useful to consider the inherent processes that enable this restructuring capability. Following a major shift in ecological thinking in the 1980s from an equilibrium paradigm to a non-equilibrium paradigm (Pickett, Pickett and White, 1985; Tarlock, 1994; Wu and Loucks, 1995), ecologists began to view the natural state of ecosystems as one of continual change rather than stasis. Ecological resilience embraced the idea that a natural system may have multiple stable states, a situation that imparts more flexibility. Natural systems achieve multiple stable states not because of intricate, ulterior design but through stochastic trial-and-error of resilience and stability (Holling, 1973), a dynamic that has not been embraced in engineering. Since Holling's landmark paper, many ecologists have come to see the resilience of natural systems to be their ability to operate in constant flux, maintaining structure, function, identity, and feedback loops (Wu & Wu, 2013); some even view resilience as incorporating transformation (Carpenter *et al.*, 2012).

The mimicry of natural systems in the human-designed world is known as biomimicry, an approach that "seeks sustainable solutions to human challenges by emulating nature's time-tested patterns and strategies" (Biomimicry Institute, 2019). However, the application of biomimicry to infrastructure has largely focused on form and process-level imita-

---

This chapter was adapted from the following journal article with publisher and lead author permissions: Alysha Helmich, Mikhail Chester, Samantha Hayes, Samuel Markolf, Cheryl Desha, and Nancy Grimm, 2020, Using Biomimicry to Support Resilient Infrastructure Design, *Earth's Future*, 8(12), e2020EF001653, pp. 1-18, doi: 10.1029/2020EF001653.

tion, which involve mimicry of the organism or organism's behavior respectively, instead of ecosystem-level mimicry (Benyus, 1997; Zari, 2018). Biomimicry, particularly at a form and process-level, does not innately result in a resilient system (Vincent, 2009). For instance, the creation of a car that has improved aerodynamics inspired by a shark (Lurie-Luke, 2014) does not produce transportation resilience even though it practices biomimicry. Therefore, we examine the 'design rules of nature,' as described by the biomimicry field, at the system level (i.e., beyond the organism and toward the ecosystem as seen in ecological organization), with the goal of identifying biomimicry principles that may support infrastructure resilience.

In contrast to engineering resilience, natural system resilience embraces the design rules of nature, which are labeled 'Life's Principles' (Biomimicry 3.8, 2013). Natural systems are resilient to known and unknown disturbances as a function of both the collective adaptive capacity of individual organisms and the capacity of communities and ecosystems to restructure in the face of changing environments. Given the multiple stable states in which natural systems can operate, ecologists and social-ecological systems scholars have developed a definition of resilience as the expected or unexpected disturbance a system can absorb while still functioning in a dynamic equilibrium (Cai et al., 2018; Holling, 1996; Yodo and Wang, 2016). This definition closely aligns with that proposed by Woods and reflects the high-level alignment between ecological resilience theory and Woods's proposed resilience theory for infrastructure.

With the development of engineering and ecological resilience concepts within a world of rapidly changing conditions, and considering biomimicry principles at the system level, this is an opportune time to rethink how infrastructure managers approach infrastructure resilience (Chester, Markolf and Allenby, 2019). Here, we seek to support infrastructure managers in innovative design for future infrastructure and retrofitting existing infrastructure. Transformational change in engineered systems, including but not limited to infrastructure, typically occurs after large events instead of incremental adjustments (Iwaniec *et al.*, 2019). Therefore, by examining how natural systems make numerous changes—both small and large—to achieve resilience, infrastructure managers may be able to adopt similar practices to prepare infrastructure for known and unknown disturbances. Furthermore, infrastructure managers may be able to design in a way that works in conjunction with nature rather than against it. Some exploration of the applicability of Life's Principles to sustainable infrastructure exists (Aitamurto, T. *et al.*, 2013; Oguntona and Aigbavboa, 2017), but there has been less exploration of their applicability to resilience, although foundations have been established by (Hayes *et al.*, 2019).

Within this context, we identify alignments, contradictions, contentions, and gaps between infrastructure resilience theory and practice and the key elements of Life's Principles, identifying opportunities for innovation and growth in resilient infrastructure design.

4.5.2. BIOMIMICRY AND RESILIENT INFRASTRUCTURE DESIGN

Here, we explore the adoption of resilience in biomimicry, infrastructure resilience theory, and infrastructure resilience practice. First we introduce biomimicry and Life's Principles. Next, we frame infrastructure resilience theory and its history. As we explore resilience theory we start with social-ecological systems and then frame the concept in engineering. Moving from academic theory to practical application, we draw on resilient infrastructure guidance documents within the United States, where resilience has been a national focus due to terrorism attacks and natural disasters (Fisher, Norman and Peerenboom, 2018). These documents span from federal to local government. Resilience in theory and practice are explored to identify core concepts of resilience in infrastructure design at large, so they may be assessed against Life's Principles.

4.5.2.1. BIOMIMICRY

In social and ecological systems, resilience appears intrinsic—in part due to the chaotic nature of these systems—but some counter that it may be a goal or trajectory of the systems (Folke *et al.*, 2010; Donohue *et al.*, 2016; Moser *et al.*, 2019). It has taken longer for resilience to be embraced by the built environment sector, where design has historically focused on creating organized and reliable (anti-chaotic) systems. Bio-inspired design terms appear in a variety of contexts within design: bionics; biomimetics; biophysics; biomechanics; animal, anthroposophic, biomorphic, ecological, evolutionary, geomorphic, organic, and zoomorphic architecture; biophilic design; regenerative design; permaculture; ecomimicry; cradle to cradle design; industrial, building, and construction ecology; and, of course, biomimicry (Zari, 2018). Biomimicry brings insights into design strategies that have proven successful within living systems toward withstanding and adapting to stresses and shocks. As Park et al. (2013) recognizes, resilience is not a designed characteristic of a system but an emergent property of a collection of design choices; therefore, by integrating characteristics of natural systems that have proven to lead to the emergence of resilience, infrastructure managers may be able to foster and reconceptualize resilience in infrastructure systems.

To promote a shift from traditional engineering resilience approaches toward a more agile, adaptive, and nature-inspired understanding of resilence, infrastructure managers must re-evaluate their design approaches. Life's Principles, derived from collective experience within academia and consulting spanning more than twenty years and six reiterations, depict common themes consistently observed in natural systems from organisms to ecosystems (Benyus, 1997; Biomimicry 3.8, 2013). These design principles have supported living systems that remain resilient to known and unknown disturbances. Recognizing resilience as an emergent system-level property, Life's Principles and associated sub-principles (Table 4.5.1) provide a system-level approach to analyzing biomimicry and infrastructure resilience.

Table 4.5.1. Life's Principles (Biomimicry 3.8, 2013)

Life's Principles	Sub-principles
1.0. Evolve to Survive *Continually incorporate and embody information for enduring performance.*	1.1. Replicate working strategies 1.2. Integrate the unexpected 1.3. Reshuffle information
2.0. Adapt to Changing Conditions *Appropriately respond to dynamic contexts.*	2.1. Incorporate diversity 2.2. Maintain integrity through self-renewal 2.3. Embody resilience through variation, redundancy, and decentralization
3.0. Be Locally Attuned and Responsive. *Fit into and integrate with the surrounding environment.*	3.1. Leverage cyclic processes 3.2. Use readily available materials and energy 3.3. Use feedback loops 3.4. Cultivate cooperation
4.0. Integrate Development with Growth. *Invest optimally in strategies that promote development and growth.*	4.1. Self-organize 4.2. Build from the bottom up 4.3. Combine modular and nested components
5.0. Be Resource Efficient *Skillfully and conservatively take advantage of resources and opportunities.*	5.1. Use low energy processes 5.2. Use multi-functional design 5.3. Recycle all materials 5.4. Fit form to function
6.0. Use Life-Friendly Chemistry *Use chemistry that supports life processes.*	6.1. Break down products into benign constituents 6.2. Build selectively with a small subset of elements 6.3. Do chemistry in water

4.5.2.2. INFRASTRUCTURE RESILIENCE THEORY

Infrastructure must be resilient to numerous potential disturbances: rapid urbanization, technical failures, emerging and disruptive technology, climate change, terrorist attacks, pandemics, cyber warfare. Expanding upon Holling's work on resilience of natural systems, researchers have begun applying these concepts to non-living systems such as transportation infrastructure (Hayes *et al.*, 2019). Existing resilient infrastructure frameworks are typically generalized and can be applied across the array of infrastructure and disturbances. The highlighted resilience frameworks are examined because of their prominence in infrastructure literature, acknowledgement of natural systems resilience, and unique contributions, reflecting several key themes of established and emergent resilience theory.

Ostrom's Social-Ecological Systems

Holling's ecological work rapidly transitioned across discipline boundaries, and the social sciences were one of the early adopters of resilience (though not without tensions, (Adger, 2000; Brown, 2014)). Elinor Ostrom made significant contributions toward defining social-ecological system (SES) resilience, systems with a complex dynamic between resources, resource users, public infrastructure, public infrastructure providers, and the surrounding environment (Ostrom, 2007; Anderies and Janssen, 2012). These systems are imbued with complexity and uncertainty sourced from both internal disturbances (e.g., quality of system relationships) and external disturbances (e.g., natural disasters). The authors define infrastructure as robust if it does not deteriorate the environment or inflict long-term social damage (Anderies, Janssen and Ostrom, 2004).

The SES framework looks closely at the relationships between social and ecological systems, recognizing that there is a large cost to embed infrastructure in society. They promote a holistic viewpoint, encouraging infrastructure managers to make decisions based on the system of systems. For instance, due to emerging and evolving relationships of infrastructure systems, the maintenance strategies that have proven to work for a sector in the past may collapse the system of systems when implemented without consideration of the other sectors. This SES framework emerged from eight identified design principles that are characteristics of robust systems: clearly defined boundaries; proportional equivalence between costs and benefits; collective-choice arrangements; monitoring; graduated sanctions; conflict-resolution mechanisms; local autonomy; and polycentric governance (Ostrom, 1990). This framework again emphasizes that resilience is an emergent characteristic of a system.

Park et al.'s Sensing, Anticipating, Adapting, Learning Framework

Park et al. (2013) developed a framework for infrastructure that expands upon Holling's resilience definition and concepts from risk management, emphasizing a need to move away from robustness, which fails to account for unknown disturbances. They define resilience as "the capacity to adapt to changing conditions without catastrophic loss of form or function" (Park *et al.*, 2013). To achieve resilience, infrastructure managers must embrace incompleteness—in other words, they must accept that infrastructure cannot be designed and minimally maintained but continuously managed and adapted. Furthermore, the authors promote transformational—instead of incremental—technical and social change since transformational change allows the system to be redesigned holistically. Incremental change can develop problems if only one component is considered at a time and is not adjusted to adequately function in the complex system.

The authors formed the sensing, anticipating, adapting, and learning (SAAL) framework to guide resilient infrastructure systems (i.e., infrastructure managers, infrastructure, and their interactions). The SAAL framework promotes design strategies that embrace diversity, adaptability, cohesion, flexibility, renewability, regrowth, innovation, and transformation to achieve resilient infrastructure (Fiksel, 2003; Mu *et al.*, 2011). The study found that resilience is not appropriately integrated into practice, and infrastructure managers would benefit from integrating resilience into risk management.

Woods's Four Concepts for Resilience

Woods describes four resilience regimes: rebound, robustness, graceful extensibility, and sustained adaptability (Woods, 2015). Capabilities described in the SAAL framework are relevant across the regimes. Rebound is the ability of an infrastructure system to return to its original functionality after a disturbance based on its own capacity. A system only enters a state of rebound if its design envelope is exceeded. Second, robustness is the ability of infrastructure to withstand an expected disturbance. To achieve this, a system must be designed for worst-case scenarios but doing so can lead to optimization and, therefore, brittleness. Third, graceful extensibility is the ability of infrastructure to respond to disturbances at the edge of its design envelope, asking, how do systems expand at a boundary? This advance preparation for stress on the system can help infrastructure prepare for cascading failure events. Lastly, sustained adaptability identifies flexible characteristics that allow infrastructure to adapt in changing social, ecological, and/or technological contexts. This last aspect of resilience encourages a holistic viewpoint with integration of social and technological solutions to achieve resilience. In all, Woods encourages engineers to move away from considering resilience as just one aspect of these concepts (e.g. robustness) but, instead, embrace all four. By investing in all four concepts of resilience, infrastructure managers can alleviate the pressures on one aspect. For instance, after an unexpected disturbance that causes system failure, there are strong initiatives to return functionality (rebound). If a system has invested not only in robustness but additionally graceful extensibility and sustained adaptability, the brittleness of the system is reduced, and it is less likely that significant robustness efforts are needed to rebound the system.

4.5.2.3. INFRASTRUCTURE RESILIENCE PRACTICE WITHIN THE UNITED STATES

Infrastructure managers currently design infrastructure to economically meet design and safety standards for delivery of intended use or service over a predetermined design life (Ayyub, 2018), and they are frequently encouraged by their employers to use past experiences and solutions to address a problem (Sydenham, 2004; Kosky *et al.*, 2006; Pahl *et al.*, 2007; Eder and Hosnedl, 2008; Park *et al.*, 2013). Resilience engineering stems from safety management (Righi, Saurin and Wachs, 2015), making the emphasis on robustness and fail-safe design unsurprising. However, to achieve resilience, infrastructure managers will have to balance innovation with operations to adapt to an uncertain world (Pahl *et al.*, 2007). Engineering resilience is often defined as the ability of infrastructure to be robust to disturbances and recover quickly; yet, existing research on recovery is very limited (Yodo and Wang, 2016). The dominant focus on robustness in the field of engineering indicates a need for resilience planning methods (Hosseini, Barker and Ramirez-Marquez, 2016). The following documents illustrate how resilience has been integrated into infrastructure design at the federal to project-level within the United States.

National Infrastructure Advisory Council's (NIAC) Critical Infrastructure Resilience Study

The integration of resilience was initially created for security purposes since infrastructure is a critical component of economic activity (NIAC, 2009). Resilience became popularized after extreme weather events, such as Hurricane Katrina, leading to an expansion of the term to cover a wide range of disturbances (Fisher, Norman and Peerenboom, 2018). The Critical Infrastructure Resiliency Study outlined the current state of resilience in infrastructure across the United States and defined resilience as "the ability to reduce the magnitude and/or duration of disruptive events" (NIAC, 2009). The report stated that a resilient system should be able to absorb, adapt to, and/or rapidly recover from a disturbance such as terrorist attacks, cyber warfare, extreme weather events, or cascading failures through the integration of three features: robustness, resourcefulness, and rapid recovery. Due to the 9/11 terrorist attack that occurred eight years prior, the report assessed infrastructure in the United States to be highly fortified (i.e., robust). It also emphasized that safety and resilience are not contradictions and could be achieved simultaneously.

The report recommended that the following actions be taken by the federal government to promote resilient infrastructure: 1) define resilience; 2) increase communication between government and infrastructure managers for consistent standards; 3) identify roles for all stakeholders during a disturbance for quick response; 4) create a public–private dialogue to improve infrastructure resilience; 5) provide market-based incentives to promote resilience; and 6) reevaluate infrastructure operation plans to identify any emerging gaps, including cross-sector planning.

Department of Homeland Security's (DHS) National Infrastructure Protection Plan

In response to the Presidential Policy Directive 21 (PPD-21) on Critical Infrastructure Security and Resilience which promotes resilient infrastructure to be robust, agile, and adaptable, the National Infrastructure Protection Plan (NIPP) was updated in 2013 to provide guidance to infrastructure managers as well as federal, state, and local governments on how to manage infrastructure risk. The PPD-21 defined resilience as the "[ability] to withstand and rapidly recover from all hazards" (The White House, 2013) and promoted integration of prevention, protection, mitigation, response, and recovery (DHS, 2013).

Notably, the NIPP acknowledged the complexity of infrastructure systems and recommended preparing for risk by recognizing disturbances, reducing vulnerabilities, and minimizing disturbance consequences (DHS, 2013). The report identified an array of feasible hazards from extreme weather; accidents of technical failure, including cascading failure; pandemics; terrorism; and cyber threats. The NIPP stated that security and resilience should be top concerns of risk management; however, it did not provide infrastructure managers with specific guidance on how to achieve resilience. Broader guidance was given to manage risk, including 1) conduct risk-informed decision making, 2) analyze complexity, 3) understand potential cascading failures, 4) promote recovery post-disturbance, 5) educate infrastructure managers, and 6) advance research on security and resilience.

American Society of Civil Engineer's Climate-Resilient Infrastructure

In 2019, the American Society of Civil Engineers (ASCE) released Climate-Resilient Infrastructure, a manual of practice (MOP) for infrastructure design in relation to climatic uncertainty. The authors adopted the same definition of resilience as the NIPP and emphasized several of the characteristics identified by the two previous reports: robustness, redundancy, resourcefulness, and rapidity (Ayyub, 2018). The authors recognized that redundancy and resourcefulness are not well defined in infrastructure design. While the text focused on climate change, population growth and development patterns were also recognized as disturbances. The MOP promoted the observational method and adaptive risk management to address known and unknown uncertainties in infrastructure design.

The observational method recommends designing to most probable, instead of least favorable, events and adjusting the infrastructure as social, ecological, and technological contexts change. Potential design modifications are identified in the design phase for every expected disturbance, and these designs are modified if they become necessary integrations to combat unfavorable events. The proposed adaptive risk management strategy is more detailed, identifying context and hazards; conducting uncertainty and extreme-value analysis; calculating failure probability estimations; performing exposure and loss analysis, economic valuation, and risk quantification; developing feasible design adaptations; computing cost-benefit and risk-informed decision analysis; monitoring hazards and risks; and, determining alternative strategies through risk-informed adaptation analysis (Ayyub, 2018). However, the MOP notes that not all information needed to complete these steps may be accessible. This method also promotes brainstorming design solutions for disturbances that are known and unknown during the design phase and states that adaptation should take place as changes appear.

Community-Level Resilience

Koliou et al. (2018a) reviewed community resilience practice, focusing on natural hazards and infrastructure. Their definition of resilience was the same as that of DHS and ASCE from PPD-21 (The White House, 2013), leading community resilience to be defined as "emergency response, preparedness and security, mitigation, risk communication, and recovery of communities from physical, economic and social disruptions" (Koliou *et al.*, 2018). Infrastructure managers tend to adopt a narrow definition of resilience, focusing on robustness and recovery in contrast to resilience work in other fields, which consider system resilience as the ability to reduce consequence, future vulnerabilities, and enhance recovery (Koliou *et al.*, 2018). Inconsistent concepts, definitions, and propositions have been identified in reviews of engineered systems resilience (Francis and Bekera, 2014) and urban resilience (Meerow, Newell and Stults, 2016). Koliou et al. (2018a) and Markolf et al. (2018) recommend integration of infrastructure, social, and economic systems perspectives into resilience frameworks as there are not general, multi-disciplinary community resilience frameworks or readily available metrics. They provide a thorough review of resilience efforts on a community scale across the power, water, and transportations sectors.

Project-Level Resilience (Rating Systems)

Leadership in Energy and Environmental Design (LEED) and Envision are popular rating systems for assessing sustainability of individual infrastructure projects, and they have both integrated concepts of resilience. LEED first began piloting resilience credits in 2015. In 2019, three revised credits were piloted, including assessment and planning for resilience, design for enhanced resilience, and passive survivability and back-up power during disruptions (U.S. Green Building Council, 2019). 'Assessment and planning for resilience' requires projects to address shocks and stresses while recognizing a non-stationarity climate. 'Design for enhanced resilience' states that design should resist natural disasters. Lastly, 'passive survivability' accounts for maintaining functionality during a disturbance. LEED does not provide a definition of resilience; but, Envision defines resilience as "the ability to successfully adapt to and/or recover readily from a significant disruption" (Institute for Sustainable Infrastructure, 2018). This rating system has identified six criteria for resilience: 1) avoid unsuitable development, 2) assess climate change vulnerability, 3) evaluate risk and resilience, 4) establish resilience goals and strategies, 5) maximize resilience, and 6) improve infrastructure integration (Institute for Sustainable Infrastructure, 2018).

Briefly looking beyond the United States, rating systems with a focus on resilience can be found in practice such as Australia's Infrastructure Sustainability (IS) Rating Scheme. The resilience category within the IS Rating Scheme references the interconnected nature of infrastructure with social and ecological systems, as well as the interdependency across systems and supporting services, that all play a role in times of disruption and disturbance. The category includes two credits: 1) resilience plan and 2) climate and natural hazards risks. Resilience approaches are informed by the 100 Resilient Cities Framework, which includes reference to robustness, redundancy, integration, and inclusivity among other principles. While remaining focused on risk management of identified and predictable risks, as opposed to building inherent flexibility and agility, it reflects an important step toward mainstreaming consideration of direct and indirect disruptions and resilience responses. Further integration of resilient infrastructure theory and practice into rating systems would increase exposure to infrastructure managers and further promote resilient design.

4.5.2.4. THEMES IN RESILIENT INFRASTRUCTURE THEORY AND PRACTICE

Infrastructure resilience theory encompasses infrastructure resilience practice, and, additionally, infrastructure resilience theory advocates agile and adaptive infrastructure. However, it fails to provide the guidance infrastructure managers need to implement such designs in practice. Therefore, infrastructure managers continue to emphasize robustness in, which has proven to work in the past. With increasing complexity and uncertainty, infrastructure managers must navigate new contexts. Infrastructure resilience practice promotes adaptability but does not define or elaborate upon what this means. Resilient infrastructure theory expands beyond robustness, looking for innovative, transformational designs that not only reinvent a single component of infrastructure but the social, ecological, and technological system (SETS, (Markolf *et al.*, 2018)) dynamics between infrastructure. In order to achieve this holistic vision, infrastructure must be adaptable to known and unknown disturbances—as do natural systems. Recognizing this, it may be useful to identify tools and frameworks for translating ecological resilience principles into tangible infrastructure design principles and strategies.

4.5.3. USING BIOMIMICRY TO SUPPORT RESILIENT INFRASTRUCTURE DESIGN

To reconcile resilient infrastructure core concepts with Life's Principles, it is necessary to assess whether each core concept aligns, contradicts, or contends with each life principle. Resilient infrastructure core concepts were drawn from the literature review of infrastructure resilience theory and practice. These core concepts are catalogued and defined in Table 4.5.2. At times, related definitions were condensed, such as adapting, adaptability, and sustained adaptability. Only a subset of core concepts—including efficiency, rebound/recover, redundancy, robustness, and status quo—represent concepts defined and implemented within resilient infrastructure practice. Many of the core concepts presented have not transitioned beyond theory to penetrate practice. Overall, these recurring themes provide a snapshot of how resilient infrastructure design is understood today. After identifying the core concepts in resilient infrastructure design, we compared them with Life's Principles, revealing alignments, contradictions, contentions, and gaps between the frameworks (Table 4.5.3). The survey was conducted by evaluating the definitions of Life's Principles (Table 4.5.1) against the infrastructure resilience core concept definitions (Table 4.5.2) through two primary inquiries:

1) Does the infrastructure resilience core concept align with the definition and/or sub-principles of the biomimicry life principle?

2) Does the infrastructure resilience core concept contradict the definition and/or sub-principles of the biomimicry life principle?

Table 4.5.2. Glossary of Resilience Core Terms.

Resilience Concept	Definition	Source
Adaptability	The ability to change in response to new shocks and stresses	(Allenby, 2012; Park *et al.*, 2013; Woods, 2015)
Agile	The ability of the physical and institutional system to maintain functionality in uncertainty	(Chester and Allenby, 2018)
Anticipating	The ability of the system to integrate the knowledge of sensing to predict potential hazardous events	(Park *et al.*, 2013)
Cohesion	The ability to utilize unifying forces and feedbacks	(Fiksel, 2003)
Diversity	The ability of a system to have multiple functions and strategies	(Fiksel, 2003)
Efficiency	The ability to provide a service with minimum resources	(Fiksel, 2003)
Flexible	The ability of infrastructure to meet changing demands	(Chester and Allenby, 2018)
Graceful Extensibility	The ability of infrastructure to respond to disturbances at the boundaries of its design envelope	(Woods, 2015)
Holistic Design	The ability to integrate multiple perspectives and infrastructure sectors in the design phase	(Anderies, Janssen and Ostrom, 2003; Koliou *et al.*, 2018)
Innovation	The ability to integrate something new	(Chester and Allenby, 2018)
Learning	The ability to learn from experience and other similar systems	(Park *et al.*, 2013)
Rebound/Recover	The ability of an infrastructure system to return to original functionality after a disturbance based on its own capacity	(Woods, 2015)
Redundancy	The ability of a system to achieve the function through alternative component pathways	(Ahern, 2011a)
Resourcefulness	The institutional ability of a system to respond to a disturbance	(NIAC, 2009)
Robustness	The ability of infrastructure to withstand a disturbance	(Woods, 2015)
Sensing	The ability of a system to recognize and understand new stresses	(Park *et al.*, 2013)
Status Quo	The ability to use previous experience and solutions to address a need	(Sydenham, 2004; Kosky *et al.*, 2006; Pahl *et al.*, 2007; Eder and Hosnedl, 2008; Park *et al.*, 2013)

Bolded concepts are prevalent in resilient infrastructure practice.

Infrastructure resilience theory frequently aligns with Life's Principles; however, infrastructure resilience in practice does not align and, at times, contradicts Life's Principles. All six of Life's Principles are addressed, aligning with at least one core concept in resilient infrastructure theory and practice (green fill in Table 4.5.3). Contradictions exist when the core concepts may not adhere to a life principle or sub-principle (orange fill in Table 4.5.3). This does not indicate that the core concept is not applicable to achieve resilience, but that tradeoffs need to be considered throughout design. For example, redundant pathways in infrastructure utilize more resources than a singular pathway; however, redundancy allows for adaptability to changing conditions.

Core concept may both align and contradict certain aspects of a life principle (yellow fill in Table 4.5.3). For instance, maintaining the status quo in design allows for replication of successes; however, it does not encourage innovation, contradicting the ability of the infrastructure to adapt. Lastly, there are significant gaps representing spaces that are not explored by a core concept but are explored by the Life's Principles (white fill in Table 4.5.3). It should be noted that while resilient infrastructure core concepts and Life's Principles are listed and analyzed individually, they are highly interconnected and interdependent. It is their ability to work in unity that creates the system's response to stresses and shocks and, eventually, resilience. The following sub-sections discuss alignments, contradictions, contentions, and gaps between resilient infrastructure core concepts and Life's Principles toward critiquing resilient infrastructure design.

Table 4.5.3. Alignments (green), Contradictions (orange), Contentions (yellow), and Gaps (white) of Resilient Infrastructure Core Concepts to Life's Principles.

Row concepts (left, with grouping numbers):

	Core Concept
1	Status Quo
1	Sensing
1	Robustness
1	Redundancy
2	Rebound/Recover
2	Learning
2	Innovation
3	Graceful Extensibility
3	Cohesion
3	Anticipating
4	Efficiency
4	Resourcefulness
4	Holistic Design
4	Diversity
6	Agile
6	Flexible
6	Adaptability

Column principles (bottom, with counts):

- **Evolve to Survive** (12)
  - Replicate strategies that work
  - Integrate the unexpected
  - Reshuffle information
- **Adapt to Changing Conditions** (11)
  - Incorporate diversity
  - Maintain integrity through self-renewal
  - Embody resilience
- **Be Locally Attuned and Responsive** (12)
  - Leverage cyclic processes
  - Use readily available materials and energy
  - Use feedback loops
  - Cultivate cooperative relationships
- **Integrate Development with Growth** (7)
  - Self-organize
  - Build from the bottom up
  - Combine modular and nested components
- **Be Resource Efficient** (5)
  - Use low energy processes
  - Use multi-functional design
  - Recycle all materials
  - Fit form to function
- **Use Life Friendly Chemistry** (4)
  - Break down products into benign constituents
  - Build selectively with a small subset of elements
  - Do chemistry in water

Darker shades represent principles and lighter tones correspond to sub-principles. Bolded core concepts are those prevalent in resilient infrastructure practice, consistent with Table 4.5.2.

4.5.3.1. EVOLVE TO SURVIVE

While natural and technological systems both replicate successful strategies, natural systems endure a test of trial-and-error through processes of mutation, inheritance, and natural selection, where unsuccessful strategies are abandoned (or don't survive). However, infrastructure systems are meticulously designed and governed to ensure service and performance for an intended life. In the built environment sector, infrastructure managers design infrastructure based on historical models and common practice, emphasizing the reliance on past experience in engineering design approaches (Sydenham, 2004; Kosky *et al.*, 2006; Eder and Hosnedl, 2008). The ability to replicate existing infrastructure increases efficiency of design resources, allows for learning from past experiences, and perpetuates the status quo. By designing in favor of the status quo, infrastructure systems are easily understood by existing institutions, increasing the infrastructure's chance of survival. This focus on implementing tried-and-true infrastructure leads to slow integration of innovations and transformation as infrastructure managers most frequently 'integrate the unexpected' by creating increasingly robust systems. Moving forward, infrastructure mangers may benefit from recognizing the shortcomings in current systems so that new design approaches can be trialed, as implementing strategies of the past limits the ability of infrastructure to become resilient to unknown and unexpected disturbances of the future.

Resilient infrastructure core concepts in theory offer additional methods to account for unknown disturbances in design, such as adaptability, flexibility, and agility. These core concepts advocate for the ability of infrastructure to sense changes in stresses, shocks, and uncertainties related to the function and demands of the system. At times, infrastructure managers will need to learn from and manage failures, which can be seen in practice through safe-to-fail design (Ahern, 2011b; Kim *et al.*, 2019). In natural systems, failure is managed by natural selection against strategies that do not work and successful mistakes arising from mutation. Resilient infrastructure theory again fills the gap in addressing this biomimicry subprinciple. First, infrastructure managers need to be innovative, realizing the status quo that has proved proficient for decades may no longer always suffice due to emerging disturbances. Knowing failures will occur, infrastructure managers have a few options, including a) continue to increase robustness, b) create management plans for controlled failure, or c) respond to failure events with learned knowledge and adapt or transform the system. Aforementioned, failures oftentimes spur change in infrastructure systems (Iwaniec *et al.*, 2019). Therefore, by considering failure — and how to manage it — in the design phase, or actively envisaging failure and the consequences during the design life, infrastructure managers can be prepared for adaptation and transformation during these chaotic times.

The importance of disturbance in altering infrastructure design places an additional emphasis on a system to be able to acknowledge reshuffled information through anticipation and graceful extensibility. Information must not only be acknowledged but integrated. For example, ecosystems are able to respond to geophysical variation in environmental parameters by transitioning between multiple stable states over time, and, occasionally, these systems undergo regime shifts. The Chihuahuan Desert will alternate between grass- and shrub-dominant vegetation depending on the temporal and spatial variation of groundwater availability, precipitation, grazing, wildfire disturbance, etc. as well as anthropogenic drivers (Bestelmeyer *et al.*, 2018; Hoover *et al.*, 2020). Learning can be used in infrastructure systems through monitoring, which can be as complex as implementing information and communication technology (ICT, a technological solution) or utilizing manpower for observations on a routine basis (a social solution) — the goal is to create a feedback loop between the infrastructure system and external conditions such as climate or cultural change. This provides infrastructure managers with a tool to address stresses to their system.

4.5.3.2. ADAPT TO CHANGING CONDITIONS

Infrastructure resilience practice emphasizes the importance of recovering functionality amongst dynamic stresses and shocks; however, the status quo prevents transformation of more adaptive infrastructure. The ability to rebound — or recover — is closely aligned with a system's ability to actively respond to disturbances. By 'incorporating diversity,' a system may have multiple functions (e.g., a park provides recreation most days but flood mitigation during storm events). Natural systems utilize diversity, self-renewal, redundancy, and decentralization to appropriately respond in these contexts (Biomimicry 3.8, 2013). Consider a soil ecosystem, which is able to host diverse plants species, store and cycle nutrients, decompose organic matter, filter water, and so on (Wagg *et al.*, 2014). Infrastructure systems can also utilize redundancy (e.g. a stormwater system may combine grey and green infrastructure methods to manage stormwater runoff). While both of these core concepts (diversity and redundancy) address adaptation to changing conditions, flexible and agile infrastructure literature promotes consideration of multi-functional infrastructure to achieve adaptability (Chester and Allenby, 2018). The current resistance to alter the status quo — or in other words, the lock-in (inability to change a system due to financial, political, technical, social, cultural, and technological barriers, Chester and Allenby, 2018) faced by infrastructure systems — directly contradicts the essence of the second Life's Principle.

In infrastructure, a cohesive integration of social-ecological-technological systems (SETS) can promote self-renewal and rival lock-in. At the ecosystem level, seed banks and stored organic matter accelerate regrowth after a disturbance (such as a wildfire) and demonstrate the self-renewal of the system. The idea of self-renewing infrastructure on a systemic scale is in early stages of innovation. The power sector is particularly interested in the ability of systems to self-heal as demand rises and cascading failures increase. This sector has looked toward real-time feedback loops to respond within localized context and be constrained by a set of pre-defined rules, and this flexibility would allow the system to determine a best course of action on its own to prevent a failure (Amin, 2001; Gu and Jiang, 2017). Research is also being conducted on self-renewal of infrastructure on a components-scale. Self-repairing concrete and asphalts are being explored for the transportation sector, where these materials are able to repair cracks in the surface without external interventions (Schlangen and Sangadji, 2013; AskNature, 2018). This adoption of self-renewal as adaptation would be a transformational switch from traditional engineering approaches of repair, which focuses on external maintenance, to the self-management similar to natural systems (Hayes *et al.*, 2019). Furthermore, the concept of recovery in resilient infrastructure practice is currently assessed as the ability of infrastructure to return to a functional state. This functional state is defined purely by the technological aspects of the infrastructure with little focus on social (beyond maximizing safety) and ecological aspects. Additionally, the pre-disruption functional state of certain social, ecological, and/or technological systems may be at undesirable levels initially. Thus, emphasizing a return to these functional states may in fact be at odds with fundamental adaptation principles and objectives. Infrastructure managers will need to be able to make tradeoffs between these concepts that expand on more than just the economics of the system but also the social and ecological effects. SETS literature (Grabowski et al., 2017; Grimm, Cook, Hale, & Iwaniec, 2015; Keeler et al., 2019; Markolf et al., 2018; Miller, Ches-ter, & Munoz-Erickson, 2018) is promoting infrastructure managers be able to take a more holistic viewpoint by considering the impacts and adaptive capacities that exist (and can emerge) between and among interconnected social, ecological, and technological systems. Infrastructure managers and institutions must be willing to transform existing practices to those that have shown to be more resilient as failure from shocks can no longer be avoided but should be managed.

Finally, the life principle advocates for 'resilience through variation, redundancy, and decentralization,' and this sub-principle is finding traction in infrastructure theory and practice. First, in natural systems, similar species do not occupy the same niche. This would increase competition and hurt chances of individual survival—instead, they evolve to effectively partition resources. Evidence suggests that loss of diversity in eco-systems is as least as important as nutrient loading and climate change to the loss of primary productivity (Hooper *et al.*, 2012), indicating the importance of species diversity to ecosystem stability. Infrastructure has historically been implemented with the consideration of economies of scale, where a larger system is seen to increase economic savings. This technique puts infrastructure in competition with an implementation of decentralized, physical systems (e.g., the power sector can host redundant sources of energy (renewables and non-renewables), and it can implement these sources in various locations across the grid rather than at one power station), allowing redundant and independent systems to respond adeptly to instability. However, renewables have struggled, historically, to be competitive with large-scale non-renewables, showing how economies of scale lock infrastructure into a non-adaptive state. The modularity of decentralized systems can allow for pilot tests before integrating at full-scale. Resilient infrastructure theory promotes graceful extensibility, and a variety of decentralized infrastructure practices can help prevent a complete failure of the technical system, providing a chance for infrastructure managers to anticipate stresses and shocks to the system.

4.5.3.3. BE LOCALLY ATTUNED AND RESPONSIVE

The imagery of being locally attuned and responsive within natural systems has previously provided inspiration to infrastructure design (e.g., industrial ecology, xeriscaping, passive design); however, practice shows one large contradiction: robustness. This life principle emphasizes the importance of cooperative relationships between infrastructure and the surrounding environment. With innovations of canals, railways, service lines, and heating and cooling ventilation systems, humans have further disentangled themselves from nature, attempting to survive independently of the surrounding environment. Now, infrastructure managers do not need to regress but should actively consider leveraging cyclic processes, readily available materials, feedback loops, and cooperative relations through holistic design (Biomimicry 3.8, 2013). In natural systems, examples of feedback loops and cooperative relationships illustrate the intricate co-evolution between ecosystems. Riparian zones, ecotones between terrestrial and aquatic environments, are one emergence. The terrestrial ecosystem provides nutrients, food supply, and canopy shade to the aquatic system; meanwhile, the aquatic system provides diverse habitats, food supply, and disturbance (e.g., flooding) to the terrestrial ecosystem (Swanson *et al.*, 1982). The riparian zone may be a refuge for organisms during disturbances and a source of propagules for post-disturbance recovery. These ecological relationships are complex while remaining highly supportive due to co-evolution. For infrastructure systems, lock-in inhibits the ability of the systems to co-evolve, limiting the use of feedback loops and cooperation. However, the water sector provides some inspiration. Water infrastructure managers have considered the natural water cycle in design through history with actions of rainwater harvesting and flood management. The first step infrastructure managers should take is to achieve closed loops, where the resource remains within the system. To truly see sustained adaptability, infrastructure managers should seek to close loops not just within their sector but between sectors (e.g., using wastewater from the water sector for cooling water in the energy sector). By creating a cohesive industrial ecology, infrastructure managers can capitalize on these sub-principles, increasing knowledge of the systems and decreasing the need for excess material and energy and, therefore, becoming more efficient. But recall, in contradiction, robustness assumes an infrastructure will be able to withstand disturbances on its own without aid from symbiosis. It is often predicated on a view of infrastructure assets as distinct technical entities as opposed to integrated components of complex SETS. By looking at the system holistically and as something that engages with SETS, infrastructure managers may be able to identify stressors and understand failure consequences of shocks more rapidly.

4.5.3.4. INTEGRATE DEVELOPMENT WITH GROWTH

Integration of development with growth is the only life principle not acknowledged by infrastructure design practice, but it is addressed in theory by adaptability, flexibility, and agility. Looking toward natural systems, they do not design with an overarching blueprint but through modularity with components evolving in conjunction—from individual species, to communities with co-evolved members, to complex ecosystems—and resilience emerges over time from the collective properties. Infrastructure systems are challenged by lock-in as seen with traditional infrastructure design approaches that oftentimes focus on robustness (Helmrich and Chester, 2020), which can send false signals of security. For example, robustness of levees has created the 'levee effect' where the strengthening of infrastructure has created a sense of security that has promoted additional development—developments now at risk for a more catastrophic failure should the levees fail (Burby, 2006; Markolf et al., 2018). The life principle prompts consideration of modularity and options for nested infrastructure components as opposed to locked-in infrastructure growth. Infrastructure throughout the United States has existed for generations and care of this system of systems is necessary to support new development. Careful consideration is required to determine when the optimal solution may be to add new infrastructure, and when it may be to adapt, retrofit, or further develop existing infrastructure. These legacy components are a source of complexity (Park et al., 2013); however, for better or worse, they are the foundation of the infrastructure system as it exists today and as it transitions to exist in the future.

Integrating development with growth is reminiscent of adaptive management strategies, approaching infrastructure design in a way that can be incrementally adjusted over time to address stresses and shocks (Allenby, 2012). This modularity allows infrastructure managers to invest in development as necessary, gracefully extending the capability of infrastructure to respond to situations beyond its original design envelope. Modular design is inherently flexible and agile, providing an outlet for diversity. Furthermore, the ability of infrastructure to 'self-organize' allows for the system to respond through social and technological means to disturbances. Socially, self-organization is present by the capacity of infrastructure managers to make decisions, or the decentralization of governance, as they are intimately familiar with the infrastructure and its surrounding environment (a holistic approach). Technologically, ICT can be embedded in infrastructure to provide real-time feedback loops as explored in the power sector (Amin, 2000; Gu and Jiang, 2017). Self-organization allows infrastructure managers to directly respond to stresses and shocks.

4.5.3.5. BE RESOURCE EFFICIENT

Although resilient infrastructure theory and practice may appear to have large gaps when addressing resource conservation, the prevalence of efficient design in practice makes this a life principle especially relevant to infrastructure systems. Resource use efficiency, in natural systems, is the 'amount of biomass produced per unit of supplied resource' (Hodapp, Hillebrand and Striebel, 2019). Hodapp et al. (2019) found that ecosystems can increase resource use efficiency by increasing diversity (species and function) given the new traits increase resource consumption—particularly in terrestrial systems (Cardinale *et al.*, 2006). Infrastructure design similarly seeks to lessen the resources and energy used to minimize costs through reuse and recycling as well as reduction. This principle, particularly the sub-principles of 'use low energy processes' and 'recycle all materials,' is represented in sustainability efforts and could be assessed through rating systems. However, sustainability—commonly understood as reducing negative environmental and social impacts—does not promise resilience; it is simply one aspect as shown in Life's Principles. For example, LEED-certified buildings, while effective at promoting sustainability, do not necessarily increase the resilience of the design (Champagne and Aktas, 2016).

Infrastructure managers are responsible for the resources consumed by infrastructure—locally, regionally, and globally—and must assess their priorities holistically when balancing efficiency. An emphasis on low energy processes can reduce the resource strain of infrastructure and allow further compatibility with being locally attuned and responsive. By reducing the dependence of infrastructure on resources (e.g., power) produced geographically far away, infrastructure is less likely to experience cascading failures due to supply-chain logistics. Furthermore, while being resource efficient advocates to 'fit form to function,' it also challenges infrastructure managers to use multi-functional design. Resilient infrastructure theory seeks to fill gaps of practice by addressing these two sub-principles. Instead of creating robust and redundant infrastructure that may utilize more resources than necessary to achieve the function, infrastructure managers may adopt adaptability through flexibility and agility by moving away from optimizing and toward satisficing. For instance, infrastructure managers could consider how infrastructure is designed to function in times of failure by asking how failure can be planned for in advance to prevent social, ecological, and technological components from failing during a disturbance. As Anderies et al. (2004) questioned, when is a system in a failed state? If social systems can be resourceful and govern failed technological systems, is the system truly failed? One such example is the Stormwater Management and Road Tunnel (SMART) in Kuala Lumpur. Exactly as the name describes, this infrastructure serves as a road tunnel to reduce traffic pressures during peak hours as well as a flood management during flash floods, allowing SMART to relieve two failure points of two infrastructure sectors, while reducing overall resource and energy consumption (Abdullah, 2004). Planning for failure events can help infrastructure managers manage shocks to the system through multi-functional design that fits form to function, and resourceful institutions provide a largely unexplored opportunity to manage information during times of stability and instability.

4.5.3.6. USE LIFE-FRIENDLY CHEMISTRY

While using life-friendly chemistry may seem out-of-scope within infrastructure resilience—only aligning with a handful of core concepts (flexibility, agility, diversity, and efficiency)—the principle ultimately promotes a reduction of complexity enforced by the use of benign constituents and modularity in design. First, minimizing or eliminating hazardous and dangerous materials can reduce ecological and technological complexities, such as pollution and disposal, and this, thereby, can also impact the social complexities by reducing the burden of individuals living near large-scale infrastructure such as landfills, fracking operations, and industrial plants. These sub-principles push infrastructure managers to consider the impacts of their material choices beyond quantity and economics (although money can be saved from reducing hazardous waste disposal) and contemplate how to make a benign system from benign components. Second, the principle of modularity can be explored with the building blocks of natural systems: hydrogen, carbon, nitrogen, oxygen, phosphorus, and sulfur. The modularity and, therefore, flexibility and agility of these six elements allows for reformation — consider the water or carbon cycle — across diverse mediums. Modularity is promoted across a few core concepts found in resilient infrastructure design, including flexible and agile infrastructure where it is listed as one of ten competencies (Chester and Allenby, 2018). By building selectively and strategically, infrastructure managers can work toward combatting the convolution of legacy systems, where future managers may not understand or have the tools necessary to fix the system. It also emphasizes the need for communication to ensure the modular components continue to work succinctly and adapt based on collective experiences. Reductions in system complexity may reveal otherwise hidden stressors.

4.5.4. CONCLUSION

These are the opportunities for infrastructure managers to rethink infrastructure resilience through the lens of biomimicry. Life's Principles are limitedly addressed by resilient infrastructure practice, which focuses on robustness and recovery; still, resilient infrastructure theory aligns closely with the Life's Principles framework. The most frequently aligned core concepts include adaptability, flexibility, and agility, which we propose are overarching themes in resilient infrastructure theory and an 'end goal' of resilient infrastructure. Resilient infrastructure design shows dominance in certain Life's Principles such as 'evolve to survive,' 'adapt to changing conditions,' and 'be locally attuned and responsive;' however, the remaining principles are addressed in part, and one — 'be resource efficient' — is more commonly evoked in practice than implied by analysis. Life's Principles provide a systematic approach to consistently and rigorously apply resilient infrastructure core concepts in theory toward practice.

Where the alignments between the frameworks provide opportunities for recognized growth, the identified contradictions and gaps provide opportunities for innovation. Infrastructure managers should address the contradictions between resilient infrastructure practice and Life's Principles in design by incorporating more core concepts that show alignment. This will expand the dominant infrastructure design strategy beyond failsafe and robust infrastructure. Furthermore, infrastructure managers can be innovative to fill the gaps between frameworks. Not all gaps are equivalent. As stated previously, resource efficiency, which appears to have many gaps with resilient infrastructure core concepts, is commonly implemented in practice; however, efficiency only provides one aspect of resilience in natural systems as seen in Life's Principles. Not every core concept will be able to address every life principle. Though exploration of these gaps may prove that the core concepts are not binary classifications but a continuum, which would strengthen a tradeoff analysis. Resilience can only be achieved through an approach that recognizes complexity, deep uncertainty, and interrelatedness of components. This means that resilient infrastructure design should seek to optimize not one but all Life's Principles through satisficed solutions.

By considering infrastructure systems and the collective of resilient infrastructure core concepts, infrastructure managers can begin assessing their designs in accordance with Life's Principles. The objective is not for every resilient infrastructure core concept to address every principle and sub-principle, but that infrastructure design addresses each of these principles and sub-principles through an agglomeration of resilient infrastructure core concepts. Life's Principles offer a succinct snapshot of design lessons and strategies that have emerged and sustained life over 3.8 billion years in the face of dynamic and often unpredictable operating conditions (Biomimicry 3.8, 2013). These strategies offer a useful and immediately practical lens for considering resilience approaches in a complex and deeply uncertain world.

4.5.5. REFERENCES

Abdullah, K. (2004) 'Stormwater management and road tunnel (SMART) a lateral approach to flood mitigation works', *International Conference on Bridge Engineering and Hydraulic Structures*, pp. 59–79.

Adger, W.N. (2000) 'Social and ecological resilience: are they related?', *Progress in Human Geography*, 24(3), pp. 347–364. Available at: https://doi.org/10.1191/030913200701540465.

Ahern, J. (2011a) 'From fail-safe to safe-to-fail: Sustainability and resilience in the new urban world', *Landscape and Urban Planning*, 100, pp. 341–343. Available at: https://doi.org/10.1016/j.landurbplan.2011.02.021.

Ahern, J. (2011b) 'From fail-safe to safe-to-fail: Sustainability and resilience in the new urban world', *Landscape and Urban Planning*, 100(4), pp. 341–343. Available at: https://doi.org/10.1016/j.landurbplan.2011.02.021.

Aitamurto, T. *et al.* (2013) 'Biomimicry: A Path to Sustainable Innovation', *Design Issues*, 29(4), pp. 1–5.

Allenby, B.R. (2012) *The Theory and Practice of Sustainable Engineering.* Hoboken, NJ: Upper Saddle River: Pearson Prentice Hall.

Amin, M. (2000) 'Toward self-healing infrastructure systems', *Computer*, 33(8), pp. 44–53. Available at: https://doi.org/10.1109/2.863967.

Amin, M. (2001) 'Toward self-healing energy infrastructure systems', *IEEE Computer Applications in Power*, 14(1), pp. 20–28. Available at: https://doi.org/10.1109/67.893351.

Anderies, J., Janssen, M. and Ostrom, E. (2003) 'Design Principles for Robustness of Institutions in Social-Ecological Systems', in. *Joining the Northern Commons: Lessons for the World, Lessons from the World*, Anchorage, AK. Available at: https://hdl.handle.net/10535/1777.

Anderies, J., Janssen, M. and Ostrom, E. (2004) 'A Framework to Analyze the Robustness of Social-ecological Systems from an Institutional Perspective', *Ecology and Society*, 9(1). Available at: https://doi.org/10.5751/ES-00610-090118.

Anderies, J.M. and Janssen, M.A. (2012) 'Elinor Ostrom (1933–2012): Pioneer in the Interdisciplinary Science of Coupled Social-Ecological Systems', *PLOS Biology*, 10(10), p. e1001405. Available at: https://doi.org/10.1371/journal.pbio.1001405.

AskNature (2018) 'Inspired Ideas', *The Biomimicry Institute* [Preprint].

Ayyub, B.M. (2018) *Climate-Resilient Infrastructure.* Edited by Committee on Adaptation to a Changing Climate. Reston, VA: ASCE Press. Available at: https://doi.org/10.1061/9780784415191.

Benyus, J.M. (1997) *Biomimicry: Innovation Inspired By Nature.* New York, NY: HarperCollins.

Bestelmeyer, B.T. *et al.* (2018) 'The Grassland–Shrubland Regime Shift in the Southwestern United States: Misconceptions and Their Implications for Management', *BioScience*, 68(9), pp. 678–690. Available at: https://doi.org/10.1093/biosci/biy065.

Biomimicry 3.8 (2013) *Life's principles: Biomimicry design lens.* Available at: https://www.researchgate.net/figure/Design-lens-Biomimicry-38-lifes-principles-Biomimicry-38-2013-Permission-granted-by_fig2_339799264 (Accessed: 23 October 2023).

Biomimicry Institute (2019) *What Is Biomimicry?* Available at: https://biomimicry.org/what-is-biomimicry/.

Brown, K. (2014) 'Global environmental change I: A social turn for resilience?', *Progress in Human Geography*, 38(1), pp. 107–117. Available at: https://doi.org/10.1177/0309132513498837.

Burby, R. (2006) 'Hurricane Katrina and the Paradoxes of Government Disaster Policy: Bringing About Wise Governmental Decisions for Hazardous Areas', *The ANNALS of the American Academy of Political and Social Science*, 604. Available at: https://doi.org/10.1177/0002716205284676.

Cai, B., Xie, M., Liu, Y., Liu, Y., & Feng, Q. (2018). Availability-based engineering resilience metric and its corresponding evaluation methodology. Reliability Engineering & System Safety, 172, 216–224. https://doi.org/10.1016/J.RESS.2017.12.021

Cardinale, B.J. *et al.* (2006) 'Effects of biodiversity on the functioning of trophic groups and ecosystems', *Nature*, 443(7114), pp. 989–992. Available at: https://doi.org/10.1038/nature05202.

Carpenter, S.R. *et al.* (2012) 'General Resilience to Cope with Extreme Events', *Sustainability*, 4(12), pp. 3248–3259. Available at: https://doi.org/10.3390/su4123248.

Champagne, C.L. and Aktas, C.B. (2016) 'Assessing the Resilience of LEED Certified Green Buildings', *Procedia Engineering*, 145, pp. 380–387. Available at: https://doi.org/10.1016/j.proeng.2016.04.095.

Chester, M., Markolf, S. and Allenby, B. (2019) 'Infrastructure and the environment in the Anthropocene', *Journal of Industrial Ecology*, 23(5), pp. 1006–1015. Available at: https://doi.org/10.1111/jiec.12848.

Chester, M.V. and Allenby, B. (2018) 'Toward adaptive infrastructure: flexibility and agility in a non-stationarity age', *Sustainable and Resilient Infrastructure*, 4(4), pp. 173–191. Available at: https://doi.org/10.1080/23789689.2017.1416846.

Chester, M.V. and Allenby, B. (2019) 'Infrastructure as a wicked complex process', *Elementa: Science of the Anthropocene*. Edited by A. Iles and M.E. Chang, 7(1), p. 21. Available at: https://doi.org/10.1525/elementa.360.

DHS (2013) *NIPP 2013: Partnering for Critical Infrastructure Security and Resilience*.

Donohue, I. *et al.* (2016) 'Navigating the complexity of ecological stability', *Ecology Letters*, 19(9), pp. 1172–1185. Available at: https://doi.org/10.1111/ele.12648.

Eder, W.E. and Hosnedl, S. (2008) *Design Engineering: A Manual for Enhanced Creativity*. Boca Raton, FL: CRC Press.

Fiksel, J. (2003) 'Designing Resilient, Sustainable Systems', *Environmental Science & Technology*, 37(23), pp. 5330–5339. Available at: https://doi.org/10.1021/es0344819.

Fisher, R., Norman, M. and Peerenboom, J. (2018) 'Resilience History and Focus in the USA', in A. Fekete and F. Fiedrich (eds) *Urban Disaster Resilience and Security: Addressing Risks in Societies*. Berlin, DE: Springer International Publishing (The Urban Book Series), pp. 91–109. Available at: https://doi.org/10.1007/978-3-319-68606-6_7.

Folke, C. *et al.* (2010) 'Resilience Thinking: Integrating Resilience, Adaptability and Transformability', *Ecology and Society*, 15(4). Available at: https://www.jstor.org/stable/26268226 (Accessed: 23 October 2023).

Francis, R. and Bekera, B. (2014) 'A metric and frameworks for resilience analysis of engineered and infrastructure systems', *Reliability Engineering & System Safety*, 121, pp. 90–103. Available at: https://doi.org/10.1016/j.ress.2013.07.004.

Grimm, N., Hale, R., & Iwaniec, D. (2015). A broader framing of ecosystem services in cities. *The routledge handbook of urbanization and global environmental change (1st Ed.)*. New York: Routledge. https://www.routledgehandbooks.com/doi/10.4324/9781315849256.ch14

Gu, X. and Jiang, N. (2017) *Self-healing Control Technology for Distribution Networks*. 1st edition. Edited by China Electric Power Press. Singapore ; Hoboken, NJ: Wiley.

Hayes, S. *et al.* (2019) 'Leveraging socio-ecological resilience theory to build climate resilience in transport infrastructure', *Transport Reviews*, 39(5), pp. 677–699. Available at: https://doi.org/10.1080/01441647.2019.1612480.

Helmrich, A.M. and Chester, M.V. (2020) 'Reconciling complexity and deep uncertainty in infrastructure design for climate adaptation', *Sustainable and Resilient Infrastructure*, 7(2), pp. 83–99. Available at: https://doi.org/10.1080/23789689.2019.1708179.

Hodapp, D., Hillebrand, H. and Striebel, M. (2019) '"Unifying" the Concept of Resource Use Efficiency in Ecology', *Frontiers in Ecology and Evolution*, 6. Available at: https://doi.org/10.3389/fevo.2018.00233.

Holling, C.S. (1973) 'Resilience and Stability of Ecological Systems', *Annual Review of Ecology, Evolution, and Systematics*, 4(1), pp. 1–23. Available at: https://doi.org/10.1146/annurev.es.04.110173.000245.

Holling, C.S. (1996) 'Engineering Resilience versus Ecological Resilience', in P. Schulze and National Academy of Engineering (eds) *Engineering Within Ecological Constraints*. Washington, DC: National Academies Press, pp. 31–44. Available at: https://doi.org/10.17226/4919.

Hooper, D.U. *et al.* (2012) 'A global synthesis reveals biodiversity loss as a major driver of ecosystem change', *Nature*, 486(7401), pp. 105–108. Available at: https://doi.org/10.1038/nature11118.

Hoover, D.L. *et al.* (2020) 'Traversing the Wasteland: A Framework for Assessing Ecological Threats to Drylands', *BioScience*, 70(1), pp. 35–47. Available at: https://doi.org/10.1093/biosci/biz126.

Hosseini, S., Barker, K. and Ramirez-Marquez, J.E. (2016) 'A review of definitions and measures of system resilience', *Reliability Engineering & System Safety*, 145, pp. 47–61. Available at: https://doi.org/10.1016/j.ress.2015.08.006.

Institute for Sustainable Infrastructure (2018) *Envision: Sustainable Infrastructure Framework Guidance Manual*. 3rd edn. Washington, DC: Institute for Sustainable Infrastructure. Available at: https://sustainable-infrastructure.org/wp-content/uploads/EnvisionV3.9.7.2018.pdf.

Iwaniec, D.M. *et al.* (2019) 'The Framing of Urban Sustainability Transformations', *Sustainability*, 11(3), p. 573. Available at: https://doi.org/10.3390/su11030573.

Keeler, B. L., Hamel, P., McPhearson, T., Hamann, M. H., Donahue, M. L., Meza Prado, K. A., et al. (2019). Social-ecological and technological factors moderate the value of urban nature. *Nature Sustainability*, 2(1), 29–38. https://doi.org/10.1038/s41893-018-0202-1

Kim, Y. *et al.* (2019) 'The Infrastructure Trolley Problem: Positioning Safe-to-fail Infrastructure for Climate Change Adaptation', *Earth's Future*, 7(7), pp. 704–717. Available at: https://doi.org/10.1029/2019EF001208.

Koliou, M. *et al.* (2018) 'State of the research in community resilience: progress and challenges', *Sustainable and Resilient Infrastructure*, 5(3), pp. 131–151. Available at: https://doi.org/10.1080/23789689.2017.1418547.

Kosky, P. *et al.* (2006) *Exploring Engineering: An Introduction for Freshmen to Engineering and to the Design Process*. Amsterdam, NL: Elsevier Science.

Lurie-Luke, E. (2014) 'Product and technology innovation: What can biomimicry inspire?', *Biotechnology Advances*, 32(8), pp. 1494–1505. Available at: https://doi.org/10.1016/j.biotechadv.2014.10.002.

Markolf, S.A. *et al.* (2018) 'Interdependent Infrastructure as Linked Social, Ecological, and Technological Systems (SETSs) to Address Lock-in and Enhance Resilience', *Earth's Future*, 6(12), pp. 1638–1659. Available at: https://doi.org/10.1029/2018EF000926.

Meerow, S., Newell, J.P. and Stults, M. (2016) 'Defining urban resilience: A review', *Landscape and Urban Planning*, 147, pp. 38–49. Available at: https://doi.org/10.1016/j.landurbplan.2015.11.011.

Moser, S. *et al.* (2019) 'The turbulent world of resilience: interpretations and themes for transdisciplinary dialogue', *Climatic Change*, 153(1), pp. 21–40. Available at: https://doi.org/10.1007/s10584-018-2358-0.

Mu, D. *et al.* (2011) 'A resilience perspective on biofuel production', *Integrated Environmental Assessment and Management*, 7(3), pp. 348–359. Available at: https://doi.org/10.1002/ieam.165.

NIAC (2009) *Critical Infrastructure Resilience: Final Report and Recommendations*. National Infrastructure Advisory Council, p. 54. Available at: https://www.cisa.gov/sites/default/files/publications/niac-critical-infrastructure-resilience-final-report-09-08-09-508.pdf.

Oguntona, O.A. and Aigbavboa, C.O. (2017) 'Biomimicry principles as evaluation criteria of sustainability in the construction industry', *Energy Procedia*, 142, pp. 2491–2497. Available at: https://doi.org/10.1016/j.egypro.2017.12.188.

Ostrom, E. (1990) *Governing the Commons: The Evolution of Institutions for Collective Action*. Cambridge, MA: Cambridge University Press.

Ostrom, E. (2007) 'A diagnostic approach for going beyond panaceas', *Proceedings of the National Academy of Sciences*, 104(39), pp. 15181–15187. Available at: https://doi.org/10.1073/pnas.0702288104.

Pahl, G. *et al.* (2007) *Engineering Design: A Systematic Approach*. London, UK: Springer Science & Business Media.

Park, J. *et al.* (2013) 'Integrating Risk and Resilience Approaches to Catastrophe Management in Engineering Systems', *Risk Analysis*, 33(3), pp. 356–367. Available at: https://doi.org/10.1111/j.1539-6924.2012.01885.x.

Pickett, S.T.A., Pickett, S.T. and White, P.S. (1985) *The Ecology of Natural Disturbance and Patch Dynamics*. Cambridge, MA: Academic Press.

Righi, A.W., Saurin, T.A. and Wachs, P. (2015) 'A systematic literature review of resilience engineering: Research areas and a research agenda proposal', *Reliability Engineering & System Safety*, 141, pp. 142–152. Available at: https://doi.org/10.1016/j.ress.2015.03.007.

Schlangen, E. and Sangadji, S. (2013) 'Addressing Infrastructure Durability and Sustainability by Self Healing Mechanisms - Recent Advances in Self Healing Concrete and Asphalt', *Procedia Engineering*, 54, pp. 39–57. Available at: https://doi.org/10.1016/j.proeng.2013.03.005.

Swanson, F.J. *et al.* (1982) 'Land-water interactions: the riparian zone', in *Analysis of coniferous forest ecosystems in the western United States.* UK: Hutchinson Ross Publishing, pp. 267–291.

Sydenham, P.H. (2004) *Systems Approach to Engineering Design.* London, UK: Artech House.

Tarlock, A. (1994) 'The Nonequilibrium Paradigm in Ecology and the Partial Unraveling of Environmental Law', *Loyola of Los Angeles Law Review*, 27(3), pp. 1121–1144.

The White House (2013) *Presidential Policy Directive -- Critical Infrastructure Security and Resilience, whitehouse.gov.* Available at: https://obamawhitehouse.archives.gov/the-press-office/2013/02/12/presidential-policy-directive-critical-infrastructure-security-and-resil (Accessed: 24 October 2023).

U.S. Green Building Council (2019) 'Leed Resilient Design Pilot Credits', p. 2.

Vincent, J.F.V. (2009) 'Biomimetics — a review', *Proceedings of the Institution of Mechanical Engineers, Part H: Journal of Engineering in Medicine*, 223(8), pp. 919–939. Available at: https://doi.org/10.1243/09544119JEIM561.

Wagg, C. *et al.* (2014) 'Soil biodiversity and soil community composition determine ecosystem multifunctionality', *Proceedings of the National Academy of Sciences*, 111(14), pp. 5266–5270. Available at: https://doi.org/10.1073/pnas.1320054111.

Woods, D.D. (2015) 'Four concepts for resilience and the implications for the future of resilience engineering', *Reliability Engineering & System Safety*, 141, pp. 5–9. Available at: https://doi.org/10.1016/j.ress.2015.03.018.

Wu, J. and Loucks, O.L. (1995) 'From balance of nature to hierarchical patch dynamics: A paradigm shift in ecology', *Quarterly Review of Biology*, 70(4), pp. 439–466. Available at: https://doi.org/10.1086/419172.

Wu, J., & Wu, T. (2013). Resilience in ecology and urban design: Linking theory and practice for sustainable cities. Ecological resilience as a foundation for urban design and sustainability (Vol. 3, 1st Ed.). Netherlands: *Springer*. https://doi.org/10.1007/978-94-007-5341-9

Yodo, N. and Wang, P. (2016) 'Engineering Resilience Quantification and System Design Implications: A Literature Survey', *Journal of Mechanical Design*, 138(111408). Available at: https://doi.org/10.1115/1.4034223.

Zari, M.P. (2018) *Regenerative Urban Design and Ecosystem Biomimicry.* London, UK: Routledge, Taylor & Francis Group.

Section 5

COGNITIVE ECOSYSTEM AND ARITIFICAL INTELLIGENCE

The preceding sections of this book have by and large been broadly functional: governance, resilience, and the Anthropocene, for example. In this last section, however, we focus on a new and unique infrastructure, artificial intelligence and, more broadly, the rise of the cognitive ecosystem.

We do so for several reasons. First, AI and cognitive functionality is increasingly being integrated into the traditional infrastructure with which engineers are concerned, from energy and the electric grid to water and power systems to intelligent buildings and road, rail, air, and freight transportation networks. While if properly done this trend enables safer and more efficient infrastructure, it also creates new and potentially significant vulnerabilities, especially in areas such as cybersecurity where traditional engineering domains are weak.

Second, AI is a unique infrastructure in that it operates not just as an infrastructure in itself, but at three different levels. As just mentioned, it is an increasingly important input to many if not most familiar infrastructure systems. Second, however, it is an infrastructure in itself, and although most people only think of software when AI is mentioned, it is in fact platformed on a highly complex, integrated, physical network of servers, chips, transmission technologies, supported by global supply chains and manufacturing facilities. In being both an enabling infrastructure and an infrastructure in itself, it is somewhat like electricity or the Internet.

But AI is also an enabling component of a metainfrastructure, the cognitive ecosystem. This cognitive ecosystem, emerging from the integrated functionality of global technologies – the Internet of Things, wired and wireless communication networks, massive server farms that support functions from memory to information processing, and so forth - could not function without AI, but it is much more than just AI. This level is, in practice, neither understood nor appreciated. Thus, for example, the EU, the US, China, and other political entities are rushing to regulate AI, but none of the proposals to date show any consideration for impacts of proposed regulation on the functioning of the cognitive ecosystem.

It is perhaps understandable that neither policymakers nor the public are able to grasp, much less manage, the complexity of AI as it operates across these three functional domains: as enabling infrastructure, as an infrastructure in itself, and as a critical component of the emerging cognitive ecosystem. But it is important for engineering professionals and technologists to understand this terrain, complex and rapidly evolving as it might be, if they are to continue to provide the built environment, and infrastructure, that the Anthropocene will demand.

5.1

RISE OF THE COGNITIVE ECOSYSTEM

5.1.1. INTRODUCTION

In the beginning of the movie 2001: A Space Odyssey, an ape, after hugging a strange monolith, picks up a bone and randomly begins playing with it ... and then, as Richard Strauss's Also sprach Zarathustra rings in the background, the ape realizes that the bone it is holding is, in fact, a weapon. The ape, the bone, and the landscape remain exactly the same, yet something fundamental has changed: an ape casually holding a bone is a very different system than an ape consciously wielding a weapon. The warrior ape is an emergent cognitive phenomenon, neither required nor deterministically produced by the constituent parts: a bone, and an ape, in a savannah environment.

Cognition as an emergent property of techno-human systems is not a new phenomenon. Indeed, it might be said that the ability of humans and their institutions to couple to their technologies to create such techno-human systems is the source of civilization itself. Since humans began producing artifacts, and especially since we began creating artifacts designed to capture, preserve, and transmit information—from illuminated manuscripts and Chinese oracle bones to books and computers—humans have integrated with their technologies to produce emergent cognitive results.

And these combinations have transformed the world. Think of the German peasants, newly literate, who were handed populist tracts produced on then-newfangled printing presses in 1530: the Reformation happened. Thanks to the printers, information and strategies flowed between the thinkers and the readers faster, uniting people across time and space. Eventually, the result was another fundamental shift in the cognitive structure: the Enlightenment happened.

This chapter was adapted from the following article with publisher permission: Braden Allenby, 2021, World Wide Weird: Rise of the Cognitive Ecosystem, *Issues in Science & Technology*, 37(3).

In the 1980s Edwin Hutchins found another cognitive structure when he observed a pre-GPS crew navigating on a naval vessel: technology in the form of devices, charts, and books were combined with several individuals with specialized skills and training to produce knowledge of the ship's position (the "fix"). No single entity, human or technological, contained the entire process; rather, as Hutchins observed: "An interlocking set of partial procedures can produce the overall observed pattern without there being a representation of that overall pattern anywhere in the system." The fix arises as an emergent cognitive product that is nowhere found in the constituent pieces, be they technology or human; indeed, Hutchins speaks of "the computational ecology of navigation tools."

Fast forward to today. It should be no surprise that at some point techno-human cognitive systems such as social media, artificial intelligence (AI), the Internet of Things (IoT), 5G, cameras, computers, and sensors should begin to form their own ecology — significantly different in character from human cognition. Perceiving the rise of such a cognitive ecosystem is a different matter, however: in periods of technological, social, and political upheaval, there is always a tension between feeling that despite superficial appearances, things are much as they always were, or, alternatively, that the world has gotten fundamentally weirder. Today, in one important way, it is increasingly apparent that the latter perspective is, in fact, correct, and that the weirdness is arising from deep roots that transcend our everyday frameworks. It is arising from an evolutionary leap foreshadowed throughout human history, but now, after a long ramp-up, emerging explosively across social, military, political, and cultural landscapes: what we will describe as a global cognitive ecosystem. Understanding this emergence is one of the principal challenges of our age, because the cognitive ecosystem undermines many, if not all, of our existing institutions and assumptions.

Although it is obvious we live in a period of dramatic technological, political, social, economic, institutional, geopolitical, and cultural weirdness, it is nonetheless difficult to independently perceive unfamiliar and unexpected emergent behaviors, especially when they involve unprecedented levels of complexity and cut against ways in which people and institutions have learned to parse their world. We are, in a sense, like the navigators on Hutchins's warship: each aware of only the parts, not the whole. So, while we may be aware of elements of technological infrastructure, it isn't surprising that the emergence of a cognitive ecosystem that includes them, and other technologies, institutions, and academic disciplines among its subsystems, is both unperceived and unremarked.

As historical examples suggest, the emergence of the cognitive ecosystem has the power to transcend and radically reshape everything from individual psychologies to institutions to societies and geopolitics, and indeed the world. Thus, it's necessary to understand how such evolving distributed cognition draws capability and capacity from across a number of apparently unrelated infrastructure, services, institutions, and technologies, driven by economic and geopolitical competition, tied together by AI and various institutional structures and networks ranging from private firms to military and security organizations. The characteristics of this evolving system are already clear:

- It contains the functional components of cognition and ever more powerful networks linking them together operationally.

- It is multi-scalar, both in scope and in complexity.

- It is globally distributed.

- It is evolving emerging systemic and behavioral capabilities.

- It includes learning and information-processing functionality at all levels that may include, but is not moderated by, humans.

- It is driven forward by powerful competitive forces at state and corporate levels.

REFLEXIVE CONTROL FROM THE 1970S TO TODAY

To understand what this looks like today, and how it differs from the past, remember an example from the Cold War. During the 1970s, the Soviet Union developed a theory of "reflexive control," which involved structuring narratives and disinformation campaigns causing people, such as activists in the United States, to act in ways that they would believe were voluntary, but in fact were predetermined by the Soviets to benefit their country's interests. Such a strategy, while seductive, proved difficult to implement given the technological and geopolitical environment of the time.

Now consider the 2016 US election campaign, where Russia appears to have used social media, weaponized narratives, bots, and other techniques to gain reflexive control of at least some American voters across the political spectrum. What seemed outlandish before the election came to seem so obvious afterward that activists of all stripes are now freely making use of weaponized narratives to implement reflexive control over, for example, their political bases.

What had changed were the conditions: the powerful networks, replication of functional components of cognition in global technological networks at multiple scales, dramatic increases in knowledge about human decision-making, psychology, and behavior—driven forward by powerful competitive forces involving government and corporate adversaries and competitors. The implications of continued application of cognitive ecosystem power to civilizational conflict, and social, cultural, and governance systems, will be profound.

Now, without recognizing it, we are building a functionally integrated global cognitive ecosystem, and we are doing so rapidly and at global scale. That's why the world seems to be weirder: it is. And that weirdness comes with a bite: the countries and companies that can work with the new capabilities and powers that the cognitive ecosystem supports will succeed, and those that can't will fail.

5.1.2. DEFINING THE COGNITIVE ECOSYSTEM

Definitions of "cognition" and related terms such as intelligence, consciousness, free will, and mind quickly get vague. Existing definitions do, however, tend to fall into two clusters: those that are anthropocentric, and those that are not. This dichotomy reveals our tendency to view human cognitive activity as the sin qua non of any sort of intelligence or mental function. Even the term "artificial intelligence" follows this view—as if human intelligence is the real thing, and anything software or machines do merely artificial.

There are, however, some significant drawbacks to equating human cognition with all forms of cognition. First, recent advances in such diverse fields as personal psychology, behavioral economics, and neuroscience have revealed just how dependent on heuristics, unconscious rules of thumb, kludges, and shortcuts human cognition and decision-making really are. For example, it is doubtful that nonhuman cognitive systems will need to use emotion the same way we do, as a convenient decision-making shortcut that reduces the need to depend on applied rationality. Even today, AI systems can make many decisions more rapidly than humans, which doesn't mean that the human brain isn't an amazing computational device, but it does suggest that as the cognitive ecosystem matures, human cognition, which has always been a part of the techno-human structures underlying the cognitive ecosystem, will assume a different role.

It is true historically, of course, that much human cognition has been integrated with institutions, cultural practices, information repositories such as books, and technologies in the form of "congealed cognition," which systematically enhance the scope, power, and creativity of real-time human cognition. Certainly, the navigation charts and tools that Hutchins observed, and which are so important to the process of navigating, represent such congealed cognition modules. Humans in such structures still provide goals and agency, while congealed cognition (practices, standards, navigation charts and devices, and so forth) provides enhanced cognitive functionality such as data acquisition, computation, memory, communication, and monitoring. Human cognition at the individual, institutional, and cultural level, then, has never been and is not today outside the cognitive ecosystem.

What is different, then, is not that the cognitive ecosystem is new. It is that the performance of the cognitive ecosystem is reaching a tipping point, where active learning and networked global techno-human cognitive processes evolve and function in information environments that involve levels of information volume, velocity, and complexity beyond anything that humans and their institutions are accustomed to.

When a functional rather than anthropocentric definition of cognition is considered, this becomes even clearer. Typical functional elements of cognition include perception, learning, differentiation, reasoning and computation, problem-solving and decision-making, memory, information processing, and communication with other cognitive systems (language, media, video, or machine to machine, for example). While the elements that taken together constitute cognition may themselves be somewhat difficult to define, they nonetheless offer a way to begin to visualize and understand the cognitive ecosystem.

For purposes of explication, the cognitive ecosystem can schematically be broken down into three large domains: the data economy, infrastructure that provides cognition, and the infrastructure of institutions and services. In practice, of course, these domains overlap, but this map provides a relatively easy way to order its components.

5.1.3. THE COGNITIVE ECOSYSTEM

The first domain, that of the data economy, is a vibrant marketplace that is growing to be as large and as complex as the money economy that it parallels. This domain thus includes data generation and distillation services ranging from Internet of Things devices and networks, and fleets of vehicles learning to be autonomous, to social media platforms, payment systems, and facial recognition technologies. The data economy requires sensors to generate the data in the first place, massive memory storage capability, and clever algorithms and the processing power to structure underlying data into meaningful patterns and products that are used for commercial, military, and security purposes. Although the physical capabilities of the expanding data economy are far beyond human cognitive capability, humans still control what kinds of data are developed, and how they are aggregated and used.

Figure 5.1.1: The Cognitive Ecosystem

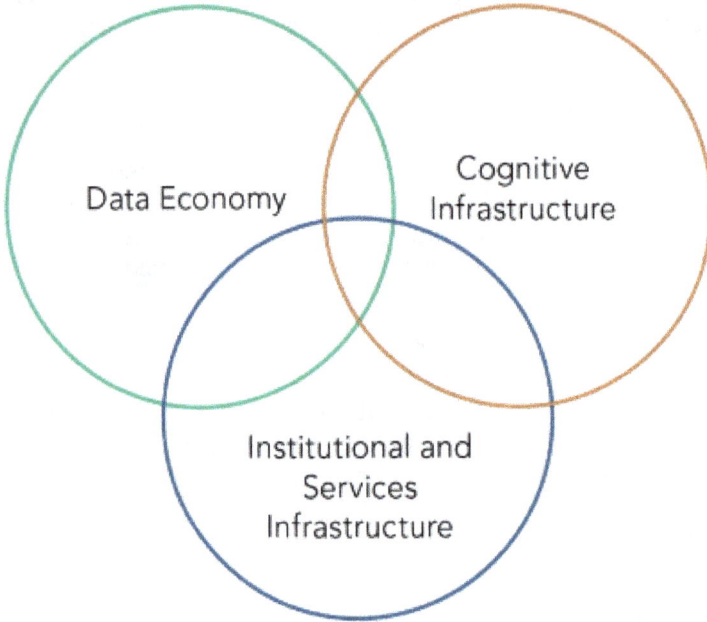

The other two domains include many diverse forms of infrastructure. Cognitive infrastructure consists of those institutions, technologies, services, and products that provide the functional elements of cognition, from perception to constructions of intelligibility to applications such as problem-solving. Sensors in mobile phones, IoT products, point of sale payment technologies, facial recognition cameras, autonomous vehicles, and other devices provide information streams that are then processed by the cognitive infrastructure of AI neural net technologies for various purposes—from security to marketing to disinformation campaigns. Institutional and services infrastructure comprises not only the platforms of such global social media firms as Facebook, X (Twitter), Weibo, WeChat, and TikTok, but also the rapidly deepening intellectual capital supporting the cognitive ecosystem, including behavioral economics, personal and evolutionary psychology, cultural studies, and neuroscience.

There is considerable overlap among these domains, particularly as they rely on massive stocks and flows of data, and thus require and enable evolution of the data economy. The firms that create social media platforms, for example, have substantial roles in shaping the cognitive ecosystem, and they in turn are affected by the different framework of policies, practices, and cultural beliefs characterizing different jurisdictions.

CHINA'S SOCIAL CREDIT SYSTEM

The People's Republic of China's social credit system (SCS), a cognitive ecosystem application par excellence, was authorized in 2014 and is currently being implemented. The SCS, still in its early stages, is a mechanism by which data regarding many aspects of an individual's private and public behavior, from jaywalking to prompt payment of debts, are integrated into a single numerical score that indicates how trustworthy, and how good a citizen, that person is. Where implemented, the SCS score controls whether a person can get on trains or planes, what dating sites they can use, whether they can get a loan, whether they can get into college, what friends they can have, and much else.

Operating such a system is a massive technological challenge and involves all three domains of the cognitive ecosystem, requiring everything from facial recognition technology to AI/big data/analytics, to networked high-speed communications technologies such as 5G, to data compiled and provided by Chinese social media firms. It creates not just cultural challenges, as its capabilities are in essence negotiated between the state and Chinese citizens, but significant institutional challenges, as it requires validating and integrating inputs from many different firms and government entities at many political levels with a degree of granularity—the individual citizen—that is possible only with the rise of the cognitive ecosystem.

At its heart, an effective and well-designed SCS represents the most significant challenge to pluralistic governance systems since the beginning of the Enlightenment. (Of course, it is not yet clear how competently China's version will be designed and implemented). At an extreme, it enables a reflexive relationship between the citizenry and the government that obviates the lack of legitimacy required by traditional authoritarianism, while at the same time providing a mechanism for governments to make rapid, adaptive responses in a shifting environment. For the first time, this powerfully ubiquitous tool offers governments the ability to design social and cultural stability with reasonable efficiency and cost.

The geopolitical struggle for ascendancy between a more assertive, social-credit-system-powered China, and an increasingly divided and hyper-partisan America is thus not occurring in a technological vacuum, but in the context of a rapidly emerging and coevolving part of the global cognitive ecosystem. It is a new technological reality that the Chinese are using well, and the Americans and the West, so far, are not.

5.1.4. UNDERSTANDING THE SCALE OF TODAY'S COGNITIVE ECOSYSTEM

Across each of the three domains constituting the cognitive ecosystem, there is rapid, accelerating development and deployment of technologies, services, and cultural and social practices that taken together are forming the constituent components of cognition, but at networked global scale. This technology is auto-catalyzing, and can evolve much more rapidly than institutional, legal and regulatory, or cultural systems, especially if the evolution is distributed at all scales throughout the cognitive ecosystem.

A snapshot of technology as of the beginning of 2020 provides a sense of the scale of the cognitive ecosystem today. As of 2020, between 25 billion and 50 billion objects — refrigerators, microwaves, microphones, cars, and airplanes, among many others — are linked to the internet. They contain an estimated billion sensors, which are designed to be sensitive to some inputs while disregarding others that are not relevant to their function — and they in turn feed information into machines and systems that further integrate the data, rejecting some and highlighting other information. Increasingly, these systems talk to each other, and to learning systems that then reprogram them to function more efficiently and effectively based on data and assessment across networks of devices; each Tesla teaches other Teslas. Thus, machine-to-machine connections at all scales are exploding: Cisco notes they increased from 17.1 billion in 2016 to 27.1 billion by 2021.

Perhaps the two most critical functions required to enable these networks to evolve and prosper across vastly different scales are AI and memory. Importantly, AI is becoming ubiquitous, enabling rapid and accelerating functionality across the ecosystem as a whole. AI's evolution is constrained not by installation of physical facilities but by reflexive modifications operating in software systems. Memory, which supports this reflexive process, is also expanding. In 2025 global stored data was estimated to stand at 200 zettabytes, roughly a factor of 6 larger than in 2018.

None of these technologies, from AI to sensors to servers, could by itself create a historical tipping point. A single neural network AI, working with a simple data set but unconnected to broader systems, is not going to fundamentally change the world. But that is not what the cognitive ecosystem represents. Rather, it is a step change in the cognitive capability of techno-human systems, with learning and cognitive capability increasingly diffused across many different networks at many different levels, but all interconnected. It is this rapidly developing structure that enables entirely new functionality, with profound implications for institutions, governments, and cultures.

Thus, China is integrating powerful data-generating technologies, such as the facial recognition and financial credit technologies, with data processing capabilities at many different scales, which are themselves integrated into vast multidomain networks such as the social credit system. This cognitive ecosystem, like human cognition, generates levels of processing networks that float on lower-level sensor and model-building functions, and in turn inform higher-level cognitive function.

Through these levels, and by design, the cognitive ecosystem is not a replacement for human cognition, but rather integrates human cognition into its operation in many ways. In general, for example, motivation, goals, and ethics are provided by the human components of techno-human cognitive systems. But performance in complex, rapidly shifting environments increasingly calls on the technological side of cognitive ecosystem capabilities.

This is not, however, a stable relationship: like the navigators on a 1980s warship, humans and organizations work on individual elements of the larger cognitive infrastructure, without knowing the capacity of the whole. As a result, human goals and desires tend to reflect local conditions rather than the state of the global cognitive ecosystem, with the predictable result that actions taken at a subsystem level, which may seem perfectly appropriate at that level, may result in undesirable behaviors of the system taken as a whole. For example, a European Union initiative that is widely embraced in America, the General Data Protection Regulation, supports privacy at the national and EU level. At the level of global geopolitics and the cognitive ecosystem, however, the GDPR serves primarily to restrict the data available to grow and train Western AI systems, thus providing a significant advantage to the Chinese, who not only do not share the privacy fetish of the West, but in fact are generating a vast data flow from their SCS.

The weirdness we mentioned early on is real, but it is far deeper than mere angst arising from cultural change. At least in part, it reflects a profound change between human "naïve cognition" and a world increasingly structured by the emergent behaviors of the cognitive ecosystem, a world where the relationship between humans and cognition is being continually redefined in ways that few people understand.

GROWING COMPLEXITY IN THE COGNITIVE ECOSYSTEM: CASE STUDIES

To better understand what the emergence of the cognitive ecosystem means, it helps to observe the way it behaves at various scales and levels of complexity.

General Electric, like many jet engine manufacturers, equips its engines with sensors, software systems, and automatic reporting technology to monitor engine health and predict potential problems in real time so they can be more easily and efficiently addressed. Data on such factors as engine operating temperatures, vibration, and flight conditions are fed into machine-learning systems to be analyzed. Prior to the pandemic, for example, GE's Middle East Technology Center in Dubai analyzed 10 gigabytes of data produced by the engines in Emirates airline's Boeing 777s every few seconds. Using these data, Emirates was able to reduce unscheduled maintenance by 50% and increase engine "time on wing" by 20%, thus lowering costs while improving safety and reliability. At a higher level, GE and other engine manufacturers integrate data feeds from engines on the many types of airplanes flying for many different carriers, enabling the identification of systemic issues, and continually improving maintenance and operation services, as well as fundamental engine design.

At a superficial level, this sounds like what Tesla, Waymo, and other firms introducing autonomous vehicle technology are doing. Equipped with sensors and information processors, every vehicle on the road learns—and, because each vehicle sends its data back to an integrated machine-learning platform, vehicles learn from each other as their software is updated by the companies' machine-learning systems. This is significant for at least three performance domains: computer vision, vehicle prediction of immediate future states, and driving policy generation and validation.

Both the airplane and the driverless vehicle systems make use of a cognitive ecosystem, but there are some fundamental differences. These systems operate at very different levels of organizational and technological complexity—as well as social and economic scale. For all its complicated technology and reams of data, the GE system is a bounded, explicitly designed, simple system operating according to deterministic and known principles controlled by a single entity.

By contrast, autonomous vehicle technology is a complex adaptive system beset by wicked complexity and tightly coupled domains that range from the highly technical software and hardware technologies required for autonomous function to a constantly changing environment full of low-probability events such as children on bicycles, all set within cultural, legal, regulatory, and ethical expectations that are themselves changing in reaction to the technology's development.

Considered from the standpoint of governance and institutional complexity, the jet engine learning system is under the explicit control of a single firm, GE. But autonomous vehicles operate within a space that includes insurance companies, federal and state regulators, shipping firms, individual consumers, and people with whom the technology shares the road. The output of the GE system is predictable: a safer, more efficient, jet engine. To the contrary, the implications of global-scale autonomous vehicle technology are profoundly unknowable: Massive unemployment? Unpredictable culture change? Fundamental shifts in urban design, energy consumption, and quality of life? Safer cars? Collapse of the automobile insurance industry? An end to suburbs? No wonder that, while the GE system is already implemented and effective, autonomous vehicle technology is proving far harder and more complex than predicted.

Measured by cognitive function, the GE engine is basic while that of autonomous vehicle is far more open-ended. In the GE case, single ownership of the cognitive process means that a relatively small number of entities—GE and its customers—understand fairly precisely the costs and benefits of implementation. In the autonomous vehicle case, the technological domain is (reasonably well) understood, but neither the firms nor their customers have any remit to consider the meta-level cognitive system impacts, except as they may impact competition between firms. And under these pressures, the highly competitive and essentially unbounded environment of autonomous vehicle deployment will also accelerate fundamental advances in machine learning/AI/big data function with spillover effects across many domains, from military systems to education. Cognitive ecosystem technology that can effectively manage autonomous transportation systems at scale, for example, will be able to do the same thing on a battlefield, or in cyberwar. The ape picks up the bone, the German peasant reads the tract, social media firms enable the subversion of American democratic practices—the world changes in unexpected ways, but faster and more unpredictably.

5.1.5. WHAT IS TO BE DONE?

It is too early, and the changes are coming too fast and too broadly, to be able to predict how humans will interact over time with the distributed techno-human cognitive structures that are growing increasingly complex and powerful around us.

This is especially fraught because the role of human cognition in distributed techno-human cognitive processes continues to fundamentally change: as functions such as memory, computation, and communication shift to technologies, humans increasingly migrate to alternative functions including defining goals, exercising agency, and supervising the training and algorithmic structure of AI. Human cognition is still critical, but increasingly it is a smaller part of larger and more complex techno-human systems platformed on a rapidly evolving cognitive ecosystem.

Moreover, one of the implications of weaponized narrative and the increasing success of Russia's reflexive control disinformation campaigns is that human identity and behavior are themselves becoming design spaces—and battle spaces. There are challenges presented by the evolution of the cognitive ecosystem not just at the level of the individual, but at the level of community, of institutions, and of governance systems themselves. China's social credit system doesn't offer just the potential for better management of a large and diverse population—the soft authoritarianism most Western analysts focus on—but a solution to the problem that has plagued large authoritarian systems for centuries: how does a remote, and authoritarian, government keep track of what the population is doing and thinking so that potential problems can be addressed before they become challenges to the legitimacy of authority? After all, the East German secret police, the Stasi, needed 90,000 full-time employees, assisted by 170,000 full-time unofficial collaborators, and a budget estimated at a billion dollars a year, to keep track of a population of some 16 million. By comparison, China's SCS is far cheaper and potentially more effective.

Responses to the deep challenges posed by the cognitive ecosystem cannot be simple or superficial—antitrust initiatives won't stop the ill effects of social media from fragmenting society. In fact, probably the best possible response is to continue incremental responses to immediate problems, while recognizing that any longer-term strategy must involve enhancing the agility and adaptability of humans and their institutions in the face of unpredictable, foundational, and accelerating change. And in both cases, explicitly embracing the cognitive ecosystem, and trying to work within it, are necessary preconditions to ethical and rational responses to its challenges.

But there is a far more fundamental cosmic reflexivity at play in the rise of the cognitive ecosystem. Think of how the big, AI-powered search engines and service firms such as Amazon, Alibaba, Tencent, and Google; and the social and financial AI-powered evaluation platforms such as the Chinese social credit system; and the powerful governments such as the U.S., China, and the European Union increasingly know what you want, and track you in detail. Then, recall Mathew 10:29 (New Living Translation): "What is the price of two sparrows—one copper coin? But not a single sparrow can fall to the ground without your Father knowing it."

In medieval Europe, it was believed that God knew every detail of your life; today, the cognitive ecosystem we are building all around the world

increasingly actually does. Indeed, at heart the cognitive ecosystem project is nothing less than the construction of the Mind of God. And we
would do well to remember nineteenth century orator Robert Green Inger-
soll's admonition: "An honest God is the noblest work of Man." He meant
it metaphorically. But as the original Enlightenment gives birth to a world
profoundly different than any humans have experienced in history—an
anthropogenic world with emerging cognitive capabilities we are not even
perceiving, much less managing—it appears less metaphor than road
map.

5.2

INFRASTRUCTURE AND THE COGNITIVE ECOSYSTEM

5.2.1. INTRODUCTION

Disruption of legacy infrastructure systems by novel digital and connected technologies represents not simply the rise of cyberphysical systems as hybrid physical and digital assets but, ultimately, the integration of legacy systems into a new cognitive ecosystem. This cognitive ecosystem, an ecology of massive data flows, artificial intelligence, and connected technologies, is poised to alter how humans and artificial intelligence understand and control our world (Smart, Heersmink and Clowes, 2017; Allenby, 2021). To be prepared for this paradigm shift does not simply mean that infrastructure managers need better tools to integrate cybertechnologies and protect against new types of vulnerabilities, but that, more significantly, their systems are becoming increasingly absorbed into an emerging suite of data, analytical tools, and decision-making technologies that will fundamentally restructure how legacy systems behave and are controlled, how decisions are made, and most importantly what the systems fundamentally are.

Consider the rise of web mapping, such as Google Maps and Apple Maps, and the control they exert over transportation services. These services now route a significant and increasing portion of vehicle traffic. Routing is based on a cognitive overview of the system only available to the mapping providers, which provides insights into the network and its use that far exceeds the information streams available to legacy transportation agencies – from in-pavement loop detectors and traffic cameras on a limited portion of the network. Google and Apple Maps - informed by

This chapter was adapted from the following journal article with publisher permission: Mikhail Chester and Braden Allenby, 2023, Infrastructure and the Cognitive Ecosystem: An Irrevocable Transformation, *Environmental Research: Infrastructure and Sustainability*, 3(3), 033002, doi: 10.1088/2634-4505/aced1f.

millions of mobile phone users - can see problems unfold in real time (Herrera *et al.*, 2010), predict traffic ahead of time using AI (Herring *et al.*, 2010; Lau, 2020), and adjust trips to avoid the problem (Powelson and Stoltzman, 2016), effectively optimizing how the system is used based on their objectives (Tseng and Ferng, 2021). Objectives range from reducing travel time to reducing fuel use (Dicker, 2021; Google, 2023), making such mapping software one of the most powerful and least recognized environmental technologies released to the public. These mapping platforms are not simply changing how traffic moves but are more so redefining what transportation systems are, an emerging ecosystem of novel technologies controlled by novel data streams accessible to increasing numbers of stakeholders steered by algorithms, layered on and changing the dynamics of existing transportation systems (Chester and Allenby, 2021).

To reduce the integration of legacy services with the cognitive ecosystem as one of technology integration is wholly insufficient, especially since the cognitive ecosystem itself includes not just technology and data but also the cultural, institutional, and disciplinary dimensions that are an integral part of it. The rise of the cognitive ecosystem is just beginning to transform how knowledge management (creating, sharing, using, and managing knowledge in an organization, Girard, Girard, and Sagology, 2015) and sensemaking (a shared comprehension of how to interpret the environment that is implemented through the application of tacit and explicit knowledge, Weick, 1995; Cook and Brown, 1999) happens in infrastructure systems and services are consumed, and how this knowledge can be used as a foundation for increasingly integrated and global systems with far greater diversity of control. The infrastructure community must accept and embrace this shift and reimagine how their organizations make sense of increasingly complex interactions with their environments and within their systems (i.e., internal and external complexity, Simon, 1996).

5.2.2. RISE OF THE COGNITIVE ECOSYSTEM

The cognitive ecosystem is an emerging and highly complex feature of an increasingly anthropogenic planet. It integrates functionality from several more traditional technologies and infrastructure from communications systems such as 5G and backbone ICT networks; to computational capabilities in everything from home devices and automobiles to vast server farms; to increasingly powerful AI tools such as generative AI; to the rules, regulations, venture capitalists, and social media systems that co-evolve with the technologies. Figure 5.2.1 illustrates the convergence of these capabilities as three domains: the data economy (data generation and distillation services), the cognitive infrastructure (institutions, technologies, services, and products that supply the cognitive components, including perception, intelligibility, and problem-solving), and the institutional and services infrastructure (institutions and platforms that support the cognitive ecosystem, including ancillary fields such as behavioral economics and neuroscience) (Chapter 5.1). Within each domain, legacy physical infrastructure systems (assets that have typically been disconnected from cybertechnologies) are becoming integrated with novel digital (non-physical) systems (including sensors, communication technologies, cloud-based systems that store and analyze data, and increasingly make decisions about how the asset should operate) that expand the capabilities of the cyberphysical systems, decentralize control by increasing the number of stakeholders able to access information from the systems and affect service use, and increasingly allow AI to make decisions.

Figure 5.2.1: Physical and Non-physical Domains of the Cognitive Ecosystem

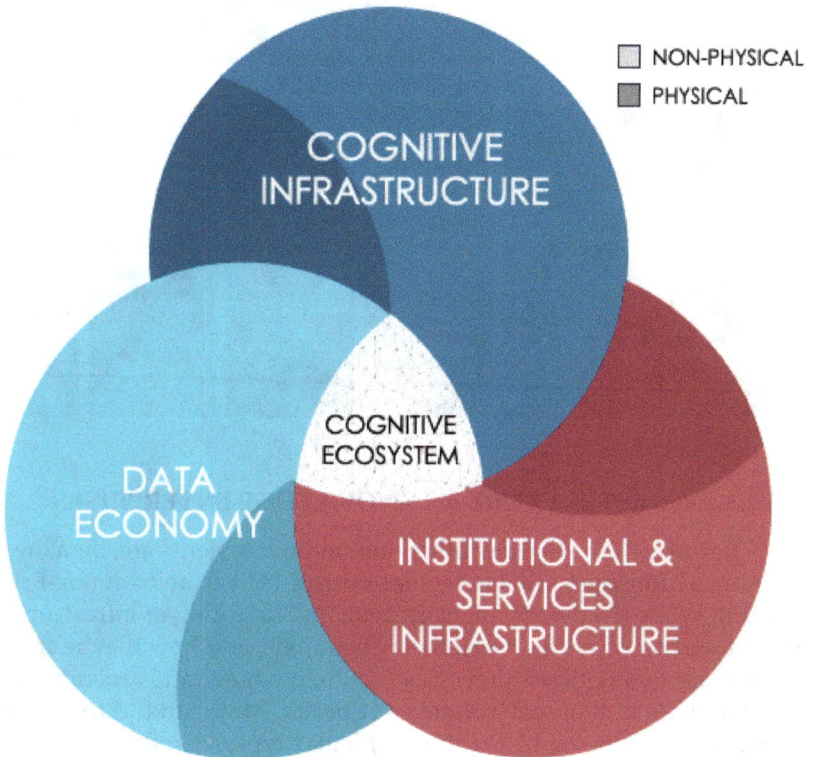

Lighter-colored areas represent non-physical (digital) systems, and darker-colored areas physical and often legacy systems. The relative area of each circle (domain) shaded as physical or non-physical is arbitrary.

This emerging ecosystem is cognitive because it increasingly displays the functions associated with cognition at multiple scales, including perception, information processing, internal and external communication, conceptualization, learning, reasoning, problem-solving, and memory. This evolving distributed cognition system increasingly integrates several apparently unrelated infrastructure, services, institutions, and technologies tied together by technical tools such as AI and a vast array of institutional structures and networks. The important defining characteristics of the cognitive ecosystem are shown in Figure 5.2.2.

Figure 5.2.2: Characteristics of the Cognitive Ecosystem

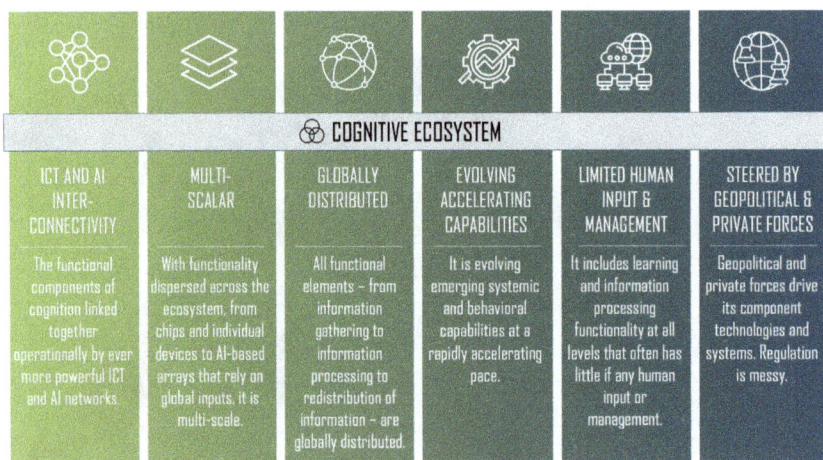

ICT AND AI INTER-CONNECTIVITY	MULTI-SCALAR	GLOBALLY DISTRIBUTED	EVOLVING ACCELERATING CAPABILITIES	LIMITED HUMAN INPUT & MANAGEMENT	STEERED BY GEOPOLITICAL & PRIVATE FORCES
The functional components of cognition linked together operationally by ever more powerful ICT and AI networks.	With functionality dispersed across the ecosystem, from chips and individual devices to AI-based arrays that rely on global inputs, it is multi-scale.	All functional elements – from information gathering to information processing to redistribution of information – are globally distributed.	It is evolving emerging systemic and behavioral capabilities at a rapidly accelerating pace.	It includes learning and information processing functionality at all levels that often has little if any human input or management.	Geopolitical and private forces drive its component technologies and systems. Regulation is messy.

5.2.3. INFRASTRUCTURE AS KNOWLEDGE ENTERPRISES

Infrastructure - like any other human-built system - are, in many ways, first and foremost knowledge enterprises (Miller and Muñoz-Erickson, 2018). While the design, management, and financing of infrastructure often focus on physical assets, their operations, and how they're managed, there are foundational processes that define how infrastructure agencies make sense of their environments (Chester, Miller and Muñoz-Erickson, 2020) and attenuate or amplify particular types of information between layers of management to make decisions (Chester and Allenby, 2022). An organization (infrastructure or otherwise) is intentionally designed to take in a select stream of information about the environment (Beer, 1985). The specific information taken in is an artifact of both the legacy organizational goals established at the inception of the agency (often decades ago) and how the organization has modernized in response to the perceived environment and mission change. If the organization has not felt compelled to change, that their environments are stable and their services uncompetitive, they likely maintained knowledge-making processes that reflect legacy priorities.

Remaining locked-in to legacy knowledge-making processes means that infrastructure systems are increasingly decoupled from their changing, expanding, integrating, uncertain, and increasingly complex environments. This decoupling implies that infrastructure agencies will be increasingly marginalized by the cognitive ecosystem and the introduction of a growing number of players who are able to exert control over their systems. Complexity in the context of infrastructure is both external and internal (Simon, 1996). External complexity refers to rapidly changing and increasingly complex environments in which infrastructure must function – driven by climate change, uncertain financing of aging assets, the evolution of coupled technologies, and social and cultural change that affect demand. Internal complexity describes infrastructure themselves, the rapid cyber integration of digital and connected technologies that creates both new possibilities as well as vulnerabilities (Chester and Allenby, 2020). Infrastructure agencies must constantly modernize their systems to engage with this complexity, adapting their systems to have a sufficient repertoire of responses relative to what the environment introduces, i.e., requisite complexity (Chester and Allenby, 2022). If climate change is producing more frequent and extreme weather events, then infrastructure agencies need to design for deep uncertainty and novel resilience strategies that they wouldn't have before. As cyber-attacks escalate, agencies will need new knowledge and new competencies to protect their systems (Chester and Allenby, 2020). Failure to change how the organization takes in the knowledge of these changing conditions and makes sense of this knowledge towards appropriate action results in systems being increasingly intervened in by new players. Our need for water, energy, and mobility won't disappear anytime soon, but who controls - or deliberately disrupts - how those services are used and delivered might.

New knowledge structures are needed for infrastructure systems that recognize the increasing fuzziness of their changing system boundaries in the cognitive ecosystem. As cybertechnologies become integrated into infrastructure, there is an accelerating emergence of cross-system functionality (Chester and Allenby, 2020). One infrastructure will integrate with another in novel ways, enabled by information and communication technology functionality that previously didn't exist. The demand for one service can be tied to the availability of another, e.g., the spinning up and down of water treatment technologies to reduce energy demand (Murray *et al.*, 2018). Novel monitoring technologies are being developed for co-located underground assets (Daulat *et al.*, 2022). An abundance of research has emerged on managing electric vehicle charging within supply constraints and reducing emissions (Sundstrom and Binding, 2012; Hoehne and Chester, 2016). These functionalities represent the emergence of trans-infrastructure systems supported by a foundation of cybertechnologies with their underlying data, analytical infrastructure, and artificial intelligence decision-making capabilities. Infrastructure also become a critical input to the data systems that support apparently unrelated functions, therefore becoming feedstock for behaviors that may arise elsewhere but are in part shaped by infrastructure. Whereas in the past knowledge- and sensemaking could be designed for and within a single infrastructure, this no longer appears to be the case. Infrastructure are being rapidly integrated into a cognitive ecosystem.

5.2.4. INFRASTRUCTURE AND THE COGNITIVE ECOSYSTEM

Positioning infrastructure managers to be better capable of navigating their systems as part of the cognitive ecosystem will fundamentally require a restructuring of organization and cross-organizational sensemaking. The cognitive ecosystem is reaching a tipping point, where learning and networked technologies have accelerated to the point where legacy institutional knowledge-making paradigms are unable to keep up. This hasn't been a problem in the past because systems were largely independent, controlled by a handful of agencies, and operating in environments with greater predictability and relatively slow change cycles. A transportation agency could largely operate without great concern for the dynamics of the energy system or navigation systems routing traffic. With the growth of the cognitive ecosystem, all systems converge on interconnectedness – enabled by cyberinfrastructure layers that support data flows and analysis – and become part of much more extensive systems in terms of both connections and global scale. The legacy knowledge-making models that pervade infrastructure agencies are simply not structured to make sense of this new paradigm. The Anthropocene is creating these major global systems. Sensemaking will differ across scales. On the ground, legacy knowledge will always be needed about how to design and operate assets and subsystems (although it may be generated increasingly by artificial intelligence). However, at broader scales of both functionality and geography, new sensemaking models will be needed to recognize the changing nature and dynamics of systems and be able to more quickly recognize and negotiate roles in terms of how the systems are being steered and to what end.

The potential for accelerating cycle times within cognitive ecosystem technologies means that infrastructure agencies must root the management of their systems in principles of agility (ability to maintain function in a non-stationary future) and flexibility (ability to meet changing demands) (Chester and Allenby, 2018). We refer to cycle times as the time between new generations of technologies, and in the Anthropocene, we can expect cognitive ecosystem cycle times to decrease as economies of scale are realized in data collection and information processing. Over the course of only a few months in early 2023, the world was exposed to ChatGPT, whose functionality was rapidly integrated into existing software. Within months of its release, the ChatGPT engine was quickly upgraded with greater imagination, sophisticated reasoning, multilingual proficiency, and the capacity to process visual information (Scharth, 2023). Multiple competing AI systems were released, and the world was suddenly thrust into a dialogue of what these now seemingly mature and rapidly evolving technologies mean. Whether AI, integrated technologies (such as vehicle-to-grid turning cars into home energy systems), or distributed technologies (such as wifi, rooftop solar, or mobility as a service), the implications of rapidly evolving technologies are serious for infrastructure agencies that have been structured around environments that have been slow to change and perceptions of no competition for service. The divide between how fast environments are changing and the capabilities of infrastructure organizations to respond is referred to as a decoupling. Accelerating organizational change starts with the structure and function of infrastructure governance, creating the capabilities to scan for future game changers (horizon or future scanning), practice horizontal governance, commit to sustained adaptation, and design for loose fit (allowing elements within the organization to restructure easily) (Chester and Allenby, 2022).

Fundamentally, the cognitive ecosystem is poised to alter what infrastructure are, producing a need for managers to self-reflect on the roles and behaviors of their systems. Reflection on the changing nature of infrastructure systems is paramount to their success, and any narrative that assumes that legacy systems will persist in slightly modified forms should be viewed with skepticism. While societies may always need water, energy, mobility, shelter, and other basic services, in some manifestations, the underlying structures and functions of legacy infrastructure systems that have delivered those services appear to be about to change significantly. This change appears to be increasingly driven by new technologies (including AI) and players that are able to harness data and agile and disruptive technologies. To this end, infrastructure managers should embark on creative exercises to reimagine the roles of their services and organizations in the deep future. They should support efforts to train their workforce to be capable of navigating this new frontier.

5.2.5. PREPARING INFRASTRUCTURE LEADERS

The continuing evolution of the Anthropocene and the cognitive ecosystem has profound implications for education in general and infrastructure education in particular. Engineering of infrastructure is essentially structured problem-solving that relies heavily on rational, usually quantitative, models and tools. As cognition diffuses across highly complex information structures, increasingly containing AI capability at all scales from the chip to the infrastructure component to the overall infrastructure system, and as the increasing complexity of the social, political, cultural, and economic structures become ever more tightly bound to built systems, reductionist, disciplinary frameworks which characterized the early Enlightenment approach to education fail (Allenby *et al.*, 2009; Allenby, 2011). Modern problem-solvers, whether they rely on quantitative or qualitative analytics and methodology, must become more agile, adaptive, and sophisticated in their approaches, and the education provided to them needs to adapt to support those capabilities. The experience with generative AI, such as ChatGPT, the most rapidly adopted and diffusing technology in history, with 100 million users in the first two months, illustrates the need for a more flexible, agile, and adaptive education system. It is true that education is both inherently conservative – tenure, for example, being a good way to ensure significant inertia – and institutionally rigid, relying on such reactionary mechanisms as peer review and accreditation programs such as ABET. Nonetheless, a few obvious ideas suggest themselves. For one, given the new capabilities which technology offers by the week, it is critical to develop institutionalized mechanisms that routinely zero-base the curriculum, asking what needs to be kept, what needs to be added, and what can be dropped. This will require a degree of flexibility and imagination that the academy has yet to display. For another, every course should, directly or indirectly, embrace the implications of the cognitive ecosystem, in particular, the need for students to be able to implement and work with AI and the constant need for attention to cybersecurity issues and vulnerabilities.

5.2.6. CONCLUSION

By asking how organizations should be structured for the challenges of the future, organizations take a critical step to reflect on bureaucratic norms, legacy priorities, and gaps between what the organization is capable of and what it needs to do going forward (Jessop, 2002; Chester, Miller and Muñoz-Erickson, 2020). However, simply reflecting is not enough. Infrastructure organizations will need to commit to sustained change that restructures and reorients the organization within a cognitive ecosystem where knowledge is generated, and control of services is wielded by myriad stakeholders. It may be that the successful infrastructure agency of the future is one that positions itself to build consensus on how services are produced and consumed (Muñoz-Erickson, Miller and Miller, 2017) and away from legacy models that emphasize control over services (Chester *et al.*, 2023), and is in a constant state of reinvention keyed to the accelerating cycle times of relevant technologies and the cognitive ecosystem itself.

5.2.7. REFERENCES

Allenby, B. *et al.* (2009) 'Sustainable engineering education in the United States', *Sustainability Science*, 4(1), pp. 7–15. Available at: https://doi.org/10.1007/s11625-009-0065-5.

Allenby, B. (2011) 'Rethinking engineering education', in *Proceedings of the 2011 IEEE International Symposium on Sustainable Systems and Technology. Proceedings of the 2011 IEEE International Symposium on Sustainable Systems and Technology*, pp. 1–5. Available at: https://doi.org/10.1109/ISSST.2011.5936869.

Allenby, B.R. (2021) 'World Wide Weird: Rise of the Cognitive Ecosystem', *Issues in Science and Technology*, 37(3), pp. 34–44.

Beer, S. (1985) *Diagnosing the system for organizations*. Chichester, UK: Wiley (Managerial cybernetics of organization). Available at: https://bac-lac.on.worldcat.org/oclc/300591269 (Accessed: 5 November 2023).

Chester, M.V. *et al.* (2023) 'Sensemaking for entangled urban social, ecological, and technological systems in the Anthropocene', *npj Urban Sustainability*, 3(1), pp. 1–10. Available at: https://doi.org/10.1038/s42949-023-00120-1.

Chester, M.V. and Allenby, B. (2018) 'Toward adaptive infrastructure: flexibility and agility in a non-stationarity age', *Sustainable and Resilient Infrastructure*, 4(4), pp. 173–191. Available at: https://doi.org/10.1080/23789689.2017.1416846.

Chester, M.V. and Allenby, B. (2021) 'Transportation for the Anthropocene', *Transfers Magazine*. Available at: https://transfersmagazine.org/magazine-article/issue-7/transportation-for-the-anthropocene/ (Accessed: 5 November 2023).

Chester, M.V. and Allenby, B. (2022) 'Infrastructure autopoiesis: requisite variety to engage complexity', *Environmental Research: Infrastructure and Sustainability*, 2(1), p. 012001. Available at: https://doi.org/10.1088/2634-4505/ac4b48.

Chester, M.V. and Allenby, B.R. (2020) 'Perspective: The Cyber Frontier and Infrastructure', *IEEE Access*, 8, pp. 28301–28310. Available at: https://doi.org/10.1109/ACCESS.2020.2971960.

Chester, M.V., Miller, T. and Muñoz-Erickson, T.A. (2020) 'Infrastructure governance for the Anthropocene', *Elementa: Science of the Anthropocene*, 8(1), p. 78. Available at: https://doi.org/10.1525/elementa.2020.078.

Cook, S.D.N. and Brown, J.S. (1999) 'Bridging Epistemologies: The Generative Dance Between Organizational Knowledge and Organizational Knowing', *Organization Science*, 10(4), pp. 381–400. Available at: https://doi.org/10.1287/orsc.10.4.381.

Daulat, S. *et al.* (2022) 'Challenges of integrated multi-infrastructure asset management: a review of pavement, sewer, and water distribution networks', *Structure and Infrastructure Engineering*, 0(0), pp. 1–20. Available at: https://doi.org/10.1080/15732479.2022.2119480.

Dicker, R. (2021) *3 new ways to navigate more sustainably with Maps*, *Google*. Available at: https://blog.google/products/maps/3-new-ways-navigate-more-sustainably-maps/ (Accessed: 23 May 2023).

Girard, John, Girard, Joann, and Sagology (2015) 'Defining knowledge management: Toward an applied compendium', *Online Journal of Applied Knowledge Management*, 3(1), pp. 1–20.

Google (2023) *2023 Environmental Report*. Mountain View, CA: Google. Available at: https://sustainability.google/reports/google-2023-environmental-report/ (Accessed: 5 November 2023).

Herrera, J.C. *et al.* (2010) 'Evaluation of traffic data obtained via GPS-enabled mobile phones: The Mobile Century field experiment', *Transportation Research Part C: Emerging Technologies*, 18(4), pp. 568–583. Available at: https://doi.org/10.1016/j.trc.2009.10.006.

Herring, R. *et al.* (2010) 'Using Mobile Phones to Forecast Arterial Traffic through Statistical Learning', in. *Transportation Research Board 89th Annual MeetingTransportation Research Board*. Available at: https://trid.trb.org/view/910552 (Accessed: 5 November 2023).

Hoehne, C.G. and Chester, M.V. (2016) 'Optimizing plug-in electric vehicle and vehicle-to-grid charge scheduling to minimize carbon emissions', *Energy*, 115, pp. 646–657. Available at: https://doi.org/10.1016/j.energy.2016.09.057.

Jessop, B. (2002) *Governance and Metagovernance: On Reflexivity, Requisite Variety, and Requisite Irony*. Lancaster, UK: Lancaster University,.

Available at: https://www.lancaster.ac.uk/fass/resources/sociology-online-papers/papers/jessop-governance-and-metagovernance.pdf.

Lau, J. (2020) *Google Maps 101: How AI helps predict traffic and determine routes, Google.* Available at: https://blog.google/products/maps/google-maps-101-how-ai-helps-predict-traffic-and-determine-routes/ (Accessed: 23 May 2023).

Miller, C.A. and Muñoz-Erickson, T.A. (2018) *The Rightful Place of Science: Designing Knowledge.* Tempe, AZ: Consortium for Science, Policy & Outcomes.

Muñoz-Erickson, T.A., Miller, C.A. and Miller, T.R. (2017) 'How Cities Think: Knowledge Co-Production for Urban Sustainability and Resilience', *Forests*, 8(6), p. 203. Available at: https://doi.org/10.3390/f8060203.

Murray, A. *et al.* (2018) 'Dynamic Demand Project (2018) | Water Projects', 1 October. Available at: https://waterprojectsonline.com/custom_case_study/dynamic-demand-project/ (Accessed: 5 November 2023).

Powelson, L.H. and Stoltzman, W. (2016) 'Dynamic rerouting during navigation'. Available at: https://patents.google.com/patent/US9360335/en (Accessed: 5 November 2023).

Scharth, M. (2023) *Evolution not revolution: why GPT-4 is notable, but not groundbreaking, The Conversation.* Available at: http://theconversation.com/evolution-not-revolution-why-gpt-4-is-notable-but-not-groundbreaking-201858 (Accessed: 5 November 2023).

Simon, H.A. (1996) *The Sciences of the Artificial.* Cambridge, MA: MIT Press.

Smart, P., Heersmink, R. and Clowes, R.W. (2017) 'The Cognitive Ecology of the Internet', in S.J. Cowley and F. Vallée-Tourangeau (eds) *Cognition Beyond the Brain: Computation, Interactivity and Human Artifice.* Cham, NL: Springer International Publishing, pp. 251–282. Available at: https://doi.org/10.1007/978-3-319-49115-8_13.

Sundstrom, O. and Binding, C. (2012) 'Flexible Charging Optimization for Electric Vehicles Considering Distribution Grid Constraints', *IEEE Transactions on Smart Grid*, 3(1), pp. 26–37. Available at: https://doi.org/10.1109/TSG.2011.2168431.

Tseng, Y.-T. and Ferng, H.-W. (2021) 'An Improved Traffic Rerouting Strategy Using Real-Time Traffic Information and Decisive Weights', *IEEE Transactions on Vehicular Technology*, 70(10), pp. 9741–9751. Available at: https://doi.org/10.1109/TVT.2021.3102706.

Weick, K.E. (1995) *Sensemaking in Organizations*. Thousand Oaks, CA: SAGE.

5.3

INFRASTRUCTURE'S CYBER FRONTIER

5.3.1. INTRODUCTION

The benefits of cybertechnologies integrated into infrastructure are be-coming clearer. In 2017 the City of San Diego saw energy use drop by 60% after LEDs were installed downtown in conjunction with optical, auditory, and environmental sensors. In 2018 the Arizona Department of Transpor-tation reported that more than a dozen wrong-way drivers were pre-vented from entering freeways by new thermal cameras and warning sys-tems. California in late 2019 released an early warning system, providing residents with precious additional seconds to find safety before an earth-quake.

The increasing integration of cybertechnologies into infrastructure is also creating vulnerabilities that we haven't ever experienced. A few days before Christmas in 2015 operators in the Prykarpattya Oblenergo electric utility of western Ukraine watched as their supervisory control and data acquisition (SCADA) system mouse pointer moved across the screen, no longer under their control, disabling substation after substation shutting down power across Ukraine. In 2017 hackers were able to access and trans-fer a casino's data using a vulnerability exploit in a Wi-Fi-connected fish tank sensor used to regulate water temperature, food, and water quality. In 2019 a ransomware attack brought the City of Baltimore's data manage-ment systems to a halt, suspending critical services related to real estate and communications.

Our increasingly connected systems are a new frontier for infrastruc-ture, one that offers remarkable capabilities to deliver new or augmented services and lower costs, while on the other end it creates radically new vulnerabilities that have never been faced or even conceived. The integra-tion of cyber and physical systems is accelerating. Yet the tools that we

This chapter was adapted from the following journal article with publisher permission: Mikhail Chester and Braden Allenby, 2020, The Cyber Frontier and Infrastructure, *IEEE Access*, 8(1), pp. 28301-28310, doi: 10.1109/ACCESS.2020.2971960.

have at our disposal to manage this integration and the outlook that we have about what this integration means remain rooted in the past century.

The number of devices that are now connected is exploding, and infrastructure is part of the trend. Estimates vary but generally show acceleration of growth in both the number of connected devices and the amount of data being transferred. Devices and data are growing faster than the global population and number of internet users (Cisco, 2019a). In 2022 there were around 22 billion connected devices (approximately 3 per planetary citizen) with expectations of roughly 30 billion by 2030.

The growth in information traffic is outpacing that of devices. Mobile traffic has grown 17-fold between 2012 and 2017, and mobile devices are projected to average 10.7 gigabytes of data traffic per month by 2022, up from 2.3 gigabytes in 2017 (Cisco, 2019b). The amount and quality of data (e.g., video resolution) being transmitted is increasing (Ericsson, 2017; Cisco, 2019a). Specific to infrastructure, the growth in machine-to-machine technologies (M2M) are of particular interest, with a projected 34% annual growth rate to 2022 (Cisco, 2019a). M2M refers to the direct communication between devices, which has been transitioning from closed network models to open. This allows devices to avoid communications hub and instead communicate directly with a centralized system or users (creating the potential for new technologies such as autonomous connected vehicle fleets). This category of interconnected devices has seen the largest growth, more than smart phones and personal computers. These devices are projected to drive much of the interconnectedness of smart cities and their infrastructure.

Viewing the embedding of smart technologies in infrastructure as simply an interconnectedness of systems is insufficient, if not irresponsible. The accelerating of the coupling may represent a singularity, a profound shift in the relationships between humans and their services (Kurzweil, 2004). It lays the groundwork for explosions of artificial intelligence, new capacities for services, radical changes in efficiency, and, with those, new vulnerabilities.

At the infancy of this shift, our comprehension of the implications of an accelerating cyberphysical world remains limited, and as such our ability to manage the implications and protect against vulnerabilities is likely woefully lacking. This unpreparedness has major implications for infrastructure managers and engineering education. It raises questions as to whether the next generation of leaders have the appropriate competencies to steer infrastructure as it transitions.

It's important to understand the context in which the acceleration of the interconnectedness between cyber and physical systems is happening. The demand for services delivered by infrastructure is one side of the story. Physical infrastructure systems (water, power, transportation, etc.) have largely been built to provide services that have for decades been relatively stable. We want water from a faucet the same way we did a century ago. How we demand electricity hasn't changed much from 1882 when Thomas Edison begin providing power through his Pearl Street Station to lower Manhattan. And over the past 70 or so years we (particularly those in the United States) have largely demanded automobility and its associated transportation infrastructure, which hasn't radically changed in technology (but certainly extent) in this time.

As such, the technologies that make up the backbone of our physical infrastructure systems have remained relatively stable for decades, if not centuries (Chester and Allenby, 2018). Certainly new technologies have been added, and efficiencies introduced, but water mains, pumps, transmission lines, transformers, and asphalt continue to dominate the core structure and functioning of these systems. If we were to bring Edison to today in a time machine he'd largely understand the power grid. But if we were to show Alexander Graham Bell a modern smart phone he'd be flummoxed by the black mirror. The acceleration of cyber technologies means that the cycle time (how quickly a past generation is replaced by a new generation) is now outpacing that of infrastructure. This is part of the challenge, working with cyberphysical systems that can't be treated as traditional coupled systems, given that cyber is cycling faster than the physical.

Concurrent with the technological change and increasing coupling of cyber and physical systems, there has been rapid acceleration in other fields, as well as social and political structures. Massive advances in computational power, data storage, and data analytics are driving advances in artificial intelligence and social media. At the same time we've seen a shift in military policy with a rise in asymmetric warfare strategies by nation states with weaker hardware, smaller armies, or less prepared armies that engage in cyberattacks to affect the strategic balance of power (Fritz, 2008; Chester and Allenby, 2018). Nation-states have adopted explicit strategies of civilizational conflict which make targets of all of society's systems, from finance to infrastructure to health (Allenby, 2015).

The combination of rapid advancement of digital technologies, increasing interconnectedness of cyber and physical systems, different outlooks on humans, and differing approaches to warfare, represents a radically new paradigm, and infrastructure is at the center. We can't ignore this context as we design and manage infrastructure going forward.

Toward providing insights into the design and management of infrastructure in a future with potentially new demands for services, vulnerabilities, and relationships between people, the environment, and technologies, we explore the changing cyberphysical dynamics and its implications. We start by exploring technological acceleration theory and what that means for infrastructure. Next, we describe how transitions from physical to cyberphysical infrastructure will create new capabilities along with vulnerabilities. We consider the changing relationship between people and their services as mediated by infrastructure, as we accelerate the cyber integration of physical systems. We conclude by recommending how infrastructure education and management must shift from models that emphasize systems as they've traditionally existed to systems that will be controlled by cyber technologies.

We discuss three coupled but conceptually different systems: 1) cyber and information and communication technology (ICT) infrastructure; 2) physical infrastructure, which increasingly includes ICT functionality and technology; and 3) the "institutional context" of infrastructure (including education and management). We don't view this chapter as an exhaustive exploration or summary of all of the issues relevant to cyberphysical systems, but more as an effort to elucidate new thinking about the rapidly changing relationship between technologies, infrastructure, and people.

5.3.2. ACCELERATING INTELLIGENCE

In 1999 the futurist Ray Kurzweil noted that many technologies tend to grow exponentially and as such the twenty-first century can be expected to yield 20,000 historical years of relative progress (Kurzweil, 2004). He branded this phenomenon the Law of Accelerating Returns, which if true is accelerating humankind toward technological change so radical and profound that we cannot comprehend the implications.

The Law of Accelerating Returns is a theory of change acceleration, which in general describes the increasing rate of technological progress that ultimately results in profound social and cultural change; many theories of change acceleration exist (Nottale, Chaline and Grou, 2000; Coren, 2003; Kurzweil, 2004; Teilhard de Chardin, 2008). While technology has always moved forward, the rate at which technology has changed up until recent times has mirrored population growth, meaning that essentially all of the world's population remained at subsistence levels of production and consumption (Clark, 2007).

But technological change is now increasingly exponential, representing a new paradigm for humans and the systems they operate (Syvitski, 2012). This acceleration creates remarkable new opportunities, and also hazards and vulnerabilities, and implies a future that is difficult to meaningfully comprehend. How long such an acceleration can proceed has been the subject of much debate (Rennie, 2010). However, as we look forward at the coming century there is accumulating evidence to warrant a critical examination of the implications of technological acceleration (Nagy *et al.*, 2011). Technological acceleration is attributed to positive feedback loops, and trends of increasing integration of cyber into physical systems raise questions of whether the perceived benefits of cyber result in integration in infrastructure that thereby changes services and vulnerabilities creating new cyber-integration demands.

MATURATION OF CYBER TECHNOLOGIES

Infrastructure have, for decades, operated as either purely physical systems or with limited and often isolated computing capabilities, frequently included in system design as mechanical devices (inertial and centrifugal governors on steam engines might be regarded as a form of computational device, for example). Sensors, software, and digital controls (including SCADA) have been increasingly used since the latter half of the twentieth century, but these digital systems largely functioned to augment the core underlying infrastructure, which—for decades if not longer—have been largely driven by physical systems and hardware.

But recent advances in hardware and software are increasing the accessibility, usability, and affordability of cyber technologies. Computer scientist Ragurathan Rajkumar provides a useful synthesis of the factors that are pushing and pulling cyber technologies leading to their ubiquity (Rajkumar, 2012). Sensors are now available to measure the properties at nano to macro scales. Actuators have become ubiquitous (again across scales). Alternative energy sources are maturing. Satellite and wireless communications are available across the globe, and internet connectivity is growing. At the same time, computing and storage capabilities are improving, and appearing in ever smaller form factors. Demand for these technologies is also growing. Building and environmental controls, critical infrastructure monitoring, process control, factory automation, healthcare, aerospace, and defense are all advancing cyberphysical systems as industries strive for radical new capabilities and efficiencies.

The result is a new paradigm of infrastructure that includes hardware, software, firmware, and wetware (people) integrated into new techno-human infrastructure. These systems are now smart and connected, delivering the ability to measure system, natural, and human dynamics in ways that weren't feasible a short time ago. They are able to generate, see, and make sense of massive data streams (often using integrated AI and human capabilities), and send data to users in real time, in ways that heretofore have not been possible. This changing paradigm will shift how we interact with infrastructure and what we ask infrastructure to do.

The proliferation of lower cost, smaller, more efficient, and more powerful computing technologies coupled with data transfer, storage, and
management technologies, and supported by emerging techniques (including AI) to make sense of voluminous and federated datasets represents profound new capabilities and efficiencies for physical systems (Rajkumar, 2012). This confluence of technologies represents an important
transition period for infrastructure. Prior to this maturation we typically
think of infrastructure as largely "dumb" physical systems (Allenby,
2004). The proliferation of the internet and augmentation of communication capacities (including bandwidth and communication protocols) has
resulted in radical new possibilities for physical systems. These technologies represent new capabilities for how we understand and interact with
natural and human systems.

BENEFITS OF CYBERPHYSICAL SYSTEMS

The integration of cyber and physical systems creates new capabilities
that didn't exist before. But it is what these capabilities enable that drive
the accelerating integration of the systems. Prior to discussing infrastructure at broad scales it is useful to examine a parallel but smaller technology and its integration of cyber and physical systems: the automobile.

Until the 1950s automobile technologies had no cyber technologies;
they were purely mechanical systems linked to each other via other mechanical systems or controlled through the cognitive capacities of the
driver. The first sensors integrated into cars simply alerted drivers to
problematic conditions such as low oil pressure via a dashboard light
(Jurgen, 1998). Critical system functioning was controlled by valves and
other mechanical devices that responded directly to the driver's input.

In 1968 Volkswagen introduced the first microchip into a car to control
fuel injection and minimize emissions, a device now known as an electronic fuel injector (EFI). Today's EFIs take in readings from dozens of
subsystem sensors, perform millions of calculations per second, and adjust the spark timing and how long the fuel injector is open, ensuring the
lowest emissions and highest fuel economy (Chips Etc., 2014). By 1999 cars
had dozens of microprocessors (Turley, 1999).

Today, sensors, processors, cameras, accelerometers, and other technologies result in 65Mb/s of data transferred throughout a vehicle,
roughly 2 miles of cabling within the vehicle, and 280 connections to manage power and that data (Katwala, 2017). It's naive to think that pushing
a gas pedal directly engages the engine. Instead, a computer determines —
based on your past behavior, environmental conditions, and readings
from the vehicle — how to give you the best ride.

But this is just one scale of the system. Navigation software (e.g., Google Maps) now takes into account thousands of other drivers and routes you based not simply on the shortest travel time in an unloaded network but with consideration of how all other users of the system are traveling. Hybrid electric vehicles can learn your frequent destinations and automatically switch to electric power as you approach those destinations, thereby saving gas (Amick, 2013). The integration of cyberphysical systems across scales as they relate to the automobile, the efficiencies they introduce, and the new capabilities radically alter our relationship with mobility services.

Cyberphysical infrastructure allow for new capabilities to optimize systems across broad scales and time frames, generate new efficiencies, and create multifunctionality where it didn't exist before. Fundamentally, the integration of cyber into physical systems creates new cognitive capacity about the system by shifting it toward relying more critically on information. New insights are created about not only the internal functioning and relationships between subsystems but also the demands (needs) being placed for the services. New optimization techniques are created with the integration of cyber creating the potential for efficiency gains. Sensors that detect ambient light can be used to control whether traffic lights are on or off, and their intensity when they're on, thereby reducing the need for electricity. Real-time information driving forecasts for electricity demand (power is perhaps the most historical major cyberphysical infrastructure) allows operators to deploy supply as needed. And in the case of mobility, a large scale connected vehicle fleet (possibly through Google Maps and emerging vehicle-to-vehicle communication technologies) offers the potential to shave peak demand, thereby reducing the need for new infrastructure and changing the kind of infrastructure that transportation engineers need to think about. Parking lots become less important; charging stations become more important.

These are possibilities that we can comprehend. With the deployment of artificial intelligence into cyberphysical systems the capabilities with full autonomy and humans not in control are beyond comprehension. But even today and in the near future, before the advent of fully autonomous AI, we must recognize that each new capability brings with it the potential for vulnerabilities and exploits.

5.3.3. VULNERABILITIES

With great promise comes the potential for radically new vulnerabilities, the likes of which we have never seen with infrastructure. These vulnerabilities arise not simply because new exploits are created, but are largely due to the new capabilities for exploiting operators, users, control systems, distributed software, and hardware. These vulnerabilities arise at a time when cyberattack tools have become available to low-expertise hackers and nation-states have established and tested strategies and devoted resources for asymmetric warfare.

TAXONOMIES OF THREATS

To understand how cyber threats emerge in infrastructure a taxonomy is helpful. Many taxonomies exist for cyber threats, differing depending on the phase of the hacking process (data collection, storage, processing, etc.), target, actors, methods, techniques, or capabilities (Harry and Gallagher, 2018). There is no preeminent taxonomy for threats to cyber-infrastructure systems.

It is helpful to first take a perspective within an infrastructure risk model. The National Institute of Standards and Technology's Guide for Conducting Risk Assessments provides a helpful model for framing risk factors for infrastructure management. Cyber threat taxonomies can map to the risk processes in NIST's model, providing a roadmap for analyzing threats as they relate to infrastructure. Starting with the threat source, taxonomies describe the types of actors involved in attacks, including professional criminals, state actors, terrorists, cyber vandals, hacktivists, internal actors, and cyber researchers (Bruijne *et al.*, 2017). They may focus on the threat event and techniques used, including degree of automation, exploited weakness, source address validity, possibility of characterization, attack rate dynamics, impact on the victim, victim type, and persistence of agent (Mirkovic and Reiher, 2004). The attack vector, vulnerability, and exploit have also been the focus of taxonomies.

Analysts Simon Hansman and Ray Hunt catalog and map the types of attacks to targets (including hardware and software exploits) and the corresponding vulnerabilities (Hansman and Hunt, 2005). Security researchers Charles Harry and Nancy Gallagher focus their taxonomy on the impacts of attacks by describing the outcomes of disruptions of operations and illicit acquiring of information (Harry and Gallagher, 2018). Figure 5.3.1 shows how organizational risk is a result of different sources of attacks, event types, impacts, and vulnerabilities.

Although the level of sophistication and number of attacks has increased over time, the intruder expertise has decreased. This trend reflects the growing availability of tools to cyberattackers. While in the past considerable expertise and resources were needed to conduct an attack, it is becoming more and more common for a small number of expert hackers to make their tools available to a broader community of novice hackers. This growing body of cybercriminals has the capability to deploy an arsenal at ever-increasing scales, diversity, and sophistication with increasingly devastating effects (Lipson, 2002; Hansman and Hunt, 2005; NRC, 2013). This trend reflects a new reality of cyberattacks.

As National Intelligence Director James Clapper testified, "Rather than a 'Cyber Armageddon' scenario that debilitates the entire U.S. infrastructure, we envision something different. We foresee an ongoing series of low-to-moderate level cyberattacks from a variety of sources over time, which will impose cumulative costs on U.S. economic competitiveness and national security." ('DNI Clapper Statement for the Record, Worldwide Cyber Threats before the House Permanent Select Committee on Intelligence', 2015). The ongoing low- to moderate-level attacks may reflect a Death by 1000 Cuts civilizational conflict strategy (Allenby, 2015), or simply that the vulnerabilities inherent in today's cyberdesigns attract an ever-increasing number of unrelated attacks.

Figure 5.3.1: Risk Model, Key Risk Factors and Associated Exemplary Taxonomies

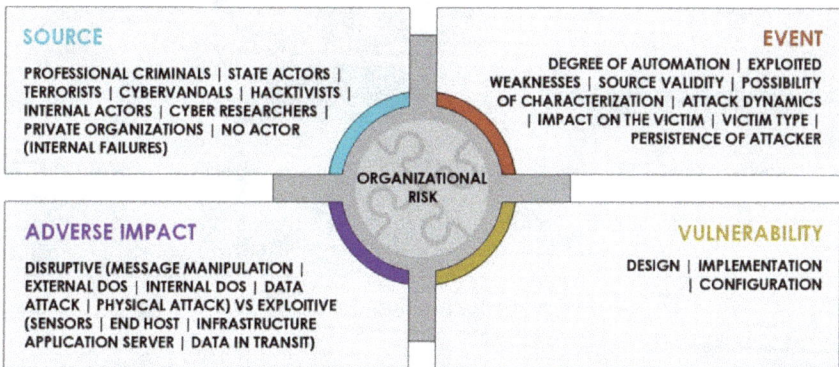

Adapted from: Howard and Longstaff (1998); Mirkovic and Reiher (2004); Hansman and Hunt (2005); Joint Task Force Transformation Initiative (2012); Bodeau and Graubart (2017); Bruijne *et al.* (2017); Harry and Gallagher (2018).

The level of sophistication of the attacker directly informs the strategies that should be developed when preparing for a cyberattack (Figure 5.3.2). Conventional threats include cyber vandalism and incursion, often involving disgruntled or suborned insiders, denial-of-service attacks, and hackers who have obtained legitimate user credentials (Bodeau and Graubart, 2017). Conventional threats can be approached by practice-driven risk management strategies that largely focus on basic hygiene and critical information protection (protocols for password changes, software updates, hardware updates, software installations, limiting users, and backing up data).

Advanced threats represent cyber adversaries that learn and evolve, such that compliance and good-practice driven strategies are insufficient and new competencies and threat-specific knowledge are needed (actors capable of such sophisticated cyber campaigns, such as Russia's Cozy Bear, are known as Advanced Persistent Threats) (Bodeau and Graubart, 2017). While conventional threats can often be handled by a properly trained IT department, advanced threats may require sustained and directed resources for cybersecurity and management of corresponding initiatives, and appropriate staff, tools, and strategic planning. Agility may be required to consider the goals of attackers, the techniques the attacker may use, and the appropriate anticipatory and reactive responses an organization can deploy to protect itself.

Figure 5.3.2: Sophistication of Cyber Threats and Risk Management Strategies

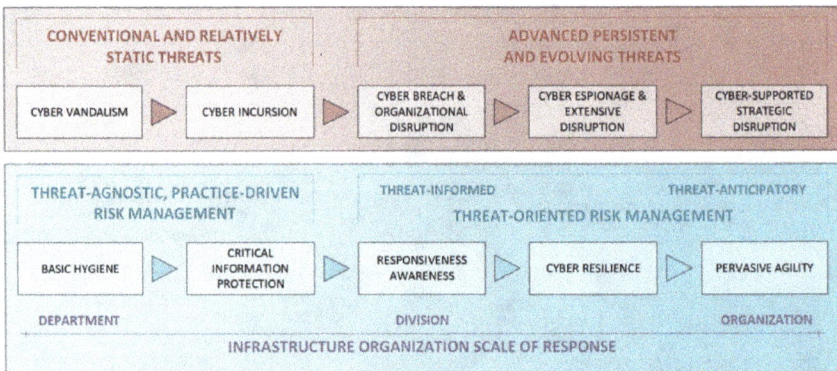

Adapted from Bodeau and Graubart (2017).

COMPLEXITY AND VULNERABILITY

As our infrastructure systems evolve toward greater complexity, in many ways defined by the increasing coupling of cyber and physical systems, vulnerabilities will need to be managed differently. We define infrastructure complexity here as the changing technical, environmental, and social context that engineers and managers must navigate to deliver and evolve services (Chester and Allenby, 2019). What is particularly interesting about the complexity associated with infrastructure is the speed and scale of which other systems are being integrated. The changing relationship of infrastructure users with the systems they rely on (e.g., the availability and price of parking spaces, the real-time arrival of the next transit vehicle, how to reroute to avoid traffic, the timed use of low-cost electricity by home appliances, the number of infrastructure elements that are offline in their region) through apps and internet connected services is exploding, fundamentally altering people's understanding and thereby use of services (Sarwar, Emirates and Soomro, 2013).

With the new possibilities created through cyberphysical systems comes vulnerabilities and exploits that didn't exist before, some of which transcend the cyberphysical system. It's possible to conceive of cyberattackers no longer needing to target the cyberphysical system itself, but instead conditioning operators with targeted disinformation. Most attacks on infrastructure occur from within, generally disgruntled employees with internal access (Cárdenas *et al.*, 2009). In 2006 engineers sabotaged intersection controls in Los Angeles, and in 2000 an ex-employee disabled critical SCADA systems with the hopes of being re-hired to fix the problem (Slay and Miller, 2008).

With new means for engaging with these operators, for instance through social media, we can conceive of a new method for inciting sabotage without directly engaging with the infrastructure. In 2019 utility operators were targeted with emails impersonating their accreditation society baiting them to open malware attachments masked as notifications that their professional credentials were being revoked (Raggi and Schwartz, 2019). The attachments contained the LookBack virus that would give the cyberattackers access to the utility's systems.

Another vulnerability that is receiving considerable attention is the controlling of outgoing information about an attack to distort facts and condition a particular response. Reflexive control (the means of conveying to a partner or an opponent specially prepared information to incline him to voluntarily make the predetermined decision desired by the initiator of the action), a principle developed by Russia since the 1960s, is particularly well-suited for the hyper-connected and information rich era (Thomas, 2004). The case of the 2015 Ukrainian power grid cyberattack, for example, was part of a broader Russia strategy that involved denial of service attacks, disinformation campaigns (including social media, mass media, and internet trolls), and energy diplomacy (involving coercion that forced Ukraine to pay market prices for oil and gas), that together sowed disinformation across international outlets. The strategy allowed Russia to deploy minimal physical forces, thereby staying below the threshold for international intervention, while achieving their objective of stopping a revolution that threatened to overturn the pro-Russian administration (King IV, 2017).

The possibilities for impact are no longer limited to the systems themselves, but span the interconnected systems in which our technical systems function. A challenge remains that those who understand the threat landscape and the complex tools being deployed are largely disconnected from those making day-to-day decisions about infrastructure. While in the

United States the National Institute of Standards and Technology and Department of Homeland Security issue valuable guidelines and recommendations for how to prepare for and protect against cyberattacks; (Joint Task Force Transformation Initiative, 2012; CISA, 2018) the reality of infrastructure at the ground level is one of limited resources and governing institutions that are structured to operate toward reliability principles that in many ways are designed to deliver services as they've been delivered in the past (and the existing engineering education structure reflects this).

CYBERWARFARE NORM

That cybersecurity has become a major challenge for engineered systems is neither new nor particularly surprising. The roots of the challenge lie deep in recent geopolitical history. Partially because the United States was the strongest country left standing after World War II and the collapse of the Soviet Union, and partially because defense expenditures by the United States have consistently been far greater than those of any rival, the conventional military forces of the United States are generally understood to be stronger than those of any other power (Qiao and Wang, 1999; Kissinger, 2014; Allenby, 2016; McFate, 2019).

This dominance has driven adversaries, especially state adversaries such as Russia and China, to adopt asymmetric warfare strategies that redefine conflict away from traditional military engagement to longer term "civilizational conflict," which among other things elevates information warfare, disinformation and subversion techniques, and weaponized narrative to priority attack mechanisms (Allenby, 2015). In perhaps the most cited military strategy article of the past decade, General Valery Gerasimov, Chief of the General Staff of the Russian Federation, notes that in the twenty-first century there has been "a tendency toward blurring the lines between the states of war and peace," and that "a perfectly thriving state can, in a matter of months and even days, be transformed into an arena of fierce armed conflict, become a victim of foreign intervention, and sink into a web of chaos, humanitarian catastrophe, and civil war." Writing before the successful Russian invasion of Crimea and eastern Ukraine in 2014, General Gerasimov emphasizes:

> *The very "rules of war" have changed. The role of nonmilitary means of achieving political and strategic goals has grown, and, in many cases, they have exceeded the power of force of weapons in their effectiveness…. The focus of applied methods of conflict has altered in the direction of the broad use of political, economic, informational, humanitarian, and other nonmilitary measures – applied in coordination with the protest potential of the population. All this is supplemented by military means of a concealed character, including carrying out actions of information conflict and the actions of special*

operations forces. The open use of forces – often under the guise of peacekeeping and crisis regulation – is resorted to only at a certain stage, primarily for the achievement of final success in the conflict (Gerasimov, 2016).

Russia is not alone in developing civilizational conflict strategies as an asymmetric response to American conventional dominance. Shocked by the success of allied forces in Desert Storm (1990–1991), Chinese strategists have developed a strategy of "Unrestricted Warfare" that contemplates conflict across the entire domain of a civilization, from financial markets to all forms of infrastructure:

There is reason for us to maintain that the financial attack by George Soros on East Asia, the terrorist attack on the U.S. embassy by Usama bin Laden, the gas attack on the Tokyo subway by the disciples of the Aum Shinri Kyo, and the havoc wreaked by the likes of Morris Jr on the Internet, in which the degree of destruction is by no means second to that of a war, represent semi-warfare, quasi-warfare, and sub-warfare, that is, the embryonic form of another kind of warfare (Qiao and Wang, 1999).

Iran, North Korea, and others are following in Russian and Chinese footsteps, although not as part of such a structured and formal geopolitical conflict strategy.

The complacency of academic engineering education institutions in light of active cyberwarfare directed at essentially all engineered systems within American, and Western, society is remarkable – and untenable. Engineering students in disciplines including civil and environmental, biomedical, and industrial engineering are taught to include ever-more advanced sensor, computing, communication, and data processing systems in their designs because of concomitant dramatic improvements in function and efficacy. But they are taught next to nothing about information and cybersecurity, both because of the inertia of engineering curricula to any proposed change, and because their professors were never trained in the subject, are not versed in it, and completely fail to perceive, much less understand, relevant geopolitical shifts.

The result is that American engineering education is optimally designed to create a generation of engineering professionals who will, among other things, unknowingly design ever more vulnerability and frailty into the built environment and infrastructure systems that are critical to our society. We should recognize that the integration of cyber technologies into infrastructure is altering the relationships between people and their services.

5.3.4. HUMANS, THEIR SERVICES, AND THE ENVIRONMENT, MEDIATED BY SOFTWARE

Edwin Hutchin's 1995 book *Cognition in the Wild* describes, through the lens of U.S. Navy pilots and sailors, the differences in cognitive approaches between individuals with no technology (the first sailors) and groups with technology (Hutchins, 1995). Hutchins argues that cognition in modern society is composed of multiple agents and their technologies. While sailors on a modern Navy vessel cannot necessarily navigate like early sailors with no technology, they are able to accomplish remarkably more, by compartmentalizing tasks, communicating effectively, and utilizing technology. Technology creates new opportunities for understanding the world around us, and as it accelerates is likely to create radical new relationships between people and their environments.

The rapid integration of cyber technologies into infrastructure and the implications for how humans interact with and demand services may represent a fundamentally new relationship that remains difficult if not impossible to comprehend. Whereas in the past new technologies often represented new capabilities and efficiencies, the hyperconnected and information-driven reality represents a radical change in how we see and experience the world. And artificial intelligence that mediates our interactions with other people, information, and services is positioned to fundamentally alter human experience. Infrastructure is at the center of this change.

Physical infrastructure systems will remain the backbone for cyber-technologies, but how they're used is poised to radically change. Several key dynamics may reshape our relationships with infrastructure:

- *Physical Systems as the Cyber Backbone*: Despite shifts from hardware to software functionality that reduces the need for physical assets (Shih, 2015), core physical systems will be needed to enable information transfer, analytics, and storage. And who controls the core physical systems will be strategically positioned, both economically and politically (see Google and Facebook's efforts to deploy fiberoptic lines around the world and recent concern over 5G hardware security) (Burgess, 2018).

- *Insights into Infrastructure Services*: Next, people are gaining and will continue to gain new insights about infrastructure that they didn't have before, thereby changing how they demand infrastructure services. The advent of smartphones created an industry of location tracking and traffic analysis firms that now deliver products and insights to travelers about the conditions of roads, how to route to minimize delays, and how to change their travel behaviors to reduce trip times (Wang and Fesenmaier, 2013; Kim and Coifman, 2014; Hu, Chiu and

Shelton, 2017). While still in its infancy, the possibilities of software making sense of the complexity of the transportation system has remarkable implications for how we use the system based on how the software understands it. Imagine similar insights and software-driven intelligence behind water and energy use, for example. And we are already heavily debating and seeing the implications of such intelligence driving how we consume news and media (Zeynep, 2015).

- *Evolving Demands for Infrastructure Services*: While it's easy to imagine how new and improved information can make our interactions with infrastructure more efficient (e.g., saving us travel time or managing our appliances to run at low-cost electricity times of day), it's likely that the possibilities offered by cyber technologies will result in demands for new services. The emergence of car, bicycle, and scooter sharing, which is resulting in major changes to how people travel in many major cities (Clewlow and Mishra, 2017), would not have been possible without smartphones and cellular networks. Furthermore, combining modalities with autonomous vehicles means that you end up needing to redefine the urban transportation network completely.

- *Adaptive Capacity*: The integration of sensing technologies coupled with analytical capabilities and software-based intelligence is likely to create new adaptive capacities for infrastructure. Sensors of various forms that can detect the conditions of assets in the system, both in terms of structure and function, are already being deployed and utilized in new ways. This information will likely drive algorithms that make sense of the overall state of the system, and decisions about how to manage assets to ensure integrity and efficiency. Imagine a SCADA system deciding to triage a portion of a water distribution network where a pipe is expected to fail to ensure that a cascading failure does not ensue. This capability will likely increase the agility and flexibility of infrastructure services to meet rapid changes in conditions and respond to hazards. Google Maps may already be showing us this adaptivity, by rerouting users with considerations of larger systems dynamics when there is a traffic accident.

These changes represent just a few of the possibilities of how cyber technologies may change our relationship with infrastructure. Preparing infrastructure managers and engineers for these shifts is critical to ensuring the integrity and safety of cyberphysical systems. As the technologies that define infrastructure change, so must education and governance for these systems.

5.3.5. PREPARING FOR CYBERINFRASTRUCTURE

Several critical and immediate efforts are needed to ensure that the integration of cyber-infrastructure results in systems that continue to support society's needs and are safe and secure. While there are certainly hardware and software changes that are needed, we focus these efforts on the institutional management of infrastructure and the training of future managers.

The training around integrated cyberphysical systems at universities is essentially non-existent and should immediately be developed as a core competency, a Fifth Column that can change the status quo of how we view and manage infrastructure (Senge, 1990). Engineers, architects, planners, and other infrastructure managers will still need knowledge around fundamentals of design principles, underlying science, and operations. However, they will need to be trained with new competencies that support a new norm for infrastructure, one where systems are increasingly focused on information management (Allenby, 2019).

Currently, disciplines such as civil, environmental, and mechanical engineering; planning; and architecture (domains largely responsible for the physical systems) mostly train independently of computer science, computer engineering, information sciences, and military or security domains. This fragmentation of knowledge is likely to lead to unintended consequences, both in the relevancy of disciplines and who and what decides how infrastructure services are managed. Cyber technology, information management, and security must become central to the training of infrastructure managers.

Cyber security competencies must become central to the training of infrastructure managers. To attempt to manage infrastructure today without such training is an ethical and professional failure, particularly in light of the increasing cyberattacks on infrastructure. New managers must have at least basic competencies to know why different actors might want to target their systems, what techniques they can use to exploit vulnerabilities, and strategies that can be deployed to protect systems (Figure 5.3.2) (Bodeau and Graubart, 2017).

It has become remarkably easy (both in terms of technology and cost) to layer new and connected technologies into old and new systems, without a comprehensive understanding of the implications, risk, and vulnerabilities. Infrastructure managers must be trained with the tools to understand how to vet hardware and software on devices, encrypt and secure communications, manage access to information, and thwart inside and outside attacks.

Infrastructure managers will need to develop roadmaps that guide the planning and development of their cyber systems into physical systems. This will require translating federal insights to their locales. It is difficult to find cyber planning and cybersecurity plans for state, regional, and local infrastructure agencies. These plans should immediately be developed and serve as a roadmap for how infrastructure agencies plan on integrating cyber into their systems and protect their systems against threats. Much of the cybersecurity literature identified was developed by federal agencies (namely NIST and DHS) and there is good reason to assume that intelligence agencies are also central to making sense of the challenge.

However, when it comes to day-to-day decisions about infrastructure assets, limited guidance exists. This information is sorely needed. It should be specific to region (considering local needs and hazards), describe threats across scales (from foreign to local actors), guide managers in how to access vulnerabilities in hardware and software, and provide strategies for protecting systems.

A remarkably difficult challenge will be steering infrastructure as artificial intelligence comes online. Software developers that are developing artificial intelligence appear to be doing so largely independent from those that design and manage infrastructure. The implications of this lack of coordination are very unclear. Will the software manage services in ways that infrastructure managers hadn't intended? Will it drive infrastructure development in an unplanned direction? Will it monopolize resources beyond the capacity of the system?

For every question that we think of there's probably two more that are beyond our comprehension, given the complexity of an AI-managed system and its potential for restructuring how we understand and interact with human systems. What is clear is that the tools and techniques that we currently train and deploy are badly out of date, and that the acceleration of the integration of cyber and physical systems is not likely one that we will be able to control. Instead, we'll need to accept that our new role is one of understanding and guiding the emerging complexity.

5.3.6. CONCLUSION

New approaches to how we think about goals and structure of infrastructure, what those systems do, and how they are operated are immediately needed to ensure that societal needs are met into the future. The cyber technologies that are increasingly integrated with physical systems are being developed faster than the infrastructure, resulting in an increasing mismatch between the new capabilities delivered by the cyber technology and the obdurate backbone physical system's capabilities. This is likely to lead to unintended consequences in how infrastructure is used and their reliability.

Furthermore, the acceleration of technologies, their pervasive use, and a dearth of knowledge and training among infrastructure managers is creating major vulnerabilities that are already being exploited. Education of infrastructure managers must include cyber technology. The growing complexity of human systems and their relationships with natural and social systems appears to be accelerating. The sooner we accept that the approaches we use to manage the core infrastructure systems that support human activities are rooted in the past century, the sooner we can reinvent infrastructure management for the coming centuries.

5.3.7. REFERENCES

Allenby, B. (2004) 'Infrastructure in the Anthropocene: Example of Information and Communication Technology', *Journal of Infrastructure Systems*, 10(3), pp. 79–86. Available at: https://doi.org/10.1061/(ASCE)1076-0342(2004)10:3(79).

Allenby, B. (2019) '5G, AI, and big data: We're building a new cognitive infrastructure and don't even know it', *Bulletin of the Atomic Scientists*, 19 December. Available at: https://thebulletin.org/2019/12/5g-ai-and-big-data-were-building-a-new-cognitive-infrastructure-and-dont-even-know-it/ (Accessed: 13 October 2023).

Allenby, B.R. (2015) 'The paradox of dominance: The age of civilizational conflict', *Bulletin of the Atomic Scientists*, 71(2), pp. 60–74. Available at: https://doi.org/10.1177/0096340215571911.

Allenby, B.R. (2016) 'In an Age of Civilizational Conflict', *Jurimetrics*, 56(4), pp. 387–406.

Amick, M. (2013) *First drive: Lincoln's 2013 MKZ hybrid makes going green easy, but lacks luxury, Digital Trends*. Available at: https://www.digitaltrends.com/cars/first-drive-lincolns-2013-mkz-hybrid-makes-going-green-easy-but-lacks-luxury/ (Accessed: 5 November 2023).

Bodeau, D. and Graubart, R. (2017) 'Cyber Prep 2.0: Motivating Organizational Cyber Strategies in Terms of Threat Preparedness'. Available at: https://www.mitre.org/sites/default/files/2021-08/15-0797-cyber-prep-2-motivating-organizational-cyber-strategies.pdf (Accessed: 9 November 2023).

Bruijne, M. de *et al.* (2017) 'Towards a new cyber threat actor typology', *Delft* [Preprint]. Available at: https://repository.wodc.nl/handle/20.500.12832/2299 (Accessed: 9 November 2023).

Burgess, M. (2018) 'Google and Facebook are gobbling up the internet's subsea cables', *Wired UK*, 18 November. Available at: https://www.wired.co.uk/article/subsea-cables-google-facebook (Accessed: 9 November 2023).

Cárdenas, A. *et al.* (2009) 'Challenges for Securing Cyber Physical Systems', in. *Workshop on Future Directions in Cyber-Physical Systems Security,* Department of Homeland Security.

Chester, M.V. and Allenby, B. (2018) 'Toward adaptive infrastructure: flexibility and agility in a non-stationarity age', *Sustainable and Resilient Infrastructure*, 4(4), pp. 173–191. Available at: https://doi.org/10.1080/23789689.2017.1416846.

Chester, M.V. and Allenby, B. (2019) 'Infrastructure as a wicked process', *Elementa: Science of the Anthropocene*. Edited by A. Iles and M.E. Chang, 7(1), p. 21. Available at: https://doi.org/10.1525/elementa.360.

Chips Etc. (2014) *Computer Chips inside Cars, Chips Etc.* Available at: https://www.chipsetc.com/computer-chips-inside-the-car.html (Accessed: 5 November 2023).

CISA (2018) *Department of Homeland Security's Cybersecurity and Infrastructure Security Agency, Department of Homeland Security.* Available at: https://www.cisa.gov/news-events/alerts/2018/11/19/cybersecurity-and-infrastructure-security-agency (Accessed: 9 November 2023).

Cisco (2019a) *Cisco Visual Networking Index: Forecast and Trends, 2017–2022.* C11-741490–00. Cisco Systems, Inc., p. 38.

Cisco (2019b) *Cisco Visual Networking Index: Global Mobile Data Traffic Forecast Update, 2017–2022.* C11-738429–01. Cisco Systems, Inc.

Clark, G. (2007) *A Farewell to Alms: A Brief Economic History of the World.* Princeton: Princeton University Press (Princeton economic history of the Western world). Available at: http://hdl.handle.net/2027/heb.30993 (Accessed: 5 November 2023).

Clewlow, R.R. and Mishra, G.S. (2017) *Disruptive Transportation: The Adoption, Utilization, and Impacts of Ride - Hailing in the United States.* UCD-ITS-RR-17-07. Davis, CA: University of California Davis Institute of Transportation Studies, p. 37. Available at: https://steps.ucdavis.edu/wp-content/uploads/2017/10/ReginaClewlowDisuptiveTransportation.pdf.

Coren, R.L. (2003) *The Evolutionary Trajectory: The Growth of Information in the History and Future of Earth.* New York, NY: CRC Press.

'DNI Clapper Statement for the Record, Worldwide Cyber Threats before the House Permanent Select Committee on Intelligence' (2015).

Washington, DC. Available at: https://www.dni.gov/index.php/news-room/congressional-testimonies/congressional-testimonies-2015/item/1251-dni-clapper-statement-for-the-record-worldwide-cyber-threats-before-the-house-permanent-select-committee-on-intelligence (Accessed: 9 November 2023).

Ericsson (2017) *Ericsson Mobility Report*. EAB-17:005964. Stockholm, Sweden: Ericsson.

Fritz, J. (2008) 'How China Will Use Cyber Warfare to Leapfrog in Military Competitiveness', *Culture Mandala: The Bulletin of the Centre for East-West Cultural and Economic Studies*, 8(1).

Gerasimov, V. (2016) *The Value of Science Is in the Foresight: New Challenges Demand Rethinking the Forms and Methods of Carrying out Combat Operations*. Military Review, translated and republished from Military-Industrial Kurier, p. 7. Available at: https://www.armyupress.army.mil/portals/7/military-review/archives/english/militaryreview_20160228_art008.pdf.

Hansman, S. and Hunt, R. (2005) 'A taxonomy of network and computer attacks', *Computers & Security*, 24(1), pp. 31–43. Available at: https://doi.org/10.1016/j.cose.2004.06.011.

Harry, C. and Gallagher, N. (2018) 'Classifying Cyber Events: A Proposed Taxonomy', *Journal of Information Warfare*, 17(3), pp. 17–31.

Howard, J.D. and Longstaff, T.A. (1998) *A common language for computer security incidents*. SAND98-8667. Sandia National Lab. (SNL-NM), Albuquerque, NM (United States); Sandia National Lab. (SNL-CA), Livermore, CA (United States). Available at: https://doi.org/10.2172/751004.

Hu, X., Chiu, Y.-C. and Shelton, J. (2017) 'Development of a behaviorally induced system optimal travel demand management system', *Journal of Intelligent Transportation Systems*, 21(1), pp. 12–25. Available at: https://doi.org/10.1080/15472450.2016.1171151.

Hutchins, E. (1995) *Cognition in the wild*. Cambridge, MA: MIT Press (Bradford Book Ser). Available at: https://openli-brary.org/books/OL1096941M (Accessed: 9 November 2023).

Joint Task Force Transformation Initiative (2012) *Guide for Conducting Risk Assessments*. NIST Special Publication (SP) 800-30 Rev. 1. National Institute of Standards and Technology. Available at: https://doi.org/10.6028/NIST.SP.800-30r1.

Jurgen, R.K. (ed.) (1998) *History of Automotive Electronics: The 1980's*. Warrendale, PA: Society of Automotive Engineers.

Katwala, A. (2017) *Connected cars are 'driving microchip development'*, *Institution of Mechanical Engineers*. Available at: https://www.imeche.org/news/news-article/connected-cars-are-'driving-microchip-development' (Accessed: 5 November 2023).

Kim, S. and Coifman, B. (2014) 'Comparing INRIX speed data against concurrent loop detector stations over several months', *Transportation Research Part C: Emerging Technologies*, 49, pp. 59–72. Available at: https://doi.org/10.1016/j.trc.2014.10.002.

King IV, F.J. (2017) *Reflexive Control and Disinformation in Putin's Wars*. University of Colorado. Available at: https://scholar.colorado.edu/downloads/ht24wj65j.

Kissinger, H. (2014) *World Order*. New York, NY: Penguin Publishing Group.

Kurzweil, R. (2004) 'The Law of Accelerating Returns', in C. Teuscher (ed.) *Alan Turing: Life and Legacy of a Great Thinker*. Berlin, Heidelberg: Springer, pp. 381–416. Available at: https://doi.org/10.1007/978-3-662-05642-4_16.

Lipson, H. (2002) *Tracking and Tracing Cyber-Attacks: Technical Challenges and Global Policy Issues*. CMU/SEI-2002-SR-009. Pittsburgh, PA: Carnegie Mellon University. Available at: https://doi.org/10.1184/R1/6585395.v1.

McFate, S. (2019) *The New Rules of War: Victory in the Age of Durable Disorder*. New York, NY: William Morrow-HarperCollins.

Mirkovic, J. and Reiher, P. (2004) 'A taxonomy of DDoS attack and DDoS defense mechanisms', *ACM SIGCOMM Computer Communication Review*, 34(2), pp. 39–53. Available at: https://doi.org/10.1145/997150.997156.

Nagy, B. *et al.* (2011) 'Superexponential long-term trends in information technology', *Technological Forecasting and Social Change*, 78(8), pp. 1356–1364. Available at: https://doi.org/10.1016/j.techfore.2011.07.006.

Nottale, L., Chaline, J. and Grou, P. (2000) *Les arbres de l'évolution: univers, vie, sociétés*. Paris, FR: Hachette littératures (Collection Sciences (Hachette (Firm))).

NRC (2013) *The Resilience of the Electric Power Delivery System in Response to Terrorism and Natural Disasters: Summary of a Workshop*. Washington, DC: National Academies Press. Available at: https://doi.org/10.17226/18535.

Qiao, L. and Wang, X. (1999) *Unrestricted Warfare*. Beijing, CN: PLA Literature and Arts Publishing House.

Raggi, M. and Schwartz, D. (2019) *LookBack Malware Targets the United States Utilities Sector with Phishing Attacks Impersonating Engineering Licensing Boards US | Proofpoint US, Proofpoint*. Available at: https://www.proofpoint.com/us/threat-insight/post/lookback-malware-targets-united-states-utilities-sector-phishing-attacks (Accessed: 9 November 2023).

Rajkumar, R. (2012) 'A Cyber–Physical Future', *Proceedings of the IEEE*, 100(Special Centennial Issue), pp. 1309–1312. Available at: https://doi.org/10.15779/Z38R11D.

Rennie, J. (2010) *Ray Kurzweil's Slippery Futurism - IEEE Spectrum*. Available at: https://spectrum.ieee.org/ray-kurzweils-slippery-futurism (Accessed: 5 November 2023).

Sarwar, M., Emirates, U. and Soomro, T.R. (2013) 'Impact of Smartphone's on Society', *European Journal of Scientific Research*, 98(2), pp. 216–226.

Senge, P.M. (1990) *The Fifth Discipline: The Art and Practice of the Learning Organization*. 1st edn. New York, NY: Doubleday/Currency.

Shih, W.C. (2015) 'Does Hardware Even Matter Anymore?', *Harvard Business Review*, 9 June. Available at: https://hbr.org/2015/06/does-hardware-even-matter-anymore (Accessed: 9 November 2023).

Slay, J. and Miller, M. (2008) 'Lessons Learned from the Maroochy Water Breach', in E. Goetz and S. Shenoi (eds) *Critical Infrastructure Protection*. Boston, MA: Springer US (IFIP International Federation for Information Processing), pp. 73–82. Available at: https://doi.org/10.1007/978-0-387-75462-8_6.

Syvitski, J. (2012) 'Anthropocene: An epoch of our making - IGBP', *Global Change Magazine*, (78), p. 12.

Teilhard de Chardin, P. (2008) *The Ohenomenon of Man*. First Harper Perennial Modern Thought edition. New York, NY: HarperCollins.

Thomas, T. (2004) 'Russia's Reflexive Control Theory and the Military', *The Journal of Slavic Military Studies*, 17(2), pp. 237–256. Available at: https://doi.org/10.1080/13518040490450529.

Turley, J. (1999) 'Embedded Processors by the Numbers', *EE Times*, 1 May. Available at: https://www.eetimes.com/embedded-processors-by-the-numbers/ (Accessed: 5 November 2023).

Wang, D. and Fesenmaier, D.R. (2013) 'Transforming the Travel Experience: The Use of Smartphones for Travel', in L. Cantoni and Z. (Phil) Xiang (eds) *Information and Communication Technologies in Tourism 2013*.

Berlin, DE: Springer, pp. 58–69. Available at: https://doi.org/10.1007/978-3-642-36309-2_6.

Zeynep, T. (2015) 'Algorithmic Harms beyond Facebook and Google: Emergent Challenges of Computational Agency', *Colorado Technology Law Journal*, 2, pp. 203–218.

5.4

ARTIFICIAL INTELLIGENCE

5.4.1. INTRODUCTION

The need for infrastructure to adapt, transform, and perform competently under conditions of complexity and accelerating change is increasingly being met by integrating infrastructure and information systems (including various artificial intelligence (AI) capabilities) into infrastructure design, construction, operation, and maintenance. However, successfully implementing this strategy requires a clear and concise understanding of relevant information, communication, and computational frameworks, as well as how they functionally couple together in practice—a particularly difficult task in today's environment. Therefore, it is not surprising that the rise of a new global infrastructure with profound implications for humans, their institutions, and their planet has gone both unperceived and unremarked. This is the cognitive infrastructure, and it already permeates virtually every aspect of our world (Allenby, 2019). In particular, each infrastructure system and sector has its own companies, experts, investors, and users. But what is often not recognized is that many of these infrastructure and technologies are not only coherent entities themselves, but also being integrated into an emergent infrastructure that includes integrated functionality from many sources, the "cognitive infrastructure."

Taking a functional definition of "cognition" (i.e., information processing, reasoning, remembering, learning, problem-solving, decision-making, etc.) (Squire, 2009), the accelerating rise of cognitive infrastructure becomes evident. For example, machine-to-machine connections are anticipated to increase from 6.1 billion in 2018 to 14.7 billion in 2023 (Cisco, 2020). Similarly, spending on sensors and other technologies related to the

---

This chapter was adapted from the following journal article with publisher and lead author permissions: Samuel Markolf, Mikhail Chester, and Braden Allenby, 2021, Opportunities and Challenges for Artificial Intelligence (AI) Applications in Infrastructure Management during the Anthropocene, *Frontiers in Water*, 2(551598), pp. 1-10, doi: 10.3389/frwa.2020.551598.

Internet-of-Things (IoT) is expected to reach $1.2 Trillion in 2022 (Columbus, 2018). Most of these sensors and devices will generate vast amounts of data and integrate some cognitive capability via accelerating deployment of AI technology such as neural nets (Lee and Li, 2018). In short, accelerating capability and capacity across a number of apparently unrelated infrastructure and technologies is generating an infrastructure, tied together by AI and a vast array of institutional structures, that 1) contains the functional components of cognition and ever-more powerful networks operationally linking them together, 2) is distributed around the world, and 3) contains evolving and emergent systemic and behavioral capabilities. Simply put, we are building a pervasive cognitive infrastructure without fully recognizing it, and we are doing so rapidly and at global scale.

Cognitive infrastructure offers challenges that more traditional infrastructure systems do not. For one, it operates at a level that humans can neither fully understand nor perceive—people are relatively low bandwidth cognitive mechanisms in a world where even contemporary cognitive infrastructure operates at far higher bandwidths, much faster speeds, and higher levels of complexity than individuals can access. This can unfortunately be seen in the tragic Lion Air Flight 610 and Ethiopian Airlines Flight 302 incidents. Although many factors appeared to have been at play, the disconnect between the development of the automated flight control systems in the Boeing 737-MAX planes and the training and implementation by the pilots was a key element in the accidents (Gelles, 2019; Johnston and Harris, 2019; Wise, 2019; U.S. House Committee on Transportation and Infrastructure, 2020). Thus, determining how to effectively integrate human and machine cognition into infrastructure systems becomes a significant professional challenge that, so far, appears to have not been adequately and effectively considered.

Integrating cognitive infrastructure is a critical capability as engineers, technologists, and policymakers try to develop infrastructure systems that are as resilient, agile, and adaptive as current (and future) conditions demand. But knowing that incorporating sensor and AI-driven adaptability into infrastructure can make it more efficient and responsive to changing conditions is only the beginning. Understanding the cognitive infrastructure as a whole is required to fully and responsibly meet the demands for better infrastructure. For example, designers of IoT devices embed sensors and communication capabilities in their products as a matter of required functionality. But, absent a systemic perspective on security and the devices' place within the overarching cognitive infrastructure, there is the potential for underappreciating/misunderstanding issues like the vulnerability to adversarial attacks that the embrace of AI technologies can create. These potential drawbacks are ultimately a symptom of understanding a few of the constituent technologies (e.g., AI) in isolation, but failing to understand that it is the cognitive infrastructure, not just those individual technologies, that their infrastructure design is integrating.

It is premature to consider tantalizing questions such as how humans should respond as critical cognitive functions migrate to higher level techno-human systems embedded in a global cognitive infrastructure. However, it is not premature to recognize that this new infrastructure, itself a reflection and driver of the complexity and challenges of the Anthropocene, is already emergent. Additionally, trying to perceive and understand some of these implications is an increasingly imperative and necessary professional responsibility. Without that first step, ethical, rational, and appropriate infrastructure design, construction, operation, maintenance (as well as the educational and institutional structures to support them) will remain beyond reach. As such, this chapter provides a broad discussion about what AI is and how it relates to infrastructure. We then explore various tasks and services within infrastructure systems that may be enhanced and/or replaced by AI. Finally, we conclude with a discussion of some of the broader implications that may emerge as AI and infrastructure systems become increasingly entwined in the coming decades.

5.4.2. AI AND INFRASTRUCTURE LEADERSHIP IN THE CONTEXT OF COMPLEXITY

"AI" is a fuzzy term. As the U. S. National Science and Technology Council says in its 2016 report, "There is no single definition of AI that is universally accepted by practitioners. Some define AI loosely as a computerized system that exhibits behavior that is commonly thought of as requiring intelligence. Others define AI as a system capable of rationally solving complex problems or taking appropriate actions to achieve its goals in whatever real-world circumstances it encounters." Herein, we use "AI" to include big data and analytics dimensions, but ultimately describe the leadership and intelligence capabilities that are needed to replace or augment people. We envision a future where humans employ AI to make sense of an increasingly complex world.

In managing dynamic and complex systems and environments, several leadership capabilities are needed to address continually changing conditions (Uhl-Bien, Marion and McKelvey, 2007). Administrative Leadership, what we largely practice today, is well-suited for stable conditions and is made up of bureaucracies that formalize the structure and function of organizations. However, in the changing or chaotic conditions that define complex environments, Adaptive Leadership is preferred. Under this approach, adaptability, creativity, and learning are emphasized to make sense of and navigate complex and uncertain conditions. Perhaps of most importance is Enabling Leadership, the ability to shift between Administrative and Adaptive Leadership practices as conditions shift from stable to chaotic. Enabling Leadership involves creating structural, financial, and knowledge conditions for flexibility (Uhl-Bien, Marion and McKelvey, 2007). In assessing the AI landscape, evaluating which techniques are best positioned to support each leadership style is increasingly useful.

Given this context, there are several tasks for which AI applications in infrastructure are well suited, including pattern recognition, classification, clustering, categorization, system control, function approximation (e.g., regression analysis), optimization, and prediction/forecasting (Chen, Jakeman and Norton, 2008; Brynjolfsson and McAfee, 2017; Eggimann *et al.*, 2017). In order to accomplish these tasks, a variety of techniques and approaches can be applied, such as rule-based systems (RBS), genetic algorithms, cellular automata, Fuzzy Systems, Multi-agent systems, Swarm Intelligence, Case-based reasoning (CBR), and Artificial Neural Networks (ANN) (Chen, Jakeman and Norton, 2008). For example, AI (particularly genetic algorithms, Artificial Neural Networks, and Deep Learning) has been applied in a variety of civil engineering contexts including optimum design of structures (Hajela and Berke, 1991; Adeli and Park, 1995; Camp, Pezeshk and Hansson, 2003; Hadi, 2003), concrete strength modeling (Yeh, 1999; Ni and Wang, 2000), predicting geotechnical settlement and liquefaction (Shahin, Maier and Jaksa, 2002; Young-Su and Byung-Tak, 2006), earthquake engineering (Lee and Han, 2002; Arslan, 2010; Yilmaz, 2011), concrete design mix (Jayaram, Nataraja and Ravikumar, 2009), prediction and forecasting of water resources (Maier and Dandy, 2000), sediment modeling (Nagy, Watanabe and Hirano, 2002; Zhang, Wai and Jiang, 2010), irrigation and water-delivery scheduling (Nixon, Dandy and Simpson, 2001; Karasekreter, Başçiftçi and Fidan, 2013), rainfall-runoff modeling (Minns and Hall, 1996; Tokar and Johnson, 1999; Cheng, Wu and Chau, 2005; Jeong and Kim, 2005; Abrahart and See, 2007), and evapotranspiration modeling (Tabari, Marofi and Sabziparvar, 2010; Genaidy, 2020). The scope and purpose of this chapter are not to provide a comprehensive overview and discussion of these different techniques. For that, we refer the readers to works by Flood and Kartam (1994a; 1994b; 1997), Flood (2001), Adeli (2001), Flintsch and Chen (2004), Chandwani et al. (2013), Ye et al. (2019), and Falcone et al. (2020). Nonetheless, a brief discussion about the ways in which various AI techniques may (or may not) support infrastructure leadership in stable and chaotic environments appears warranted and is included below.

Some AI techniques may be well suited for enhancing operations during stable conditions, while others may be more appropriate for supporting leadership during unstable times (e.g., extreme events, funding uncertainty, pandemics, etc.). For example, techniques that establish algorithms to solve novel problems by recalling and referencing similar problems from the past (e.g., CBR) are particularly suitable for the well-defined and stable conditions endemic of Administrative Leadership. In this context, these approaches can be particularly useful for applications related to system control, planning, prediction, and diagnosis (Chen, Jakeman and Norton, 2008). Conversely, techniques that mimic the manner in which human brains process information via a series of layered and interconnected processing units (e.g., ANN) are increasingly well-suited for the complex, data-intensive, multivariable, and dynamic conditions (i.e., instability) that warrant Adaptive Leadership. In this context, AI can help make predictions (based on a series of input patterns) and/or intuit relationships between various inputs—even in situations where the underlying rules and structure of the problem may be unknown or hard to express (Chen, Jakeman and Norton, 2008). Overall, various forms of AI appear poised to greatly complement (or even in some cases replace) Administrative and Adaptive Leadership activities and roles within our infrastructure systems. In turn, the humans and institutions that interact with and govern our infrastructure systems may play an increasingly important role as the primary source of Enabling Leadership within our systems. Thus, it will be crucial for humans and institutions to recognize the benefits and tradeoffs among the different types of leadership, roles, and services provided by various AI. Perhaps most importantly, additional consideration appears warranted regarding the frameworks, resources, structures, and knowledge systems that may be needed to facilitate the smooth and agile transition between leadership approaches as future conditions continually fluctuate between stable and chaotic. The following section explores this issue further by examining some of the various roles and tasks AI may fill in infrastructure systems moving forward.

5.4.3. AI INTELLIGENCES AND TASKS WITHIN INFRASTRUCTURE

Evaluating the potential for AI to augment or replace existing capabilities requires a critical examination of the intelligences involved. Huang and Rust (2018) assert that AI job replacement fundamentally occurs at the task level, and that 'lower' intelligence tasks (e.g., repetitive, routine tasks) are easier for AI to replace than 'higher' intelligence tasks (e.g., highly emotional/empathetic tasks). Given that, at their core, infrastructure systems are service providers, we adapt Huang and Rust's framework to 1) link various infrastructure services to the four types of intelligences described by Huang and Rust (i.e., Mechanical, Analytical, Intuitive, and Empathetic), and 2) outline cases (and examples where possible) of how AI has and/or could potentially replace various infrastructure-related tasks at each level of intelligence.

5.4.3.1. MECHANICAL INTELLIGENCE

The 'lowest' level of intelligence is Mechanical, which is defined by routine and repeated tasks, minimal creativity, and an emphasis on efficiency and consistency (Huang and Rust, 2018). AI at this level are rule-based and are well suited for homogenous tasks that are repetitive, performed often, and unsophisticated (Sawhney, 2016; Huang and Rust, 2018). As a result, AI at this level often have an advantage over humans with respect to consistency, reliability, and work-rate (Huang and Rust, 2018).

One of the primary challenges associated with Mechanical AI is that it can be difficult to scale to the systems level, which in turn can limit its applicability to the large-scale and dynamic infrastructure systems typical of modern cities. Mechanical tasks are typically conducted by a single unit (or small, tightly integrated group of components). As a result, this type of AI is best suited for well-bounded and tightly constrained situations. Thus, increasing the network, scale, and/or state of operations adds complexity that can eventually overwhelm the system — under these circumstances AI at higher levels of intelligence will likely be more appropriate and effective.

5.4.3.2. ANALYTICAL INTELLIGENCE

The second level of intelligence is Analytical, which relies on the ability to process information, make decisions, problem solve, and adjust to new information (Huang and Rust, 2018). Analytical intelligence is defined by tasks that can be complex (often data-intensive), yet consistent and predictable. AI at this level use algorithms to iteratively learn and gain insights from large and/or continuous data sets. Analytical AI increasingly consist of networked units rather than a stand-alone machine. Human interpretation and intuition are still vital complements to AI at this level. AI provides increasingly varied and valuable decision support, but humans are still the ones ultimately making the decision.

One of the biggest potential challenges with Analytical AI is that it is likely not well suited for problems that do not have similar analogs from the past (Chen, Jakeman and Norton, 2008). This drawback is particularly important to consider in the context of managing infrastructure systems under a changing climate. Non-stationarity, the concept that past conditions and data are not indicative of future trends and conditions, is increasingly a reality for urban and infrastructure systems (Milly *et al.*, 2008; Koutsoyiannis, 2011; Lins, 2012). Thus, Analytical AI should not be treated as an 'off-the-shelf' or 'plug-and-play' solution for a wide range of problems. Engineers and infrastructure managers should take great care to understand the nuances, strengths, and weaknesses of AI when applying it to infrastructure that has significant interaction with climatic variables (e.g., weather prediction, stormwater systems, flood management systems, etc.).

5.4.3.3. INTUITIVE INTELLIGENCE

The next level of intelligence is Intuitive, which relies on experience-based thinking and creativity. Tasks related to intuitive intelligence are contextual, chaotic, complex, and idiosyncratic (Huang and Rust, 2018). AI at this level function in a more human-like manner by learning and adapting based on previous experience and new information. Understanding a problem or situation based on context and prior experience is a hallmark characteristic of intuitive intelligence in both humans and AI.

One potential challenge with Intuitive AI is that the problems to which it may be applied are often 'wickedly complex' and do not have one 'right' solution (e.g., the allocation and management of natural resources) (Chester and Allenby, 2019). The algorithms supporting this type of AI often learn from human-defined data as to what the outcome should be. Thus, the training of and learning by the AI can be severely inhibited in situations where the outcome/solution is not clear (Meserole, 2018). Under these circumstances, AI can still be very helpful in generating, exploring, and analyzing various scenarios. However, human stakeholders will ultimately be responsible for deciding on the final outcomes or course of action.

Another potential challenge associated with Intuitive AI is that there can be a 'black-box' element to the analysis and outcomes due to the fact that it provides solutions and insights with minimal knowledge of the underlying systems and processes (Chen, Jakeman and Norton, 2008). For example, the AI may produce outputs that are non-intuitive and/or fail to converge on a solution, and it may be difficult to ascertain why. Ultimately, some level of this 'black box' is likely unavoidable. Presumably, one of the main reasons to deploy Intuitive AI is because the system in question is already operating at a scale and/or level of complexity beyond human cognitive capabilities. If total understanding and mastery of system dynamics and complexity (i.e., elimination of the 'black box') is achievable, then Intuitive AI was likely not needed in the first place. Thus, the critical question is not 'how do we eliminate the black-box?', but rather, 'what degree of black-box are we comfortable with?' As AI systems continue to evolve and become further embedded in our infrastructure systems, we may be implicitly or explicitly releasing control of our infrastructure systems to software and algorithms. The potential benefits of this arrangement may very well outweigh the drawbacks in certain circumstances. However, it is important for communities, policy-makers, and infrastructure managers to have open and candid discussions about the potential implications of this shift in control and whether or not those implications are desirable.

5.4.3.4. EMPATHETIC INTELLIGENCE

The 'highest' level of intelligence is Empathetic, which relies on empathy, social interaction, and communication. Empathetic tasks relate to the ability to understand emotions, appropriately respond to emotions in others, and influence other's emotions (Huang and Rust, 2018). AI at this level "relates to, arises from, or influences emotions (Picard, 1995)", and behaves as if it has feeling. Empathetic AI are still in the nascent stages of development, with initial applications tending to relate to emotional analytics (Abou-Zeid and Ben-Akiva, 2010; Quercia, Schifanella and Aiello, 2014). Nonetheless, the high level of social and communication skills needed for empathetic intelligence seem to indicate that humans will remain integral at this level for the foreseeable future.

Similar to Intuitive AI, aspects of wicked complexity and wicked problems can be especially challenging for Empathetic AI. One of the elements of a wickedly complex problem is the presence of a wide degree of norms and values among the various stakeholders within the system. These values/interests may not always be clearly stipulated or coded in anyway. Additionally, they can shift and fluctuate over time. As a result, it is very difficult for the AI to understand the different (and often conflicting) values among the stakeholders, let alone 'train' the AI around a centrally agreed upon solution/outcome (Baum, 2020).

Related to the issue above, Empathetic AI can be particularly susceptible to various biases. The biases may be implicit or explicit, and can be the result of the individuals who wrote the algorithms or the data from which the algorithm was trained (Tomer, 2019). For example, facial recognition AI has been found to contain racial bias (Grother, Ngan and Hanaoka, 2019). It is unlikely that biases can fully be eliminated from Empathetic (and other) AI systems. Thus, similar to the 'black box' issue, perhaps the best approach is for citizens, decision makers, and AI developers to have open and candid discussions about the appropriate applications of Empathetic AI given the potential unintended consequences that may result from these biases.

Table 5.4.1 provides a summary of the key elements of each intelligence, examples from infrastructure systems, and current/potential applications of AI in infrastructure across each level of intelligence.

Table 5.4.1: Four AI Intelligences and their applications to infrastructure systems.

Intelligence	Overview for Infrastructure
Mechanical	Task Description: Routine, repeated tasks; Minimal creativity; Consistency and efficiency are paramount. Examples: System components (e.g., traffic lights, water pumps, etc.) Traditional thermostats; 'Simple' supply/ demand systems based on time-of-day (e.g., time of day timing of traffic lights, time-of-day pricing and provision of water); Fixed schedule airport trams/people movers. AI Capabilities: AI with minimal learning and adapting; Actions are efficient, consistent, and precise; Actions and reactions are repetitive and based on observation. Applications: Simple rule-based controllers (e.g., automated diversion of water through pipe network); Drones/ Robots for infrastructure inspection Limited (Level 3) vehicle automation (e.g., Smart Circuit Bus in Columbus, OH); Wastewater treatment control and operation (Zhang et al, 2017; Corominas et al., 2018); Water quality and pollution control (Liu et al., 2016; Mounce et al., 2016).
Analytical	Task Description: Process information; Problem- solving; Decision-making; Learn from new information; Data/ information intensive; Tasks can be complex, yet systematic, consistent, and predictable. Examples: Intelligent Transportation Systems (e.g., cameras and loop detectors to alter timing and sequence of traffic lights); Load Management and Demand Forecasting (e.g., real-time pricing, real-time demand management in power sector); 'Smart' Thermostats and Irrigation systems; Navigation/ Traffic Routing (e.g., incorporation of network conditions and travel behavior in system routing). AI Capabilities: AI learn/adapt systematically based on data; Actions are logical, rational, and rule-based. Applications: Automated maintenance/replacement scheduling based on factors such as system age, system use, projected demand, etc.; Navigation and traffic management informed by citywide cell-phone data and environmental sensors; 'Highly' (Level 4) autonomous transportation systems (e.g., autonomous freight, autonomous segments in specific locations); Forecasting of water demand (Yang et al., 2017); Estimating and controlling water quality (Amanollahi et al., 2017; Pisa et al., 2019); Predicting precipitation and water levels (Rezaeianzadeh et al., 2015).
Intuitive	Task Description: Experience- based thinking; Creativity/ Creative problem solving; Tasks can be complex, chaotic, contextual, idiosyncratic. Examples: Long-term forecasting (e.g., 10-year plans by Metropolitan Organizations, regional water suppliers, and electric utilities); User and/or system manager improvisation during abnormal/ surprise conditions (e.g., Captain Sullenberger emergency landing on Hudson River, 'improvised' solutions to spillway breach at Oroville Dam); Project/location specific design changes based on local variables and past experiences (e.g., designing a bridge foundation more robust than plans given local extremes). AI Capabilities: AI learn and adapt intuitively based on understanding; AI function in more flexible and 'human-like' manners; Unlikely that the same mistake is made twice. Applications: Intelligent Air Traffic Control Systems; "Fully" (Level 5) autonomous transportation systems (i.e., all roads and vehicles are autonomous and connected); Automated allocation and dispatch of resources (e.g., dynamic public transportation supply and routing; anticipation of which water mains are on the verge of breaking and dispatch of resources to prevent disruption); Disaster evacuation/ preparedness (e.g., predictive and automated shut down of some portions of the network and diverted capacity/resources to others); Optimization and planning of water infrastructure (Beh et al., 2017; Zhang et al, 2018; Rastegaripour et al., 2019); Water demand forecasting (Ghalehkhondabi et al., 2017).
Empathetic	Task Description: Recognize/ understand emotions; Respond (appropriately) to other's emotions; Influence other's emotions; Relies on communication, empathy, and social interaction. Examples: Planning based on 'quality of life'/ sense of place/ community; Incorporating equity and social justice issues into planning/ implementation; Understanding of how people feel about comfort/ safety/ etc. of different infrastructure system elements. AI Capabilities: AI learn and adapt empathetically based on experience; Decisions/ outputs incorporate emotions; Signified by affective computing, emotion recognition, and highly advanced communication. Applications: Communicating service quality/ outages; Social nudges for behavioral change/demand management (Conniff, 2017; Matias, 2017; Suh, 2019; Baum, 2020; Altering service provision based on emotional analytics (e.g., traffic management based on emotional needs) (Abou-Zeid & Ben-Akiva, 2010; Quercia et al., 2014); Automated disaster planning and recovery based on community emotions, needs, and capabilities; AI for negotiating wicked problems across various stakeholders.

5.4.3.5. HOW MIGHT AI DISRUPT INFRASTRUCTURE SERVICES AND INTRODUCE NEW CAPABILITIES?

Exploration of the four levels of intelligences in the context of infrastructure systems reveals a few key insights. First, it appears that AI (or at least automation) has already been widely implemented for Mechanical tasks. Although there is still some potential for AI growth and evolution at this level, it appears that we may have already reached a saturation point, thereby making fundamental transformations less likely. This outcome further underscores the potential for AI to complement and supplement Administrative Leadership roles within infrastructure systems. On the other hand, Analytical tasks are where AI appears poised to have the largest disruption (at least in the near-to-medium term). As AI capabilities continue to improve (especially due to the combination of ever-increasing data availability, ever-decreasing computing costs, and advancements in techniques like ANNs), analytical tasks (and Adaptive Leadership roles) will increasingly be accomplished by AI. Considering that the vast majority of engineering and infrastructure jobs are analytical by nature, the augmentation and/or replacement of analytical tasks by AI is likely to have a fundamental, profound, and transformative impact on infrastructure systems as we know them. Thus, moving forward, a key adaptation for engineers and infrastructure managers will be to strengthen and place increasing emphasis on Intuitive and Empathetic tasks/intelligences, which in turn should strengthen Enabling Leadership capabilities. This is particularly important, because even though humans exhibit much higher levels of Intuitive and Empathetic intelligence than AI (and are likely to remain that way for quite a while), there is still room for improvement. Human error is always a concern when operating under both mundane and surprise conditions. Similarly, Empathetic intelligence currently does not appear to be widely incorporated or considered in the development of engineered/infrastructure systems. Thus, in order to most effectively balance the Mechanical (i.e., Administrative Leadership) and Analytical (i.e., Adaptive Leadership) advantages of AI with the Intuitive and Empathetic (i.e., Enabling Leadership) advantages of humans, we (humans) will need to continually learn from past mistakes and develop skills to make effective decisions under surprise conditions. Additionally, substantial and continual efforts should be made toward enhancing our ability to incorporate social, emotional, and equity dynamics into engineering and infrastructure planning and implementation.

5.4.4. DISCUSSION AND CONCLUSION

It is useful to consider how AI technologies in infrastructure are likely to create new capabilities that, if leveraged correctly, can help us adapt to the rapidly changing conditions in which infrastructure systems must thrive. As evidence emerges of the accelerating and increasingly uncertain conditions that characterize infrastructure environments (Steffen *et al.*, 2015), design and management must be able to respond to these conditions with agility and flexibility (Chester and Allenby, 2018; Gilrein *et al.*, 2019). With any new technology, control processes are created to harness and guide the new capabilities toward the goals of the managing institution (Beniger, 1989). For example, the advent of engines and novel processes during the industrial revolution released energy at rates and scales never before seen. In turn, these technological advancements required new institutions and processes to channel this power. Whether AI follows historical patterns of technological control is questionable. AI technologies are fundamentally focused on augmenting and replacing cognition. Cognitive infrastructure that learns and makes decisions for us implies that control may not be fully attainable (like it was for the steam engines in the industrial era). Instead, our control efforts may need to focus on establishing relationships with AI that recognize that cyber-technologies will be guiding us in ways that we may not always fully understand.

AI may be uniquely positioned to help us learn about and navigate increasingly complex environments. In designing knowledge systems, institutions enable sensing and analytical capabilities (coupled with different leadership styles) to operate in both calm and chaotic environments (Miller and Muñoz-Erickson, 2018). As our systems, and the environments in which they operate, become increasingly complex and beyond the cognitive understanding of any group of individuals or institutions, AI may offer critical cognitive insights to ensure that systems adapt, services continue to be provided, and needs continue to be met.

The mapping of AI applications to intelligences and leadership roles appears to support the varying approaches needed to address domains of complexity. The Cynefin framework classifies systems as simple, complicated, complex, or chaotic, and as we transition from one domain to another, disorder governs (Snowden and Boone, 2007; Chester and Allenby, 2019). Each domain requires a fundamentally different approach to address challenges. Infrastructure have historically been complicated systems and are now increasingly viewed as complex (Chester and Allenby, 2018). A complicated system calls for data collection, analyzing and decision-making, while a complex system shifts towards probing, testing, and a commitment to adaptability and transformation. The intelligence mapping presented in Table 5.6.1 provides a useful set of AI applications that can be applied to infrastructure in complicated and complex environments. The Mechanical and Analytical intelligences appear to align well with complicated situations where the emergent behaviors of systems are predictable and their environments somewhat stable. The Intuitive and

Empathetic intelligences appear to align with complex systems, where perturbations can result in unpredictable emergent behaviors, and 'satisficing' is needed to manage wicked problems across technical and social requirements (Chester and Allenby, 2019). While all intelligences are needed at various times during the operation of a system, the development and deployment of Intuitive and Empathetic intelligences (and Enabling Leadership) in humans and institutions, as well as the development and deployment of Administrative and Adaptive Leadership via AI appears necessary to address the growing complexity and non-stationarity of our systems and the environments in which they operate.

Ultimately, we are in the nascent stages of AI development and application to infrastructure systems. The topics in this chapter are intended to be an initial discussion of some of the key opportunities and challenges associated with AI in infrastructure systems—especially in the context of the leadership and skills needed to face the complex challenges of the Anthropocene. Avenues for future work that can build on this endeavor include interviews and surveys aimed at gaining a better understanding of infrastructure practitioner's current thoughts and expectations about the possible benefits and downsides of AI. Additionally, it would be beneficial to further explore which level of intelligence appears most appropriate for specific problems/contexts, as well as a more detailed assessment of the specific AI techniques likely to be most effective/appropriate in these circumstances. Finally, in conjunction with (if not prior to) these efforts, open, candid, and iterative discussions are required amongst society writ large to debate what level of cognitive infrastructure we are comfortable with and the level of 'control' (or at least perceived control) we are comfortable offloading to cognitive infrastructure. By doing so, engineers and infrastructure users/managers can hopefully ensure that they are striking the right balance between human and AI capabilities required to effectively and equitably navigate our increasingly complex world.

5.4.5. REFERENCES

Abou-Zeid, M. and Ben-Akiva, M. (2010) 'A Model of Travel Happiness and Mode Switching', in S. Hess and A. Daly (eds) *Choice Modelling: The State-of-the-art and The State-of-practice.* Emerald Group Publishing Limited, pp. 289–305. Available at: https://doi.org/10.1108/9781849507738-012.

Abrahart, R.J. and See, L.M. (2007) 'Neural network modelling of non-linear hydrological relationships', *Hydrology and Earth System Sciences*, 11(5), pp. 1563–1579. Available at: https://doi.org/10.5194/hess-11-1563-2007.

Adeli, H. (2001) 'Neural Networks in Civil Engineering: 1989–2000', *Computer-Aided Civil and Infrastructure Engineering*, 16(2), pp. 126–142. Available at: https://doi.org/10.1111/0885-9507.00219.

Adeli, H. and Park, H.S. (1995) 'A neural dynamics model for structural optimization — Theory', *Computers & Structures*, 57(3), pp. 383–390. Available at: https://doi.org/10.1016/0045-7949(95)00048-L.

Allenby, B. (2019) '5G, AI, and big data: We're building a new cognitive infrastructure and don't even know it', *Bulletin of the Atomic Scientists*, 19 December. Available at: https://thebulletin.org/2019/12/5g-ai-and-big-data-were-building-a-new-cognitive-infrastructure-and-dont-even-know-it/ (Accessed: 13 October 2023).

Arslan, M.H. (2010) 'An evaluation of effective design parameters on earthquake performance of RC buildings using neural networks', *Engineering Structures*, 32(7), pp. 1888–1898. Available at: https://doi.org/10.1016/j.engstruct.2010.03.010.

Baum, S.D. (2020) 'Social choice ethics in artificial intelligence', *AI & SOCIETY*, 35(1), pp. 165–176. Available at: https://doi.org/10.1007/s00146-017-0760-1.

Beniger, J.R. (1989) *The Control Revolution: Technological and Economic Origins of the Information Society*. Cambridge, MA: Harvard University Press.

Brynjolfsson, E. and McAfee, A. (2017) 'The Business of Artificial Intelligence', *Harvard Business Review*, 18 July. Available at: https://hbr.org/2017/07/the-business-of-artificial-intelligence (Accessed: 4 December 2023).

Camp, C.V., Pezeshk, S. and Hansson, H. (2003) 'Flexural Design of Reinforced Concrete Frames Using a Genetic Algorithm', *Journal of Structural Engineering*, 129(1), pp. 105–115. Available at: https://doi.org/10.1061/(ASCE)0733-9445(2003)129:1(105).

Chandwani, V., Agrawal, V. and Nagar, R. (2013) 'Applications of Soft Computing', *Civil Engineering: A Review. International Journal of Computer Applications*, 81(10), pp. 13–20. Available at: https://doi.org/10.5120/14047-2210.

Chen, S.H., Jakeman, A.J. and Norton, J.P. (2008) 'Artificial Intelligence techniques: An introduction to their use for modelling environmental systems', *Mathematics and Computers in Simulation*, 78(2), pp. 379–400. Available at: https://doi.org/10.1016/j.matcom.2008.01.028.

Cheng, C.-T., Wu, X.-Y. and Chau, K.W. (2005) 'Multiple criteria rainfall–runoff model calibration using a parallel genetic algorithm in a cluster of computers / Calage multi-critères en modélisation pluie–débit par un algorithme génétique parallèle mis en œuvre par une grappe d'ordinateurs', *Hydrological Sciences Journal*, 50(6), p. 1087. Available at: https://doi.org/10.1623/hysj.2005.50.6.1069.

Chester, M.V. and Allenby, B. (2018) 'Toward adaptive infrastructure: flexibility and agility in a non-stationarity age', *Sustainable and Resilient Infrastructure*, 4(4), pp. 173–191. Available at: https://doi.org/10.1080/23789689.2017.1416846.

Chester, M.V. and Allenby, B. (2019) 'Infrastructure as a wicked complex process', *Elementa: Science of the Anthropocene*. Edited by A. Iles and M.E. Chang, 7(1), p. 21. Available at: https://doi.org/10.1525/elementa.360.

Cisco (2020) *Cisco Annual Internet Report (2018–2023) White Paper*. C11-741490–01. San Jose, CA: Cisco Systems, Inc., pp. 1–41. Available at: https://www.cisco.com/c/en/us/solutions/collateral/executive-perspectives/annual-internet-report/white-paper-c11-741490.pdf (Accessed: 26 October 2023).

Columbus, L. (2018) '2018 Roundup Of Internet Of Things Forecasts And Market Estimates', *Forbes*, 13 December. Available at: https://www.forbes.com/sites/louiscolumbus/2018/12/13/2018-roundup-of-internet-of-things-forecasts-and-market-estimates/ (Accessed: 26 October 2023).

Eggimann, S. *et al.* (2017) 'The Potential of Knowing More: A Review of Data-Driven Urban Water Management', *Environmental Science & Technology*, 51(5), pp. 2538–2553. Available at: https://doi.org/10.1021/acs.est.6b04267.

Falcone, R., Lima, C. and Martinelli, E. (2020) 'Soft computing techniques in structural and earthquake engineering: a literature review', *Engineering Structures*, 207, p. 110269. Available at: https://doi.org/10.1016/j.engstruct.2020.110269.

Flintsch, G.W. and Chen, C. (2004) 'Soft Computing Applications in Infrastructure Management', *Journal of Infrastructure Systems*, 10(4), pp. 157–166. Available at: https://doi.org/10.1061/(ASCE)1076-0342(2004)10:4(157).

Flood, I. (2001) 'Neural networks in civil engineering: a review', in B.H.V. Topping (ed.) *Civil and structural engineering computing: 2001*. Kippen, UK: Saxe-Coburg Publications, pp. 185–2019.

Flood, I. and Kartam, N. (1994a) 'Neural Networks in Civil Engineering. I: Principles and Understanding', *Journal of Computing in Civil Engineering*, 8(2), pp. 131–148. Available at: https://doi.org/10.1061/(ASCE)0887-3801(1994)8:2(131).

Flood, I. and Kartam, N. (1994b) 'Neural Networks in Civil Engineering. II: Systems and Application', *Journal of Computing in Civil Engineering*, 8(2), pp. 149–162. Available at: https://doi.org/10.1061/(ASCE)0887-3801(1994)8:2(149).

Gelles, D. (2019) 'Boeing 737 Max: What's Happened After the 2 Deadly Crashes', *The New York Times*, 22 March. Available at: https://www.nytimes.com/interactive/2019/business/boeing-737-crashes.html, https://www.nytimes.com/interactive/2019/business/boeing-737-crashes.html (Accessed: 26 October 2023).

Genaidy, M.A. (2020) 'ESTIMATING OF EVAPOTRANSPIRATION USING ARTIFICIAL NEURAL NETWORK', *Misr Journal of Agricultural Engineering*, 37(1), pp. 81–94. Available at: https://doi.org/10.21608/mjae.2020.94971.

Gilrein, E.J. *et al.* (2019) 'Concepts and practices for transforming infrastructure from rigid to adaptable', *Sustainable and Resilient Infrastructure*, 6(3–4), pp. 213–234. Available at: https://doi.org/10.1080/23789689.2019.1599608.

Grother, P., Ngan, M. and Hanaoka, K. (2019) *Face Recognition Vendor Test (FRVT) Part 3: Demographic Effects*. NISTIR 8280. Washington, DC: US Department of Commerce-National Institute of Standards and Technology, p. 82. Available at: https://doi.org/10.6028/NIST.IR.8280.

Hadi, M.N.S. (2003) 'Neural networks applications in concrete structures', *Computers & Structures*, 81(6), pp. 373–381. Available at: https://doi.org/10.1016/S0045-7949(02)00451-0.

Hajela, P. and Berke, L. (1991) 'Neurobiological computational models in structural analysis and design', *Computers & Structures*, 41(4), pp. 657–667. Available at: https://doi.org/10.1016/0045-7949(91)90178-O.

Huang, M.-H. and Rust, R.T. (2018) 'Artificial Intelligence in Service', *Journal of Service Research*, 21(2), pp. 155–172. Available at: https://doi.org/10.1177/1094670517752459.

Jayaram, M.A., Nataraja, M.C. and Ravikumar, C.N. (2009) 'Elitist Genetic Algorithm Models: Optimization of High Performance Concrete Mixes', *Materials and Manufacturing Processes*, 24(2), pp. 225–229. Available at: https://doi.org/10.1080/10426910802612387.

Jeong, D.-I. and Kim, Y.-O. (2005) 'Rainfall-runoff models using artificial neural networks for ensemble streamflow prediction', *Hydrological Processes*, 19(19), pp. 3819–3835. Available at: https://doi.org/10.1002/hyp.5983.

Johnston, P. and Harris, R.L. (2019) 'The Boeing 737 MAX Saga: Lessons for Software Organizations', *Software Quality Professional Magazine*, 21(3), pp. 4–12.

Karasekreter, N., Başçiftçi, F. and Fidan, U. (2013) 'A new suggestion for an irrigation schedule with an artificial neural network', *Journal of Experimental & Theoretical Artificial Intelligence*, 25(1), pp. 93–104. Available at: https://doi.org/10.1080/0952813X.2012.680071.

Kartam, N. *et al.* (eds) (1997) *Artificial Neural Networks for Civil Engineers: Fundamentals and Applications.* New York, NY: American Society of Civil Engineers.

Koutsoyiannis, D. (2011) 'Hurst-Kolmogorov Dynamics and Uncertainty1', *JAWRA Journal of the American Water Resources Association*, 47(3), pp. 481–495. Available at: https://doi.org/10.1111/j.1752-1688.2011.00543.x.

Lee, K.-F. and Li, K. (2018) *AI Superpowers: China, Silicon Valley, and the New World Order.* Boston, MA: Houghton Mifflin Harcourt.

Lee, S.C. and Han, S.W. (2002) 'Neural-network-based models for generating artificial earthquakes and response spectra', *Computers & Structures*, 80(20), pp. 1627–1638. Available at: https://doi.org/10.1016/S0045-7949(02)00112-8.

Lins, H.F. (2012) 'A Note on Stationarity and Nonstationarity', in *Commission for Hydrology, Advisory Working Group. Fourteenth Session of the Commission for Hydrology*, World Meteorological Organization.

Maier, H.R. and Dandy, G.C. (2000) 'Neural networks for the prediction and forecasting of water resources variables: a review of modelling issues and applications', *Environmental Modelling & Software*, 15(1), pp. 101–124. Available at: https://doi.org/10.1016/S1364-8152(99)00007-9.

Meserole, C. (2018) *Artificial Intelligence and the Security Dilemma, Brookings.* Available at: https://www.brookings.edu/articles/artificial-intelligence-and-the-security-dilemma/ (Accessed: 4 December 2023).

Miller, C.A. and Muñoz-Erickson, T.A. (2018) *The Rightful Place of Science: Designing Knowledge.* Tempe, AZ: Consortium for Science, Policy & Outcomes.

Milly, P.C.D. *et al.* (2008) 'Stationarity Is Dead: Whither Water Management?', *Science*, 319(5863), pp. 573–574. Available at: https://doi.org/10.1126/science.1151915.

Minns, A.W. and Hall, M.J. (1996) 'Artificial neural networks as rainfall-runoff models', *Hydrological Sciences Journal*, 41(3), pp. 399–417. Available at: https://doi.org/10.1080/02626669609491511.

Nagy, H.M., Watanabe, K. and Hirano, M. (2002) 'Prediction of Sediment Load Concentration in Rivers using Artificial Neural Network Model', *Journal of Hydraulic Engineering*, 128(6), pp. 588–595. Available at: https://doi.org/10.1061/(ASCE)0733-9429(2002)128:6(588).

Ni, H.-G. and Wang, J.-Z. (2000) 'Prediction of compressive strength of concrete by neural networks', *Cement and Concrete Research*, 30(8), pp. 1245–1250. Available at: https://doi.org/10.1016/S0008-8846(00)00345-8.

Nixon, J.B., Dandy, G.C. and Simpson, A.R. (2001) 'A genetic algorithm for optimizing off-farm irrigation scheduling', *Journal of Hydroinformatics*, 3(1), pp. 11–22. Available at: https://doi.org/10.2166/hydro.2001.0003.

Picard, R.W. (1995) *Affective Computing. MIT Media Laboratory Perceptual Computing Section Technical Report No. 321, MIT Media Lab*. Available at: https://hd.media.mit.edu/TechnicalReportsList.html (Accessed: 4 December 2023).

Quercia, D., Schifanella, R. and Aiello, L.M. (2014) 'The shortest path to happiness: recommending beautiful, quiet, and happy routes in the city', in *Proceedings of the 25th ACM conference on Hypertext and social media*. New York, NY: Association for Computing Machinery (HT '14), pp. 116–125. Available at: https://doi.org/10.1145/2631775.2631799.

Sawhney, M. (2016) 'Putting Products into Services', *Harvard Business Review*, 1 September. Available at: https://hbr.org/2016/09/putting-products-into-services (Accessed: 4 December 2023).

Shahin, M.A., Maier, H.R. and Jaksa, M.B. (2002) 'Predicting Settlement of Shallow Foundations using Neural Networks', *Journal of Geotechnical and Geoenvironmental Engineering*, 128(9), pp. 785–793. Available at: https://doi.org/10.1061/(ASCE)1090-0241(2002)128:9(785).

Snowden, D.J. and Boone, M.E. (2007) 'A Leader's Framework for Decision Making', *Harvard Business Review*, 1 November, pp. 68–76, 149.

Squire, L.R. (2009) 'Cognition', in *Encyclopedia of Neuroscience*. Amsterdam, NL: Elsevier/Academic Press.

Steffen, W. *et al.* (2015) 'The trajectory of the Anthropocene: The Great Acceleration', *The Anthropocene Review*, 2(1), pp. 81–98. Available at: https://doi.org/10.1177/2053019614564785.

Tabari, H., Marofi, S. and Sabziparvar, A.-A. (2010) 'Estimation of daily pan evaporation using artificial neural network and multivariate non-linear regression', *Irrigation Science*, 28(5), pp. 399–406. Available at: https://doi.org/10.1007/s00271-009-0201-0.

Tokar, A.S. and Johnson, P.A. (1999) 'Rainfall-Runoff Modeling Using Artificial Neural Networks', *Journal of Hydrologic Engineering*, 4(3), pp. 232–239. Available at: https://doi.org/10.1061/(ASCE)1084-0699(1999)4:3(232).

Tomer, A. (2019) *Artificial Intelligence in America's Digital City, Brookings*. Available at: https://www.brookings.edu/articles/artificial-intelligence-in-americas-digital-city/ (Accessed: 4 December 2023).

Uhl-Bien, M., Marion, R. and McKelvey, B. (2007) 'Complexity Leadership Theory: Shifting leadership from the industrial age to the knowledge era', *The Leadership Quarterly*, 18(4), pp. 298–318. Available at: https://doi.org/10.1016/j.leaqua.2007.04.002.

U.S. House Committee on Transportation and Infrastructure (2020) *The Boeing 737 MAX Aircraft: Costs, Consequences, and Lessons from it Design, Development, and Certification*. SuDoc: Y 4.T 68/2:B 63. Washington, DC: U.S. House Committee on Transportation and Infrastructure, p. 14. Available at: https://www.govinfo.gov/app/details/GOVPUB-Y4_T68_2-fb0f3812fefe3515ebcf3f4170fce64b.

Wise, J. (2019) 'The Boeing 737 Max and the Problems Autopilot Can't Solve', *The New York Times*, 14 March. Available at: https://www.nytimes.com/2019/03/14/opinion/business-economics/boeing-737-max.html (Accessed: 26 October 2023).

Ye, X.W., Jin, T. and Yun, C.B. (2019) 'A review on deep learning-based structural health monitoring of civil infrastructures', *Smart Structures and Systems*, 24(5), pp. 567–585. Available at: https://doi.org/10.12989/sss.2019.24.5.567.

Yeh, I.-C. (1999) 'Design of High-Performance Concrete Mixture Using Neural Networks and Nonlinear Programming', *Journal of Computing in Civil Engineering*, 13(1), pp. 36–42. Available at: https://doi.org/10.1061/(ASCE)0887-3801(1999)13:1(36).

Yilmaz, S. (2011) 'Ground motion predictive modelling based on genetic algorithms', *Natural Hazards and Earth System Sciences*, 11(10), pp. 2781–2789. Available at: https://doi.org/10.5194/nhess-11-2781-2011.

Young-Su, K. and Byung-Tak, K. (2006) 'Use of Artificial Neural Networks in the Prediction of Liquefaction Resistance of Sands', *Journal of Geotechnical and Geoenvironmental Engineering*, 132(11), pp. 1502–1504. Available at: https://doi.org/10.1061/(ASCE)1090-0241(2006)132:11(1502).

Zhang, F.X., Wai, O.W.H. and Jiang, Y.W. (2010) 'Prediction of sediment transportation in deep bay (Hong Kong) using genetic algorithm', *Journal of Hydrodynamics*, 22(1), pp. 582–587. Available at: https://doi.org/10.1016/S1001-6058(09)60260-2.

5.5

COGNITIVE ECOSYSTEM AND THE GOVERNANCE OF ARITIFICAL INTELLIGENCE

In the movie "Her," Theodore Twombly, played by Joaquin Phoenix, falls in love with the AI-powered virtual assistant in his mobile phone, Samantha, voiced by Scarlett Johansson, only to find out to his distress that "she" is more different than he ever imagined. He has mistaken a profoundly alien meta-cognitive entity for something familiar, in this case a girl.

Today, everyone from countries to firms to academics to the U.N. is desperate to be seen as responsive to the many challenges of generative AI. Hysteria abounds. The policy world is awash with regulatory proposals. But we need to think harder about Theodore and his heartbreak, for most of these proposals are, in essence, assuming a girl.

In particular, we are making two implicit, fatally flawed, assumptions. First, we are assuming that we know enough, at this very initial stage of perhaps the most potent technology humans have come up with since printing and literacy, to regulate effectively. This is hubris at a cosmic scale. We further assume that we are basically confronting a simple system rather than a complex adaptive system (CAS). Policy tools that assume knowledge and control that we can't possibly have simply fail under such circumstances.

A simple system can be effectively regulated only because you know enough about it, and how it works, to be able to predict the outcome of your regulatory action: "if I implement Regulation A, I will almost certainly get Output B". If you regulate the use of asbestos in buildings, for example, you can reliably predict that you will reduce exposure to asbestos particles in buildings over time, and thus protect human health. But if you are facing a CAS, you don't have the information or knowledge to

---

This chapter was adapted from the following article with publisher permission: Braden Allenby, 2023, "Riding the AI Tiger," *The Future of Being Human Blog*, https://andrewmaynard.substack.com/p/riding-the-ai-tiger.

know a priori what the outcome of your regulation will be. So, your regulatory paradigm is, instead, "if I implement Regulation A, who the heck knows what will happen?".

Moreover, CASs are dynamic and constantly changing, and can neither be known, nor controlled, by a central authority. Remember that one of the causes of the collapse of the Soviet Union was its inability to run a modern economy, a quintessential CAS, using the centralized authority of Gosplan, the State Planning Committee.

We are frantically trying to build a Gosplan for AI. Indeed, we can hardly help it, for our tools and institutions assume adequate knowledge and simple systems. Accordingly, in our haste and ignorance we are making a profound category mistake. If Theodore's heartbreak is inadequate warning, the fall of the Soviet Union should at least be cautionary.

Yet to date China and the West, the two major civilizational players in the AI arena, are both making that familiar category mistake (although, to be fair, they both appear to be beginning to realize it). The approach of the West fails because the core assumption required for regulation to be an effective tool, that the output of a regulatory system can be predicted from the input, is wrong. Chinese approaches have an additional problem. Because the Party must always be able to exercise centralized and explicit control, it is trying to require predictable results from AI. This fails because in a CAS control mechanisms and knowledge cannot be centralized. Rather, they are diffused throughout dynamic networks. This is patently obvious from the experience with generative AI to date, which has if anything reinforced the reality of AI as black box.

Basically, there are three obvious challenges for a successful approach to managing AI:

The "cycle time problem": No regulatory structure or institution has the capability to match the rate at which these technologies are evolving.

The "knowledge problem": No one today has any idea of the myriad ways in which AI generally, and generative AI specifically, are now being used across global societies and economies.

The "scope of effective regulation problem": Such potent technologies are most rapidly adapted by fringe elements of the global economy, especially the pornography industry and global criminal enterprises; such elements are precisely those which will pay no attention to regulation anyway.

Even if, however, reinventing Gosplan for AI is not a viable strategy, there are three obvious responses which could support far more effective management of AI.

The first and most important step is to create informal, broad-based, networked working groups - say, at the level of the E.U., or the U.S., or China - which would have as their only task perceiving current realities on an on-going, close to real time, basis. Such entities wouldn't suggest laws or policies or regulations, make investments, impose ideological mandates, or critique databases, because any such task would bias the core task of perception. It would only seek to know what is out there, and what is going on. Knowledge comes before management – at least, it should.

Second, slow regulatory and institutional responses have already decoupled from the cycle time of AI evolution. They cannot hope to keep up, nor should they, for doing so might undermine their current effectiveness (law works as a critical cultural and civilizational infrastructure precisely because it is stable and predictable over time). Rather, an additional focus should be to create agile, adaptive capability, especially in critical areas such as security. This is not impossible; to some extent it's how the U.S. Food and Drug Administration and the European Medicines Agency operate today, albeit with far more developed institutional frameworks and far deeper knowledge resources. Moreover, while broad regulatory initiatives are likely to be dysfunctional, there will undoubtedly be specific issues that can be addressed en passant. The failure of Gosplan does not mean that societies do not routinely impose laws and regulations on economic behavior.

Indeed, many companies and governments have recognized that formal institutions, policies, laws, regulations, and other tools are too inflexible when faced with rapid and unpredictable change. Thus, new approaches such as "soft law" – think of industry codes of conduct, for example - and institutional initiatives such as "skunk works" have long been used to provide flexibility and rapid adaptability even within more structured and formal institutions.

Managing AI is unquestionably a fundamental challenge to today's governments and institutions. It is not insurmountable, but there is little to suggest that we are yet deploying the imagination, and the creativity, required.

Section 6

CONCLUSION:

INFRASTRUCTURE FIRST PRINCIPLES FOR THE ANTHROPOCENE

Anthropocene dynamics represent a wicked and complex environment that necessitates fundamentally new infrastructure models. These models – fundamental assumptions that drive societal and organizational representations of a changing reality – should be based in principles that reflect emerging infrastructure technologies and goals, the increasingly diffusive nature of infrastructure systems, and the increasingly complex environments that infrastructure must adapt and transform to such that future generations can thrive (Myntti, 2024). We propose a set of first principles for infrastructure as they attempt to disrupt legacy norms and innovate in response to future challenges, their changing natures, and their uncertainties. We then place these principles within a proposed System of Systems of novel emerging relationships and hierarchies of infrastructure. These first principles are not intended to be specific recommendations but instead fundamental guidance on how to navigate infrastructure. They are informed by novel infrastructure research that often leverages insights from other domains that appear better equipped to engage with uncertain and complex environments.

This section was adapted from the following article with publisher permission: Mikhail Chester and Braden Allenby, 2024, Infrastructure First Principles for the Anthropocene, *Environmental Research: Infrastructure and Sustainability*, 4(4), 043001, doi: 10.1088/2634-4505/ad8834.

Figure 6.1: Infrastructure First Principles for the Anthropocene

6.1. FIRST PRINCIPLES

The first principles are guided by the challenges of innovating infrastructure technologies and governance systems to engage with increasingly complex conditions. They use concepts from emerging infrastructure and other discipline theory that seek to describe how systems can adapt and transform as the environments in which they are accustomed change. The principles span technologies, physical networks, operational and governmental processes, information and knowledge networks, and physical and cyber control systems.

PRINCIPLE 1: PLAN FOR COMPLEX CONDITIONS AND SURPRISE

Infrastructure are caught between growing system complexity and increasingly complex environments yet are designed for simpler conditions. To start, it is helpful to recognize that the approaches to design and manage infrastructure will differ depending on a starting frame that normalizes complicated versus complex systems and environments (Snowden and Boone, 2007). Infrastructure design is largely rooted in predictability where expert diagnosis is required to assess cause and effect relationships and produce a right answer. This frame fundamentally differs from one that recognizes the growing complexity in the systems themselves as well as their environments, where there are unpredictable emergent behaviors and therefore no right answers (Chester and Allenby, 2019a). Consider the NYC subway as an example of infrastructure internal complexity, a system that has experienced technology accretion over a century, where hardware and controls from the early 1900s must interact with modern hardware and controls. You must approach design and management of the NYC subway and its internal complexity differently than a new subway system that may be less complex because it relies on only one generation of technology. Changing external environments intersect with this internal complexity creating wicked challenges for infrastructure managers. Consider adapting the NYC subway to increased pluvial flooding while having to adapt sub-systems that span a century of technologies. The design and management of infrastructure must plan for these complex conditions and the increasing frequency of surprise. Predictable envelopes of volatility and normative scenarios that assume stable futures will be inimical to the ability to deliver reliable services. Infrastructure managers should adopt perspectives that their systems will fail and when they do will need to extend themselves (Kim et al., 2019; Woods, 2015, 2018), will need to constantly adapt and transform to new and unforeseen conditions (see Principle 2), and should be capable of enabling governance transitions between stable and unstable conditions (see Principle 3). Novel data streams, software, and AI appear poised to make better sense of this complexity than legacy sensemaking processes, while at the same time adding complexity (as interdependencies, vulnerabilities, and new means of control (see Principle 4)) to the systems themselves.

PRINCIPLE 2: RECOUPLE WITH AGILITY AND FLEXIBILITY

The rigidity inherent in the technologies and governance that characterize infrastructure work well when conditions are relatively stable, but with rapidly changing environments agility (physical structure and the rules, policies, norms, and actors who manage and operate it, will need to be able to maintain function in a non-stationary future.) and flexibility (ability to meet changing demands in the face of both predictable and unpredictable challenges) are needed (Chester and Allenby, 2019b). Recoupling is the set of agile and flexible processes used for changing infrastructure at pace and scale to environment change (Chester et al., 2021). Agile and flexible processes are not simply technical (e.g., compatibility, connectivity, modularity, software-for-hardware substitution) and governmental (e.g., roadmapping, design for obsolescence, organic cultures that emphasize change) competencies but more so describe a means by which infrastructure organizations produce adaptation repertoires, an output of a system generating complexity to navigate between stability and instability (Chester et al., 2023; Chester and Allenby, 2022; Lambert, 2020; Langton, 1997; Levy, 2000; Packard, 1988; Porter, 2006). Rigid processes are needed for times of stability, when conditions are relatively stable and organizations can emphasize efficiency over exploration (Markolf et al., 2022). However, when organizations recognize that their environments are routinely going to introduce instability that threatens to push their systems beyond what they are designed to tolerate then agility and flexibility are needed to adapt. There is the potential for many configurations of technologies and governance processes to introduce agility and flexibility. Some technologies and processes will be stable while others will change rapidly and perhaps unpredictably. This duality across system scales, technologies, and process is a way of building agility and adaptability into the parts that are rapidly evolving, while still maintaining the stability of the system taken as a whole. Consider how iOS is relatively stable while apps built on the operating systems evolve at breakneck speed. This has operational implications, as it means that the design and management of rapidly evolving modules within more stable interconnected systems becomes crucial.

PRINCIPLE 3: GOVERN FOR EXPLORATION AND INSTABILITY

Little attention is given to the governance structures of infrastructure organizations, how they came to be, what they emphasize in terms of knowledge generation and sensemaking, and ultimately how they support or constrain novel goals (e.g., sustainability, resilience, equity and justice, cybersecurity). Governance structures can take on many different forms and these forms may support or constrain adaptation (Chester et al., 2020; Muñoz-Erickson et al., 2017; Muñoz-Erickson et al., 2016). Key to future infrastructure governance is the creation of space to explore how to respond to increasingly complex systems and environments (including satisficing as per Principle 4). Organizational theory and leadership communities have described the capabilities needed to innovate and adapt at the edge of chaos, the boundary between stability and instability. First, organizations should scan for weak signals that despite low likelihoods of occurring would change the conditions under which the infrastructure functions (Iwaniec et al., 2019), e.g., new environmental conditions that exceed designs, changes in service consumption to new models, and financial conditions that undermine operational requirements. The identification of weak signals should spark planning efforts for how to respond. At the edge of chaos governance must pivot from organizational structures, rules, resource allocation, and goals that emphasize efficiency under stability (administrative leadership) to models that emphasize adaptation (adaptive leadership). The pivot is possible because additional planning efforts have already described how to restructure teams and resources (enabling leadership). Adaptive leadership creates space for teams to self-organize responsive to the particular hazard, implements resourcing models so those teams have the appropriate knowledge, finances, and physical resources, and imbues those teams with the authority to make decisions on behalf of the organization. Horizontal governance structures during chaos empowers dynamic team building through frontline workers that are in the best position to see change and how best to use limited organizational resources. This comports with the concept of near decomposability, that when a system is perturbed the organizational units at the lowest level of the system can be expected to reach equilibrium faster than middle and higher levels of the organization (Simon, 2002) and with the governance theory of "subsidiarity," as practiced by, e.g., the European Union, where decisions are to be made at the most local level possible. Additionally, infrastructure governance must be able to pivot on what systems and sub-systems are deemed critical (i.e., dynamic criticality) (Hoff et al., 2023). The emerging hazard landscape is simply too complex and unknown for static framings that are inflexible to the specific challenges that may present with each hazard.

PRINCIPLE 4: BUILD CONSENSUS AS CONTROL DECENTRALIZES

As legacy infrastructure integrate smart and connected devices, shift data analytics to the cloud (and software), and are increasingly steered by third parties who can harness novel data streams then control will decentralize. Currently, infrastructure agencies are built around governance structures that emphasize two parties: the producers and consumers. The controls that they've implemented reflect this two-party history (e.g., regulation of service providers, loop detectors to sense traffic volumes, demand management strategies for electricity such as peak pricing). Yet recently there has been increasing disruption by third parties who can produce services (solar, rainwater capture), independent technologies that control services (smart thermostats and irrigation controllers), and unregulated products that steer behavior (Google and Apple Maps). This disruption only seems poised to accelerate and become more shocking to a system that is designed around two parties and an assumption that control of services is almost entirely in the hands of producers. The integration of smart and connected technologies into infrastructure isn't simply the introduction of novel capabilities, but the absorption of infrastructure into a cognitive ecosystem defined by massive data flows, artificial intelligence, and connected technologies, that is poised to alter how humans and artificial intelligence understand and control our world (Allenby, 2021; Chester and Allenby, 2023; Smart et al., 2017). Cognition and control of infrastructure services is increasingly yielding to software and third parties capable of harnessing these dynamics far better than legacy infrastructure agencies whose governance structure predates those technologies. This decentralization of control, which is a fundamental characteristic of complex systems, represents a shift from the two-party system where third parties – largely technology companies – can decide or steer how services are consumed, and sometimes generated. And AI appears to be well positioned to diffuse decision making away from centralized managers and organizations. Decentralization of control coupled with increasingly complex conditions are together poised to produce more volatility, and therefore are harbingers for a reassessment of the goals of infrastructure agencies. As public institutions that deliver public services, infrastructure agencies should recognize the pressing need for consensus building, that a greater array of parties will have a say in what infrastructure do, and the role of infrastructure governance must increasingly fill the challenge of ensuring the public's well-being. Governance processes should move away from legacy approaches that emphasize optimized outcomes and towards satisficing, where good enough is the goal combined with a commitment to sustained reassessment of what systems are doing and whether adjustments are needed (i.e., sustained adaptation) (Chester and Allenby, 2019a; Simon, 1957, 1947; Woods, 2015).

PRINCIPLE 5: RESTRUCTURE TO ENGAGE WITH POROUS BOUNDARIES

The boundaries that define our infrastructure are quickly diffusing as cyber technologies create new capabilities and markets that can leverage services across existing infrastructure to create new capabilities. As we reach the limits (aging assets and insufficient funding, ideological polarization, extreme weather events, steering of services by third parties, targeting of civilian infrastructure by geopolitical adversaries, etc.) of our ability to manage legacy infrastructure in more complex environments, we will naturally look to other systems to augment capabilities. This isn't simply linking two technologies, e.g., electricity and electric vehicles through vehicle-to-grid, but more so an awareness of our ability to work across dynamics of social, ecological, and technological systems in new ways, either out of necessity or for new capabilities. For example, electric vehicles are not simply transportation technologies but also energy systems, capable of disrupting the norms of energy use both in charging and with vehicle-to-grid discharging. With intelligent controls Kuala Lumpur's SMART tunnel can pivot between traffic and stormwater management (Isah, 2016). Cybertechnologies are inserting themselves as novel infrastructure DNA that is restructuring the relationships of systems. Whereas in the past an infrastructure agency could largely disassociate itself from other infrastructure, this is increasingly impossible today not simply because systems are becoming more integrated through cybertechnologies but also because of necessity. Infrastructure managers seem to be increasingly confronting the limits of their systems and it makes sense that they then look to other co-dependent systems to maintain reliability. A transportation agency combating increasing wildfire and post-fire debris flow risk with climate change may find that it is less costly and more effective to invest in forest management practices than to try to harden their system. To engage with increasingly porous boundaries infrastructure agencies should look to co-management governance practices rooted in the social-ecological systems (SES) communities. Co-management – the joint management of commons – is a problem-solving process and governance approach that organizes and distributes power-sharing functions across government agencies, private entities, and local communities. Instead of taking a perspective that infrastructure have independent goals and processes, co-management could help recognize that functions across systems (commons) are increasingly intertwined and novel power-sharing arrangements are needed (Olsson et al., 2004).

PRINCIPLE 6: CYBERTHREAT PLANNING IS MISSION CRITICAL

As infrastructure have become connected new vulnerabilities have emerged requiring cyber resilience competency throughout infrastructure organizations. With the integration of cyber technologies into infrastructure new capabilities have emerged at a cost of opening up systems to a vast spectrum of threats (Chester and Allenby, 2020). At one end of this spectrum are well acknowledged threats including nuisance (e.g., botnets and spam), data theft (e.g., ransomware), and hacktivism (e.g., website defamation). At the other end of the spectrum is a far more complex threat landscape that is treating infrastructure as cyber battlespaces that allow foreign nations to bypass militaries and directly sow chaos among populations (i.e., civilizational conflict) (de Bruijne et al., 2017; Fritz, 2008; Hjortdal, 2011). These threats are not manageable by the usual geopolitical means; novel strategies are needed that embed infrastructure within national security strategy (Allenby, 2016, 2015). Recent examples include Russia disabling Ukraine's power grid prior to their 2015 invasion (Zetter, 2016), Russia's denial of service and data wiping attacks on key Ukrainian government websites (Lyngaas, 2022), and China's intrusion and prepositioning in critical infrastructure to disrupt critical services if conflict over Taiwan were to escalate (Fairclough, 2024). The cyber threat landscape is evolving rapidly adding complexity that must directly be engaged. The legacy roles of IT departments do not appear sufficient to counter these new threats. Instead, infrastructure governance should adopt pervasive cyber agility competencies that are central to what the organization is and does (Bodeau and Graubart, 2017). This pervasive cyber agility should be structured around co-governance models that build knowledge and responses across federal agencies and private technology companies in addition to the infrastructure agency itself. For example, Microsoft at times may be in the best position to make sense of strategic threats. Power sharing over how an infrastructure organization prepares for and responds to that threat may benefit from an open relationship with the company. At the highest levels of the organization governance should emphasize cyber capabilities and vulnerabilities, not as ancillary to the mission of delivering legacy services but as part of the service itself. To deliver water, power, mobility, or any other critical service will require navigating a cognitive ecosystem of cyber and legacy technologies that define the critical infrastructure systems.

6.2. NOVEL DYNAMICS AND ECOLOGIES

The legacy boundaries that have defined infrastructure planning, design, and management appear increasingly problematic against a backdrop of deeply uncertain environments and the emergence of novel cybertechnology and control systems that are integrating conventional infrastructure. Whereas in the past infrastructure could be managed as non-cyber independent or directly interdependent systems, this no longer is the case. Infrastructure appear increasingly steered by cybertechnologies and emerging cyber cognition, with diffuse boundaries, and controlled by a diversity of players. This change is driven by an expanding ecology of massive data flows, artificial intelligence, institutional and intellectual structures, and connected technologies (referred to as the Cognitive Ecosystem) (Allenby, 2021; Chester and Allenby, 2023). As such, the first principles should be contextualized and practiced within this emerging system of systems (Jamshidi, 2008) that includes four cores sub-systems that are respectively increasingly dynamic and together wicked and complex:

1)   Technological assets (e.g., the water or electric infrastructure) – the physical assets that makeup infrastructure;

2)   Operational/Organizational – which would include not just the organization running the infrastructure network, but also the politics, natural systems, stakeholders, geopolitics, cybersecurity and criminal networks that exert power;

3)   Information networks that connect all of these – including the information that an infrastructure agency gets from long-term weather forecasts, their people at the state capitol, and consulting firms, and federal agencies on security threats;

4)   Controls systems – which may be formal (e.g., a water utility controls water flow) or informal (e.g., state politics and regulation, the natural availability of the infrastructure resource such as water resources, and technology companies steering demand).

This system of systems is an emerging, evolving, dynamic and complex ecosystem and the first principles reflect novel practices, technologies, and knowledge creation processes, to engage with the complexity. In the Anthropocene the four sub-systems are increasingly coupled and internally dynamic to produce complexity (i.e., emergent system behavior that is volatile and functionality that is dispersed across sub-systems) that is difficult to approach directly, and perhaps even to understand directly. The sub-systems are themselves becoming difficult enough to parse, and the system of systems might be complex enough so that any framing we might understand is necessarily partial and somewhat arbitrary. As such, the first principles represent practices to: i) perceive the complexity in ways that are useful; and ii) learn enough about it to track the dynamics that our interventions in the sub-systems, and perhaps the system of systems itself, generate or contribute to.

6.3. CONCLUSION

For an enterprise to thrive in the face of chaos and complexity it has two options: attempt to reduce that complexity or adapt to it (Boisot and McKelvey, 2011). Attempts by infrastructure agencies to reduce complexity – e.g., disregarding deeply uncertain climate futures for assumptions of predictable futures, pretending that all types of infrastructure aren't active cyberconflict targets, assuming that traffic can be controlled disregarding Apple Maps -- appear ill-fated. Infrastructure appear to be near or at tipping points where accelerating technology integration, distribution of control, unpredictable environments, and other factors are quickly upending the norms that have allowed infrastructure to persist relatively unchanged for the past century. The six principles serve as guidance for engaging with complexity across technologies, governance, and higher order emerging systems. The principles themselves are perhaps less important than the emergent characteristic of the novel systems they open up. The principles should serve as guidance towards producing the requisite complexity to adapt to complexity. Through this adaptation the ultimate goal should be to recouple, i.e., create novel infrastructure system (as technologies and governance) that are able to thrive in the face of chaos and complexity.

6.4. REFERENCES

Allenby, B., 2021. World Wide Weird: Rise of the Cognitive Ecosystem. Issues Sci. Technol. 37.

Allenby, B., 2016. In An Age of Civilizational Conflict. Jurimetrics 56, 387–406.

Allenby, B.R., 2015. The paradox of dominance: The age of civilizational conflict. Bull. At. Sci. 71, 60–74. https://doi.org/10.1177/0096340215571911

Bodeau, D., Graubart, R., 2017. Cyber Prep 2.0: Motivating Organizational Cyber Strategies in Terms of Threat Preparedness. The MITRE Corporation, Bedford, MA.

Boisot, M., McKelvey, B., 2011. Complexity and Organization–Environment Relations: Revisiting Ashby's Law of Requisite Variety, in: The

Sage Handbook of Complexity and Management. SAGE Publications Ltd, London, pp. 278–298. https://doi.org/10.4135/9781446201084

Chester, M., Miller, T., Muñoz-Erickson, T., Helmrich, A., Iwaniec, D., McPhearson, T., Cook, E., Grimm, N., Markolf, S., 2023. Sensemaking for entangled urban social, ecological, and technological systems in the Anthropocene. Npj Urban Sustain. 3. https://doi.org/10.1038/s42949-023-00120-1

Chester, M., Underwood, B.S., Allenby, B., Garcia, M., Samaras, C., Markolf, S., Sanders, K., Preston, B., Miller, T.R., 2021. Infrastructure resilience to navigate increasingly uncertain and complex conditions in the Anthropocene. Npj Urban Sustain. 1, 1–6. https://doi.org/10.1038/s42949-021-00016-y

Chester, M.V., Allenby, B., 2023. Infrastructure and the cognitive ecosystem: an irrevocable transformation. Environ. Res. Infrastruct. Sustain. 3, 033002. https://doi.org/10.1088/2634-4505/aced1f

Chester, M.V., Allenby, B., 2022. Infrastructure autopoiesis: requisite variety to engage complexity. Environ. Res. Infrastruct. Sustain. 2, 012001. https://doi.org/10.1088/2634-4505/ac4b48

Chester, M.V., Allenby, B., 2019a. Infrastructure as a wicked complex process. Elem Sci Anth 7, 21. https://doi.org/10.1525/elementa.360

Chester, M.V., Allenby, B., 2019b. Toward adaptive infrastructure: flexibility and agility in a non-stationarity age. Sustain. Resilient Infrastruct. 4, 173–191. https://doi.org/10.1080/23789689.2017.1416846

Chester, M.V., Allenby, B.R., 2020. The Cyber Frontier and Infrastructure. IEEE Access 8, 28301–28310. https://doi.org/10.1109/ACCESS.2020.2971960

Chester, M.V., Miller, T., Muñoz-Erickson, T.A., 2020. Infrastructure governance for the Anthropocene. Elem. Sci. Anthr. 8. https://doi.org/10.1525/elementa.2020.078

de Bruijne, M., van Eeten, M., Hernández Gañán, C., Pieters, W., 2017. Towards a New Cyber Threat Actor Typology. TU Delft, Delft, Netherlands.

Fairclough, N.M. and G., 2024. China Is 'Prepositioning' for Future Cyberattacks—and the New NSA Chief Is Worried [WWW Document]. Wall Str. J. URL https://www.wsj.com/politics/national-security/china-is-prepositioning-for-future-cyberattacksand-thenew-nsa-chief-is-worried-5ede04ef (accessed 7.2.24).

Fritz, J., 2008. How China will use cyber warfare to leapfrog in military competiitiveness. Cult. Mandala Bull. Cent. East-West Cult. Econ. Stud. 8.

Hjortdal, M., 2011. China's Use of Cyber Warfare: Espionage Meets Strategic Deterrence. J. Strateg. Secur. 4, 1–24.

Hoff, R., Helmrich, A., Dirks, A., Kim, Y., Li, R., Chester, M., 2023. Dynamic criticality for infrastructure prioritization in complex environments. Environ. Res. Infrastruct. Sustain. 3, 015011. https://doi.org/10.1088/2634-4505/acbe15

Isah, N., 2016. Flood occurrence, smart tunnel operating system and traffic flow: a case of Kuala Lumpur smart tunnel Malaysia. Universiti Tun Hussein Onn Malaysia, Parit Raja, Malaysia.

Iwaniec, D.M., Cook, E.M., Barbosa, O., Grimm, N.B., 2019. The Framing of Urban Sustainability Transformations. Sustainability 11, 573. https://doi.org/10.3390/su11030573

Jamshidi, M. (Ed.), 2008. Systems of Systems Engineering: Principles and Applications. CRC Press, Boca Raton. https://doi.org/10.1201/9781420065893

Kim, Y., Chester, M.V., Eisenberg, D.A., Redman, C.L., 2019. The Infrastructure Trolley Problem: Positioning Safe-to-fail Infrastructure for Climate Change Adaptation. Earths Future 7. https://doi.org/10.1029/2019EF001208

Lambert, P.A., 2020. The Order - Chaos Dynamic of Creativity. Creat. Res. J. 32, 431–446. https://doi.org/10.1080/10400419.2020.1821562

Langton, C.G., 1997. Artificial Life: An Overview. MIT Press.

Levy, D., 2000. Applications and Limitations of Complexity Theory in Organization Theory and Strategy, in: Handbook of Strategic Management. Taylor and Francis.

Lyngaas, S., 2022. Key Ukrainian government websites hit by series of cyberattacks [WWW Document]. CNN. URL https://www.cnn.com/2022/02/23/europe/ukraine-government-commercial-organizations-data-wiping-hack/index.html (accessed 7.2.24).

Markolf, S., Helmrich, A., Kim, Y., Hoff, R., Chester, M., 2022. Balancing Efficiency and Resilience Objectives in Pursuit of Sustainable Infrastructure Transformations. Curr. Opin. Environ. Sustain. 56. https://doi.org/10.1016/j.cosust.2022.101181

Muñoz-Erickson, T.A., Campbell, L.K., Childers, D.L., Grove, J.M., Iwaniec, D.M., Pickett, S.T.A., Romolini, M., Svendsen, E.S., 2016. Demystifying governance and its role for transitions in urban social–ecological systems. Ecosphere 7, e01564. https://doi.org/10.1002/ecs2.1564

Muñoz-Erickson, T.A., Miller, C.A., Miller, T.R., 2017. How Cities Think: Knowledge Co-Production for Urban Sustainability and Resilience. Forests 8, 203. https://doi.org/10.3390/f8060203

Myntti, C., 2024. Infrastructure and well-being. Environ. Res. Infrastruct. Sustain. 4, 033001. https://doi.org/10.1088/2634-4505/ad6cf0

Olsson, P., Folke, C., Berkes, F., 2004. Adaptive Comanagement for Building Resilience in Social–Ecological Systems. Environ. Manage. 34, 75–90. https://doi.org/10.1007/s00267-003-0101-7

Packard, N.H., 1988. Adaptation Toward the Edge of Chaos. University of Illinois at Urbana-Champaign, Center for Complex Systems Research.

Porter, T., 2006. Coevolution as a Research Framework for Organizations and the Natural Environment. Organ. Environ. 19, 479–504. https://doi.org/10.1177/1086026606294958

Simon, H.A., 2002. Near decomposability and the speed of evolution. Ind. Corp. Change 11, 587–599. https://doi.org/10.1093/icc/11.3.587

Simon, H.A., 1957. Models of Man: Social and Rational; Mathematical Essays on Rational Human Behavior in Society Setting. Wiley, Berkeley, CA.

Simon, H.A., 1947. Administrative Behavior. Macmillan Company, Basingstoke, UK.

Smart, P., Heersmink, R., Clowes, R.W., 2017. The Cognitive Ecology of the Internet, in: Cowley, S.J., Vallée-Tourangeau, F. (Eds.), Cognition Beyond the Brain: Computation, Interactivity and Human Artifice. Springer International Publishing, Cham, pp. 251–282. https://doi.org/10.1007/978-3-319-49115-8_13

Snowden, D., Boone, M., 2007. A Leader's Framework for Decision Making. Harv. Bus. Rev.

Woods, D., 2015. Four concepts for resilience and the implications for the future of resilience engineering. Reliab. Eng. Syst. Saf., Special Issue on Resilience Engineering 141, 5–9. https://doi.org/10.1016/j.ress.2015.03.018

Woods, D.D., 2018. The theory of graceful extensibility: basic rules that govern adaptive systems. Environ. Syst. Decis. 38, 433–457. https://doi.org/10.1007/s10669-018-9708-3

Zetter, K., 2016. Inside the Cunning, Unprecedented Hack of Ukraine's Power Grid. Wired.

Acknowledgements

The chapters of this book have been supported by many funding agencies, organizations, and people. However, the collective and final work has been produced with direct support from the U.S. National Science Foundation, specifically the Urban Resilience to Extremes Sustainability Research Network (Award No. 1444750) and Growing Resilience Convergence Research project (Award No. 1934933).

The first edition of this book was supported by ASU's Consortium for Science Policy and Outcomes (CSPO) program under their *Rightful Place of Science* book series. CSPO provided invaluable support in launching the first edition.

We thank Dr. Mindy Kimball (editing), Brennan Liu (referencing), and ASU's Spring 2025 Urban Infrastructure Anatomy class (consistency reviewing) for their support.

Index

L

Lock-in, 19, 32, 35, 47, 69, 92, 93, 94,
95, 107, 172, 201, 204, 218, 220,
221, 232, 258, 299, 315, 316, 318,
329, 330, 331, 332, 336, 337, 368,
382, 404, 405, 406, 407, 415

R

Resilience, 11, 16, 22, 23, 29, 55, 59,
63, 82, 84, 88, 101, 116, 119, 120,
121, 122, 123, 125, 126, 128, 143,
148, 165, 166, 167, 168, 169, 170,
171, 172, 174, 205, 206, 207, 216,
217, 219, 223, 228, 230, 233, 235,
236, 238, 242, 245, 251, 252, 253,
254, 255, 256, 257, 258, 259, 260,
261, 269, 277, 278, 279, 284, 287,
289, 301, 302, 313, 314, 318, 333,
334, 336, 341, 350, 351, 353, 356,
363, 364, 365, 366, 368, 369, 370,
371, 372, 373, 374, 382, 383, 384,
389, 390, 391, 392, 393, 394, 395,
396, 397, 398, 399, 400, 401, 404,
405, 407, 408, 409, 410, 411, 414,
415, 419, 439, 467, 498, 501, 506

S

Safe-to-Fail, iii, 63, 138, 291, 318,
327, 328, 330, 333, 334, 335, 337,
347, 348, 351, 352, 353, 354, 355,
358, 359, 360, 361, 362, 363, 364,
366, 368, 369, 373, 403

W

Warfare, iii, 5, 10, 22, 62, 77, 102,
107, 216, 269, 283, 392, 395, 424,
449, 454, 455, 458, 459, 467, 501
Wickedness, iii, 2, 23, 29, 31, 33, 34,
35, 36, 37, 38, 40, 41, 42, 43, 44, 45,
46, 48, 62, 77, 78, 110, 195, 199,
200, 201, 210, 242, 278, 430, 478,
479, 482, 495, 496, 502

About the Authors

MIKHAIL CHESTER

Mikhail Chester is a professor of Civil, Environmental, and Sustainable Engineering, and the director of the Metis Center for Infrastructure and Sustainable Engineering at Arizona State University. He runs a research program focused on innovation and disruption of infrastructure systems for the challenges of the coming century including technologies, governance, and education. His work spans climate adaptation, disruptive technologies, innovative financing, cybersecurity, transitions to agility and flexibility, and modernization of infrastructure management. He is broadly interested in the changes needed in infrastructure governance, design, and education for the Anthropocene, an era marked by acceleration and uncertainty. He was awarded the American Society of Civil Engineer's early career researcher Huber prize (2017). He has also published *Urban Infrastructure: Reflections for 2100* (with Sybil Derrible, 2020).

BRADEN ALLENBY

Brad Allenby is President's Professor of Engineering and Lincoln Professor of Engineering and Ethics at Arizona State University. He moved to ASU from his previous position as the environment, health, and safety vice president for AT&T in 2004. Allenby received his BA from Yale University, his JD and MA (economics) from the University of Virginia, and his MS and PhD in environmental sciences from Rutgers University. He has served as president of the International Society for Industrial Ecology; chair of the AAAS Committee on Science, Engineering, and Public Policy; and chair of the IEEE Presidential Sustainability Initiative. He is also a AAAS Fellow and a Fellow of the Royal Society for the Arts, Manufactures & Commerce. Recent books include *The Techno-Human Condition* (with Daniel Sarewitz, 2011), *The Theory and Practice of Sustainable Engineering* (2011), *Future Conflict & Emerging Technologies* (2016), and *The Applied Ethics of Emerging Military and Security Technologies* (2016).

Revision History

Date	Ed.	History
March 2021	1	Published.
June 2025	1	Unpublished.
	2	Published.
August 2025	2	Added Library of Congress Control Number to frontmatter. Table of Contents corrections for chapter titles.

v250821

www.ingramcontent.com/pod-product-compliance
Lightning Source LLC
Chambersburg PA
CBHW061229220326
41599CB00028B/5380